Industrial Wastewater Treatment Technology, Second Edition

Industrial Wastewater Treatment Technology, Second Edition

James W. Patterson

BUTTERWORTH PUBLISHERS
Boston • London
Sydney • Wellington • Durban • Toronto

Library of Congress Cataloging in Publication Data

Patterson, James William, 1940–
Industrial wastewater treatment technology.

Rev. ed. of: Wastewater treatment technology. c1975.
Includes bibliographies and index.
1. Sewage—Purification. 2. Water—Purification.
I. Patterson, James William, 1940– Wastewater treatment technology. II. Title.
TD745.P32 1985 628.3 84-29314
ISBN 0-409-90002-8

Butterworth Publishers
80 Montvale Avenue
Stoneham, MA 02180

10 9 8 7 6 5 4 3 2 1

Printed in the United States of America

Contents

Preface

Wastewater Treatment Technology was published in 1975 and has since gone through several printings. I suspect that everyone involved in the production of that volume, including my publisher, has been surprised by the steady demand for the text. This demand, hopefully reflecting the text's usefulness to our profession, is gratifying.

Since the publication of the first edition of *Wastewater Treatment Technology*, a significant body of supplemental information on industrial wastewater pollution control has been generated. The preparation of this second edition was undertaken, in part, to incorporate the new material regarding modern industrial practices in water pollution abatement. In addition, a number of industrial pollutants not included in the first edition have become serious environmental concerns. Thus, the second objective of this revised edition is to address the technologies appropriate for these additional pollutants. Four chapters have been added to accomplish this latter goal.

A chapter on industrial sources and treatment technologies for aluminum has been added. Aluminum salts have been used for many years in the purification of drinking water. However, there is recent information which suggests that aluminum may exhibit chronic toxicity effects. Chapters have been added on sources and control of organic and ammonia-nitrogen and of nitrite and nitrate-nitrogen. The fourth new chapter addresses industrial sources and treatment technologies for the 114 organic priority pollutants. These organic toxicants have only recently come under regulatory scrutiny and requirements for their control. In addition to these new chapters, the chapters brought forward from the first edition have been updated and expanded.

In 1975, the industry was still two years away from the 1977 national deadline to implement the level of water pollution control termed "best practicable control technology currently available," as required by federal legislation. Although that deadline has passed, industry now faces a new series of targets for water pollution control. These include the implementation by 1984 of a higher level of pollution abatement termed "best available technology economically achievable" (BATEA) as well as the achievement of pretreatment standards based upon BATEA concepts, for indirect industrial dischargers to municipal sewer systems. The accomplishment of these future objectives will depend on an understanding of the technical alternatives available, their associated capital and operating economics, and the design and operating factors which influence the performance of those technologies.

This second edition of *Wastewater Treatment Technology* includes 24 chapters, each devoted to a specific pollutant or category of pollutants. The volume is almost twice the size of the previous edition, and the number of references cited has more than doubled.

The overall objectives of this work remain as stated in the preface to the first edition:

1. Identify sources and typical levels of the pollutant discussed.
2. Describe available treatment technology with regard to:
   a) how each technology operates and its limitations,
   b) what levels of treatment have been accomplished in full-scale treatment systems (preferably), or in pilot-scale systems if no full-scale data are available,
   c) the economics of each technique, with regard to comparison of both capital investment requirement, and operating and maintenance costs.
3. Summarize concisely the major types of technology and effluent levels of pollutant achievable.

As before, I have relied for treatment information on that reported for full-scale pollution control facilities, except in instances where pilot or laboratory scale data would assist in understanding the behavior of the full-scale systems.

The literature of our field can sometimes be confusing and contain vague or even conflicting information regarding the effectiveness and economics of alternative treatment technologies. Where such conflicts exist in the literature, I have attempted to note this. However, I have also stated my opinion and reasons regarding which of the two opposing viewpoints was valid.

Although I did not dedicate the first edition of this text, other than to the many pollution control engineers and scientists who through their writings have established the technical literature basis for our field, I have elected to dedicate this new edition to a very special lady. That lady is my grandmother, Mrs. W. C. Rahn, who on July 13, 1984 celebrated her ninety-fourth birthday. From her historical perspective, our field is surely in its infancy.

J. W. Patterson

1

Aluminum

Aluminum is quite soluble in water at both high and low pH extremes, due to its amphoteric nature (Figure 1.1). Its aqueous chemistry is solely that of the plus 3 oxidation state (Al^{3+}), but in solution a range of positive and negative charged complex hydroxyalumino species exist, including polynuclear (with respect to the aluminum) ions. However, in the intermediate pH range, aluminum is quite insoluble, with reported minimum solubility at pH 5–6 [1]. Aluminum is not considered toxic to humans at levels found in fluids used for general consumption. In fact, aluminum salts are frequently used as wastewater treatment chemicals (coagulants) for other industrial contaminants, and have a long and successful history of use in the treatment of potable water by chemical coagulation. At pH extremes, the acid or base aspects of an aluminum wastewater would normally overshadow the potential for aluminum toxicity. High aluminum content in soils is of concern with respect to plant toxicity and has been cited in assessing problems of revegetation of lands disturbed by strip mining for coal and other minerals.

INDUSTRIAL SOURCES

Principal sources of aluminum wastes include the manufacture of aluminum from bauxite ores, as well as industrial processes which produce aluminum parts from ingots, or which clean or etch aluminum parts (generally with hydrofluoric acid) prior to other steps such as lithography. Acid or alkaline treatments for oxide removal during working or finishing of the metal product will produce aluminum wastewaters. Hydrofluoric acid etching is a common process that leads to aluminum-contaminated rinse wastewater. Typically, however, greater concern arises from the fluoride content than from the aluminum content of the wastewater, since aluminum is frequently used in fluoride wastewater treatment (see Chapter 9).

Aluminum parts are often coated, for both protective and decorative purposes. The coating process for aluminum is an electrolytic treatment called *anodizing*. The anodizing process yields an artificial oxide coating on the metal or its alloy. The anodizing operation produces spent baths and rinsewaters containing aluminum plus other metallic pollutants. Dyes may also be present as a result of their use to impart color to the oxide coating. The use of aluminum

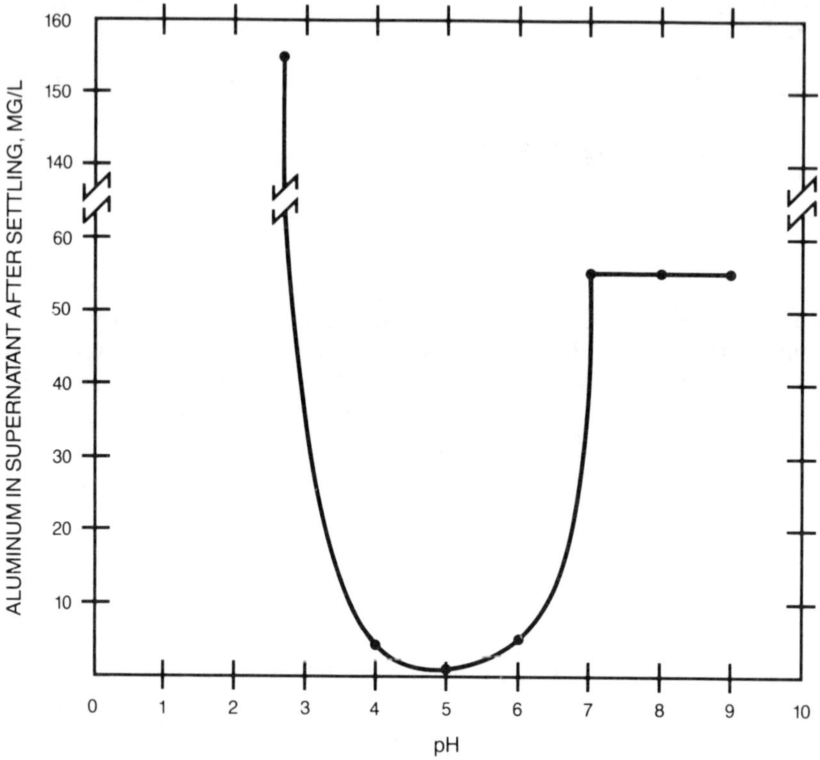

Figure 1.1 Influence of treatment pH on aluminum for an anodizing wastewater [1].

parts is rapidly expanding, and anodizing wastewaters are increasingly significant in terms of pollution control.

Because aluminum has traditionally not been cast in the role of a pollutant, data on wastewater aluminum concentrations are limited. An aluminum concentration of 185 mg/l has been reported in the oily waste resulting from rolling of aluminum sheet, plate, and foil [2]. Chemical rinsewaters resulting from the acid or base treatment of forged aluminum pieces have been reported to contain a mean concentration of 183 mg/l aluminum. Variations in pH of 8.6–12 were noted [3]. Wire-treatment plant wastes (essentially iron wire) with a pH of 2.5 contained only 0.5 mg/l aluminum [4]. One aluminum parts manufacturing plant's waste was reported to contain 22.5 mg/l aluminum [5], while a cast carburetor chromating line rinsewater contained 4–13 mg/l [6]. A gray iron foundry wastewater has been reported to contain 16–110 mg/l aluminum, with an average 60.3 mg/l concentration [7]. Aluminum-forming wastewaters are also high in aluminum content. A survey of 64 facilities revealed an average concentration of 1,283 mg/l aluminum [8]. Spent aluminum etching and anodizing baths are highly concentrated [9]. Acidic coal-mine drainage also con-

Table 1.1 Aluminum Concentrations Reported for Industrial Wastewaters

Source	*Concentration (mg/l)*	*Reference*
Aluminum sheet-rolling wastewater	185	2
Forging rinsewater	183	3
Iron wire mill rinsewater	0.5	4
Cast chromating line rinsewater	4–13	6
Gray iron foundry wastewater	16–110 (range) 60.3 (avg)	7
Aluminum forming wastewaters	1,283	8
Spent anodizing bath	7,500	9
Spent etching bath	43,380	9
Acidic coal mine drainage	0.5–18 (range) 7.6 (avg)	10
Acidic coal-cleaning leachate	370	11

tains aluminum. One study reported a range of 0.5–18 mg/l and an average 7.6 mg/l aluminum [10]. An acidic coal-cleaning waste leachate contained 370 mg/l aluminum [11]. Table 1.1 summarizes aluminum concentrations reported for industrial wastewaters. Typical characteristics of aluminum ore-reduction plant wastes are presented in Table 1.2. Table 1.3 presents data on etching plus anodizing wastewater characteristics.

TREATMENT TECHNOLOGY

Treatment technology for aluminum removal is essentially limited to chemical precipitation by pH adjustment. The practice is well established in potable

Table 1.2 Characteristics of Aluminum Ore-Reduction Plant Wastes [12]

	Plant A		*Plant B*	
Parameter[a]	*Avg*	*Range*	*Avg*	*Range*
Total residue	464	407–540	358	356–361
Filterable residue	420	332–487	214	193–236
Nonfilterable residue	44	16–75	144	125–163
Total aluminum	9.8[b]	0.8–22	14.5[c]	7.8–26
Fluoride	145	—	107	93–116
pH	—	3.6–3.9	—	2.9–3.2

[a] All concentrations except pH in mg/l.
[b] Mean percent particulate = 64.
[c] Mean percent particulate = 81.

Table 1.3 Characteristics of Aluminum Etching and Anodizing Wastewaters [1]

Parameter[a]	*Range*
pH	2.0–12.0[b]
Total suspended solids	100–11,000
Total dissolved solids	1,500–3,500
Total aluminum	30–2,500
Total hexavalent chromium	0.2–30
Total trivalent chromium	1.0–400
Total copper	0.2–2.0

[a] All concentrations except pH in mg/l.
[b] Anodizing baths are typically acidic and etching baths, typically alkaline.

water treatment, where aluminum is commonly employed as a treatment chemical. Aluminum hydroxide is precipitated, typically yielding a white, flocculant, gelatinous solid which is difficult to settle and dewater. Process control measures are normally directed at reducing the final sludge volume that must be disposed.

In many aluminum-bearing wastewaters, other metallic pollutants may also be present and simultaneously coprecipitated, although the aluminum is typically the principal constituent. Therefore, the proper precipitation of aluminum becomes a primary factor in the overall effectiveness of the precipitation/solids removal process.

Although the chemistry of aluminum hydrolysis and precipitation has been studied extensively, particularly with regard to coagulation treatment for drinking water supplies, there still remain uncertainties regarding the optimum pH for precipitation treatment. One author and his associates [1] have reported minimum aluminum solubility to occur at pH 5. However, other performance characteristics influenced by pH, such as sludge settleability and handling properties, may be more important than solubility alone. Table 1.4 compares the results of two studies on lime precipitation of aluminum wastewater. In one study [13] the wastewater was treated by lime addition to various pH levels,

Table 1.4 Solubility of Lime-Precipitated Aluminum Solutions

Settled Supernatant (mg/l) [13]			*Filtered Supernatant (mg/l) [11]*		
Initial Concentration	*Treatment pH*	*Residual Concentration*	*Initial Concentration*	*Treatment pH*	*Residual Concentration*
97	3.9	97	370	2.3	370
	5.9	78		5.8	0.46
	7.2	9		6.5	<0.10
	8.8	4		8.1	<0.10
	10.0	73		10.2	1.1

and the settled supernatant measured. Best results were obtained at pH 8.8. In the second study [11], lime was added but after reaction the solution was filtered. Lowest concentration of filterable aluminum occurred at pH 6.5 to 8.1. The results suggest that, while minimum solubility is achieved at pH 6.5, best settling precipitate is achieved at a pH near 9.0. To achieve good removal at treatment pH near that of minimum solubility, extended clarification or effluent filtration may be required. In a pilot evaluation of a lime-treated radiator manufacturing wastewater, Knapp and Paulson reported that gravity filtration reduced the average clarifier effluent aluminum from 0.1 to 1.3 (average 0.5) to 0.3 mg/l [14]. These data would seem to refute a claim that lime precipitation plus sedimentation alone can achieve an effluent aluminum concentration of 0.2 mg/l [8].

Olver et al [1] have discussed the importance of the chemical form of the precipitated aluminum, in terms of its solubility, settleability, and sludge-dewatering characteristics. They note that the aluminum precipitate can exist in either an amorphous or a crystalline form, with the latter being desired in terms of ease of precipitate removal and sludge volume reduction. In a treatability study performed on an anodizing waste (see Table 1.3), the precipitation of the aluminum was reported to be influenced by the initial and final solution pH, temperature and time of reaction, initial concentration of aluminum, mixing intensity, and presence of other wastewater constituents [1]. Optimum treatment was reported in the pH range of 4.5–5.5. Optimum reaction time for formation of a proper hydroxide precipitate was approximately one hour. In the treatability study, instantaneous (i.e. rapid) pH adjustment resulted in the formation of an amorphous gel with poor settling and thickening properties. Generally, higher initial aluminum concentrations resulted in greater ease of precipitation of the crystalline hydroxide form.

These design criteria were translated into a treatment plant handling 40,000–45,000 gpd (2 shifts/day) of anodizing wastewater, and having a capital cost in the range of $200,000–250,000 (1978 basis). This represents a capital investment of $4.44 6.25/gpd treatment capacity. Costs include sludge thickening and filter press dewatering, plus process control automation [1].

Saunders et al [9] also investigated the influence of treatment conditions on the precipitate properties of concentrated etching and anodizing wastes. They reported no effect of treatment temperature on the precipitate settling velocity. However, there was a significant pH effect, at a treatment temperature of 80°C. In contrast to the results reported by Olver et al [1], lowest settling velocities were observed at precipitation pH 5.5, and highest velocities at pH 8.5. These results are consistent with the data reported in Table 1.4. Sludge dewaterability and filter cake solids were also better at the higher treatment pH [9].

There is limited information available on the performance of aluminum wastewater precipitation treatment systems. Aluminum is usually precipitated in the pH range of 6–7, which is near the pH of minimum solubility. The actual treatment pH selected may be chosen to avoid an unduly acidic (below pH 6) effluent, or to avoid the need to further adjust the treated effluent pH prior to discharge. At near-neutral treatment pH, residual soluble aluminum levels below 0.1 mg/l appear possible. The hydroxide precipitate formed can be segregated by sedimentation or filtration. The latter process is common in aluminum treatment because of the poor settleability of most aluminum sludges.

Osantowski and Ruppersberger [7] reported the results of a pilot treatability study on a gray iron foundry wastewater. Treatment involved lime neutralization to pH 7.0, clarification, and dual media filtration. Influent pH ranged from 5.7 to 6.8. Influent aluminum ranged from 16.2 to 110 mg/l, and averaged 60.3 mg/l. Clarifier effluent aluminum averaged 2.5 mg/l, and the filter effluent contained 2.4 mg/l aluminum. Estimated costs for a treatment system are presented in Table 1.5.

The disadvantages of precipitation treatment include the possibility of inefficient precipitate formation due to nonoptimal treatment pH or the presence of complexing agents, and difficult-to-settle particles which are not removed in a sedimentation step. The major advantage of the precipitation process is that the equipment used is relatively simple and the process is relatively easy to operate. Treatment costs are low compared to those of other processes. Capital costs for granular media filters are given in Table 1.6. The

Table 1.5 Estimated Costs for Precipitation Treatment of Foundry Wastewater [7]

	Plant Size, MGD		
Cost	*0.1*	*0.5*	*2.5*
Capital, $1,000			
Neutralization/clarification	269	377	834
filtration	114	221	736
Total	383	598	1,570
Capital, $/gpd capacity			
Neutralization/clarification	2.69	0.75	0.33
filtration	1.14	0.44	0.29
Total	3.83	1.19	0.62
Operating, $1,000			
Neutralization/clarification	60	108	313
filtration	26	33	51
Total	86	141	372
Operating, $/1,000 gal			
Neutralization/clarification	1.64	0.59	0.34
filtration	0.71	0.18	0.06
Total	2.35	0.77	0.40

Table 1.6 Capital Cost of Granular Media Filtration, 1979 Basis [15]

Filter Area *(ft^2)*	*Capital Cost* *($1,000)*
30	84
40	98
100	134
500	330
1,000	610
5,000	2,250

Costs in this table were originally reported in 1973 dollars, and have been updated by the Marshall & Swift Equipment Cost Index. 1973 = 344.1; 1979 = 599.4.

solids can possibly be recovered since they are not destroyed during precipitation [16].

One aluminum company used ultrafiltration followed by reverse osmosis for treating oily wastes in its rolling operation [2]. The raw waste contained 185 mg/l aluminum, 80,000 mg/l hexane-extractable (oily) substances, 207 mg/l suspended solids, and 6,400 mg/l sulfate. Ultrafiltration alone reduced the aluminum to 1.7 mg/l, and further treatment by reverse osmosis yielded 0.06 mg/l aluminum. The extent of aluminum removal by ultrafiltration suggests that the aluminum in this wastewater was principally in particulate form. Operating costs were $3.76/1,000 gal for ultrafiltration and $3.53/1,000 gal for the ultrafiltration plus reverse osmosis. The latter cost advantage results from reuse of the reverse osmosis permeate. Both cost figures include oil recovery credit.

Wilmoth et al [10] have reported a pilot ion-exchange study for a coal acid-mine drainage wastewater. The pilot facility involved a two-resin system. The cation exchanger was operated on the hydrogen cycle, and the anion exchanger on the hydroxyl cycle. Treatment results are presented in Table 1.7.

Table 1.7 Performance of Ion Exchange Pilot Plant on Coal Acid Mine Drainage [10]

Parameter[a]	*Raw Wastewater*	*Cation Effluent*	*Anion Effluent*
pH	5.0	1.5	9.4
Aluminum	8.6	0.6	0.3
Calcium	360	32	26
Iron, total	207	18	2.3
Iron, ferrous	193	16	0.1
Magnesium	110	7.9	7.3
Sulfate	2,500	2,373	637
TDS[b]	3,533	2,670	1,033

[a] All concentrations except pH in mg/l.
[b] Total Dissolved Solids.

Although the system was effective in removing aluminum and other constituents, cation regenerant (sulfuric acid) costs alone ranged from \$1.50 to \$3.50 per 1,000 gallons treated. Anion regenerant (sodium hydroxide) costs were \$2.40–3.70/1,000 gal. Further, Pelosi and McCarthy [17] have warned that aluminum can foul cation exchange resins, by coating the resin beads with gelatinous aluminum hydroxide. This would represent a more serious risk for sodium or other than hydrogen-cycle resin operation, however.

SUMMARY

Because of its nontoxic nature, aluminum is not normally cited as a wastewater pollutant. However, it is present as a contaminant in many industrial wastes. Precipitation treatment is most frequently employed, although other processes have been utilized when wastewaters contain other pollutants as well as aluminum. One example is the application of ultrafiltration for oily aluminum-bearing wastewaters, such as result from acid cleaning of preoiled rolled aluminum stock.

When precipitation is utilized, effective treatment of aluminum is achieved at pH 6.5–8.5. Concern for an unacceptably acid effluent, or attempts to concurrently precipitate or coprecipitate other contaminants such as fluoride, may result in treatment at nonoptimum (for aluminum) pH.

Aluminum precipitates are notorious for their poor settleability, and the sludges produced are voluminous and difficult to dewater. There is some evidence that gradual pH adjustment to induce precipitation, and a reaction time of approximately one hour, will yield a more crystalline precipitate form, with improved sedimentation and sludge volume reduction characteristics.

REFERENCES

1. Olver, J., W. Kreye, R. Roberts, and H. Sutton. "Anodizing Wastewater Treatment and Sludge Disposal: A Total Systems Design," *Proceedings of the 3rd Annual Conference on Treatment and Disposal of Industrial Wastewaters and Residues*, Houston, TX (April 18–20, 1978) pp 94–101.
2. Sonksen, M.K., F. M. Sittig, and E.F. Mariarz. "Treatment of Oily Wastes by Ultrafiltration/Reverse Osmosis. A Case History," Paper presented at the 33rd Annual Purdue Industrial Waste Conference, West Lafayette, IN (May 1978).
3. Hockenbury, M.R. and A.W. Loven. "Treating Metal Forging and Processing Wastewater," *Ind. Wastes* (May/June 1977): 45–49.
4. Fuller, J.K. "The Collection, Neutralization and Disposal of Rinse Water," *Wire J.* (January 1977): 60–63.
5. "Aluminum Manufacturer Removes Chromium," *Ind. Wastes* (January/February 1978): 24, 34.
6. Scott, M.C. "Sulfide Process Removes Metals, Produces Disposable Sludge," *Ind. Wastes* (July/August 1979): 34–38.

7. Osantowski, R. and J. Ruppersberger. "Upgrading Foundry Wastewater Treatment," presented at the 39th Annual Purdue Industrial Waste Conference, West Lafayette, IN (May 1984).
8. Carlton, J., G. Aronberg, D. Washington, J. Gopodwin, and E. Hall. "Handling of Aluminum Forming Wastewater," presented at the 37th Annual Purdue Industrial Waste Conference, West Lafayette, IN (May 1982).
9. Saunders, F., M. Sezgin, and J. Medero. "Segregated Neutralization for Treatment of Concentrated Aluminum Finishing Wastes," *Proceedings of the 4th Conference on Advanced Pollution Control for the Metal Finishing Industry*, USEPA 600/9-82-022 (December 1982).
10. Wilmoth, R., R. Scott, and E. Harris. "Application of Ion Exchange to Acid Mine Drainage Treatment," presented at the 32nd Annual Purdue Industrial Waste Conference, West Lafayette, IN (May 1977).
11. Wangen, L. and J. Williams. "Control by Alkaline Neutralization of Trace Elements in Acidic Coal Cleaning Waste Leachates," *J. Water Poll. Control. Fed.*, 54 (1982): 1302–1310.
12. Sylvester, R.O., R.T. Oglesby, D.A. Carlson, and R.F. Christman. "Factors Involved in the Location and Operation of an Aluminum Reduction Plant," *Proceedings of the 22nd Industrial Waste Conference, Purdue University* (May 1967) p. 441.
13. Netzer, A., A. Bowers, and J. Norman. "Removal of Heavy Metals from Wastewater by Lime," *Proceedings of the 16th Conference on Great Lakes Research*, Huron, OH (1973), pp 260–265.
14. Knapp, R. and E. Paulsen. "Gravity Filtration Reduces Suspended Metals in a Lime Precipitation Treatment System," presented at the 37th Annual Purdue Industrial Waste Conference, West Lafayette, IN (May 1982).
15. Edwards, A.W., W.W. Eckenfelder, Jr., and J.A. Roth. *Industrial Water Cost Studies*, Vol. 1, Office of Research and Development, USEPA Contract 68-01-2485 (1975).
16. Fukuyama, Y., Y. Misaka, and K. Kato. "Recovery of Aluminum Hydroxide from Fabricating Plants of Aluminum Products," *Proceedings of the 29th Industrial Waste Conference, Purdue University* (May 1974), pp 704–715.
17. Pelosi, P. and J. McCarthy. "Preventing Fouling of Ion-Exchange Resins—I," *Chem. Eng.* (1982): 75–78.

2 Arsenic

The principal aqueous forms of inorganic arsenic are the arsenite ion (AsO_2^-) and the arsenate ion (AsO_4^{-3}). In the presence of dissolved oxygen, arsenite is ultimately oxidized to arsenate, a less toxic form [1]. The oxidation state of the anion appears to have a significant effect upon the efficiency of some treatment techniques for arsenic removal.

Arsenic and arsenical compounds have been reported in wastewaters of the metallurgical industry, of glassware and ceramic production, of tannery operations, and of dye and pesticide manufacture [2]. Other industrial sources include the organic and inorganic chemicals- and petroleum-refining industries [3] and the rare-earth industry [4]. Arsenic is also widely associated with the manufacture of herbicides and pesticides [5]. In recent years, with the development of organic arsenicals, many of the inorganic arsenic-based forms have declined in use. The treatment of organic arsenical wastewaters can present unique difficulties.

Other industrial applications of arsenic include wood and hide preservatives, sheep dip, decolorizers in glass manufacturing, pigment formulations, lead shot manufacture, phosphate detergent builders and presoaks, and as an impurity of calcined phosphate used in many fertilizers.

Arsenic is a common constituent of most coals, at levels of up to 100 mg/kg in some Appalachian coal [6]. As a result, arsenic typically occurs in coal-fired power-plant fly-ash pond waters. Arsenic is present in the pond water as both arsenite and arsenate, although at near-neutral pH and in well-aerated systems the arsenite slowly converts to the more stable pentavalent arsenate form.

Arsenic wastes from nonferrous smelting and refining have been reported [3, 7–10]. A major form appears to be as extremely small (1–3 μ) particles from which arsenic may be profitably recovered [7]. The soluble concentration of arsenic is dependent upon both chemical speciation and wastewater pH. For example, one gold ore-extraction process waste stream, at pH 9.5, contained 910 mg/l particulate arsenic and only 10.1 mg/l soluble arsenic [11]. The soluble fraction, representing only 1% of the total arsenic present, was in the arsenite form. For a second waste stream, at pH 3.1, particulate arsenic was 880 and soluble arsenic was 132 mg/l (13%). Acid mine drainage of metal mines in the Rocky Mountains has been found to contain 6.0–22.0 mg/l soluble arsenic [12].

Table 2.1 Arsenic Concentrations Reported for Industrial Wastewaters

	Arsenic (mg/l)		
Source	*Total*	*Soluble*	*Reference*
Insecticide manufacture	362	—	14
Gold ore extraction	910	10.1	11
	1012	132	11
Acid mine drainage	—	6.0–22.0	12
	2.3	—	15
Sulfuric acid manufacture	200–500	—	13
Zinc ore extraction	—	0.1–0.68	9
Copper ore—slag granulation	—	0.05–5.70	10
acid leaching	—	0.15–19.0	10
acid leaching	230	—	15
Arsenic trioxide plant	310	—	10
Electrolytic copper refining	—	0.001–51	10
Boric acid production	—	0.04–0.92	16
Ammonia manufacture	430	—	17
Wood products processing	—	13–50	18
Timber products processing	0–14	—	19
Geothermal water	0.03–3.0	—	6
Geothermal power plant condensate	11	—	21
Coal-fired power plant ash pond water	0.001–1.0	—	6
Steam electric plant cleaning	0.0–310	—	20
Coal-cleaning leachate	—	0.76	22
Petroleum refining	0.0–0.48	—	20
Paint and ink formulation	0.0–0.80	—	20
Commercial laundry	0.0–1.60	—	20
Porcelain enameling	0.0–2.80	—	20

In one instance, wastewater from a sulfuric acid production plant was reported to contain 200–500 mg/l total arsenic [13]. Sulfur dioxide used in the plant was obtained from a sulfide-ore roasting process, and the high arsenic level presumably resulted from arsenic constituents of the ore. Table 2.1 summarizes reported industrial wastewater concentrations of arsenic.

High arsenic levels have also been encountered in raw municipal water supplies, necessitating arsenic removal by the water treatment industry. Many groundwater supplies in Central and South America contain excessive arsenic [23, 24]. Deep-well waters in Taiwan, which contain up to 2 mg/l arsenic, are believed to be responsible for an endemic illness in the area called *blackfoot disease* [25, 26]. Deep-well waters in desert areas of the southwest United States contain arsenic at levels exceeding 0.1 mg/l [27].

TREATMENT TECHNOLOGY

Limited information is available on current arsenic-wastewater treatment processes and removals obtained. Much of the literature describing treatment of arsenic wastes is 30 years or more old. More up-to-date information is available on the removal of arsenic from drinking water, and in fact the methods for treatment of both drinking water and industrial wastes are similar. The treatment methods and arsenic removal efficiencies discussed in detail below are summarized in Table 2.2.

Common treatment methods for arsenic include lime or sulfide precipitation or coprecipitation with iron or aluminum hydroxide, plus adsorption onto coagulant floc, with enmeshment of particulate arsenic. This second process is typical of the traditional coagulation process used in the water treatment

Table 2.2 Summary of Arsenic Treatment Methods and Removals Achieved

Treatment	*Initial Arsenic (mg/l)*	*Final Arsenic (mg/l)*	*% Removal*	*Reference*
Precipitation with sulfide	—	0.05	—	28
	132.0	26.4	80	11
	—	—	99	20
Ferric sulfide filter bed	0.8	0.05	94	25
Precipitation with lime	0.2	0.03	85	29
	0.5	0.03	95	11
	—	0.01	73	9
Precipitation with lime plus iron	—	0.05	—	4
	—	—	69–99	20
			85–92	
Coprecipitation with alum	0.35	0.003–0.005	85–92	30
	430	0.023	99+	18
Coprecipitation with ferric sulfate	0.31–0.35	0.003–0.006	98–99	30
	25	5	80	31
Coprecipitation with ferric chloride	3.0	0.005	98	32
	0.58–0.90	0.0–0.13	81–100	26
Coprecipitation with ferric salt	—	0.6	—	33
	362.0	15–20	94–96	14
Charcoal bed filtration	0.2	0.06	70	29
Activated carbon adsorption	0.5	0.3	40	11
	—	—	21–99	20
Ion exchange	2.3	0.52	77	15

industry. Other processes used, with varying degrees of success, include adsorption onto activated carbon or alumina, and ion exchange.

Precipitation

It has been reported that effluent arsenic levels of 0.05 mg/l are obtainable by precipitation of arsenic as the sulfide, with addition of sodium or hydrogen sulfide at pH 6–7 recommended [28]. Sulfide precipitation has been found partially effective for arsenate, but ineffective for arsenite, in gold ore-extraction plant effluent [11]. At an optimal sulfide treatment pH of 7 and optional sulfide:arsenic dosage ratio of 0.5, arsenate was reduced from 132 to only 26.4 mg/l. No precipitate resulted from sulfide treatment of arsenite wastewater. Table 2.3 presents results from sulfide as well as other precipitating chemicals investigated [11]. These results indicate that for this wastewater, lime precipitation yielded equal or better treatment than other chemicals, at lower cost. It was effective on both arsenite and arsenate waste streams, since arsenite rapidly converts to the pentavalent form at extremely high pH.

The primary disadvantage of lime precipitation treatment is the high required treatment pH. The data of Table 2.3 indicate that a treatment pH of 12

Table 2.3 Precipitation Treatment of Arsenite and Arsenate in Gold Ore Extraction Wastewaters [11]

Precipitation Chemical	*Arsenic Form*	*Optimal pH*	*% Removal*	*Optimal Ratio, Precipitating Ion: Arsenic Ion*
Sodium sulfide	AsO_4^{-3}	7	80	$\frac{S^{-2}}{As} = 0.5$
	AsO_2^-	No precipitation	0	
Lime	AsO_4^{-3}	12	95	$\frac{Ca^{+2}}{As} = 9.8$
	AsO_2^-	12	95	
Caustic	AsO_4^{-3}	10	80	$\frac{Na^+}{As} = 3.8$
	AsO_2^-	No precipitation	0	
Ferric sulfate	AsO_4^{-3}	8	94	$\frac{Fe^{+2}}{As} = 1.5$
	AsO_2^-	No precipitation	0	
Ferric chloride	AsO_4^{-3}	9	90	$\frac{Fe^{+3}}{As} = 4.0$
	AsO_2^-	8	95	
Alum	AsO_4^{-3}	7–8	90	$\frac{Al^{+3}}{As} = 4.0$
	AsO_2^-	7–8	95	

is necessary for optimum arsenic removal by this method. This requirement has been confirmed in a separate study, which reported best treatment by lime precipitation at pH 11.9 [34]. In that study, 20 g/l slaked lime was required to adjust the sulfuric acid production wastewater (initial pH 0.6) to the desired treatment pH. In addition to the disadvantages of high treatment chemical demand and, for this waste, concurrent precipitation of voluminous quantities of gypsum ($CaSO_4$), the treated effluent would normally require reneutralization to more moderate pH prior to discharge.

Efficient removals of arsenic in lime precipitation treatment systems can occur at pH values far below the optima presented for lime in Table 2.3. This removal, however, is usually associated with coprecipitation with other ions present in the raw wastewater. Coprecipitation, discussed in the next section, is primarily an adsorption phenomenon wherein a trace constituent such as arsenic is adsorbed onto a bulk-precipitated solid such as ferric hydroxide. Wangen and Williams [22] have reported lime treatability results for an acidic coal-cleaning waste leachate which demonstrate this effect. Their results (all soluble concentration values) are presented in Table 2.4. Arsenic removal occurs at a pH of 2.7, concurrent with partial iron (III) precipitation; and is completed by pH 5.8 by which pH both aluminum and iron (III) coprecipitation are essentially complete. Variations in soluble calcium concentration reflect both lime addition and precipitation of calcium sulfate.

Although standard municipal sanitary sewage treatment processes do not remove arsenic, one common water treatment method is at least partially effective in arsenic treatment. Use of the lime water-softening process was found to reduce an initial arsenic level of 0.2 mg/l down to approximately 0.03 mg/l, yielding 85% removal [26].

Lime treatment has been recommended as the best practicable technology for arsenic wastewater from phosphorus trichloride manufacture [35], and for copper and zinc smelting and refining segments of nonferrous metals manufacture [9,10]. Heavy-metal coprecipitation would occur concurrently for lime treatment of these wastewaters, yielding enhanced treatment effectiveness. For example, lime treatment of one zinc plant waste, at pH 8.2, yielded an arsenic reduction of 73% and an effluent arsenic level below 0.1 mg/l [9]. The primary

Table 2.4 Lime Treatment Results for an Acidic Coal-Cleaning Waste Leachate [22]

	Soluble Concentration (mg/l)				
pH	*Calcium*	*Aluminum*	*Iron (II)*	*Iron (III)*	*Arsenic*
2.3	350	370	1,680	1,630	0.76
2.7	540	370	1,800	160	0.26
5.8	430	0.46	1,350	<0.02	<0.02
6.5	450	<0.1	620	<0.02	<0.02
8.1	500	<0.1	2.2	<0.02	<0.02

mechanism of removal was adsorption and coprecipitation with precipitated heavy-metal hydroxides.

Coprecipitation

Arsenic in an industrial waste which also contains heavy metals in solution can be concurrently coprecipitated upon precipitation of the heavy metals. Coprecipitation can be considered to involve both adsorption of the soluble ion onto another, bulk-precipitated solid, and coagulation or enmeshment of fine solids by the bulk precipitate. Approximately 90% reduction of arsenic is to be expected [5]. One patented waste recycle process coprecipitates arsenic and cadmium simultaneous with the precipitation of iron as ferric hydroxide [36].

In arsenic removal from wastewaters of the rare-earth industry, arsenic has been effectively removed by lime addition, but only in the presence of iron [4]. Treatment levels below 0.05 mg/l were obtained. Similar experience was described by Nilsson [37], who was unable to achieve effluent arsenic levels below 9.7 mg/l through lime treatment of a low-iron water.

The kinetics of arsenic coprecipitation with iron hydroxide are rapid, with a report of 90% completion of arsenate removal within 10 minutes, and stable residual arsenic levels achieved within one hour of contact time [6]. The principal mechanism of soluble arsenate removal by ferric hydroxide appears to involve adsorption, since Leckie [6] reports equivalent removals with both preformed and actively precipitating ferric hydroxides. Ferric hydroxide coprecipitation is also effective for organic arsenic removal.

There is a strong pH effect on the efficiency of arsenate removal by ferric hydroxide coprecipitation, with declining efficiency of removal at pH above 9. This adverse effect is accentuated when the mechanism of arsenate removal is through adsorption onto preformed ferric hydroxide [6]. After removal, a shift in pH to a more alkaline value will result in slow desorption of the sorbed arsenate.

Table 2.3 indicates that the optimum dosage of ferric chloride for arsenic removal was fourfold the tested wastewater arsenic concentration. Experience with a concentrated arsenic waste from a sulfuric acid plant indicated improved treatment at Fe^{+3}:As ratios up to 8, but no improvement at greater iron dosages [13]. Data reported by Leckie [6] for arsenate removal at pH near 9 and Fe^{+3}:As ratios ranging up to 2,000 support this latter dosage optimum. Increased coagulant dosage will result in greater sludge yield for ultimate disposal.

Extensive pilot plant studies on physical-chemical removal processes for arsenic in municipal waste involved addition of a coagulant, followed by flocculation and settling, dual media filtration, and carbon adsorption [38]. Types and levels of coagulants employed are described in Table 2.5. Treatment results for an initial arsenic concentration of 5 mg/l are reported in Table 2.6. For this wastewater, filtration provided little or no improvement, although carbon column polishing was somewhat effective in all cases.

Table 2.5 Treatment Systems for Arsenic Removal [38]

System	*Coagulant*	*pH*
Iron	Ferric sulfate at 45 mg/l Fe	6.0
Low-lime	Lime at 260 mg/l	
	Ferric sulfate at 20 mg/l Fe	10.0
High-lime	Lime at 600 mg/l	11.5

Arsenic removal has been observed in a conventional water treatment plant employing coagulation, settling, and filtration [31]. Ferric sulfate was employed as a coagulant, and at a coagulant dosage of approximately 500 mg/l it was possible to reduce the arsenic content of the water from 25 mg/l to 5 mg/l or less. Chemical oxidation of arsenite to arsenate, prior to coagulation treatment, improved arsenic removal, suggesting that arsenate forms the more stable sorbed complex with the precipitating metal.

Ferrous hydroxide coprecipitation has also been employed to effectively remove arsenic from water supplies [21]. The treatment sequence involved addition of ferrous sulfate for precipitation, and subsequent filtration. An effluent arsenic level less than 0.05 mg/l was the goal of the treatment process. Ferrous coagulant treatment of arsenic-contaminated water has also been reported by others [23, 33].

Logsdon et al [30] studied arsenate removal by precipitation with ferric sulfate and alum in pilot plant tests. For initial arsenic concentrations of 0.31–0.35 mg/l, ferric sulfate treatment yielded 91–94% removal. Dual media filtration achieved an additional 5–7% removal, for overall removals of 98–99% and effluent levels of 0.003–0.006 mg/l arsenic. Alum treatment was less effective, yielding 75–79% removal without filtration and 85–92% removal with filtration [30]. Arsenite removals of only 10–25% were achieved by alum and 40–60% by ferric chloride. However, when chlorination preceded coagulant addition, removals equivalent to those reported for arsenate were observed [30]. Chlorine apparently oxidized the arsenite to arsenate. For ferric sulfate precipitation, treatment efficiency remained constant from pH 5.5–8.5 (see also Table 2.4). Above this pH, efficiency declined. With alum, optional precipitation pH was at 7.0, with slight reduction in efficiency at pH below 7.0, and rapid deterio-

Table 2.6 Pilot Plant Arsenic Removal [38]

	Cumulative Percent Removal			*Effluent Concentration (mg/l)*
System	*Settling*	*+ Filtration*	*+ Carbon*	
Iron	90	89	96–98	0.06
Low-lime	79	79	82–84	0.92
High-lime	73	75	84–88	0.77

Table 2.7 Effectiveness of Mechanical Dewatering Devices for Coprecipitated Arsenic Sludge [39]

Process	*Cake Solids (%)*
Filter press	25
Belt filter press	13–16
Centrifuge	10
Vacuum filter	10–15
Precoat vacuum filter	10–15

ration in removal efficiency at pH above 7.0. Similar pH effects for arsenate treatment by iron and aluminum coagulant salts have been observed in other studies [1, 6]. In one pilot application of alum, chlorination followed by alum addition was reported to reduce an ammonia-manufacture wastewater arsenic content from 430 to 0.023 mg/l [17]. Treatment pH was 5.0, and the final effluent was gravity clarified, followed by filtration.

Experience with full-scale treatment of deep-well water to remove arsenic indicates that both potassium permanganate and chlorine are effective oxidants of arsenite to arsenate [26]. Copper sulfate coagulant performed poorly, as did alum, in precipitating the arsenic, while ferric sulfate and ferric chloride provided good removals. During four months of plant operation, arsenic levels of 0.35–0.56 mg/l were reduced to below 0.01 mg/l by chlorination to a chlorine residual of 0.1–2.2 mg/l, and ferric chloride dosages of 30–60 mg/l [26].

Sludge management from arsenic treatment by coprecipitation may be difficult. Wilson et al [39] report that ferric chloride ($Fe^{+3}:As = 2.5:1$) plus lime coprecipitation of a gold ore leachate yields a fine precipitate with poor settling properties and a voluminous hydrous sludge. Initial arsenic levels of 300–600 mg/l are reduced below 0.5 mg/l, and polymer addition to the clarifier enhances sludge settleability but does not improve the sludge-dewatering characteristics. Bench-scale mechanical dewatering studies on the sludge yielded the results presented in Table 2.7. A filter press installation for the facility, capable of handling approximately 10,000 gpd sludge at 5% initial solids, was estimated to have a 1984 capital cost of $776,000 [39].

Other Processes

Activated carbon adsorption has been tested for removal of arsenic from a potable water supply. Simple filtration through a charcoal bed yielded an effluent containing 0.06 mg/l of arsenic [29]. This represented 70% removal of the original 0.2 mg/l. Rosehart and Lee reported 40% removal of arsenite by activated carbon, from an initial concentration of 0.5 mg/l–0.3 mg/l [11]. This and other evidence [30, 40] indicate that carbon is not as effective as other means of treating arsenic.

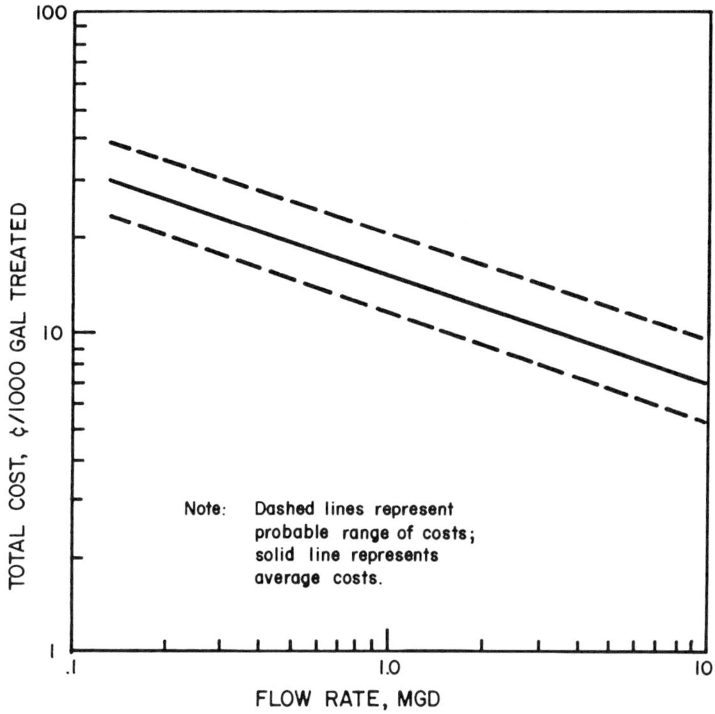

Figure 2.1 Total cost of treatment by coagulation, sedimentation, and rapid sand filtration. Cost (for 1968) includes capital investment for 30 years at 4%, labor, power, chemicals, maintenance and repair, and heating of buildings [42].

Both weak and strong base ion exchange resins appear effective in removing arsenate and arsenite from wastewater. Calmon [41], treating an arsenate waste containing 68 mg/l arsenic at pH 6.9 with a weak-base anion exchange resin, reported 82-100% removal. Medium and strong base resins were less effective. Rosehart and Lee reported successful treatment of arsenite wastes below pH 3 with a weak base resin and at pH 4-13 with a strong base resin. Tests with an intermediate base resin showed optimum removal at pH 4.5-6, but overall removal was less than either the strong or weak base resins [11].

SUMMARY

Arsenic is used as a raw material and produced as a waste in a variety of industrial processes. Despite this wide usage, there is limited information available in the current literature on the performance of full-scale industrial waste

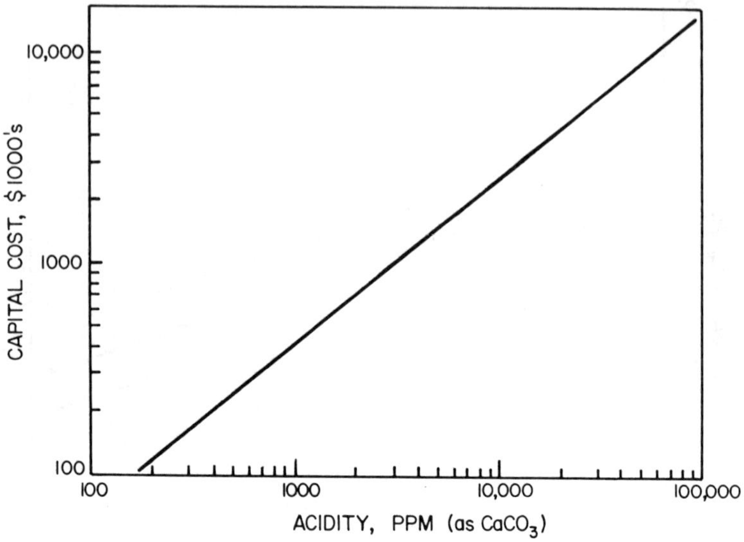

Figure 2.2 Capital cost (in 1970) for a 1 MGD line neutralization plant, including sludge disposal [43].

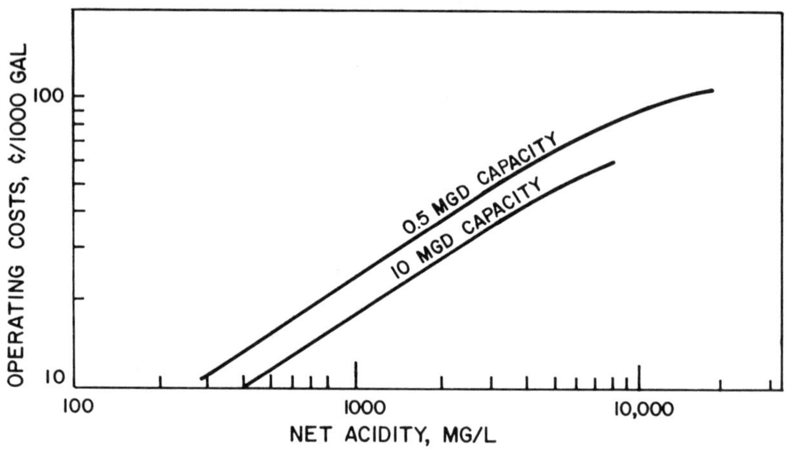

Figure 2.3 Operating costs (in 1970) for lime neutralization, including sludge dewatering by vacuum filtration. [43]

Table 2.8 Criteria for Chemical Selection [44–46]

Chemical Process	*Dosage Range (mg/l)*	*Resulting pH*	*Sludge Dewatering*
High-lime	150–600	11.5–12.0	Easy
Low-lime	75–250	9.5–10.5	Easy
Very low-lime	50–100	8.0–8.5	Fair
Solid alum	75–250	4.5–7.0	Difficult
Liquid alum	75–250	4.5–7.0	Difficult
Ferric chloride	35–150	4.0–7.0	Fair
Ferrous sulfate	70–200	4.0–7.0	Fair

treatment processes. Removal of arsenic from municipal waters has been more widely reported. Arsenic treatment processes for both municipal waters and industrial wastes, insofar as the available literature indicates, are similar and commonly involve coprecipitation by addition of a polyvalent metallic coagulant, to produce a hydroxide floc.

If the arsenic is present as arsenite, effective treatment may require oxidation to arsenate by chlorine or permanganate addition prior to coprecipitation. Based upon limited experience, sulfide or lime precipitation is also an effective treatment technique. For low initial arsenic levels, these processes yield effluent arsenic concentrations of 0.05 mg/l or less.

The treatment processes employed are standard operations. Therefore, it would appear that costs associated with routine water treatment processes would provide some basis for estimating the costs of arsenic removal. Figure 2.1 provides total treatment costs for a treatment process involving addition of coagulant, settling, and sand filtration. Costs of $0.15–0.45/1,000 gal are indicated for plants treating 0.1–0.5 MGD. Figures 2.2 and 2.3 provide incremental lime costs for pH adjustment. Suggested guidelines for the selection of coagulants are presented in Table 2.8.

REFERENCES

1. O'Connor, J.T. "Trace Metals vs. Conventional Water Treatment Systems," *Water Sew. Works* 12 (1974): 72–75.
2. McKee, J.E. and H.W. Wolf. *Water Quality Criteria*, 2nd ed., California State Water Quality Control Board Publication 3-A (1963).
3. Dean, J.G., F.L. Bosqui, and K.H. Lanouette. "Removing Heavy Metals from Waste Water," *Environ. Sci. Technol.* 6 (1972): 518–522.
4. Skripach, T., V. Kagan, M. Ramanov, L. Kamer, and A. Semina. "Removal of Fluorine and Arsenic from the Wastewater of the Rare-Earth Industry," *Proceedings of the 5th International Conference on Water Pollution Research* (New York: Pergamon Press, 1971) vol. 2, pp III–34.
5. Lund, H.F. *Industrial Pollution Control Handbook* (New York: McGraw-Hill, 1971).

6. Leckie, J.O. "Trace Element Removal from Water by Co-Precipitation with Iron Hydroxide," Progress Report EPRI-RP-910, Electric Power Research Institute, Palo Alto, CA (May 1978).
7. Swain, R.E. "Waste Problems in the Nonferrous Smelting Industry," *Ind. Eng. Chem.* 31 (1939): 1358–1361.
8. Stooff, H. and L.W. Haase. "Occurrence and Removal of Arsenic in Drinking Waters," *Chem. Abstr.* 32 (1938): 6370.
9. "Development Document for Interim Final Effluent Limitations Guidelines and Proposed New Source Performance Standards for the Zinc Segment of the Nonferrous Metals Manufacturing Point Source Category," USEPA 440/1–75/032, Group I, Phase II (February 1975).
10. "Developmental Document for Interim Final Effluent Limitations Guidelines and Proposed New Source Performance Standards for the Primary Copper Smelting Subcategory and the Primary Copper Refining Subcategory of the Copper Segment of the Nonferrous Metals Manufacturing Point Source Category," USEPA 440/1–75/032-b, Group I, Phase II (February 1975).
11. Rosehart, R. and J. Lee. "Effective Methods of Arsenic Removal from Gold Mine Wastes," *Can. Min. J.* (June 1972): 53–57.
12. Larsen, H.P., J.K.P. Shou, and L.W. Ross. "Chemical Treatment of Metal-Bearing Mine Drainage," *J. Water Poll. Control. Fed.* 45 (1973): 1682–1695.
13. Ohsako, N. and H. Yoneyama. "Treatment of Arsenic-Containing Sulfuric Acid Production Plants," *Chem. Abstr.* 78 (1973): 235.
14. Cherkinski, S.N. and F.I. Ginzburg. "Purification of Arsenious Waste Waters," *Water Poll. Abstr.* 14 (1941): 315–316.
15. Welmoth, R.C. et al. "Full Scale Removal of Trace Elements from Acid Mine Drainage," undated report of the USEPA Industrial Environmental Research Laboratory, Cincinnati, OH.
16. Petrov, N.M., Y.V. Leiderman, O.V. Maidurova, A.M. Khalemskii, L.F. Kuznetsova, G.P. Kondrat'eva, and T.A. Lomakina. "Removal of Arsenic from Washing Water of a Sulfuric Acid Department," *Chem. Abstr.* 86 (1977): 336.
17. "Developmental Document for Interim Final Effluent Limitations Guidelines and Proposed New Source Performance Standards for the Significant Inorganic Products Segment of the Inorganic Chemicals Manufacturing Point Source Category," USEPA 440/1–75/037, Group I, Phase II (May 1975).
18. Hiroyuki, N. and H. Watanabe. "Removing Arsenic from Wastewaters," *Chem. Abstr.* 81 (1974): 285.
19. "Development Document for Effluent Limitations Guidelines and New Source Performance Standards for the Plywood, Hardboard and Wood Preserving Segment of the Timber Products Processing Point Source Category," USEPA 440/1-74-023-a (April 1974).
20. Cawley, W., ed. *Treatability Manual,* Vol. I, *Treatability Data,* USEPA 600-8-80-042-a (July 1980).
21. Christensen, D.C. and J.A. McNeese. "Removal of Arsenic and Boron from Geothermal Brines," paper presented at the 32nd Purdue Industrial Water Conference, West Lafayette, IN (May 1977).
22. Wangen, L. and J. Williams. "Control by Alkaline Neutralization of Trace Elements in Acidic Coal Cleaning Waste Leachates," *J. Water Poll. Control Fed.* 54 (1982): 1302–1310.

23. Trelles, R.A. and F.D. Amato. "Treatment of Arsenical Waters with Lime," *Water Poll. Abstr.* 23 (1950): 125.
24. Viniegra, G. and R.E. Marquez. "Chronic Arsenic Poisoning in the Lake Region: Section 4. Treatment of Drinking Water," *Water Poll. Abstr.* 38 (1965): 430–431.
25. Shen, Y.S. and C.S. Chen. "Relation Between Blackfoot Disease and the Pollution of Drinking Water by Arsenic in Taiwan," *Proceedings of the 2nd International Conference on Water Pollution Research, Tokyo* (Elmsford, NY: Pergamon Press, 1964) 1:173–190.
26. Shen, Y.S. "Study of Arsenic Removal from Drinking Water," *J. Am. Water Works Assoc.* 651 (1973): 543–548.
27. Bellack, E. "Arsenic Removal from Potable Water," *J. Am. Water Works Assoc.* 63 (1972): 454–458.
28. Curry, N.A. "Philosophy and Methodology of Metallic Waste Treatment," *Proceedings of the 27th Industrial Waste Conference, Purdue University* (May 1972).
29. Magnusen, L.M., T.C. Waugh, O.K. Galle, and J. Bredfeldt. "Arsenic in Detergents: Possible Danger and Pollution Hazard," *Science* 168 (1970): 389–390.
30. Logsdon, G.S., T.J. Sorg, and J.M. Symons. "Removal of Heavy Metals by Conventional Treatment," *Proceedings of the 16th Water Quality Conference*, University of Illinois, Urbana (February 13–14, 1974): 11–133.
31. Buswell, A.M., R.C. Gore, H.E. Hudson, H.C. Wiese, and T.E. Larson. "War Problems in Analysis and Treatment," *J. Am. Water Works Assoc.* 35 (1943): 1303–1311.
32. Irukayama, K. "Discussion—Relation between Blackfoot Disease and the Pollution of Drinking Water by Arsenic in Taiwan," *Proceedings of the 2nd International Conference on Water Pollution Research, Tokyo* (Elmsford, NY: Pergamon Press, 1964) Vol. 1, pp 185–187.
33. Berezman, R.I., "Removal of Inorganic Arsenic from Drinking Water under Field Conditions," *Water Poll. Abstr.* 29 (1965): 185.
34. Ohtsubo, K. "Treatment of Arsenic Containing Waste Water," *Chem. Abstr.* 85 (1976): 379.
35. "Development Document for Effluent Limitations Guidelines and New Source Performance Standards for the Phosphorus Derived Chemicals Segment of the Phosphate Manufacturing Point Source Category," USEPA Publication 440/1-74-006-a (January 1974).
36. Electrolytic Metal Corporation. "Recovery of Manganese from Solutions Containing Iron, Manganese, Nickel and Cobalt—Patent," *Chem. Abstr.* 54 (1960): 12960 (c).
37. Nilsson, R. "Removal of Metals by Chemical Treatment of Municipal Waste Water," *Water Res.* 5 (1971): 51–60.
38. Maruyama, T., S.A. Hannah, and J.M. Cohen. "Removal of Heavy Metals by Physical and Chemical Treatment Processes," *J. Water Poll. Control Fed.* 47 (1975): 962–975.
39. Wilson, R., K. Murphy, and J. Ramshaw. "Arsenic Sludge Dewatering," paper presented at the 39th Annual Purdue Industrial Waste Conference, West Lafayette, IN (May 1984).
40. Lee, J.Y. and R.G. Rosehart. "Arsenic Removal by Sorption Processes from Waste Waters," *Can. Min. Metallurg. (CIM) B.* 65 (1972): 33–37.
41. Calmon, C. "Comment," *J. Am. Water Works Assoc.* 651 (1973): 568–569.

42. Dorr-Oliver, Inc. "Cost of Waste Water Treatment Processes," Robert A. Taft Water Research Center Report TWRC-6, USDI, Washington, DC (1968).
43. *The Economics of Clean Water*, vol. III. *Inorganic Chemicals Industry Profile*, (Washington, DC: USDI, 1970).
44. Tofflemire, T.J. and L.J. Hetling. "Chemicals and Clarifiers: Which are Best?" *Water Wastes Eng*. 10 (1973): F24–26.
45. Edwards, A.W., W.W. Eckenfelder, Jr., and J.A. Roth. *Industrial Water Cost Studies*, vol. 1, (Washington, DC: Office of Research and Development, USEPA, 1975).
46. Van Note, R.H., P.V. Hebert, R.M. Patel, C. Chupek, and L. Feldman. "A Guide to the Selection of Cost-Effective Wastewater Treatment Systems," USEPA 430/9-75-002 (July 1975).

3
Barium

Barium is utilized in metallurgy; in glass, ceramics, and dyes manufacture; and in vulcanizing rubber [1]. Barium (as barite, barium sulfate) is used industrially as a white pigment for paints [2]. Of values reported for paint manufacturing wastewaters (five plants) only one plant had appreciable wastewater barium levels (0.77–5.7 mg/l). The remaining plants ranged from <1 to 1.67 mg/l [3]. Barium has been reported present in explosives-manufacturing wastewater [4] and in midwestern groundwaters at concentrations up to 10 mg/l [5]. The presence of barium in ink-manufacturing wash water is also reported [6]. Additional processes that utilize barium are zinc oxide formulation by the wet process, and potassium iodide manufacturing [7].

TREATMENT TECHNOLOGY

In treatment of an explosives-manufacturing wastewater, barium was removed by precipitation as barium sulfate upon addition of sodium sulfate [4]. Coagulation of suspended solids in the explosives wastewater (including fine, nonsettleable barium sulfate particles) with ferric sulfate enhanced solids removal. The process of precipitation of barium sulfate by addition of sodium sulfate is used by barite producers to recover barium from solution [2]. Others have also recommended sulfate precipitation for barium treatment, based upon the relative solubility products of barium sulfate and barium hydroxide [5]. Barium sulfate is relatively insoluble, having a minimum theoretical solubility of approximately 1.4 mg/l as barium at stoichiometric concentrations of barium and sulfate [8]. The solubility level of barium can be reduced by addition of sulfate in excess. Alternately, in waters of moderate to high carbonate alkalinity, barium carbonate precipitation may be preferable [9].

Barium removal has been investigated in 4 gpm pilot plant studies [10, 11]. Continuous-flow pilot plants received municipal wastewater dosed with barium and other metals, and removals by each unit process were determined. The unit processes were coagulation followed by settling, mixed media filtration, and activated carbon adsorption. Table 3.1 lists the coagulants and operating pH values for each of the four pilot plants. The ferric sulfate dosages for the iron and low-lime systems, applied to initial barium concentrations of 5 mg/l, are equivalent to respective sulfate dosages of 92.2 and 36.5 mg/l. The

Table 3.1 Pilot Treatment Systems for Barium Removal [10, 11]

System	*Coagulant*	*pH*	*Sulfate Dosed (mg/l)*
Iron	Ferric sulface at 45 mg/l Fe	6.0	92.2
Alum	Aluminum sulfate at 200 mg/l	6.4	164.5
Low-lime	Lime at 260 mg/l Ferric sulfate at 20 mg/l Fe	10.0	36.5
High-lime	Lime at 600 mg/l	11.5	0.0

alum (aluminum sulfate) dosage, applied to an initial barium concentration of 0.5 mg/l, is equivalent to a sulfate dosage of 164.5 mg/l. Thus, sulfate was present in excess of the barium. Excess sulfate will reduce barium solubility, and thereby improve the effectiveness of the barium sulfate precipitation process [12].

The results of the pilot plant studies are presented in Table 3.2. Treatment levels of 90% or greater were found for the processes using ferric sulfate or alum, while the high-lime process at pH 11.5 was much less efficient. The data of Table 3.2 indicate that increasing sulfate dosage may not be the only controlling treatment parameter, since the low-lime system at pH 10, which employed sulfate at 36.5 mg/l, was slightly more effective than the iron system at pH 6.4 and a sulfate dosage of 92.2 mg/l. The major fraction of removal was achieved by settling. Carbon adsorption did not improve treatment efficiency. The inefficiency of activated carbon in removing barium has also been noted by others [5, 13].

The lack of simple correlation found in the pilot study between sulfate dosage and barium removal appears to be partially explainable in terms of the kinetics of barium sulfate precipitate formation. In batch studies on barium sulfate precipitation, high initial sulfate dosages apparently resulted in supersaturated barium sulfate solutions, which came to equilibrium with the precipitant solid phase very slowly [5]. Coagulation, with aluminum sulfate and ferric

Table 3.2 Pilot Plant Barium Removal Results [10, 11]

	Cumulative Percent Removal			*Effluent Barium Concentration (mg/l)*
System	*Settling*	+ *Filtration*	+ *Carbon*	
Iron	97	95	95	0.27[a]
Alum	87	79	93	0.04[b]
Low-lime	99	99	99	0.03[a]
High-lime	80	78	81	0.94[a]

[a] Initial barium concentration = 5 mg/l.
[b] Initial barium concentration = 0.5 mg/l.

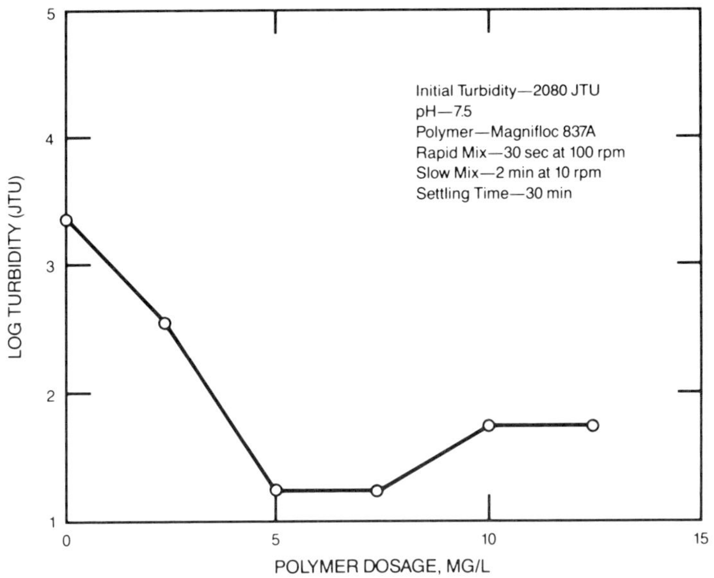

Figure 3.1 Effect of polymer dosage on the removal of turbidity in supernatant from system employing direct neutralization with barium carbonate [12].

sulfate coagulant dosages up to 120 mg/l, yielded barium removals of 20–40% for initial barium concentrations of 7.0–8.0 mg/l at treatment pH 7.5–8.0. However, initial coagulant dosages of 100 mg/l, with subsequent (up to 24 hours later) coagulant dosages of 20 mg/l, yielded overall removals of 70% [5]. These results suggest a very slow approach to precipitation equilibrium at high sulfate levels for barium sulfate precipitation, requiring long wastewater detention times. Barium sulfate forms a colloidal precipitate and may require additional coagulation and/or filtration to improve colloidal barium precipitant removal. Figure 3.1 demonstrates the effect of polymer addition on residual turbidity after barium sulfate precipitation by barium carbonate treatment chemical addition to a high-sulfate wastewater.

Carter [14] reported bench-scale treatability results for a piston ring-manufacturing wastewater containing oil, barium, and heavy metals. The barium presence resulted from the use of barium carbonate in purifying chrome-plating baths by sulfate precipitation plus use of a proprietary barium salt to precipitate lead and other heavy metals in a pretreatment stage. The treatment process tested involved acid addition to pH 3.0, for oil demulsification (see Chapter 17), followed by lime precipitation and clarification. Initial barium concentration was 2,100 mg/l. Use of hydrochloric acid in the demulsification step yielded a final settled barium concentration of 377 mg/l. Use of sulfuric acid yielded 1.78 mg/l effluent barium, through the precipitation of barium sulfate.

Table 3.3 Equipment Costs for Chemical Treatment Systems—1979 Basis [3]

	Cost (\$/1000 gpd)[a]			
Capacity (MGD)	*Reaction Tanks*	*Clarifiers and Thickeners*	*Sludge Filters or Centrifuges*	*Total Costs*[b]
0.01	2,600	2,600	4,400	10,500
0.1	450	700	450	1,600
1.0	66	130	350	550
10.0	9	35	130	175

[a] These costs, originally reported in 1973 dollars, have been updated by the Marshall & Swift Equipment Cost Index. 1973 = 344.1; 1979 = 599.4.

[b] Includes engineering, land preparation and installation only.

Efficient barium removal has been achieved by lime-softening treatment of hard, alkaline water with an initial barium level of 7–8 mg/l [5]. Precipitation of barium carbonate upon pH adjustment with lime is the most probable barium removal mechanism. Treatment efficiency was pH dependent, increasing to an optimum of 98% at pH 10.5, and dropping off at higher pH. These results are comparable to those described in Tables 3.1 and 3.2. At the reported optimum pH of 10.5, essentially all bicarbonate alkalinity has been converted to carbonate. The reduced treatment efficiency at pH above 11 may be due to preferential formation of a more soluble form of barium salt.

Generalized capital costs for chemical precipitation systems are given in Table 3.3. Chemical costs depend upon chemical requirements (neutralization, polymer or coagulant feeds, etc.). Detailed capital cost information for individual unit operations for chemical treatment (according to the treatment schematic of Figure 3.2) are given in Figures 3.3–3.10. Additional cost would be incurred for sand or mixed media filtration of the effluent for solids control. Figure 3.11 presents capital and operating costs (1973 basis) for effluent filtration systems.

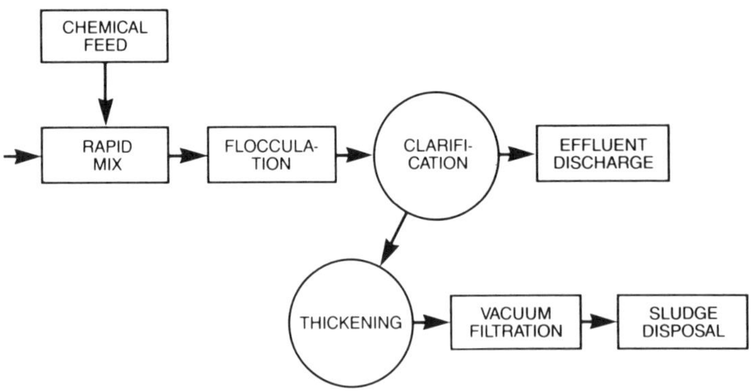

Figure 3.2 Basic chemical treatment flow diagram [15].

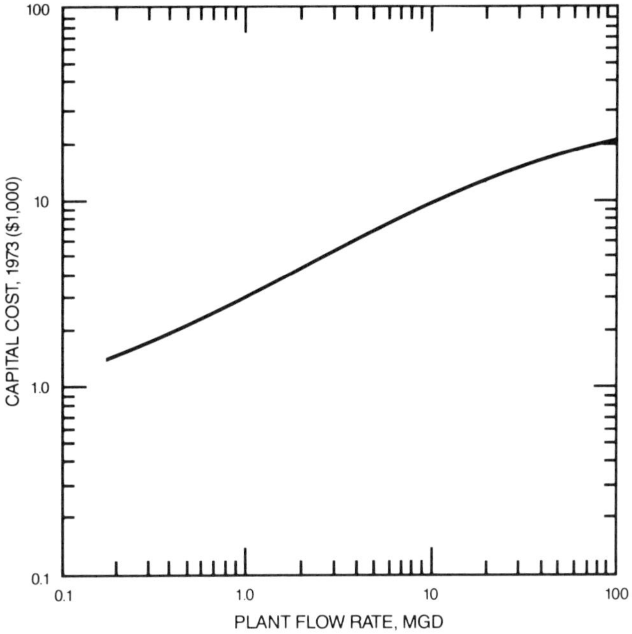

Figure 3.3 Capital cost (in 1973) of polymer feed facilities [15].

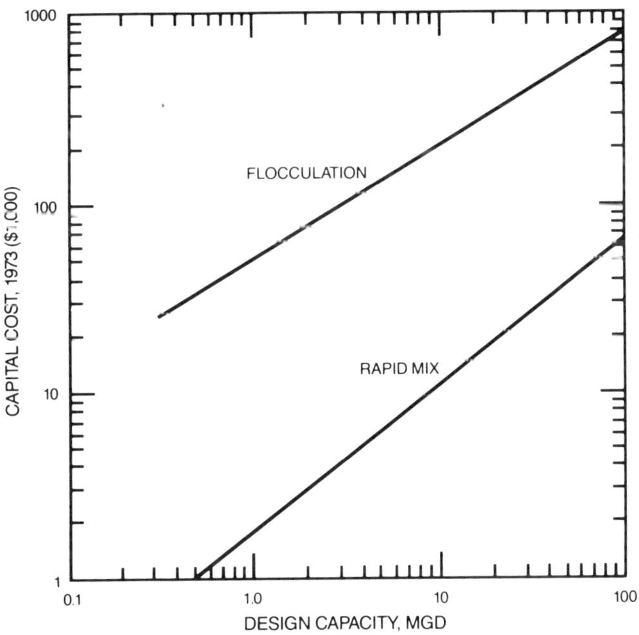

Figure 3.4 Capital cost (in 1973) of rapid mix and flocculation [15].

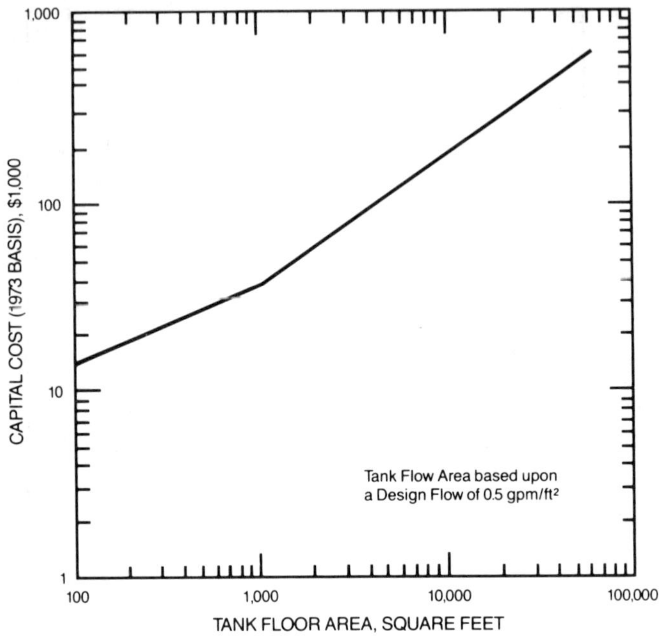

Figure 3.5 Capital cost (in 1973) of clarification [15].

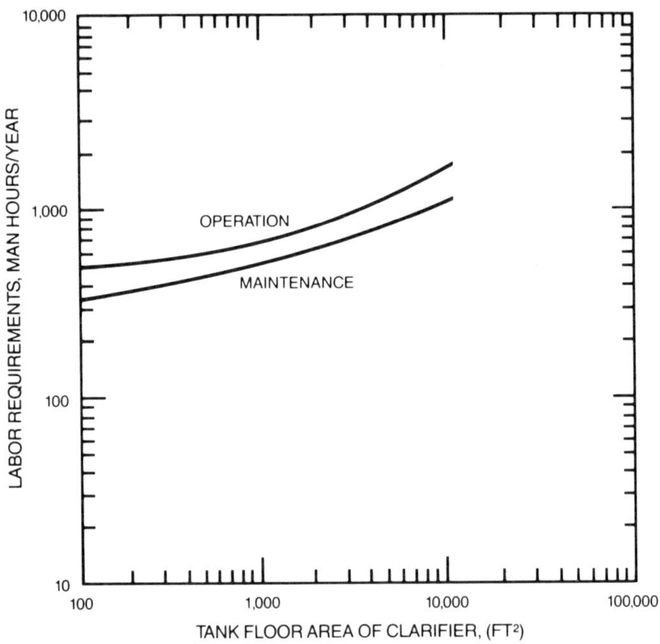

Figure 3.6 Labor requirements for operation and maintenance of chemical treatment facilities [15].

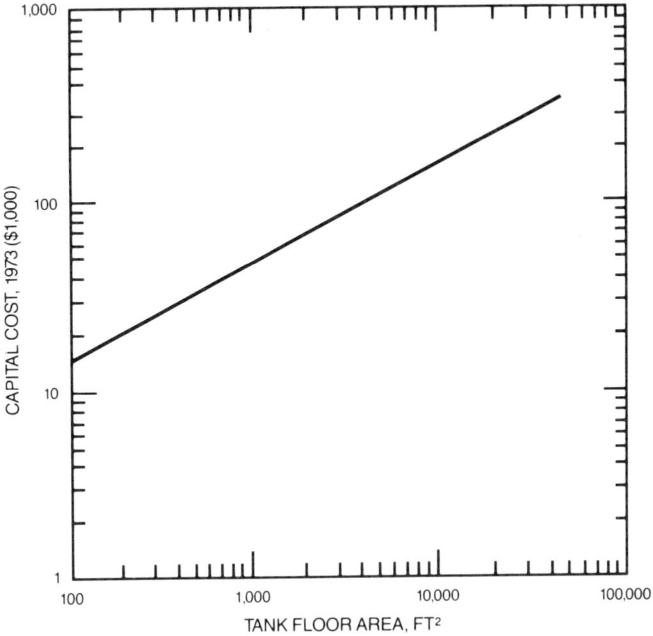

Figure 3.7 Capital cost (in 1973) of gravity thickeners [15].

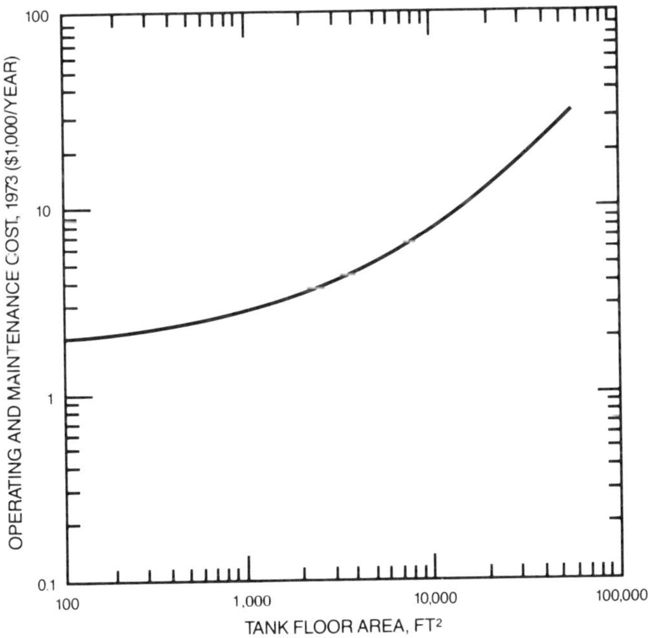

Figure 3.8 Operation and maintenance cost (in 1975) for gravity thickening [15].

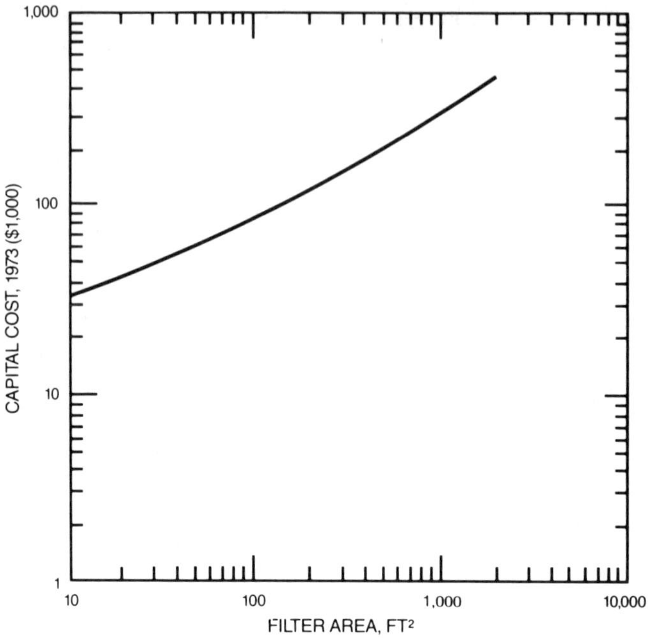

Figure 3.9 Capital cost (in 1973) of vacuum filtration [15].

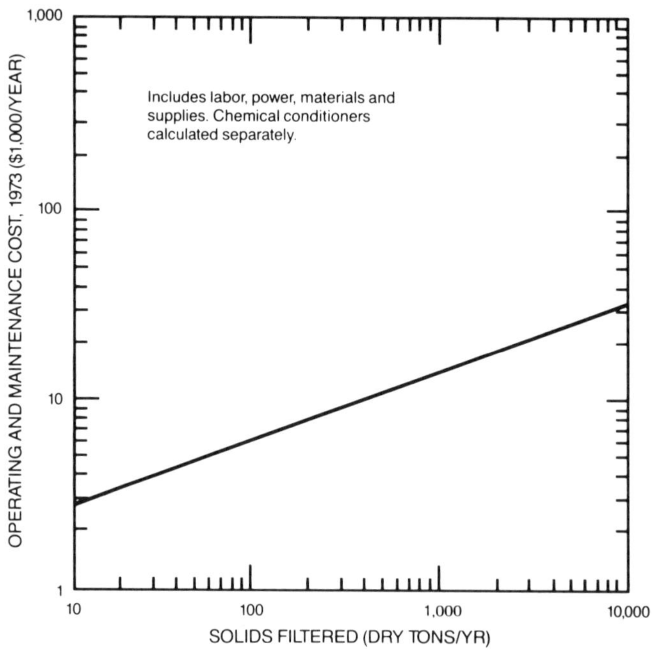

Figure 3.10 Operation and maintenance cost (in 1973) for vacuum filtration [15].

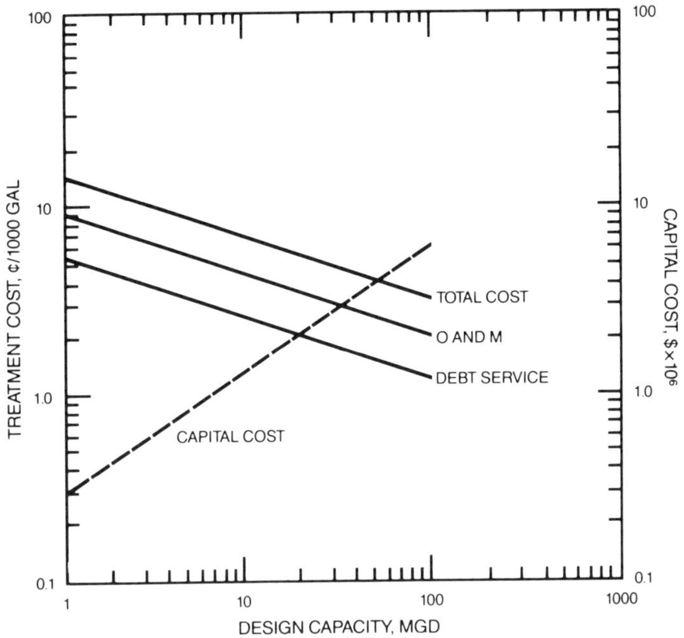

Figure 3.11 Capital and treatment costs (in 1973) for mixed media or sand filtration. (From Monti and Silbermann [16], courtesy of *Water Wastes Engineering*.)

Another reported barium treatment process is ion exchange. Successful ion-exchange treatment of barium has been reported for both nuclear wastes [17, 18], and groundwater softening [5]. Ayres [17] cited extremely high removals (>99.9%) for radioactive barium-140 by ion exchange, and Pressman et al [18] report 99% removal of nonradioactive barium in a U.S. Army mobile ion-exchange unit. In a full-scale ion-exchange groundwater softening plant, barium reduction from 11.7 to 0.17 mg/l, a 98.5% removal efficiency, has been observed [5]. These results indicate that ion exchange may be a more effective barium treatment technique than either sulfate or carbonate precipitation. However, successful ion-exchange treatment would likely require use of general, nonspecific water-softening type resins, with consequent removal of constituents not removed by the precipitation technique. Further, ion exchange may represent a four-to-nine-fold greater treatment cost than chemical precipitation plus filtration [19].

SUMMARY

Methods of barium removal are summarized in Table 3.4, and include precipitation as barium sulfate upon addition of ferric, aluminum, or sodium sulfate;

Table 3.4 Summary of Barium Treatment Methods

Method	*pH*	*Barium Concentration (mg/l)* *Initial*	*Final*	*% Removal*	*Reference*
Precipitation					
Sulfate	6.0	5.0	0.27	95	10
	6.4	0.5	0.04	93	11
	10.0	5.0	0.03	99	10
	—	—	0.5	11	20
	—	2,100	1.78	99+	19
	7.5–8.0	7.0–8.0	2.1–2.6	70	5
Carbonate	10.5	7.0–8.0	0.15	98	5
Hydroxide	11.5	5.0	0.94	81	10
Ion exchange	—	11.7	0.17	98	5
Ion exchange	—	—	—	99.9+	17
Ion exchange	—	—	—	99	18

precipitation as barium carbonate at pH 10–10.5 with lime used for pH adjustment; and ion exchange. Extremely high removal efficiencies were reported for the third process. Based upon relative economics, however, and the low solubility of barium sulfate and barium carbonate, the precipitation method of barium removal would appear to be the method of choice.

REFERENCES

1. McKee, J.E. and H.W. Wolf. *Water Quality Criteria*, 2nd ed., California State Water Quality Board 3-A (1963).
2. *The Economics of Clean Water*, vol. III. *Inorganic Chemicals Industry Profile*, (Washington, DC: USDI, 1970).
3. "Development Document for Effluent Limitations Guidelines and New Source Performance Standards for the Major Inorganic Products Segment of the Inorganic Chemicals Manufacturing Point Source Category," USEPA 440/1-74-007-a (March 1974).
4. "Taming Explosives," *Ind. Eng. Chem.* 46 (1954): 20A.
5. Logsdon, G.S., T.J. Sorg, and J.M. Symons. "Removal of Heavy Metals by Conventional Treatment," *Proceedings of the 16th Water Quality Conference*, University of Illinois, Urbana (February, 13–14, 1974).
6. "Development Document for Effluent Limitations Guidelines and New Source Performance Standards for the Oil Base Solvent Wash Subcategories of the Paint Formulating and the Ink Formulating Point Source Category," USEPA 440/1-75/050-a, Group II (July 1975).
7. "Development Document for Interim Final Effluent Limitations Guidelines and Proposed New Source Performance Standards for the Significant Inorganic Prod-

ucts Segment of the Inorganic Chemicals Manufacturing Point Source Category," USEPA 440/1-75/037, Group I, Phase II (May 1975).

8. Sillen, L.G. and A.E. Martel. *Stability Constants of Metal-Ion Complexes*, 2nd ed., Special Publication 17 (London: Chemical Society, 1964).
9. Culp, G.L. and R.L. Culp. *New Concepts in Water Purification* (New York: Van Nostrand Reinhold, 1974).
10. Maruyama, T., S.A. Hannah, and J.M. Cohen. "Removal of Heavy Metals by Physical and Chemical Treatment Processes," *J. Water Poll. Control. Fed.* 47 (1975): 962–975.
11. Hannah, S.A., M. Jelus, and J.M. Cohen. "Removal of Uncommon Trace Metals by Physical and Chemical Treatment Processes," *J. Water Poll. Control. Fed.* 49 (1977): 2297–2309.
12. King, P.H., L.R. Norris, and C.W. Randall. "Sulfuric Acid Neutralization and Concurrent Sulfate Removal by Precipitation with Barium," paper presented at the 30th Annual Purdue Industrial Waste Conference, West Lafayette, IN (May 1975).
13. Sigworth, E.A. and S.B. Smith. "Adsorption of Inorganic Compounds by Activated Carbon," *J. Am. Water Works Assoc.* 64 (1972): 386–391.
14. Carter, J.L. "Treatment for Removal of Heavy Metals and Oil from a Piston Ring Manufacturing Waste," paper presented at the 37th Annual Purdue Industrial Waste Conference, West Lafayette, IN (May 1982).
15. Edwards, A.W., W.W. Eckenfelder, Jr., and J.A. Roth. *Industrial Water Cost Studies*, vol. 1, Office of Research and Development, USEPA Contract 68-01-2485 (1975).
16. Monti, R.P. and P.T. Silbermann. "Wastewater System Alternatives: What Are They and What Cost? *Water Wastes Eng.* 11 (1974): 40–48, 75.
17. Ayres, J.A. "Treatment of Radioactive Waste by Ion Exchange," *Ind. Eng. Chem.* 43 (1951): 1526–1531.
18. Pressman, M., D.C. Lindsten, and R.P. Schmitt. "Removal of Nuclear Bomb Debris; Strontium-90-Yttrium-90, and Cesium-137-Barium-137 from Water with Corps of Engineers Mobile Water Treatment Equipment," *Water Poll. Abstr.* 36 (1963): 42–43.
19. Middlebrooks, E.J., J.H. Reynolds, G.L. Shell, and N.B. Jones. "Modifying Wastewater Treatment Plants to Meet New Standards," *Public Works* 103 (1972): 80–85.
20. Curry, N.A. "Philosophy and Methodology of Metallic Waste Treatment," *Proceedings of the 27th Industrial Waste Conference, Purdue University* (May 1972).

4
Cadmium

Potential sources of cadmium wastewaters include metallurgical alloying, ceramics and inorganic pigments manufacture, electroplating, textile printing, mine drainage, and some photographic wastes [1, 2]. Of the total industrial cadmium use, 90% is utilized in electroplating, pigments, plastics stabilizers, alloying, and batteries. Most of the remaining 10% is used for television tube phosphors, golf course fungicides, rubber-curing agents, and nuclear reactor shields and rods [3].

High concentrations of cadmium have been reported in wastewaters from plating, nonferrous metals smelting and refining, and acid mine drainage (Table 4.1). Cadmium plating provides a corrosion-protective coating over the base metal (commonly iron or steel). Cadmium cyanide is a popular electroplating bath solution, due to the good adherence properties and fine-grained deposit formed. The typical cadmium cyanide bath contains cadmium at 26 g/l, with cyanide concentration nearly four times as great [12]. The major source of cadmium plating waste is from plating bath dragout into rinsewaters. Dumping of cadmium baths is not normally practiced [12].

Cadmium discharges from inorganic pigment-industry plants range from 2 to 120 grams per day, with a plant average of 8 g/day [21]. The principal pigments produced are cadmium sulfide and cadmium selenide. A cadmium concentration of 15 mg/l has been reported for one municipal incinerator scrubber water [22].

TREATMENT TECHNOLOGY

Methods of cadmium wastewater treatment currently in use are essentially those employed for most heavy-metal wastes encountered in the plating and metal-processing industries. Common methods include precipitation as the hydroxide or sulfide, and ion exchange. Carbonate precipitation has also been described. Although not often mentioned in the cadmium treatment literature, electrolytic and evaporative recovery processes, as used for copper, silver, and nickel wastes, are technically feasible, for a concentrated cadmium waste or after a preconcentration step such as ion exchange. Effluent values for cadmium treatment processes are summarized in Table 4.2. Additional cadmium

Table 4.1 Cadmium Concentrations Reported for Industrial Wastewaters

Process	*Cadmium Concentration (mg/l)*	*Reference*
Plating rinsewaters		
Automobile heating control manufacturing	14–22	4
Automatic barrel Zn and Cd plant	10–15	5
Mixed plating, manual Cd	0.9	6
Mixed manual barrel and rack	7–12	5
Large installations	15 avg, 50 max	7
Rinse dragout	48	8
	158	9
	0.1–6.0	10
	0.4	11
	58	11
Large job shop	3.1	12
Recirculating rinse	1,000–3,330	10
Metal finishing plant	2–8	13
Bright dip and passivation plating baths	2,000–5,000	5
Nonferrous Metals Manufacture		
Copper smelting	0.09–1.08	14
Lead smelting and refining	0.08–1.20	15
Zinc smelting and refining	0.02–33.0	16
Copper and zinc smelting	15	11
Iron and steel manufacturing	0.00–80	17
Iron foundry wastewater	0.16–0.95	18
Iron foundry wastewater	0.00–0.74	17
Paint and ink formulation	0.00–0.81	17
Rubber processing	0.00–0.72	17
Porcelain enameling	0.00–9.60	17
Acid lead mine drainage	1,000	1
Acid mine drainage	400–1,000	19
Coal-cleaning acid leachate	0.21	20

data, for lime precipitation treatment of nonferrous metals manufacturing wastewaters, are present in Table 4.3.

Chemical Precipitation

Cadmium forms an insoluble and highly stable precipitate at alkaline pH [32]. Freshly precipitated cadmium hydroxide has been reported to leave approxi-

Table 4.2 Treatment Technology for Cadmium Removal

Method	*Treatment pH*	*Initial Cd (mg/l)*	*Final Cd (mg/l)*	*% Removal*	*Reference*
"Hydroxide" precipitation	8.0	—	1.0	—	23
	10.0	—	0.01	—	23
	9.0	—	0.54	—	24
	9.3–10.6	4.0	0.2	95	6
	—	5.2	0.4	92	6
	9.4–10.2	1.2	0.7	45	6
Hydroxide precipitation	10.0	0.34	0.054	84.1	18
plus filtration	10.0	0.34	0.033	90.3	18
Hydroxide precipitation	11.0	—	0.00075	—	25
plus filtration	11.0	—	0.00070	—	25
Hydroxide precipitation plus filtration	11.5	—	0.014	—	26
Coprecipitation with ferrous hydroxide	6.0	—	0.050	—	26
	10.0	—	0.044	—	26
Coprecipitation with alum	6.4	0.7	0.39	45	28
Sulfide precipitation	6.5	440–1000	0.008	99+	19
	9.2	2,200	1.2	99+	29
Reverse osmosis	11.5	—	—	78–99+	30
Freeze concentration	—	100	0.4	99.6	31

Table 4.3 Cadmium Precipitation by Lime Treatment of Nonferrous Metals Manufacturing Industry Wastewaters [14–16]

	Cadmium Concentration (mg/l)	
Treatment pH	*Influent*	*Effluent*
7.1	—	0.06
8.2	0.38	0.02
9.5	—	0.70
9.3–10.8	—	0.03–0.7
10.5	—	0.01–0.3
11.8	33	0.03

Table 4.4 Bench-Scale Lime Precipitation Results for Cadmium [33]

Treatment pH	*Cadmium Concentration (mg/l)*[a]
5.8	94.0[b]
8.2	51.0
9.0	6.3
10.4	0.15
10.8	<0.05
11.4	<0.05

[a] Concentration in settled supernatant.
[b] Initial cadmium concentration.

mately 1 mg/l residual cadmium ion in solution at pH 8 [23]. Most effective cadmium hydroxide precipitation should occur above pH 10.0 (Table 4.4). According to theoretical cadmium hydroxide solubility considerations, little additional treatment benefit should accrue from application of treatment pH above 10.0 [34]. However, many full-scale cadmium-precipitation treatment plants operating at lower pH consistently achieve effluent cadmium levels below 0.15 mg/l [35]. This anomaly has recently been explained by studies on the form of cadmium salt precipitated, upon pH adjustment [34, 36].

From these recent studies, it has been observed that in wastewaters containing moderate to high background carbonate levels, or to which supplemental carbonate has been added, cadmium can preferentially precipitate as the carbonate salt. This salt is much more insoluble than is cadmium hydroxide. As demonstrated by the theoretical solubility curves of Figure 4.1, even modest background carbonate (C_T) levels should greatly reduce the cadmium solubility. This carbonate influence is observed at pH values below pH 11–12, depending upon the carbonate concentration present.

These solubility effects have been confirmed in field tests on cadmium cyanide-plating bath rinsewaters [36]. Full-scale batch studies were performed on the effects of treatment pH and supplemental carbonate addition on treatment of recirculating rinsewater from a cadmium cyanide-plating line. The results for four batch treatments are summarized in Table 4.5, and the data plotted on Figure 4.1. Providing that the cyanide has first been adequately destroyed, significant improvements in cadmium treatment efficiency occur at near-neutral pH, in the presence of carbonate. At relatively low carbonate levels and high treatment pH values, the benefits of carbonate precipitation disappear. Osantowski and Ruppersberger [18] report results for parallel pilot treatment trains involving precipitation, clarification, and filtration. One train employed only lime, and the second, lime plus a small supplement of about 200 mg/l sodium carbonate. Both trains operated at pH 10.0, and the filtrate cadmium level from both treatment trains averaged 33–34 μg/l.

Typically, cadmium cyanide-plating rinsewaters, due to high plating-bath pH, contain significant carbonate as a result of uptake of atmospheric CO_2.

Table 4.5 Effect of Carbonate Addition on Batch Treatment of Cadmium-Plating Wastewater [36]

Batch No.	*Inorganic Carbon Added*[a] *(mg/l)*	*pH*		*Cadmium*[b] *(mg/l)*	
		Initial	*Final*	*Initial*	*Final*
1	0	11.6	8.9	2,100	6.1
2	457	11.8	8.3	1,080	3.0
3	684	11.6	11.2	840	0.42
4	1,142	11.3	8.3	850	0.37

[a] Raw wastewater inorganic carbon averaged 1,379 mg/l.
[b] Values reported are soluble concentrations.

Other types of industrial cadmium wastewaters, particularly where initially acidic, may be deficient and require supplemental inorganic carbon addition (for example, with soda ash or a bicarbonate salt) to achieve the improved performance associated with precipitation of the cadmium carbonate. Additional benefits cited for this technique, due to the crystalline nature of the carbonate precipitate formed, are improved settleability, reduced sludge volumes, and enhanced sludge dewatering characteristics [36]. An additional benefit, as demonstrated from Figure 4.1, is that treatment in the presence of carbonate can be effective at near-neutral pH.

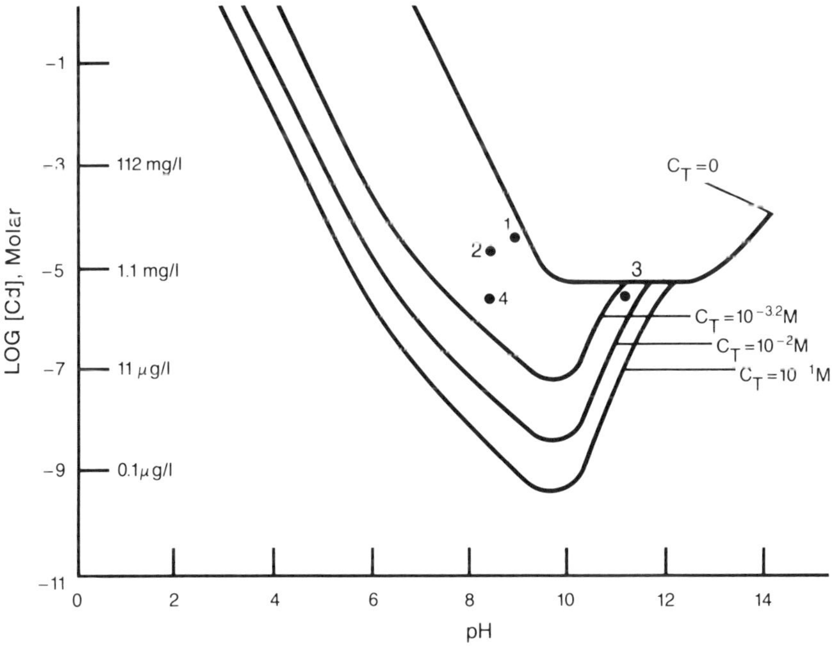

Figure 4.1 Effect of carbonate level (C_T) on cadmium solubility [36].

It has been indicated that coprecipitation with iron hydroxide at pH 8.5 improves cadmium removal [37, 38], and that coprecipitation with aluminum hydroxide is also effective for cadmium as well as for zinc and nickel [25]. Evidence for the beneficial treatment effect of coprecipitation with iron hydroxide is presented by Maruyama et al [26]. In pilot studies of metals removal in municipal wastewater, three treatment techniques were investigated. These were ferrous sulfate (45 mg/l Fe) addition at pH 6; low-lime (260 mg/l) plus ferrous sulfate (20 mg/l Fe) to pH 10; and high-lime (600 mg/l) to pH 11.5. Flocculation and settling were followed by mixed media filtration prior to passage through activated carbon columns. With iron addition only, a residual cadmium concentration of 0.05 mg/l was reported. Low-lime treatment yielded 0.044 mg/l and high-lime, 0.014 mg/l cadmium residuals. Thus, iron coprecipitation alone at pH 6 yielded near-equal results to iron plus lime treatment at pH 10. Later work, using alum coagulant (220 mg/l) pretreatment at pH 6.4 of lower initial levels of cadmium (0.7 mg/l), yielded an average cadmium removal of 43% after settling, 45% after filtration, and 98.1% cumulative removal after old carbon. Virgin carbon achieved only 55% cumulative removal [28]. The influence of previously exhausted and regenerated carbon on cadmium removal has not been determined, but could be due to an increase in either carboxy or sulfide sites on the carbon.

Coprecipitation may be an inherent phenomenon in cadmium removal, if high concentrations of other ions such as aluminum or iron are already present in the raw wastewater. Wangen and Williams [20] report achieving a soluble cadmium concentration below 3 μg/l at treatment pH 5.8, in a wastewater initially containing 370 mg/l aluminum, 1,680 mg/l ferrous iron, and 1,630 mg/l ferric iron.

In precipitation of cadmium from a waste stream at a machine-plating plant, an effluent pH of 9.0 and cadmium concentration of 0.54 mg/l was reported for the lime precipitation process [24]. Although sand filtration was not used, the process incorporated coagulant aid addition at 1–2 mg/l. The effluent contained only 2 mg/l suspended solids, indicating efficient solids removal by sedimentation.

In the presence of complexing agents (e.g. cyanide), it is difficult to precipitate the cadmium ion [38]. In the event that appreciable complexing agents are present in the waste solution, pretreatment to destroy these agents is required. Cadmium-plating wastes normally contain cyanide. Effective precipitation of cadmium from spent plating baths or rinsewaters is therefore dependent upon prior cyanide removal. Fortunately, cyanide breakdown is relatively rapid and easy, as has been demonstrated for zinc and copper cyanide baths [39, 40]. The availability of a cyanide-free cadmium-plating bath has been suggested as an answer to simplified treatment needs [41]. However, the bath has a high ammonium salt concentration (112 g/l) and also contains a proprietary chelating agent. Problems related to cadmium complexation by both the ammonia and proprietary chelating agent could be anticipated from such a system [42].

Table 4.6 Results of Cadmium Precipitation Treatment in the Electroplating Industry [12]

	Cadmium Concentration (mg/l)	
Treatment pH	*Influent*	*Effluent*
6.0–9.5	—	0.03
6.5–9.5	—	0.01
6.5–8.0	—	0.30
7.2	—	0.01
8.2	—	0.03
8.8	3.10	0.21
10.5	—	0.05
—	—	0.02
—	—	0.05
—	—	0.02[a]

[a] Gravity clarification plus sand filtration.

Lime addition for precipitation of cadmium wastewaters is widely practiced in the electroplating industry. Table 4.6 summarizes operating data for cadmium treatment of electroplating wastewaters. The treatment data of Table 4.6 likely represent the lower limit of hydroxide precipitation treatment, since the plants for which the data are reported have been identified as "exemplary" plants by the USEPA [12]. Variations in treatment performance among the 10 plants of Table 4.6 could result from (1) ineffective cyanide destruction, (2) treatment pH applied, (3) background carbonate level, and (4) efficiency of precipitate solids separation.

In the absence of or after destruction of cyanide, unit operations involved in cadmium precipitation are similar to those of the conventional tertiary treatment sequence of coagulation, settling, and sand filtration. The latter step has become of increasing importance for complete removal of nonsettling and slow-settling metal hydroxide flocs [12, 25]. Data presented in Table 4.7 for actual plant operation illustrate the distribution of effluent cadmium between soluble

Table 4.7 Distribution of Cadmium in Precipitation Treatment Effluent for Two Plating Wastes [6]

Plant A			*Plant B*		
pH	*Dissolved (mg/l)*	*Total (mg/l)*	*pH*	*Dissolved (mg/l)*	*Total (mg/l)*
9.3	0.01	17.0	9.4	0.09	0.17
10.2	0.08	21.1	9.8	0.10	4.0
10.3	0.12	4.4	9.9	0.10	1.6
10.6	0.09	1.9	10.2	0.08	0.50

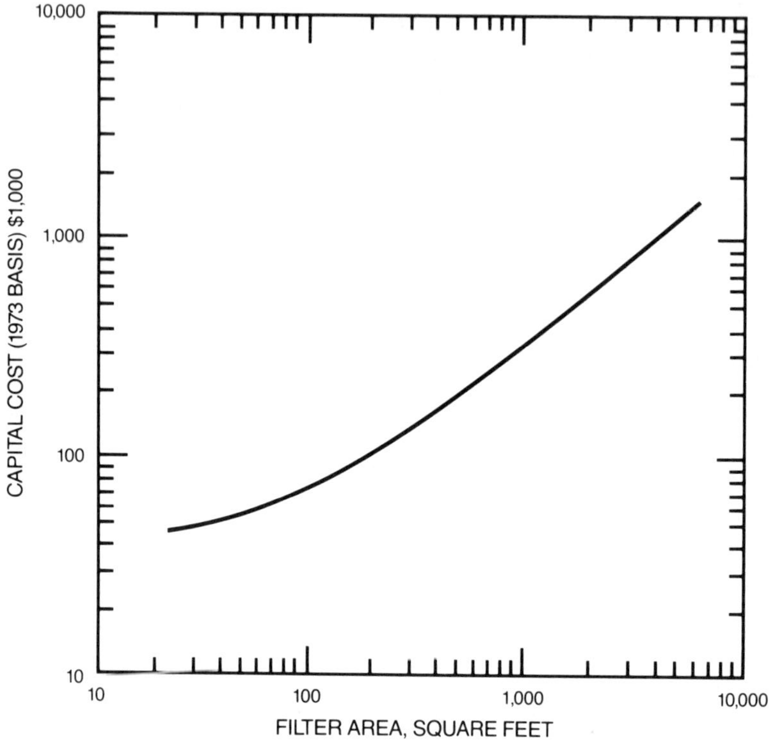

Figure 4.2 Capital cost (in 1973) of granular media filtration [43].

and particulate phases after precipitation. Figures 4.2 and 4.3 provide cost data for granular media filtration. Filtration would represent an incremental operating cost of about \$0.10/1,000 gal treated, at a waste volume of 100 gpm. Since cadmium treatment is likely to be integrated with other treatment schemes, generalized cost data for precipitation systems also apply (see Table 1.5), unless high treatment pH (i.e. above 10) is employed. Higher lime costs would be expected under such conditions.

Capital costs for precipitation treatment by the nonferrous-metals smelting and refining industry are summarized in Table 4.8. Capital costs range from \$348 to \$2,832/1,000 gpd treatment capacity. Table 4.9 presents generalized operating cost data (1974 basis) for precipitation treatment, with and without cyanide destruction, for the electroplating industry. The effect of economy of scale on treatment costs is apparent from Tables 4.8 and 4.9.

A proprietary hydrogen peroxide plus formalin oxidation precipitation system (the Kastone process) has been developed, which simultaneously oxidizes cyanide and precipitates the oxide of cadmium (rather than the hydroxide). Cadmium oxide precipitate is considered to be much easier to remove from solution than is cadmium hydroxide [44]. This process is reported as

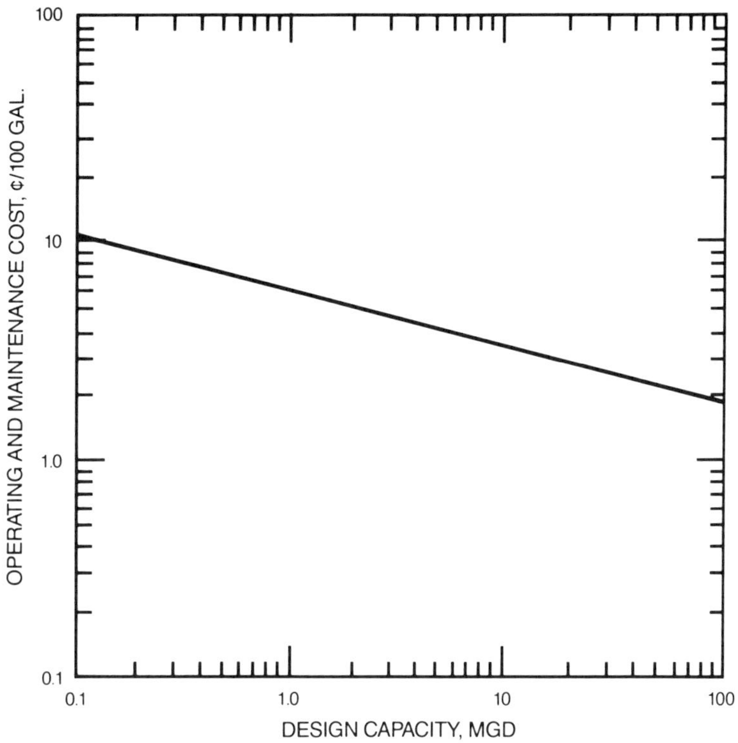

Figure 4.3 Operation and maintenance cost (in 1973) for granular media filtration [43].

Table 4.8 Summary of Cadmium Precipitation Treatment Costs (1979 Basis) for Nonferrous Metals Smelting and Refining [15, 16]

Industry	*Wastewater Volume (gpm)*	*Capital Cost*[a] *($/1,000 gpd)*
Lead[b]	100	1,942
Lead[b]	50	1,061
Zinc	500	1,263
Zinc[b]	745	534
Zinc	230–650	348–1,239
Zinc	150–325	1,306–2,832
Zinc	450	1,582

[a] Costs, originally reported in 1971 dollars, updated to 1979 dollars on the Marshall & Swift Equipment Cost Index. 1971 = 321.3; 1979 = 599.4.
[b] Costs based on design estimates.

Table 4.9 General Treatment Costs (1974) for the Electroplating Industry [12]

Wastewater Volume (gpm)	*Precipitation Treatment Operating Cost ($/1,000 gal)*	*Precipitation Plus Cyanide Treatment Operating Cost ($/1,000 gal)*
26	1.07	1.36
60	0.92	1.24
102	0.87	1.12

suitable for small plating operations, since solids removal is readily accomplished by slight modification to simple filtration apparatus already present in most shops. One potential disadvantage of the process is the production of glycolic acid in the cyanide destruction step [12]. Utilization of the process would likely be restricted to plants discharging to municipal sewers.

Hydroxide precipitation with lime may be unsatisfactory for cadmium due to the high pH required for effective treatment [45]. The use of sulfide for cadmium precipitation appears feasible, although it has had only limited application [12, 19, 46–50]. In one test application, acid mine drainage initially at pH 2.6 and containing cadmium at 440–1,000 mg/l has been treated by two-stage precipitation. In the first stage, the wastewater was neutralized with lime to pH 5.0. In the second stage, pH was further adjusted to 6.5 and sodium sulfide was added to precipitate cadmium and other heavy metals. Effluent cadmium of 0.008 mg/l was achieved [45]. The high removal efficiency may have resulted in part from cadmium coprecipitation.

Recent comparison studies on lime treatment versus lime plus sulfide treatment have been performed on several wastewaters from the electroplating and nonferrous metals industries [11]. Lime treatment alone, with and without effluent filtration, was compared to treatment where sodium sulfide was added concurrently with lime at a rate necessary to maintain 0.5 mg/l free sulfide in the reactor. The results, presented in Table 4.10, do not suggest a significant

Table 4.10 Comparison of Lime and Lime Plus Sulfide Precipitation of Cadmium [11]

Initial Concentration[a]	*Treatment pH*	*Lime-Settled*	*Lime-Filtered*	*Lime + Sulfide-Settled*	*Lime + Sulfide-Filtered*
0.40	9.0	0.098	0.011	0.055	0.029
15.00	8.5	0.080	0.007	0.050	0.002
58.00	10.0	1.130	0.923	0.026	<0.010
1.40	8.5	0.432	0.360	—[b]	—[b]
1.40	10.0	0.073	0.066	—[b]	—[b]

[a] All concentrations in mg/l.

[b] No sulfide test. Authors report that the extremely high sulfide demand (995–1,400 mg/l) required for this waste would make a full-scale process uneconomical.

Table 4.11 Capital and Operating Costs (1978) for a Lime-Precipitation Treatment System [11]

Capacity of Plant, gpd	*Capital, $/gpd*			*Operating, $/1,000 gal*		
	Clarification	+	*Filtration*	*Clarification*	+	*Filtration*
10,000	23.3		23.7	22.57		22.72
200,000	4.00		4.03	3.40		3.42
500,000	2.59		2.62	2.08		2.09

benefit associated with sulfide addition except in the instance of the most concentrated cadmium wastewater. Capital and operating costs (1978 basis) for the lime system are presented in Table 4.11. Capital cost for the sulfide system is about 10% higher than for the lime system. Operating costs would be 10–40% higher, depending upon plant treatment capacity and sulfide demand [11]. To reduce problems with free sulfide ion, use of insoluble ferrous sulfide as the sulfide source has been proposed [47, 49]. Sludge-handling problems have been noted [48], and residual sulfide ion must be destroyed prior to discharge.

Ion Exchange

There are many ion exchange resins with high specificity for cadmium [51]. Ion exchange can be used as a polishing treatment or as a recovery process. The concentrated solution obtained upon regeneration of the ion exchange resin is often suitable for economical recovery procedures. The recovery value of cadmium is estimated at $1.20–6.00/1,000 gal, at a solution concentration of 50 250 mg/l cadmium [52]. The extra capital cost of an ion-exchange treatment plant can be offset in 0.5–2 years when product recovery value is substantial, as is the case with cadmium [51]. Ion exchange is unsuitable, however, for recovery of mixed cadmium-cyanide solutions [53], due to irreversible fouling of the resin. Depending upon the ion exchange resin and regeneration configuration, installed costs may be $11–464/1,000 gpd capacity, with operating costs of $0.52–0.66/1,000 gal treated [54].

Evaporative Recovery

A single-effect 100 gph evaporator has been installed on a 5-barrel, 800-gallon-bath cadmium-plating line in one plant [55]. Economics for the cadmium system were not available but a similar zinc recovery system in the same plant had proved to be successful. Since the recovery value of cadmium-plating solution is greater than that of zinc, similar results were to be expected. Culotta and Swanton [56] have compared capital and operating costs for various evaporator systems. Single-effect evaporators for a 150 gph unit represent a capital investment of $13,889/1,000 gpd capacity (23-hour basis), and treatment costs are $17.40/1,000 gal, based on 4,000 hr/yr operation. Chemical recovery and water reuse values obviously must offset these costs for evaporative recovery to become economically feasible.

The recovery process necessitates waste stream segregation if water and chemical reuse are to offset capital and operation costs. The recovery system is not restricted to evaporation, since any method which will concentrate the waste solution to the extent that it can be reintroduced to the process chemical reservoir, while isolating water for rinse reuse, can be used.

Other Processes

Reverse osmosis has been examined for treatment of cadmium plating rinse solution [30]. Cadmium removal of 78–99+% was reported, with cyanide removals of 10–97%. Capital costs for reverse osmosis units of 10 and 100 gpm capacity have been reported as $24,000 and $180,000, respectively. This represents capital investment costs of $1,666 and $1,250/1,000 gpd. Operating costs of $0.60–1.00/1,000 gal for units of 10–100 gpm capacity have been reported [43].

Electrodialysis has been claimed to be a successful and economical means of treating wastewater in the plating industry, but data are not available in support of the claim [12, 57]. Electrolysis, in use for high-level cyanide destruction, has recently drawn more attention for dilute waste treatment. In pilot studies, an industrial-size unit using a packed bed of porous carbon fibers ($2.4 \times 10^6 cm^2/g$) was able to reduce a mixed plating waste with 0.3 mg/l Cd to <0.1 mg/l [9]. Capital cost of $50,000 (1978) for a 30 gpm unit was given. Operating costs were projected at 0.044¢/gal (80¢/hr for electrical cost of 2¢/KW hr). Freeze recovery has also been evaluated and reported effective. A 2,500 gpd pilot plant unit has achieved 99.6% reduction of cadmium, to a final concentration of 0.4 mg/l [31]. Specific treatment costs are not available for either electrodialysis or freeze recovery treatment of cadmium wastes.

SUMMARY

Removal of cadmium from a waste stream can be accomplished by precipitation as the hydroxide at elevated pH, but values of pH 10 or greater are necessary for effective removal in the absence of coprecipitation with iron or aluminum hydroxide. Carbonate precipitation of cadmium has been reported as more effective than hydroxide precipitation, with additional advantages in sludge handling. Coprecipitation with, or adsorption on, ferrous hydroxide floc may enhance cadmium removal. Sulfide precipitation, at near-neutral pH, appears feasible, although the higher capital and operation costs may not be warranted by the treatment benefits achieved. Organic or inorganic complexing agents in the waste stream will reduce the effectiveness of precipitative removal.

REFERENCES

1. McKee, J.E. and H.W. Wolf. *Water Quality Criteria*, 2nd ed., California State Water Quality Control Board, Publication 3-A (1963).
2. Santaniello, R.M. "Air and Water Pollution Quality Standards, Part 2: Water Quality Criteria and Standards for Industrial Effluents," *Industrial Pollution Control Handbook*, Herbert F. Lund, ed. (New York: McGraw-Hill, 1971).
3. "Metals Focus Shifts to Cadmium," *Environ. Sci. Technol.* 5 (1971): 754–755.
4. Gard, S.M., C.A. Snavely, and D.J. Lemon. "Design and Operation of a Metal Wastes Treatment Plant," *Sew. Ind. Wastes* 23 (1951): 1429–1438.
5. Lowe, W. "The Origin and Characteristics of Toxic Wastes with Particular Reference to the Metals Industries," *Water. Poll. Control* (London, 1970): 270–280.
6. Yost, K.J. and D.R. Masarik. "A Study of Chemical-Destruct Waste Treatment Systems in the Electroplating Industry," *Plating Surface Fin.* (January 1977): 35–40.
7. Pinkerton, H.L. "Waste Disposal. Inorganic Wastes," *Electroplating Engineering Handbook*, 2nd ed., A. Kenneth Graham, ed. (New York: Reinhold, 1962).
8. Nemerow, N.L. *Theories and Practices of Industrial Waste Treatment* (Reading, MA: Addison-Wesley, 1963).
9. Kennedy, I.F.T. and S.D. Gupta. "Electrochemical Removal of Trace Metals for Metal Wastes with Simultaneous Cyanide Destruction," *Fin. Manag.* (July 1978): 7–18.
10. Minear, R.A. and J.W. Patterson. "Carbonate Precipitation Treatment for Cadmium," paper presented at the 8th WWEMA Industrial Pollution Conference, Houston, TX (June 5, 1980).
11. Robinson, A. and I. Sum. "Sulfide Precipitation of Heavy Metals," USEPA 600/2-80-139 (June 1980).
12. "Development Document for Interim Final Effluent Limitations Guidelines and Proposed New Source Performance Standards for the Common and Precious Metals Segment of the Electroplating Point Source Category," USEPA 440/1-75/040, Group I, Phase II (April 1975).
13. McDonald, C.W. and R.S. Bajwa. "Removal of Toxic Metal Ions from Metal-Finishing Wastewater by Solvent Extraction," *Sep. Science* 12 (1977): 435–445.
14. "Development Document for Interim Final Effluent Limitations Guidelines and Proposed New Source Performance Standards for the Primary Copper Smelting Subcategory and the Primary Copper Refining Subcategory of the Copper Segment of the Nonferrous Metals Manufacturing Point Source Category," USEPA 440/1-75/032-b, Group I, Phase II (February 1975).
15. "Development Document for Interim Final Effluent Limitations Guidelines and Proposed New Source Performance Standards for the Lead Segment of the Nonferrous Metals Manufacturing Point Source Category," USEPA 440/1-75/032-a, Group I, Phase II, (February 1975).
16. "Development Document for Interim Final Effluent Limitations Guidelines and Proposed New Source Performance Standards for the Zinc Segment of the Non-

ferrous Metals Manufacturing Point Source Category," USEPA 440/1-75-032, Group I, Phase II (February 1975).

17. Cawley, W., ed. *Treatability Manual*, vol. I. *Treatability Data*, USEPA 600-8-80-042-a (July 1980).
18. Osantowski, R. and J. Ruppersberger. "Upgrading Foundry Wastewater Treatment," paper presented at the 39th Annual Purdue Industrial Waste Conference, West Lafayette, IN (May 1984).
19. Larsen, H.P., J.K.P. Shou, and L.W. Ross. "Chemical Treatment of Metal-Bearing Mine Drainage," *J. Water Poll. Control. Fed.* 45 (1973): 1682–1695.
20. Wangen, L. and J. Williams. "Control by Alkaline Neutralization of Trace Elements in Acidic Coal Cleaning Waste Leachates," *J. Water Poll. Control. Fed.* 54 (1982): 1302–1310.
21. Barrett, W.J., G.A. Morneau, and J.J. Roden. "Waterborne Wastes of the Paint and Inorganic Pigments Industries," USEPA 670/2-74-030, Cincinnati, OH (1974).
22. Iammartino, N.R. "Mercury Cleanup Routes—II," *Chem. Eng.* 82 (1975): 36–37.
23. Jenkins, S.H., D.G. Knight, and R.E. Humphreys. "The Solubility of Heavy Metal Hydroxides in Water, Sewage and Sewage Sludge, I. The Solubility of Some Metal Hydroxides," *Int. J. Air Water Poll.* 8 (1964): 537–556.
24. Hansen, N.H. and W. Zabban. "Design and Operation Problems of a Continuous Automatic Plating Waste Treatment Plant at the Data Processing Division, IBM, Rochester, Minnesota," *Proceedings of the 14th Industrial Waste Conference, Purdue University* (May 1959), pp 227–249.
25. Culp, G.L. and R.L. Culp. *New Concepts in Water Purification* (New York: Van Nostrand Reinhold, 1974).
26. Maruyama, T., S.A. Hannah, and J.M. Cohen. "Removal of Heavy Metals by Physical and Chemical Treatment Processes," *J. Water Poll. Control Fed.* 47 (1975): 962–975.
27. Chalmers, R.K. "Pretreatment of Toxic Wastes," *Water Poll. Control* (London, 1970): 281–291.
28. Hannah, S.A., M. Jelus, and J.M. Cohen. "Removal of Uncommon Trace Metals by Physical and Chemical Treatment Processes," *J. Water Poll. Control. Fed.* 49 (1977): 2297–2309.
29. Laughlin, R., H. Robey, and P. Gooderham. "Batch Hydrolysis System for the Destruction of Cyanides in Electroplating Effluents," *Proceedings of the 4th Conference on Advanced Pollution Control for the Metal Finishing Industry*, USEPA 600/9-82-022 (December 1982).
30. Donnelly, R.G., R.L. Goldsmith, K.J. McNulty, and M. Tan. "Reverse Osmosis Treatment of Electroplating Wastes," *Plating* 61 (1974): 432–442.
31. Campbell, R.J. and D.K. Emmerman. "Recycling of Water from Metal Finishing Wastes by Freezing Process," paper presented at the Reuse Treatment of Waste Water Symposium, ASME/EPA/AIChE, New Orleans, LA (March 28, 1972).
32. Cotton, F.A. and G. Wilkinson. *Advanced Inorganic Chemistry* (New York: Interscience, 1962).
33. Netzer, A., A. Bowers, and J. Norman. "Removal of Heavy Metals from Wastewater by Lime," *Proceedings of the 16th Conference on Great Lakes Research*, Huron, OH (1973).
34. Patterson, J.W., H.E. Allen, and J.J. Scala. "Carbonate Precipitation for Heavy Metals Pollutants," *J. Water Poll. Control. Fed.* 49 (1977): 2397–2410.

35. Illinois Effluent Standards Advisory Group. "Evaluation of Effluent Regulations of the State of Illinois," State of Illinois Institute for Environmental Quality Report 76/21 (June 1976).
36. Minear, R.A. and J.W. Patterson. "Carbonate Precipitation Treatment for Cadmium," paper presented at the 8th WWEMA Industrial Pollution Conference, Houston, TX (June 5, 1980).
37. Anderson, J.S., J.M. Phillips, and C.B. Schriver. "Water Contamination by Certain Heavy Metals," paper presented at the 44th Annual Meeting, Water Pollution Control Federation, San Francisco, CA (1971).
38. Weiner, R.F. "Acute Problems in Effluent Treatment," *Plating* 54 (1967): 1354–1356.
39. Eden, G.E., B.L. Hampson, and A.B. Wheatland. "Destruction of Cyanide in Waste Waters by Chlorination," *J. Soc. Chem. Ind.* 69 (1950): 244–249.
40. Lancy, L.E. "An Economic Study of Metal Finishing Waste Treatment," *Plating* 54 (1967): 157–161.
41. Beckwith, M.M. "Cyanide-Free Cadmium Plating Bath," *Plat./Fin. Pract.* 58 (1971): 413–415.
42. Minear, R.A. and J.W. Patterson. "Treatment of Metallic Wastewater," in *Water Pollution in Metropolitan Areas, Proceedings of a Symposium at Illinois Institute of Technology* (November 30, 1972) pp. 110–139.
43. Edwards, A.W., W.W. Eckenfelder, Jr., and J.A. Roth. *Industrial Water Cost Studies*, vol. I, Office of Research and Development, USEPA Contract 68-01-2485 (1975).
44. "New Process Detoxifies Cyanide Wastes," *Environ. Sci. Technol.* 5 (1971): 496–497.
45. Dean, J.G., F.L. Bosqui, and K.H. Lanouette. "Removing Heavy Metals from Waste Water," *Environ. Sci. Technol.* 6 (1972): 518–522.
46. Shimoiizaka, J. "Recovery of Xanthate from Cadmium Xanthate," *Nippon Kogyo Kaishi* (Japan) 88 (1972): 539–543.
47. Huang, S-D. and D.J. Wilson. "Foam Separation of Mercury(II) and Cadmium(II) from Aqueous Systems," *Sep. Science* 11 (1976): 214–222.
48. Metzner, A.W. "Removing Soluble Metals from Wastewater," Water Sewage Works (April 1977): 98–101.
49. Rouse, J.V. "Removal of Heavy Metals from Industrial Effluents," *ASCE J. Environ. Eng. Div.* (October 1976): EE5 929–936.
50. Feigenbaum, H.N. "Removing Heavy Metals in Textile Wastes," *Ind. Wastes* (March/April 1977): 32–24.
51. Mattock, G. "Modern Trends in Effluent Control," *Metal Fin. J.* 14 (1968): 168–175.
52. Digregorio, D. "Cost of Wastewater Treatment Processes," Robert A. Taft Water Research Center Report TWRC-6, USDI, Washington, DC (1968).
53. Oldan, K. and T.C. Hesler. "Profitable Recovery of Plating Wastes by Reconcentration of Reduced Volume Rinses," *Plating* 43 (1956): 1022–1025.
54. Zievers, J.F. and C.J. Novotny. "Recovery of Mixed Rinse Waters by Means of Ion Exchange," *Plating* 58 (1971): 482–485.
55. Gallo, B.R. and J.M. Culotta. "Save on Plating Waste System," *Water Wastes Eng.* 9 (1972): A18–A19.
56. Culotta, J.M. and W.F. Swanton. "Recovery of Plating Wastes: Selection of Lowest Cost Evaporator," *Plating* 57 (1970): 1221–1223.
57. Bovet, E.D. "Industrial Waste Desalting for Byproduct Recovery," *J. Am. Water Works Assoc.* 62 (1970): 539–542.

5

Hexavalent Chromium

Chromium occurs in aqueous systems as both the trivalent (Cr^{+3}) and the hexavalent (Cr^{+6}) ions. Chromium is present in industrial wastes primarily in the hexavalent form, as chromate (CrO_4^{-2}) and dichromate ($Cr_2O_7^{-2}$). Chromium compounds are added to cooling water to inhibit corrosion. They are employed in the manufacture of inks and industrial dyes and paint pigments, as well as in chrome tanning, aluminum anodizing, and other metal cleaning, preplating, and electroplating operations. Chromates are also contained in some preservatives and fire-retardant chemicals used in wood preservative treatment [1]. Automobile parts manufacturers are one of the largest producers of chromium-plated metal parts. Frequently the major source of waste chromium is the chromic acid bath and rinsewater used in such metal-plating operations. Table 5.1 is a compilation of typical sources and concentrations of hexavalent chromium wastewaters. Use of chromates in cooling-water systems is widespread, and blowdown from the cooling system can constitute a significant chromium-containing wastewater. Table 5.2 contains a summary of reported chromium concentrations in cooling-water blowdown.

TREATMENT TECHNOLOGY

Reduction of hexavalent chromium from a valence state of plus six to plus three, and subsequent hydroxide precipitation of the trivalent chromic ion, is the most common method of hexavalent chromium control. The treatment of trivalent chromium is presented in Chapter 6. To meet increasingly stringent effluent standards, some industries have turned to ion exchange to treat chromate and chromic acid wastes. Evaporative recovery of concentrated chromate and chromic acid wastes has also proved technically and economically feasible as a pollution abatement alternative. The application of other processes, such as electrochemical and activated-carbon adsorption techniques, is receiving increasing attention.

Reduction

The standard reduction treatment technique is to lower the waste stream pH to 2.0–3.0 with sulfuric acid, and convert the hexavalent chromium to trivalent

Table 5.1 Hexavalent Chromium Wastewater Sources and Concentrations

Industrial Source	*Chromium (VI) Concentration (mg/l)* Avg	Range	*Reference*
Leather tanning	40	—	2
Sodium dichromate production	—	560–1490	3
Sodium dichromate-chromic acid manufacture	1,300	—	4
Chromic oxide production	101 (Particulate CrO_3)	—	5
Chrome pigments production	—	17–957	6
Multiproduct pigments manufacturing	—	2–2000	7
Paint manufacturing	—	0.4–7.5	8
Dye house waste	300	—	9
Ink formulating waste	150	—	8
Municipal refuse incinerator scrubber water	0.5	—	10
Ferroalloy manufacturing	—	0.06–121	11
Aluminum manufacturing	136	—	12
Production of			
Automobile grills	700	—	13
Automobile parts	30	—	13
	11.5	—	14
Carburetors	—	46–81	15
	91	—	16
Missile parts	1	—	13
Typewriters and office machines	16	—	13
Silverware	5	—	13
Metal fasteners	52	—	13
Ornamental metal parts	9	—	14
Specific metal treatment operations			
Bright dip rinse	—	1–6	17
Bright dip bath	—	10,000–50,000	2
Etching bath	—	200–58,000	18
Anodizing bath	173	—	19
	—	15,000–52,000	20
Anodizing rinse	49	—	19
	—	30–100	20
	—	0.2–30	21
Anodizing and plating rinse	10.4	—	16

Table 5.1 (*continued*)

Industrial Source	*Chromium (VI) Concentration (mg/l)*		*Reference*
	Avg	*Range*	
Plating	1,300	—	22
	600	—	2
	—	100,000–270,000	20
	—	60–80	23
	688	—	24
Plating bath rinse	450	—	24
	2,310	—	25
	73	—	26
Electroplating	140	—	27
	41	15–75	28
	—	8–20.5	29
Diversified plating shop	—	12–60	30
Zinc bonderizing chromate bath	43	—	31
Mixed plating combined waste	12	—	25

chromium with a chemical reducing agent such as sulfur dioxide, sodium bisulfite, metabisulfite or hydrosulfite, or ferrous sulfate. The trivalent chromium is then removed, usually by hydroxide precipitation [36, 37]. The reduction of hexavalent to trivalent chromium is not completely effective, and the amount of residual nonreduced hexavalent chromium depends upon the allowed time of reaction, pH of the reaction mixture, and concentration and type of reducing

Table 5.2 Chromium Concentrations in Cooling Tower Blowdown

Industrial Source	*Chromium Concentration (mg/l)*		*Reference*
	Avg	*Range*	
Fertilizer manufacturing	31.4	—	32
	—	8–10.7	33
	—	10–60	2
Wood preserving	—	0.23–1.5	1, 2
Hydrogen gas manufacturing	28.6	—	6
Industrial inorganic gas manufacturing	—	0.65–22.6	34
Ferroalloy smelting	—	1.49–10.8	11
Petroleum refining	—	0–6	35
Ornamental metal parts	190	—	14

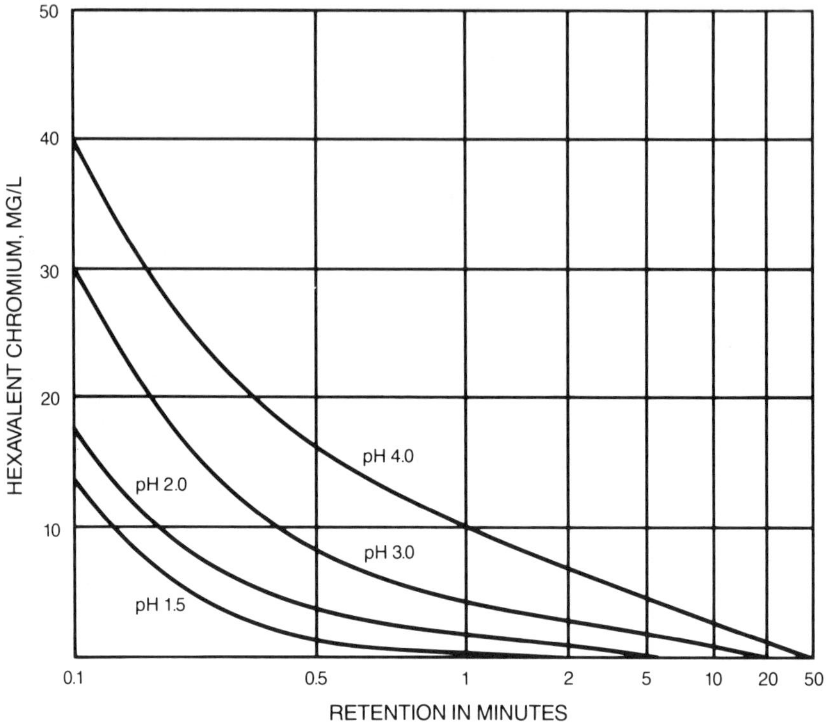

Figure 5.1 Effect of treatment pH and reaction time on conversion of hexavalent to trivalent chromium. (From Beevers [38], courtesy of *Metal Finishing Journal.*)

agent employed. The interdependence of some of these variables is shown in Figure 5.1 for chromium reduction with sulfur dioxide.

Sulfur dioxide is the most popular reducing agent used in treatment of chromium wastes, primarily because it is relatively inexpensive [39]. In one full-scale application to treat chromium in a metal-finishing waste, the wastewater is adjusted to pH 2.5 with sulfuric acid before reduction [40]. Sulfur dioxide, used to treat chromic acid-plating bath and chromic acid-etching bath wastes, at an average waste flow of 40 gpm and with a 20–30 minute treatment period, is reported to yield residual hexavalent chromium below 1.0 mg/l [41]. Sulfur dioxide has been employed in another application to treat a waste reported to contain 1,300 mg/l hexavalent chromium. This chromate waste was treated with sulfur dioxide at pH 2.0. Treatment detention time was approximately 90 minutes. Sulfuric acid was added to the waste, to maintain the proper pH [22].

Hexavalent chromium levels of 0.01 mg/l have been achieved by reduction at pH 2.5–4.0 [14, 42]. Use of H_2SO_4 and SO_2 reduced Cr (VI) from 8.75 mg/l to 0.01–0.03 mg/l, in an automotive plant wastewater containing plating wastes and cooling water blowdown [14]. A fully automated continuous-flow chromate waste treatment system has been reported, which yields hexavalent chromium levels below 0.05 mg/l [43].

The sulfur dioxide reduction process has also been used to treat wood-preserving wastewater [1, 44]. At influent chromium levels of 0.23–1.5 mg/l, effluent hexavalent chromium averaged 0.1 mg/l. Cost of the treatment plant, which included chromate reduction plus lime precipitation of chromium, copper, fluoride, and arsenic, was below $30,000. Total capital investment was approximately $50,000, including engineering and site preparation. The daily operating cost is $10.00 plus an additional $10.00/day for laboratory analyses. At a wastewater flow of 20 gpm, 24 hr/day, operating costs are $0.69/1,000 gallons.

An unusual source of sulfur dioxide has been reported in use at one plant, to treat hexavalent chromium in plating wastes [45]. Sulfur dioxide is washed out of the powerhouse stack gas and used as the chemical reducing agent for converting hexavalent chromium to the trivalent form. The process began operation in 1959. On occasions when additional reduction is necessary, sodium bisulfite is employed as a supplement. The use of sulfur dioxide for reduction imparts an oxygen demand to the waste effluent, unless the effluent is oxygenated by passing air through it [46]. A required sulfur dioxide dosage of 2.0 lb/lb hexavalent chromium reduced has been reported [28, 38].

In the treatment of an electroplating waste containing 140 mg/l hexavalent chromium, reduction was carried out at pH 2.5–2.8, employing sodium bisulfite as the reducing agent [27]. The process was reported to achieve effluent hexavalent chromium concentrations of 0.7–1.0 mg/l. One attempt to use sodium bisulfite to treat the waste from chrome plating of automobile bumpers was abandoned because of odors and corrosion hazards associated with the use of this agent. Sulfur dioxide is now employed in its place [47]. In contrast, the replacement of sulfur dioxide with sodium metabisulfite as a reductant has also been reported, due to industrial health problems associated with SO_2 gas leakage [48]. Use of sodium bisulfite to treat a chrome-plating waste has been reported to yield effluent hexavalent chromium levels of 0.05–0.1 mg/l [49]. Residual hexavalent chromium levels below 0.10 mg/l have been reported from sodium bisulfite treatment of a chromate-plating line waste initially containing up to 688 mg/l hexavalent chromium [24]. The reduction was carried out at pH 2.0, to an ORP level of 250–280 mV. Treatment costs for bisulfite reduction of $0.55/lb [28] and $0.80/lb [50] of chromium treated have been cited.

Two-stage chromate reduction has been used at several plants. In the first stage sulfur dioxide, sodium bisulfite, or metabisulfite is employed. This is followed by a second "polishing" stage of chromate reduction with hydrazine, often combined with a neutralizing chemical such as sodium hydroxide to simultaneously precipitate chromium-plating rinsewater. In one such application, the first stage, operating at pH 2.5–4.5, uses sodium bisulfite concentrations of 1,000–2,500 mg/l, followed by a hydrazine concentration of 50–200 mg/l. Chromate concentration is reported reduced from 8 to 20 mg/l down to less than 0.1 mg/l in the effluent [29]. Treatment cost is $0.39/lb chromic acid treated. Similar two-stage reduction systems, using hydrazine in the second stage, have been reported for other plants [51, 52].

The use of metabisulfite as a reducing agent has been reported for several hexavalent chromium treatment plants. One treatment plant reduced an initial chrome level of 31.4 mg/l in a cooling tower blowdown of 3,600 gpm, to concentrations below 0.5 mg/l [32]. Reduction was carried out at pH 2.0, achieved by addition of sulfuric acid. A metabisulfite dosage of 4.2 lb/lb chromium was employed. This represented a 75% excess above the theoretical stoichiometric dosage required [36]. Reductive chromate treatment of a similar blowdown waste has achieved hexavalent chromium levels of 0.025–0.05 mg/l. In another application of metabisulfite, for treatment of a chromate waste from a metal-plating operation, effluent hexavalent chromium levels of 0.1 mg/l or less were obtained at pH 2.0 [53]. An 8 gpm metabisulfite treatment process achieved effluent hexavalent chromium levels of 0.001–0.04 mg/l, at pH 3.0 [54].

Sodium hydrosulfite use has also been reported for chromate reduction. However, this treatment chemical breaks down with exposure to air and must be prepared fresh daily. At one plant its use was abandoned in favor of a two-stage bisulfite-hydrazine reduction system [29].

Sodium sulfide has been suggested for reduction of hexavalent chromium. However, batch tests at pH 8.0 on a sodium dichromate wastewater resulted in extremely slow reduction kinetic [3]. Treatment pH above neutral is required to suppress hydrogen sulfide gas evolution. Applications of sulfide at up to ten times the stoichiometric demand resulted in residual chromate levels of about 1.0 mg/l, even after 10 days of reaction time [3].

Several installations employ ferrous sulfate as a reducing agent for chromate. In three cases, treatment occurred at steel mills, where spent sulfuric acid-ferrous sulfate pickle liquor was available to use in the reduction process [17, 55, 56]. Ferrous sulfate has been reported to have the advantage of effectiveness independent of pH [19]. However, Bennett [57] claims otherwise, stating that ferrous sulfate reduction at pH 2–3 proceeds within 30 minutes to a residual hexavalent chromium level of 1 mg/l, but at pH above 4, equivalent treatment requires one hour. It has also been reported that use of ferrous sulfate yields much greater quantities of sludge than are produced through use of sulfur dioxide or bisulfite, since voluminous iron precipitates will simultaneously form in the chromic precipitation step [28, 57], resulting in about four times the sludge volume produced with other reduction/precipitation alternatives [58]. In addition, the use of ferrous sulfate to treat chromate wastes containing cyanide results in the formation of very stable ferrocyanide complexes, which will prevent subsequent effective cyanide treatment.

One plant treated wastes resulting from sodium dichromate-chromic acid manufacturing by isolation of chrome-bearing wastes and reduction with waste pickle liquor in batch operation [4]. Hexavalent chrome was reduced from 1,300 to 0.01 mg/l. Costs of the 0.348 MGD plant (operational in 1972) were $2,876/

Table 5.3 Results of Batch Treatment of Sodium Dichromate Wastewaters with Ferrous Chloride Pickle Liquor [3]

pH		*Molar Ratio*[a]	*Chromate (mg/l)*	
Initial	*Treatment*	$Fe^{+2}:Cr^{+6}$	*Initial*	*Final Filtrate*
10.8	2.8	11.6	1,475	53.5
10.5	2.8	12.1	1,418	59.0
10.2	3.2	25.5	590	50.0
10.2	3.0	26.9	560	<0.004
10.0	2.8	27.4	550	<0.004
10.1	3.0	42.0	590	<0.004
10.1	2.9	46.7	530	0.005
10.1	3.0	48.5	510	<0.004
10.3	3.4	49.8	560	<0.04
10.1	2.6	51.7	480	7.55
10.5	2.9	61.9	1,390	<0.04
10.4	3.0	61.9	1,390	<0.04
10.7	2.8	65.1	1,453	0.011
10.1	2.8	65.3	530	0.50
10.7	3.2	66.8	1,415	0.007
10.6	2.8	71.6	1,440	<0.04

[a] Theoretical molar stoichiometric ratio is 3:1.

1,000 gpd (capital) and $6.41/1,000 gal treated (operating) based only on the waste flow from the chromium-manufacturing segment of the total plant.

Batch treatability studies on sodium dichromate wastewaters yielded the results presented in Table 5.3. Treatment involved the addition of ferrous chloride pickle liquor plus sufficient hydrochloric acid to achieve a reaction pH of 2.8–3.0, reaction for 2–4 hours, neutralization with lime, aeration (to oxidize ferrous to ferric iron), clarification, and dual media filtration [3]. Although there are some exceptions, the data of Table 5.3 indicate that at pH near 3.0 and ferrous iron:chromate chromium ratios above 25.0, residual chromate levels below 0.01 mg/l are achievable. At such dosage ratios, however, extremely large sludge quantities will result, due to the precipitation of ferric hydroxide.

As an alternative approach, to reduce iron demand and consequent sludge production, Gould [59] has proposed the use of metallic scrap iron to reduce hexavalent chromium. By direct conversion of metallic to ferric iron, chromium reduction would have a theoretical molar stoichiometric iron demand of 1:1 versus the theoretical 3:1 molar demand associated with ferrous iron as the reductant, and ferric iron is a direct reaction product. The process is analogous to the "cementation" process for copper discussed in Chapter 7. Gould's studies indicated increased reduction kinetics with increased initial chromium concentration, increased iron surface area, and reduced pH. The rate constant increased sharply at ionic strength below 0.1 M, but was constant at higher ionic strength. Mixing rate also influenced the rate constant, up to a mixing rate of 400 rpm. The observed experimental molar stoichiometry was 1.3:1.0, independent of reaction pH. The stoichiometric ratio decreased, however, with increasing initial chromate concentration.

Complete reduction of a 100 mg/l Cr (VI) solution by this process has been reported within 8 minutes, at pH 2.0 [60]. In one full-scale application, hexavalent chromium at 2.5–52.0 mg/l was reduced to 0.01–1.40 mg/l by the cementation process, for a mixed metal manufacturing complex involving plating, anodizing, and brass pickle liquor [61]. Average influent hexavalent chromium was 16.3 mg/l, and average effluent was 0.01 mg/l, for an average reduction of 99.4% [62]. The process uses scrap iron, suspended in a perforated rotating drum through which the wastewater flows. The waste pH is 2–3. Though copper recovery augmented process economics, the primary intent of this application was for chromate reduction [61].

Comparative chemical requirements and resulting sludge quantities for six common chrome reduction options are presented in Table 5.4. Chemical costs (1978) for sodium bisulfite and sulfur dioxide were reported to be $260 and $170/ton, respectively [63]. Sodium metabisulfite offers stability and dry chemical feed capabilities as advantages over the use of the bisulfite [58]. Specific choice of reductant and auxiliary chemicals depends upon overall process options and localized chemical and sludge disposal costs.

For complete treatment of chromate wastes by chemical reduction, including pH adjustment, reduction, and neutralization, and removal of the trivalent chromium, treatment operating costs range from $55 to $100 per day, dependent on the cost of the treatment chemicals selected, and of municipal water to the industry [64]. The basis for these costs is a waste flow of 100 gpm at 120 mg/l chromium, and at a 16-hour day.

Additional cost curves are presented in Figures 5.2 (Monthly Operating Costs) and 5.3 (Initial Capital Costs). At a flow of 70 gpm, Figure 5.2 indicates an operating treatment cost of approximately $0.40–1.00/lb chromium reduced. These operating costs do not include the additional expense of removing the trivalent chromium. The capital costs associated with batch and continuous-mode hexavalent chromium treatment are presented in Figure 5.3. Economy of scale resulting from treating large waste volumes is indicated.

The one-step Sulfex™ process (Permutit) is carried out at pH 8–9 with solid (nearly insoluble) ferrous sulfide in excess, eliminating pH reduction required for conventional hexavalent chromium conversion [25, 66, 67]. Performance is claimed equal to or better than chemical reduction/hydroxide precipitation treatment. Pilot-plant comparative studies on a multiproduct pigment manufacturing waste with wide variations in influent Cr (VI) and pH levels (2–2,000 mg/l and 1–13, respectively), yielded Sulfex process effluent levels of 0.1–0.2 mg/l Cr (VI), while SO_2 reduction yielded 0.3–0.5 mg/l Cr (VI) [7]. Since ferrous sulfide is largely insoluble and always added in excess, its value lies with highly variable hexavalent chromium wastes where excess FeS addition does not overtreat the waste. For the pilot study cited above, use of SO_2 involved treatment

Table 5.4 Chemicals Used and Sludge Produced For Reductive Treatment of Chromium (IV) (lb Chemicals Required or lb Sludge Produced/100 lb Chromic Acid Reduced) [58]

	Sodium Metabisulfite, $Na_2S_2O_5$ (97%)	*Sodium Bisulfite, $NaHSO_3$*	*Sodium Sulfite, Na_2SO_3*	*Sodium Hydrosulfite, $Na_2S_2O_4$*	*Sulfur Dioxide, SO_2*	*Ferrous Sulfate, $FeSO_4 \cdot 7H_2O$*
Reducing agent	147	156	189	261	97	834
Acid (H_2SO_4, 66 Be)	80	80	160	—	—	316
Base						
$Ca(OH)_2$	111	111	111	—	111	444
Na_2CO_3	159	159	159	—	159	631
NaOH	120	120	120	120	120	480
Sludge produced from						
$Ca(OH)_2$	307	307	307	—	307	1,346
Na_2CO_3	103	103	103	—	103	423
NaOH	103	103	103	103	103	423

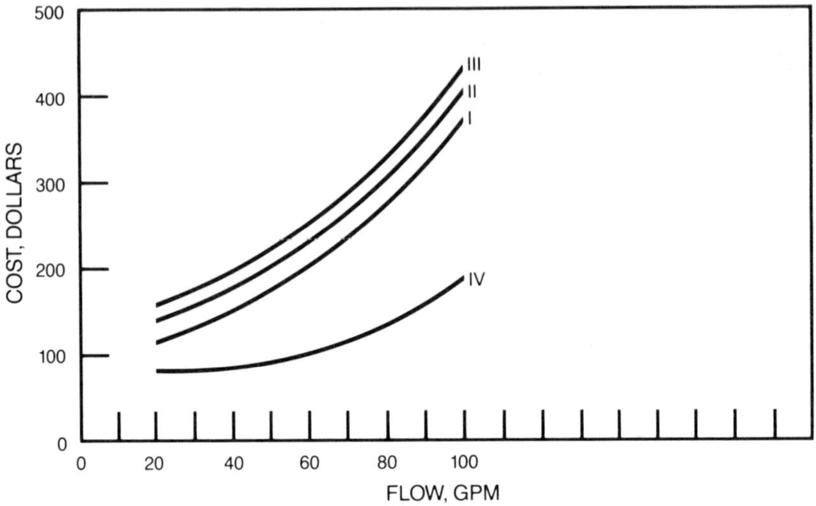

Figure 5.2 Monthly operating costs (in 1968) for two-stage chromium treatment (to trivalent), plus pH adjustment to 8.0. Depreciation plus operating costs at 50 mg/l chromium, 172 hr/month. I, no instrumentation; II, medium instrumentation; III, full instrumentation; IV, using "mist depressant" type chrome bath, no instrumentation. (From Zievers et al [65], courtesy of *Plating*.)

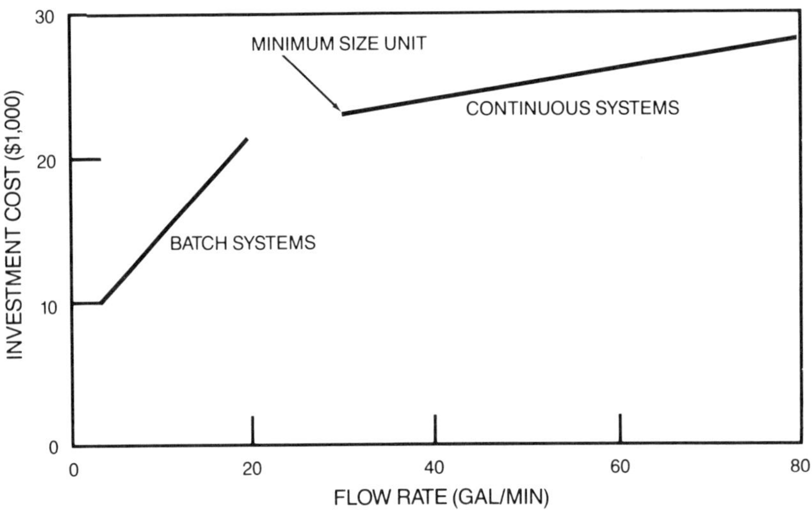

Figure 5.3 Initial capital cost (in 1968) for chromium reduction system [65].

chemical additions more than 100% above stoichiometric requirement. As a partial offset, however, the excess FeS added in the Sulfex process can increment treatment sludge yields to twice or more than that resulting from SO_2 treatment.

A cost analysis for a pigment-manufacturing waste indicated that the Sulfex process would represent an increased capital investment of $550,000 (1977 basis) and annual increment in operating (including sludge handling) cost of $165,000 above the SO_2 reduction system [7].

Ion Exchange

The ion exchange process is reported to be economical for chromium recovery and elimination of waste discharge [67, 68]. Cation exchange can be applied to remove trivalent chromium, and anion exchange employed to remove chromate and dichromate. When the anion exchange resin is exhausted it is regenerated (usually with sodium hydroxide), and sodium chromate is eluted from the ion exchange resin. The eluted sodium chromate can be passed through a cation exchange resin to recover purified chromic acid at concentrations as high as 6% [69]. If chromic acid is not recovered, the concentrated waste yielded by the resin regeneration process must be treated. This may be accomplished by reduction to trivalent chromium, followed by precipitation. Treatment by ion exchange produces a reusable water which may provide an economic benefit when water costs are high.

The earliest ion exchange systems for chromate utilized weak-base resins. However, these have generally proved to be unstable, due to the oxidizing conditions under which the exchangers operate. Strong-base gel-type resins have also been widely applied, but have experienced widespread fouling problems. More recently, weak-base macroporous resins have been utilized, with success [70]. Table 5.5 presents data on ion exchange performance in chromate removal.

In the ion exchange system, wastewater pH is a critical factor. At pH below 4, the oxidizing power of the chromic acid begins to attack the resin. At pH above 6, the ratio of chromate to dichromate in solution increases. Most anion exchange resins are less selective for dichromate, and early leakage of chromium occurs [79]. Highly basic anion-exchange resins preferentially remove chromate at pH 4.5–5 before phosphate or sulfate [80].

One significant use of ion exchange treatment is for chromate-contaminated blowdown from recirculating cooling-water systems [79–81]. With proper pH adjustment (to pH 4.0–5.0), the chromate will be removed even in the presence of several hundred mg/l sulfate and chloride [79]. The pH dependence of strong-base anion exchange resins is demonstrated in Figures 5.4 and 5.5. Numerous plants employ ion-exchange recovery systems. Based on chemical costs only (including resin replacement), reductive treatment was reported to cost twice that of ion exchange, $1.09/lb CrO_4 vs $0.51/lb CrO_4 [82].

Table 5.5 Ion Exchange Performance in Hexavalent Chromium Removal

Wastewater Source	*Chromium (mg/l)* *Influent*	*Effluent*	*Resin Capacity*[a]	*Reference*
Cooling-tower blowdown	17.9	1.8	5–6	71
	—	0.1	—	72
	10.0	1.0	2.5–4.5	73
	7.4–10.3	1.0	—	74
	9.0	0.2	2.5	75
Plating rinsewater	44.8	0.025	1.7–2.0	76
	41.6	0.01	5.2–6.3	77
Pigment manufacture	1,210	<0.5	—	78

[a] lb chromate/ft^3 resin.

Successful ion exchange treatment of a metal-finishing waste has been reported to meet a hexavalent chromium effluent standard of 0.02 mg/l [83]. The chromic acid wastes treated at this plant by ion exchange were segregated from other metal-finishing wastes. Recovery of chromic acid was practiced, and resulted in a five-year savings of $8,342 above the cost that would have been required for chromium reduction and precipitation.

After a year-long pilot plant test of ion-exchange treatment of cooling-tower blowdown containing 8–10.7 mg/l hexavalent chromium, Richardson et al [33] reported 0.0–0.9 mg/l effluent chromium, depending upon the period of operation between exchange resin regenerations.

A small plating shop has successfully employed ion exchange to reduce rinsewater chromate levels to 0.025 mg/l [84]. The concentrated eluate is treated with sodium metabisulfite to reduce chromate, followed by neutralization and

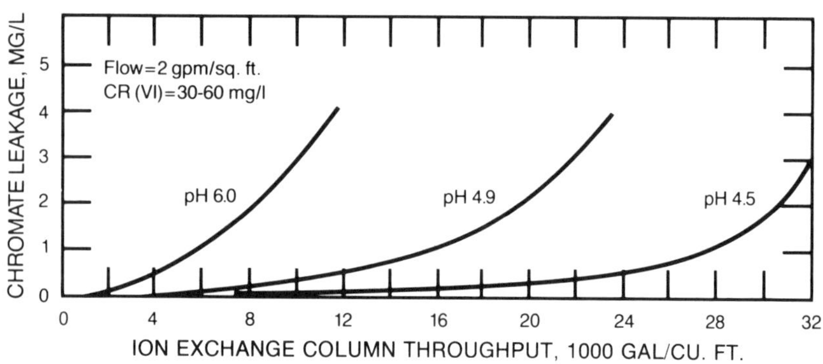

Figure 5.4 Chromate leakage through ion exchange column for wastewater at TDS of 1,400 mg/l [81].

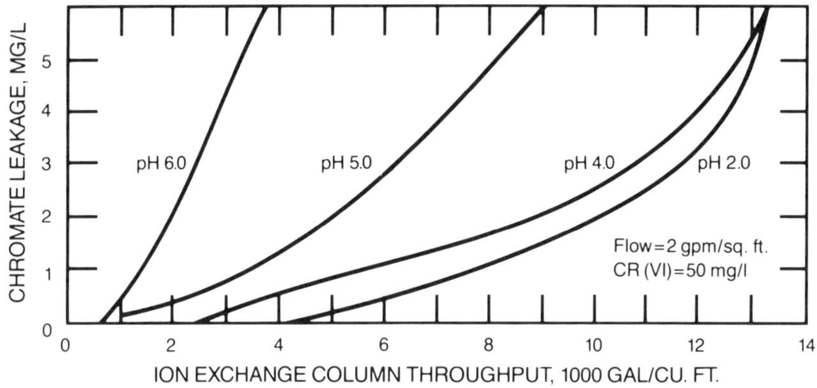

Figure 5.5 Chromate leakage through ion exchange column for wastewater at TDS of 7,600 mg/l [81].

chromic-hydroxide precipitation. Use of anion exchange directly on three chrome-plating rinsewaters has also been reported, for a complex electroplating operation [85].

The cost of ion-exchange treatment depends upon the industrial source of the waste. Waste-treatment operating costs of $0.16–0.24/1,000 gal treated have been reported [46]. Operating costs for ion-exchange waste treatment of a waste stream of 73 gpm with a chromic acid concentration of approximately 5,000 mg/l (equivalent to 2,300 mg/l hexavalent chromium) totaled $72.24/day ($0.69/1,000 gal), with a credit of $57.30/day resulting from chromic acid recovery plus reuse of purified water [68]. Net operating costs were, therefore, $12.94/day, or $0.13/1,000 gal. By contrast, reduction and precipitation would have cost $33.19/day for that waste.

Electrochemical Reduction

An electrochemical process similar to the cementation process, but employing an imposed DC electrical potential and consumable iron electrodes, has been reported to achieve chromate reduction from 3.5 mg/l to below 0.05 mg/l in cooling-tower blowdown, treating 0.036 MGD [86]. Electric current applied to the electrodes results in the release of ferrous ions into solution. These ferrous ions then reduce the Cr (VI) to yield trivalent chromium ions. Both this process and the cementation process produce some excess sludge solids upon precipitation of the trivalent chromium, as a result of the soluble iron released in the electrochemical reactions. One source has reported a fourfold increase in dry sludge solids [63].

Evaporative Recovery

This process consists of evaporating metal-plating rinsewater, to drive off the water as vapor and thus concentrate the chromic acid for recovery and reuse. Evaporative recovery and reuse are appropriate for almost all process rinse-

Table 5.6 Evaporative Recovery Treatment Costs

Treatment Capacity[a] *(gpd)*	*Capital Cost (1979 \$/gpd) [90]*	*Operating Costs (1968 \$/1,000 gal) [64]*
1,000	45	10
2,500	36	5
5,000	24	2.5

[a] One shift/day operation.

water systems, with the exception of those which chemically deteriorate with use [87]. In evaporative recovery, all nonvolatile constituents of the wastewater are retained in the concentrated product. In practice, this has been a major disadvantage of evaporative recovery, since the buildup of impurities can result in defective plating practices such as discoloration or loss of corrosion resistance [88]. Plating-waste rinsewater containing only a few mg/l chromic acid can be concentrated by evaporative treatment to above 900 mg/l [89]. Evaporative recovery treatment costs are presented in Table 5.6.

The full-scale application of evaporative recovery has been reported for a closed-loop Navy facility plating line [91]. The wastewater treated contains up to 10,000 mg/l chromic acid, plus iron and trivalent chromium. These latter two contaminants are removed by cation exchange pretreatment, and the ion exchange effluent is further concentrated to approximately 250,000 mg/l chromic acid by vapor recompression evaporation followed by waste heat evaporation. The concentrate is then returned to the plating bath. The distillate for the two evaporative units is recycled to the plating rinse tanks. The flow treated is 480 gpd. Capital cost for the evaporative system, exclusive of ion exchange pretreatment, is \$103/gpd treatment capacity. Operating cost is \$44/1,000 gal treated. However, an annual \$22,000 saving is reported in plating and water treatment chemicals, water consumption, and sludge disposal [91].

Other Processes

Some success has been reported from pilot plant work on chromate removal from dilute wastewaters by activated carbon. In one study of metal removal from municipal effluent, initial hexavalent chromium levels of 0.09–0.19 mg/l were reduced to 0.04 mg/l or less. The average effluent concentration reported was 0.017 mg/l [92]. Direct alum coagulation, sedimentation, mixed media filtration, and carbon adsorption reduced Cr (VI) from 0.7–0.018 mg/l. Coagulation/filtration alone removed 61–64% of the chromium [93]. Initial hexavalent chromium levels of 5 mg/l were reduced to 0.006–0.087 mg/l by carbon adsorption [94]. It appears that activated carbon may not be equally effective at high chromate levels, however. Effluent hexavalent chromium levels of 0.05–195 mg/l have been reported for carbon treatment on initial concentrations of

Table 5.7 Summary of Treatment Levels Reported for Hexavalent Chromium Wastes

Treatment Process	*Chromium (VI) Concentration (mg/l)* *Initial*	*Final*	*Reference*
Reduction			
Sulfur dioxide	100	<0.05	5
	—	0.3–1.3	39
	—	1.0	41
	1,300	"zero"	22
	—	0.01	42
	—	0.05	43
	0.23–1.5	0.1	44
Bisulfite	140	0.7–1.0	27
	—	0.05–0.1	49
	450–688	<0.10	24
	10.4	<0.005	16
Bisulfite plus hydrazine	8–20.5	0.1	29
Metabisulfite	70	0.5	32
	—	0.025–0.05	100
	—	0.1	53
	—	0.001–0.4	54
Ferrous sulfate	—	1.0	57
Ferrous sulfate (waste pickle liquor)	1,300	0.01	4
Ion exchange	—	0.023	83
	—	0.025	84
	8 10.7	0.0–0.09	33
Cementation	16.3	0.09	60
Electrolytic (DC-12V, Carbon bed)	100–200	<1.0	58
Activated carbon	0.09–0.19	0.017 (avg)	92
	5.0	0.017 (avg)	94
	100–820	0.006–0.087 5.05–195	28
Freeze concentration	100	0.225	99

Table 5.8 Summary of Reported Plating Waste Effluent Chromium Concentrations [after 101]

		Effluent Cr(VI) (mg/l)		*Effluent Total Cr (mg/l)*	
Type Treatment	*No. Plants*	*Range*	*Avg*	*Range*	*Avg*
Reduction plus precipitation batch treatment	9	<0.01–016	<0.15	<0.05–1.0	<0.35
continuous treatment	17	0.0–0.5	<0.10	0.11–10	<1.75
Ion exchange	2	—	0.01	0.02–1.0	<0.5
Evaporation	3	0.05–1.0	0.5	0.16–1.5	0.9

100–820 mg/l. Treatment efficiencies ranged from 50% to more than 99% [28]. Reduction of 600 mg/l Cr (VI) to below 0.05 mg/l has been reported, but the carbon bed could not be regenerated to yield subsequent chrome removal [66].

Reverse osmosis is also reported to have application for hexavalent chromium treatment, although experience with the process is limited. Improved efficiency has been reported to result upon wastewater neutralization. At pH 2.6, a high-selectivity cellulose acetate membrane yielded 92.6% rejection, with an increase to 98.6% rejection at pH 7.6 [95]. Polyetheramide membranes are also reported to perform well for chromic acid concentration and recovery, provided that a pH above 1.0 is maintained [96]. pH control is required to minimize acid hydrolysis of the reverse osmosis membrane [67]. Moreover, some membranes are not suitable for waste solutions with high oxidation potential [63]. One analysis of reverse osmosis use for the plating industry concluded that the technology is not yet satisfactory for treatment of chromic acid wastes per se [97].

Freeze concentration of concentrated plating line rinsewaters has also been claimed economical, followed by either reuse of the chemical or reduction and precipitative treatment of the concentrated slurry. In pilot plant operation, a 2,500 gpd waste containing 100 mg/l chromium yielded a product-water chromium concentration of 0.22 mg/l [98].

The direct precipitation of chromate as $BaCrO_4$ by addition of barium carbonate has been reported, and has been suggested as being useful for both barium and chromium recovery [58]. However, high barium salt costs and high effluent barium concentrations are disadvantages.

As an alternative to chromate treatment, it has been suggested that there are applications in which proprietary trivalent chromium baths may be substituted for chromic acid plating baths [99]. Most such formulations are organic solutions in which the trivalent chromium is at least partially stabilized by complexation with proprietary organic agents. The use of trivalent chromium formulations would eliminate the need for chromate treatment, but would likely impose severe treatment limitations on the traditional precipitation treatment used for trivalent chromium. For cooling-tower blowdown, an alternative to

Table 5.9 Summary of Reported Capital and Operating Costs for Total Treatment Facilities for Electroplating and Associated Metal Treatment Processes [after 101]

Process	*Waste flow (MGD)*	*Capital Cost (1974) ($/1,000 gpd)*	*Operating Cost (1974) ($/1,000 gal)*
Plating, common metals	0.168	202	1.13
	0.101	1703	4.76
	0.026	1074	1.36
Plating, common and precious metals	0.043	4651	5.37
Plating, precious metals	0.080	312	4.35
	0.154	428	1.59
	0.216	1389	5.38
	0.721	555	1.81
	0.288	382	2.08
	0.366	181	0.83
Electroplating, anodizing	0.196	510	0.98
Electrodeless plating	0.011	4120	24.61
	0.025	932	2.31
	0.661	329	0.72
	0.234	907	0.34
Anodizing	0.057	3400	5.40
	0.120	1396	1.66
	0.505	358	0.91
	0.822	189	0.34
	0.019	6579	1.91
	0.144	857	1.17
Chemical milling	0.048	364	1.74
	0.048	6250	20.90
	1.201	2421	2.27
	1.009	577	1.32
	0.043	680	1.62
Phosphating	0.364	273	1.13
Etching	0.042	7038	33.38
Immersion	0.072	819	2.34
Printed circuits	0.577	1820	2.46
Electropolishing	0.195	215	0.42

chromium treatment is discontinuation of chromate addition by substitution of other water treatment chemicals. There are many reported instances of this practice [34].

SUMMARY

Removal of hexavalent chromium from waste streams has been accomplished by reduction and precipitation, ion exchange, evaporative recovery, and other techniques. Ion exchange and evaporative recovery have the advantage that the chromium plus process water may be recovered for reuse. If waste flow rates and chromium concentrations are low (less than 50 mg/l), the total water volume for reuse and the amount of chromium involved may not warrant reclamation. In this instance, reduction to trivalent chromium, followed by precipitation, appears most feasible. Reported treatment levels for chromate waste are summarized in Table 5.7. Reported effluent levels, for Cr (VI) and total chromium, for electroplating wastewaters are presented in Table 5.8. Capital investments shown in Figure 5.3 appear to be substantiated by many installations. Operating and maintenance costs ranging from $0.15 to $0.80/lb chromium treated have been reported. Reported capital and operating costs associated with individual electroplating and associated metal-processing treatment installations are presented in Table 5.9. The wide variation of costs presented reflects site-specific conditions.

REFERENCES

1. "Development Document for Effluent Limitations Guidelines and New Source Performance Standards for the Major Inorganic Products Segment of the Inorganic Chemicals Manufacturing Point Source Category," USEPA 440/1-74-007-a (March 1974).
2. Cheremisinoff, P.N. and Y.H. Habib. "Cadmium, Chromium, Lead, Mercury: A Plenary Account for Water Pollution, Part I—Occurrence, Toxicity and Detection," *Water Sew. Works* 119 (1972): 73–85.
3. *Treatability Studies for the Inorganic Chemicals Manufacturing Point Source Category*, USEPA 440/1-80/103 (July 1980).
4. McGarvey, F.X. and S.A. Fisher, "Effluent Control in Zinc Bonderizing Process, II. The Recycle of Dilute Chromic Acid Baths via an Ion Exchange Process," *Plating* 62 (1975) 335–339.
5. "Chromates in Cooling Tower Reduced to 0.05 PPM or Less," *Chem. Process* 38 (1975): 71.
6. Iammartino, N.R. "Mercury Cleanup Routes—II," *Chem. Eng.* 82 (1975): 36–37.
7. Campbell, H.J. Jr., N.C. Scrivner, K. Batzer, and R.F. White. "Evaluation of Chromium Removal from a Highly Variable Wastewater Stream," *Proceedings of the 32nd Industrial Waste Conference, Purdue University* (May 1977), pp 102–115.

8. "Development Document for Effluent Limitations Guidelines and New Source Performance Standards for the Petroleum Refining Point Source Category," USEPA 440/1-74-014-a (April 1974).
9. Feigenbaum, H.N. "Removing Heavy Metals in Textile Waste," *Ind. Wastes* (March/April 1977): 32–34.
10. "Development Document for Effluent Limitations Guidelines and New Source Performance Standards for the Smelting and Slag Processing Segments of the Ferroalloy Manufacturing Point Source Category," USEPA 440/1-74-008-a (February 1974).
11. Patterson, J.W. and R.A. Minear. "State-of-the-Art for the Inorganic Chemicals Industry: Industrial Inorganic Gases," USEPA 600/2-74-009c (March 1975).
12. "Aluminum Manufacturer Removes Chromium" *Ind. Wastes* (January/February 1978): 24, 34.
13. "Development Document for Effluent Limitations Guidelines and New Source Performance Standards for the Textile Mills Point Source Category," USEPA 440/1-74-022-a (June 1974).
14. Cullinane, M.J. Jr. and J.D. Dietz. "Treatment of Plating Wastes from the Automotive Industry," *Ind. Wastes* (March/April 1978): 29–32.
15. Scott, M.C. "Sulfide Process Removes Metals, Produces Disposable Sludge," *Ind. Wastes* (July/August 1979): 34–38.
16. Robinson, A. and J. Sum. "Sulfide Precipitation of Heavy Metals," USEPA 600/2-80-139 (June 1980).
17. Donovan, E.J. Jr. "Treatment of Wastewater for Steel Cold Finishing Mills," *Water Wastes Eng.* (November 1970): F22–F25.
18. Soboroff, D.M., J.D. Troyer, and A.A. Cochran. "A One-Step Method for Recycling Waste Chromic Acid-Sulfuric Acid Etching Solutions," paper presented at the 33rd Annual Purdue Industrial Waste Conference, West Lafayette, IN (May 1978).
19. Germain, J., C. Vath, and C. Griffin. "Solving Complex Waste Disposal Problems in the Metal Finishing Industry," paper presented at Georgia Water Pollution Control Association Meeting (September 1968).
20. Lowe, W. "The Origin and Characteristic of Toxic Wastes, with Particular Reference to the Metal Industry," *Water Poll. Control* (London, 1970): 270–280.
21. Olver, J., W. Kreye, R. Roberts, and H. Sutton. "Anodizing Wastewater Treatment and Sludge Disposal: A Total Systems Design," *Proceedings of the 3rd Annual Conference on Treatment and Disposal of Industrial Wastewaters and Residues*, Houston, TX (April 18–20, 1978), pp 94–101.
22. Hansen, N.H. and W. Zabban. "Design and Operation Problems of a Continuous Automatic Plating Waste Treatment Plant at the Data Processing Division, IBM, Rochester, Minnesota," *Proceedings of the 14th Industrial Waste Conference, Purdue University* (May 1959) pp 227–249.
23. Crowle, V.A. "Effluent Problems as They Affect the Zinc Die-Casting and Plating-on-Plastics Industries," *Metal Fin. J.* 17 (1971): 51–54.
24. Yeats, A.R. "Ion Exchange Selectively Removes Heavy Metals from Mixed Plating Wastes," paper presented at the 32nd Annual Purdue Industrial Waste Conference, West Lafayette, IN (May 1977).
25. Kennedy, I.F.T. and S.D. Gupta. "Electrochemical Removal of Trace Metals for Metal Wastes with Simultaneous Cyanide Destruction," *Fin. Manag.* (July 1978): 7–18.

26. Jashnani, I.L. and R. Green. "Pretreatment System for a Circuit Board Plant," paper presented at the 8th National Conference on Municipal Sludge Management, Municipal and Industrial Sludge Management, Disposal and Utilization, Miami Beach, FL (March 19–21, 1979).
27. Avrutskii, P.I. "Control of Chromium (VI) Concentration in Waste Waters," *Chem. Abstr.* 70 (1969): 206–207.
28. "An Investigation of Techniques for Removal of Chromium for Electroplating Wastes," USEPA 12010 EIE 13/71 (1971).
29. Martin, J.J. Jr. "Chemical Treatment of Plating Waste for Removal of Heavy Metals," USEPA R2-73-044 (May 1973).
30. McDonald, C.W. and R.S. Bajwa. "Removal of Toxic Metal Ions from Metal-Finishing Wastewater by Solvent Extraction," *Sep. Science* 12 (1977): 435–445.
31. "Development Document for Interim Final Effluent Limitations Guidelines and Proposed New Source Performance Standards for the Significant Inorganic Products Segment of the Inorganic Chemicals Manufacturing Point Source Category," USEPA 440/1-75/037, Group I, Phase II (May 1975).
32. Landy, J.A. "Chromate Removal at a Saudi Arabian Fertilizer Complex," *J. Water Poll. Control. Fed.* 43 (1971): 2292–2293.
33. Richardson, E.W., E.D. Stobbe, and S. Bernstein. "Ion Exchange Traps Chromates for Reuse," *Environ. Sci. Technol.* 2 (1968): 1006–1016.
34. Development Document for Effluent Limitations Guidelines and New Source Performance Standards for the Oil Base Solvent Wash Subcategories of the Paint Formulating and the Ink Formulating Point Source Category," USEPA Publication 440/1-75-050-a, Group II (July 1975).
35. Leimkushler, K.C. "How NASA Removes Hexavalent Chrome," *Ind. Wastes* 21 (1975): 38–40.
36. MacDougall, H. "Waste Disposal at a Steel Plant: Treatment of Sheet and Tin Mill Wastes," *ASCE Proc.* Separate 493 (1954).
37. Besselievre, E.B. *The Treatment of Industrial Wastes* (New York: McGraw-Hill, 1969).
38. Beevers, M. "Chlorine and Sulfur Dioxide in the Treatment of Cyanide and Chromium Wastes," *Metal Fin. J.* 18 (1972): 232–235.
39. Stone, E.H.F. "Treatment of Non-Ferrous Metal Process Waste," *Metal Fin. J.* 18 (1972): 280–290.
40. Halse, B.T., R.P. Selim, and G.E. Summers. "Control of Metal Finishing Wastes Using ORP," *J. Water Poll. Control. Fed.* 32 (1960): 975–981.
41. Shink, C.A. "Plating Wastes: A Simplified Approach to Treatment," *Plating* 55 (1968): 1302–1305.
42. Curry, N.A. "Philosophy and Methodology of Metallic Waste Treatment," *Proceedings of the 27th Industrial Waste Conference, Purdue University* (May 1972).
43. Lavin, D. and C. Lavin. "A Practical Approach to Water Pollution Control for the Plating Company," paper presented at the 27th Annual Purdue Industrial Waste Conference, West Lafayette, IN (May 1972).
44. Teer, E.H. and L.V. Russell. "Heavy Metals Removal from Wood Preserving Wastewater," paper presented at the 27th Annual Purdue Industrial Waste Conference, West Lafayette, IN (May 1972).
45. Fisco, R. "Plating and Industrial Waste Treatment at the Expanded Plant of Fisher Body, Elyria, Ohio," *Proceedings of the 25th Industrial Waste Conference, Purdue University* (May 1970).

46. Lacy, W.J. and A. Cywin. "The Federal Water Pollution Control Administration's Research and Development Program: Industrial Pollution Control," *Plating* 55 (1968): 1299–1301.
47. Avrutskii, P.I. "Liquid SO_2 in New Reduction Role," *Ind. Eng.* 53 (1961): 29A–30A.
48. Kelsey, G.D. and W.E. Seeds. "Fully Automated Effluent Treatment Plant," *Metal Fin. J.* 18 (1972): 246–248.
49. "Effluent Treatment at B.S.R.," *Metal Fin. J.* 17 (1971): 201.
50. Coulter, K.R. "Pollution Control and the Plating Industry," *Plating* 59 (1970): 1197–1202.
51. Crowle, V.A. "Effluent Treatment and Materials Recovery from the Metal Finishing Industry Using the Integrated Method of Treatment," *Water Poll. Control* (London, 1972): 636–645.
52. "Zinc Plating Effluent—Treatment and Recovery," *Effluent Water Treatment J.* 12 (1973): 543.
53. Werner, H.W. "Treatment of Metal Finishing Wastes by Robertshaw Controls Company," paper presented at the 27th Annual Purdue Industrial Waste Conference, West Lafayette, IN (May 1972).
54. O'Neill, F. "Facing up to Pollution," *Plating* 57 (1970): 1211–1213.
55. Heynike, J.J.C. and F.V.K. von Reiche. "Water and Pollution Control in the Iron and Steel Industry, with Special Reference to the South African Iron and Steel Industrial Corporation," *Water Poll. Control* (London, 1969): 569–573.
56. Stoner, L.B. "Waste Treatment Facilities for Jones and Laughlin Steel Corporation, Hennepin Works," *Proceedings of the 26th Industrial Waste Conference, Purdue University* (May 1971), pp 761–765.
57. Bennett, J.R. "The Treatment of Effluents from Metal Cleaning and Finishing Processes," *Metal Fin. J.* 18 (1972): 272–276.
58. Development Document for Effluent Limitations Guidelines and New Source Performance Standards for the Fabricated and Reclaimed Rubber Segment of the Rubber Processing Point Source Category," USEPA Publication 440/1-74/030a, Group I, Phase II (December 1974).
59. Gould, J. "The Kinetics of Hexavalent Chromium Reduction by Metallic Iron," *Water Res.* 16 (1982): 871–877.
60. Chase, O.P. "Metallic Recovery from Waste Waters Utilizing Cementation," USEPA 670/2-74-008 (1974).
61. Kolesar, T.J. "The Use of Atmospheric-Evaporative Towers for Recycle and Waste Elimination," *Plating* 61 (1974): 571–574.
62. Joster, T.L. and T.H. Taylor. "Industrial Waste Treatment at Scovill Manufacturing Company," paper presented at the 28th Annual Purdue Industrial Waste Conference, West Lafayette, IN (May 1973).
63. "Economics of Wastewater Treatment Alternatives for the Electroplating Industry," USEPA Technology Transfer Report EPA 625/5-79-016 (June 1979).
64. Yoronis, D. "Metal Finishing Waste Treatment: Comparative Economics," *Plating* 55 (1968): 1071–1074.
65. Zievers, J.F., R.A. Crain, and F.G. Barclay. "Waste Treatment in Metal Finishing: U.S. and European Practices," *Plating* 55 (1968): 1171–1179.
66. Skovronek, H.S. and M.K. Stinson. "Advanced Treatment Approaches for Metal Finishing Wastewaters. Part 1 and Part 2," *Plating Surface Fin.* (October 1977): 30–38; (November 1977): 24–31.

67. Jakobsen, K. and R. Laska. "Advanced Treatment Methods for Electroplating Wastes," *Poll. Eng.* (October 1977): 42–46.
68. Ross, R.D., ed. *Industrial Waste Disposal* (New York: Reinhold, 1968).
69. Dvorin, R. "Water and Waste Treatment—A Review of Methods for the Metal Finishing Industry," *Metal Finishing Guidebook-Directory* (1960).
70. Calmon, C., J. Casana, and H. Gold. "Ion Exchange and Resin Adsorption Data for the Removal of Toxic Contaminants from Wastewaters," report to the USEPA Industrial Environmental Research Laboratory, Cincinnati, OH (September 1980).
71. Hessler, J. "Recovery and Reuse of Chromates from Cooling Tower Blowdown," *Proceedings of the International Water Conference,* Pittsburgh, PA (1964).
72. Miller, W. "Removal and Recovery of Chromates from Cooling Tower Blowdown," *Ion Exchange for Pollution Control,* CRC Press (1979) vol. II, chap. 15.
73. Kunin, R. and M. Ellis. "New Technology for Recovery of Chromates," *Proceedings of the International Water Conference,* Pittsburgh, PA (1976).
74. Yamamoto, D., K. Yabe, and O. Abe. "Recovery of Chromates from Cooling Tower Blowdown by Ion Exchange Resins," *Proceedings of the Cooling Tower Institute,* Houston, TX (1975).
75. Seward, R. "Chromate Removal by Ion Exchange," *Proceedings of the International Water Conference*, Pittsburgh, PA (1977).
76. Morrison, W. "Chromium Removal from Plating Solutions," *Ion Exchange Technology,* (New York: Academic Press, 1956), Chap. 13.
77. Schuldt, A. and R. Haber. "Chromate and Water Recovery from a Tin Plating Operation Using Ion Exchange Technologies," *Proceedings of the International Water Conference,* Pittsburgh, PA (1976).
78. Robinson, D. and H. Weisberg. "An Ion Exchange Process for Recovery of Chromate from Pigment Manufacture," USEPA 670/2-044 (1974).
79. Anderson, R.E. "Some Examples of the Concentration of Trace Heavy Metals with Ion Exchange Resins," Proceedings Traces of Heavy Metals in Water-Removal Processes and Monitoring, USEPA 902/9-74-001 (1974).
80. Calmon, C. "Trace Heavy Metal Removal by Ion Exchange," Proceedings Traces of Heavy Metals in Water-Removal Processes and Monitoring, USEPA 902/0-74-001 (1974).
81. Joster, T.L. and T.H. Taylor. "Cementation Plant Reduces Chromium, Recovers Copper," *Ind. Wastes* 21 (1975): 26–28.
82. Hesler, J.C. "Recovery and Reuse of Chromates Part 1: The Ion Exchange Recovery Process," Nalco Reprint 150, from *Proceedings, International Water Conference, Engineering Society of West Pennsylvania,* Pittsburgh, PA (September 28–30, 1964).
83. Rothstein, S. "Five Years of Ion Exchange," *Plating* 45 (1958): 835–841.
84. "Compact Ion-Exchange System Works for 'Small' Plater," *Ind. Waste* 19(1) (1973): 22–23.
85. Kelly, B.J. "Recovering Chromates," Nalco Reprint 174 from *Ind. Water Eng.* (September 1968).
86. Kraljik, J. "Practical Guide to Treatment of Chromium Waste Liquors," *Metal Fin.* 73 (1975): 49–55.
87. Klein, E.W. and J. DelPico. "RO Comes of Age for Plating Pollution Control," *Prod. Fin.* 38 (1974): 74–78.

88. Culotta, J.M. and W.F. Swanton. "Case Histories of Plating Waste Recovery Systems," paper presented at the 56th Annual Conference of the American Electroplaters Society, Detroit, MI (1969).
89. Culotta, J.M. and W.F. Swanton. "Recovery of Plating Wastes: Selection of Lowest Cost Evaporator," *Plating* 57 (1970): 1221–1223.
90. "Control Technology for the Metal Finishing Industry—Evaporators," USEPA Technology Transfer 625/7-79-002 (June 1979).
91. Yacoub, J. and K. Chacey. "Zero Discharge Chrome Recovery System Saves Energy and Recovers Valuable Resources," paper presented at Annual Conference, Water Pollution Control Federation (October 1983).
92. Maruyama, T., S.A. Hannah, and J.M. Cohen. "Removal of Heavy Metals by Physical and Chemical Treatment Processes," *J. Water Poll. Control. Fed.* 47 (1975): 962–975.
93. Hannah, S.A., M. Jelus, and J.M. Cohen. "Removal of Uncommon Trace Metals by Physical and Chemical Treatment Processes," *J. Water Poll. Control. Fed.* 49 (1977): 2297–2309.
94. Druver, J.E. "Reverse Osmosis—Where it Stands Today," *Water Sew. Works* (October 1973): 74–77.
95. Donnelly, R.G., R.L. Goldsmith, K.J. McNulty, and M. Tan. "Reverse Osmosis Treatment of Electroplating Wastes," *Plating* 61 (1974): 432–442.
96. Crampton, P. and R. Wilmoth. "Reverse Osmosis in the Metal Finishing Industry," *Metal Fin.* (March 21–27, 1982).
97. Golomb, A. "Application of Reverse Osmosis to Electroplating Waste Water Treatment, Part IV. Potential Reutilization of Chromium Plating Wastes by Other Industries." *Plating* 61 (1974): 931–934.
98. Campbell, R.J. and D.K. Emmerman. "Freezing and Recycling of Plating Rinsewater," *Ind. Water Eng.* 9 (1972): 38–39.
99. Silman, H. "The Substitution of Toxic Materials," *Plating* 61 (1974): 332–336.
100. Cheremisinoff, P.N. and Y.N. Habib. "Cadmium, Chromium, Lead, Mercury: A Plenary Account for Water Pollution," *Water Sew. Works* 119 (July 1972): 45–51.
101. "Development Document for Effluent Limitations Guidelines and New Source Performance Standards for the Copper, Nickel, Chromium and Zinc Segment of the Electroplating Point Source Category," USEPA 440/1 74-003-a (March 1974).

6
Trivalent Chromium

Chromium in industrial wastes occurs predominantly in the hexavalent form, for example, as chromate (CrO_4^{-2}). Industrial sources of and treatment methods for hexavalent chromium are described in Chapter 5. Hexavalent chromium treatment normally involves reduction to the trivalent (Cr^{+3}) form, prior to removing the chromium from the industrial waste. Thus trivalent chromium in industrial effluents may result from one step of the waste treatment process itself, that of chemical reduction of hexavalent chromium. Industries which employ trivalent chromium directly in manufacturing processes include glass, ceramics, photography, inorganic pigments (including anodizing compounds), textile dye, and animal-glue manufacture. Proprietary trivalent chromium-plating solutions have been developed as substitutes for chromic acid-plating baths, although their use is limited [1].

Trivalent chromium may be the dominant form in certain wastes. The electrolytic production of chromium metal yields only trivalent chromium due to the process design, and an effluent chromium level of 1,964 mg/l in the furnace scrubber water has been reported [2]. A municipal refuse incinerator scrubber water was reported to contain 25 mg/l trivalent chromium and only 0.5 mg/l hexavalent chromium [3]. One chromic oxide-manufacturing plant's raw waste contained 108 mg/l trivalent chromium and 10 mg/l hexavalent chromium, the latter present as particulate CrO_3 [4].

In addition to the predominant hexavalent form, some few mg/l of trivalent chromium may be encountered in chromic acid-plating wastes, even prior to chemical reduction of hexavalent chromium. For example, one plant waste contained hexavalent chromium concentrations ranging from 0.0 to 18.0 mg/l and total chromium concentrations of 2.9–31.8 mg/l [5]. By difference, a trivalent chromium concentration in this one waste stream of 2.9–13.9 mg/l is indicated. For the same plant, an acid-bath waste stream contained hexavalent chromium at 122–270 mg/l and trivalent chromium levels of 37–282 mg/l. Evaluation of plating rinse data from other plants indicates trivalent chromium levels of 2.5–15 mg/l prior to waste treatment [6–8].

Scrubber wastewater from a chrome ore-smelting plant averaged 136 mg/l total chromium and 112 mg/l hexavalent chromium, yielding 24 mg/l trivalent chromium by difference [9]. Spray-tower discharge and slag-processing waste from a ferrochrome-ferrosilicon chrome alloy smelter averaged 4.4 and 2.3 mg/l trivalent chromium, respectively [9].

Table 6.1 Trivalent and Total Chromium Content of Industrial Wastewaters

Source	*Trivalent (mg/l)*	*Total (mg/l)*	*Reference*
Ornamental metal facility			
Total waste	7.25	16	10
Cooling-tower blowdown	60	250	
Sheepskin tannery	15–60	15–60	11
Tannery	42	42	12
Tannery	47–52	47–52	13
Steel mill effluent	5–10	5–10	14
Stainless steel acid rinse	3.5	3.6	15
Metal plating	10.2	11.8	16
Circuit board chrome rinse	32	105	17
Aluminum anodizing	28	164	18
	1–400	1–430	19
Dye house wastes	300	600	20
Spent etchants	7,000–45,000	22,200–87,000	21
Piston ring coating	0.14–4.7	0.16–4.7	22
Coal-cleaning leachate	0.42	0.42	23
Titanium dioxide production	50	50	24
Sodium dichromate	240	800	24
production	10	1,500	24

Cooling-tower blowdown containing chromates may also contain an appreciable quantity of trivalent chromium, as a result of hexavalent chromium reduction by water constituents. One reported analysis indicated total chromium of 3.31 mg/l, including a trivalent chromium level of 1.82 mg/l [9]. Additional data on trivalent and total chromium content of industrial wastewaters are presented in Table 6.1.

TREATMENT TECHNOLOGY

Trivalent chromium may be removed by precipitation as the chromic hydroxide upon addition of lime or caustic, or may be concentrated by ion exchange and recovered. In one instance, precipitation as chromic sulfide has been reported [25]. The high cost of chromium as an industrial material indicates that ion exchange and recovery should be seriously considered as a waste treatment process. However, at this time precipitation of chromic hydroxide, and disposal of the resulting sludge, remains the most common practice.

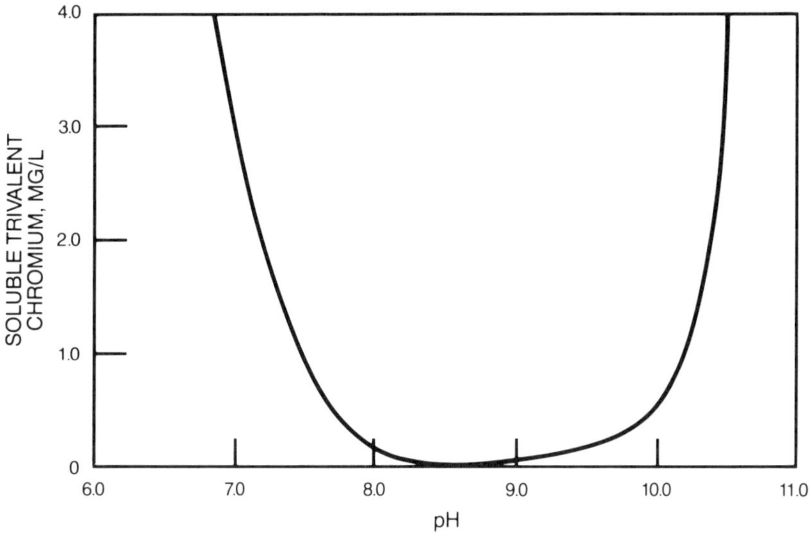

Figure 6.1 Effect of pH on solubility of trivalent chromium. (From Parsons [27], courtesy of National Lime Association.)

Precipitation

Trivalent chromium can be removed as insoluble chromic hydroxide, $Cr(OH)_3$, by precipitation with caustic soda or lime [26]. There is considerable disagreement regarding the effect of pH on the solubility of chromic hydroxide. The precipitation process is reported by one author to be most effective at pH 8.5–9.5, due to the low solubility of chromic hydroxide in that range (Figure 6.1), although best removal has been reported in one case at pH 12.2 for trivalent chromium in an animal-glue manufacturing waste (Table 6.2). However, Netzer et al [29] have reported the results presented in Table 6.3 for lime precipitation of a synthetic chromic chloride solution. These results appear to refute Figure 6.1 in two respects. First, the data of Table 6.3 indicate that minimum solubility is achieved by treatment pH 7.0. Second, solubility does

Table 6.2 Trivalent Chromium Precipitation Treatment of a Glue-Manufacturing Waste [28]

	Chromium (mg/l)		
pH	*Initial*	*Final*	*% Removal*
4.5	650	500	23
8.8	650	18	97
12.2	650	0.3	99.9+

Table 6.3 Lime Precipitation Treatment of a Synthetic Chromic Chloride Solution [29]

Treatment pH	*Chromium Concentration[a] (mg/l)*
3.5[b]	92[b]
5.6	55
7.0	<0.1
8.1	<0.1
9.6	<0.1
10.3	<0.1
11.0	<0.1
11.4	<0.1

[a] Settled supernatant chromium concentration.
[b] Initial solution values.

not increase at pH above 10. The solubility pattern of Table 6.3 has been confirmed in treatability studies on a titanium dioxide-production wastewater [24]. Treatability studies on a mixed plating waste, however, seem to support the solubility pattern given in Figure 6.1 [30].

Because a typical metal-bearing waste will include other metal ions as well as chromium to be removed by precipitation, and because the optimum pH of precipitation varies for different metal hydroxides, an average waste stream treatment pH of about 8 often seems to achieve the best overall results on a waste containing several metal ions [31]. Thus, much of the chromium treatment reported in the literature occurred at pH 7–8.

Soda ash has been used to precipitate trivalent chromium [26, 32], although the most commonly employed chemical for precipitation is lime. Reduction of the chromium content of one electroplating waste from 140 to 1.0 mg/l was achieved by neutralization and precipitation with lime, at pH 7–8 [33]. In waste treatment at an IBM plant, hexavalent chromium (at 1,300 mg/l) in one waste stream was converted to trivalent chromium and precipitated with lime to achieve an effluent containing no hexavalent and 0.06 mg/l trivalent chromium [34]. A coagulant aid was employed to improve the precipitation-sedimentation removal of chromic hydroxide. This coagulant aid undoubtedly accounted for the high degree of treatment achieved, since the influent to the sedimentation basin contained 1,000–2,000 mg/l suspended solids, while the effluent routinely contained less than 10 mg/l. Thus 99–99.5% suspended solids removal was achieved by employing the coagulant aid. Equally good effluent hydrous metallic suspended-solids levels were reported in another case where a coagulant aid was also employed to improve chromic hydroxide precipitate removal [35].

Isolation of chromium-bearing wastes from the manufacture of sodium dichromate and chromic acid yielded a raw waste containing 1,300 mg/l hexavalent chromium. Batch treatment by reduction with ferrous sulfate (waste pickle liquor) and precipitation yielded effluent levels of 0.01 mg/l hexavalent chromium and 0.01–0.31 (avg 0.14) mg/l trivalent chromium, mostly present as solids [36]. Total treatment costs of $2,876/1,000 gpd capacity and $6.41/1,000 gal treated were reported for the 0.348 MGD facility. Lime addition to wastewater from an electrolytic chrome metal facility containing 1,764 mg/l trivalent chromium yielded an effluent value of about 0.2 mg/l [2].

Filtration of effluent from the postprecipitation sedimentation basin has been employed to remove the small, nonsettleable chromic hydroxide particles not readily removable by simple sedimentation. In one instance, the use of filtration significantly improved the effluent quality of a chromium-bearing waste from a British metal alloy plant [37]. Hexavalent chromium, as sodium dichromate, was employed to produce a bright finish on copper alloy products. The plating was carried out in a mixture of the sodium dichromate and sulfuric acid. Waste chromate from the process, at 7,400 mg/l, was reduced with sulfur dioxide to trivalent chromium, and was precipitated at pH 8.5 with lime. Hydroxides of copper, zinc, and nickel coprecipitated in the process. The metal hydroxides were removed by sedimentation. During a two-week surveillance period, chromium content in the settling basin effluent was found to range from 1.3 to 4.6 mg/l. Passage of this effluent through a sand filter to remove the finely divided chromic hydroxide particles reduced the effluent chromium concentrations to 0.3–1.3 mg/l. This residual should principally be in the soluble Cr^{+6} form, rather than as trivalent chromium.

In treatability studies on a titanium dioxide wastewater, filtration was also assessed. Treatment involved lime addition to pH 9.8–10.0, clarification, and dual media filtration. The raw wastewater ranged from 32 to 65 mg/l in trivalent chromium, and averaged 49.4 mg/l. Clarifier effluent chromium was 0.17 mg/l, and filter effluent chromium was 0.05 mg/l [24].

A typical treatment process involves chromate reduction by sodium metabisulfite to below 0.1 mg/l, followed by waste neutralization with lime to pH 9.0, precipitation of chromic plus other metallic hydroxides, and settling [38]. Effluent trivalent chromium levels below 0.2 mg/l can be achieved by such treatment [39]. Lime precipitation of a dilute wood-preserving wastewater, containing trivalent chromium up to 2.2 mg/l, yielded an average effluent of 0.02 mg/l [40]. Solids separation was by settling.

High effluent chromium levels often reflect inefficient chromate reduction pretreatment. This has been demonstrated in treatability studies on a sodium dichromate wastewater [24]. The treatment involved chromate reduction with ferrous chloride, lime precipitation at pH 8.0–9.0, clarification, and dual media filtration. Average performance results are compared in Table 6.4 for treatment runs where effluent hexavalent chromium was less than 0.5 mg/l versus higher effluent concentration. The impact of ineffective chromate reduction on total chromium removal is apparent.

The use of reduction and lime precipitation to treat chromium-bearing waste from a General Electric appliance plant has been reported [5]. Sedimentation of the precipitate was improved by addition of an anionic polyelec-

Table 6.4 Evaluation of Influence of Efficiency of Hexavalent Chromium Reduction on Total Chromium Removal [after 24]

Process Stream	*Filtrate Chromate Below 0.5 mg/l*	*Filtrate Chromate Above 0.5 mg/l*
Raw wastewater[a]	1,050	958
Clarifier effluent[a]	0.29	35.2
Filter effluent[a]	0.21	33.2
Filter effluent[b]	<0.02	28.6

[a] Total chromium concentrations, mg/l.
[b] Hexavalent chromium concentrations, mg/l.

trolyte. Plant effluent was reported to contain no hexavalent chromium and an average 0.75 mg/l total chromium.

Large quantities of sludge are produced in the precipitation of metal hydroxides from industrial wastes. Furthermore, use of ferrous sulfate as the hexavalent chromium-reducing agent preparatory to trivalent chromium precipitation will produce roughly four times the sludge weight encountered with use of other reductants, upon precipitation of the chromic hydroxide [41]. The specific choice of base for pH adjustment will also influence the sludge production, with lime yielding roughly three times more sludge compared to soda ash (Na_2CO_3) or caustic soda (NaOH), due to the formation of calcium carbonate or sulfate. Typically these sludges settle rapidly but do not compact very well, and contain only 2–3% solids. Various methods of sludge concentration and disposal are presented in Table 6.5, as reported for typical industrial processes. Table 6.6 presents selected vacuum filtration sludge-concentrating data. The filtrate from sludge dewatering is normally returned to the original sludge sedimentation basin for further treatment.

There is a growing emphasis on concentration/dewatering of metal hydroxide sludges, due to increased costs of sludge disposal. Figures 6.2 and 6.3 present chromic hydroxide sludge filtration operating equipment costs. Figure 6.2 presents operating expenses as a function of sludge flow rate in gallons per

Table 6.5 Metal Hydroxide Sludge Concentration and Disposal Techniques [42]

Method	*Comments*
Lagooning	Requires large areas of land. Might contaminate groundwaters unless properly lined
Vacuum Filtration	Equipment and operating costs high. Discharge cake dry enough for haulage
Pressure Filtration	Equipment and operating costs high. Requires proper conditioning of sludge. Dry cake
Centrifugation	Method of solids removal for large volumes and large particles

Table 6.6 Vacuum Filtration Treatment of Metal Hydroxide Sludges

Initial Sludge Solids (mg/l)	*Final Sludge Solids (mg/l)*	*Reference*
30,000–60,000	100,000–120,000	34
100,000	290,000	37
—	200,000–250,000	43

minute versus monthly operating cost in thousands of dollars. Three operating conditions are presented in Figure 6.2, based upon the use of bodyfeed and/or flocculating agents to improve filtration. Bodyfeed cost was based upon use of diatomaceous earth as a filter aid. The filterability of chromium sludge precipitated by caustic is poorer than that of sludge precipitated with lime, and the former sludge frequently requires a bodyfeed filter-aid to improve dewatering characteristics, while the latter sludge does not require such aid. Figure 6.3 shows equipment costs in thousands of dollars versus sludge flow in gallons per minute. The economy of scale associated with high waste flows is evident.

In one case where vacuum filtration was not necessary, sludge from the sedimentation basin was held for short periods of time in sludge storage tanks

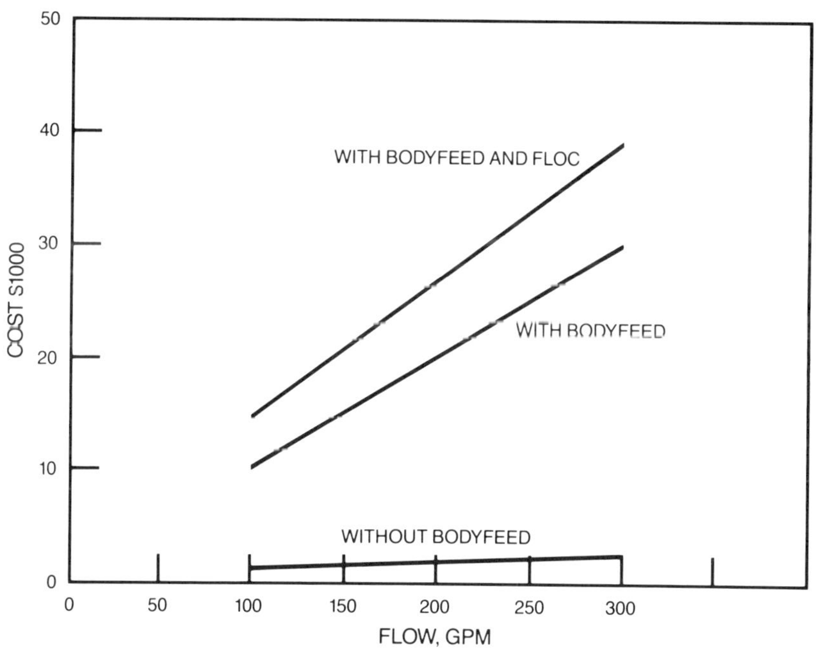

Figure 6.2 Monthly operating plus depreciation costs (12 yr straight line) for wastewater filtration, 720 hr/month. (From Zievers et al [31], courtesy of *Plating*.)

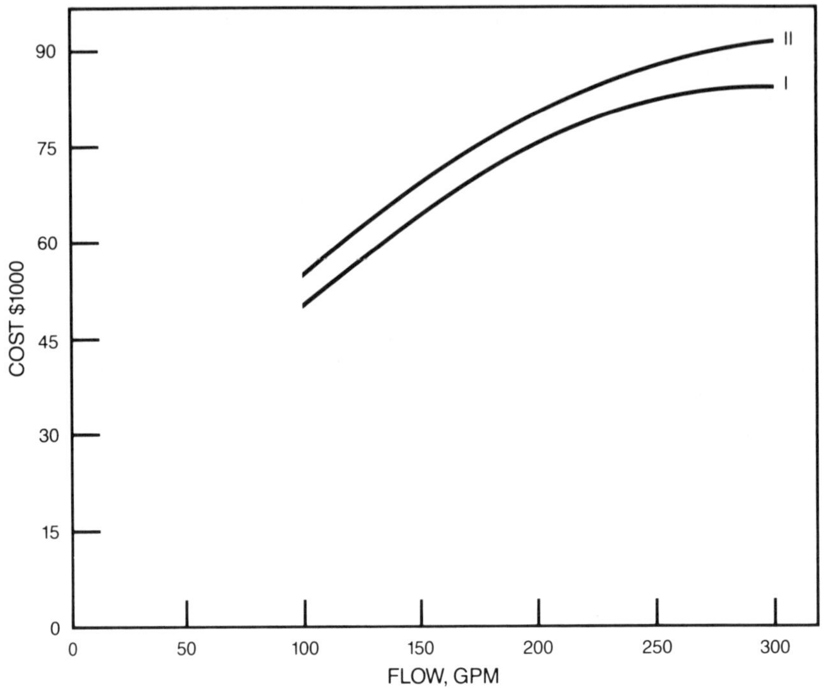

Figure 6.3 Initial capital costs (in 1968) for filtration equipment, including installation. I, manual operation; II, semiautomatic operation. (From Zievers et al [31], courtesy of *Plating*.)

[34]. The sludge stored in the holding tank thickened (or compacted) to a suspended solids concentration of 10–12%, equivalent to that obtainable by vacuum filtration. The thickening was thought to be the result of applying a coagulant aid in the sedimentation basin, although another possible explanation is that the chromic-hydroxide sludge aged and underwent transformation to the more dense chromic oxide form.

Treatment data on chromic hydroxide precipitation of aircraft engine-manufacturing wastes have been published which indicate that concentrated wastes yield most dense sludge, and dilute wastes a more voluminous sludge [44]. Results from chromic hydroxide treatment studies suggest that each volume of dilution water added to an equal initial volume of concentrated trivalent chromium waste increases the final settled sludge volume by approximately 50%. This would indicate a major benefit in final sludge handling through treating concentrated rather than diluted wastes.

The treatment costs associated with trivalent chromium removal by precipitation are due to the precipitation process and the expense of sludge concentration and disposal. At $80/ton, lime costs approximately $0.65/lb of trivalent chromium precipitated. Cost of complete treatment from the hexavalent chromium form includes chemical reduction cost, precipitation cost, and sludge handling and disposal costs. Capital costs of $12,000–$15,000 have been quoted for typical off-the-shelf automated and continuous-flow chromium treatment plants which incorporate chromate reduction, pH adjustment, and chromic-hydroxide precipitation [38].

Ion Exchange

Relatively little information is available in the technical literature on ion exchange treatment of trivalent chromium. This is in part because most industrial-waste chromium exists in the hexavalent form, and if recovery is preferred direct anion exchange of the hexavalent chromate or dichromate is employed. Ion exchange is dependent upon the electron charge on an ion. Trivalent chromium has a +3 charge, and hexavalent chromium in its usual form of chromate or dichromate has a −2 charge. Thus, different ion exchange resins must be employed to recover each form of chromium. An anionic type of resin is required to capture negatively charged forms such as the hexavalent chromate and dichromate ions. A cationic type of resin is employed for positively charge (e.g., Cr^{+3}) forms.

Where ion exchange is utilized for control of trivalent chromium, it may be employed in either of two modes. One is as a concentration mechanism to recover the trivalent chromium value. The second mode is as a polishing treatment step, after conventional treatment by precipitation. In one pilot study of ion exchange in this latter mode, however, following hexavalent chromium reduction plus precipitation treatment for a wastewater also containing copper, nickel, and zinc, the trivalent chromium breakthrough into the exchange column effluent far preceded the appearance of any other metal [30].

Two reasons have been suggested to recover trivalent chromium by ion exchange [45]. The first is the passage of concentrated chromic acid baths through a cation resin to remove metallic contaminants such as iron, aluminum, and trivalent chromium from the chromic acid. This process would purify the chromic acid solution, which could then be reused. One difficulty with this application is that chromic acid, at the concentrations typically used in plating baths (250–400 g/l) will attack and destroy exchange resin. Experience has indicated that the maximum acceptable concentration of chromic acid is approximately 250 g/l [46]. Therefore, chromic acid solutions above this concentration must be diluted before ion exchange treatment. After ion exchange, reconcentration of the purified chromic acid by evaporation in glass-lined vessels is common [46]. Application of this technique to reclaiming of weak chromating baths in the zinc bonderizing process has been reported [47].

The second reason for ion-exchange treatment for trivalent chromium removal is for the complete purification of a plating rinsewater, in which a rather dilute waste would be passed through both an anionic resin and a cationic

Table 6.7 Summary of Trivalent Chromium Treatment

Method	*pH*	*Chromium (mg/l)* *Initial*	*Final*	*Reference*
Precipitation	—	—	0.75	5
	8.8	650	18	28
	12.2	650	0.3	28
	7–8	140	1.0	33
	—	1,300	0.06	34
Precipitation	8.5	7,400	1.3–4.6	37
With sand filtration	8.5	7,400	0.3–1.3	37
Precipitation	—	2.2	0.02	40
	7.8–8.2	16.0	0.06–0.15	10
	8.5–10.5	26.0	0.44–0.86	49
	—	11.75	2.50	16
	8.8–10.1	—	0.6–30	50
	8.5	47–52	0.3–1.5	13
	—	164	1	18
Precipitation	9.8–10.0	49.4	0.17	24
With filtration	9.8–10.0	49.4	0.05	24
Electrocoagulation		15–60	5–4	11
Microflotation	—	5–10	1	14
Electrolysis	—	2.1	0.1	16
Alum coagulation				
Filtration	6.4	0.7	0.017	51
Followed by carbon absorption	—	—	0.006	51

resin to remove chromium species of both positive and negative charge [40]. The treated rinsewater could then be recycled.

No capital cost figures are available for ion exchange treatment of trivalent chromium. Lacy and Cywin [48] have reported general operating costs for ion exchange treatment of \$0.16–0.24/1,000 gal. Ross [45] has reported daily operating costs for ion exchange treatment of trivalent chromium as approximately \$28/day for a waste flow of 73 gpm. Assuming a 16-hr day, operating costs are estimated at \$0.40/1,000 gal.

SUMMARY

Most trivalent chromium in industrial waste results from reduction of hexavalent chromium as a preliminary step in waste treatment. This trivalent chromium is most often treated with lime to precipitate chromic hydroxide. Tables

Table 6.8 Summary of Treated Effluent Trivalent Chromium Concentrations for Plating and Associated Process Wastes [52]

Treatment Process	*No. Plants*	*Chromium (III) Concentration (mg/l)* Range of Values	*Chromium (III) Concentration (mg/l)* Avg for Plants
Batch Precipitation	8	0.04–1.0	<0.24
Continuous Precipitation	13	0.07–4.1	<1.17
Integrated Process	8	0.00–0.7	<0.17

6.7 and 6.8 summarize results of precipitation treatment. The chromic hydroxide sludge is dewatered by vacuum filtration or disposed of directly to landfill or otherwise.

Under special circumstances, ion exchange removal of trivalent chromium has been employed. Ion exchange is characterized by effecting complete Cr^{+3} removal, until the exchange capacity of the resin is exhausted. This exchange capacity may be recovered by regenerating the resin.

REFERENCES

1. Silman, H. "The Substitution of Toxic Materials," *Plating* 61 (1974): 332–336.
2. "Development Document for Interim Final Effluent Limitations Guidelines and Proposed New Source Performance Standards for the Electrolytic Ferroalloys Segment of the Ferroalloy Manufacturing Point Source Category," USEPA 440/1-75-038-a, Group I, Phase II (February 1975).
3. Iammartino, N.R. "Mercury Cleanup Routes—II," *Chem. Eng.* 82 (1975): 36–37.
4. "Development Document for Interim Final Effluent Limitations Guidelines and Proposed New Source Performance Standards for the Significant Inorganic Products Segment of the Inorganic Chemicals Manufacturing Point Source Category," USEPA 440/1-75/037, Group I, Phase II (May 1975).
5. Anderson, J.S. and E.H. Iobst, Jr. "Case History of Wastewater Treatment in a General Electric Appliance Plant," *J. Water Poll. Control. Fed.* 10 (1968): 1786–1795.
6. Germain, J.E., C.A. Vath, and C.F. Griffin. "Solving Complex Waste Disposal Problems in the Metal Finishing Industry," paper presented at the Georgia Water Pollution Control Association Meeting (September 1968).
7. Martin, J.J. Jr. "Chemical Treatment of Plating Waste for Removal of Heavy Metals," USEPA R2-73-044 (1973).
8. "Development Document for Interim Final Effluent Limitations Guidelines and Proposed New Source Performance Standards for the Common and Precious Metals Segment of the Electroplating Point Source Category," USEPA 440/1-75/040, Group I, Phase II (April 1975).
9. "Development Document for Effluent Limitations Guidelines and New Source Performance Standards for the Smelting and Slag Processing Segment of the Fer-

roalloy Manufacturing Point Source Category," USEPA 440/1-74-008-a (February 1974).

10. Cullinane, M.S., Jr. and J.D. Dietz. "Treatment of Plating Wastes from the Automotive Industry," *Ind. Wastes* (March/April 1978): 29–32.
11. Ramirez, E.R., L.K. Barber, and O.A. Clemens. "Physiochemical Treatment of Tannery Wastewater by Electro-Coagulation," *Proceedings of the 32nd Industrial Waste Conference, Purdue University* (May 1977), pp. 183–188.
12. Olsen, C.H., L.D. Rose, J.G. Kremer, and W. Garrison. "Control of Chromium and Sulfide Discharges at Three Tanneries," paper presented at the 51st Annual Water Pollution Control Federation, Anaheim, CA (October 1978).
13. Bishop, P.L. "Physiochemical Pretreatment of Wastes from a Secondary Tannery," paper presented at the 33rd Annual Purdue Industrial Waste Conference, West Lafayette, IN (May 1978).
14. Cottrell, W.R.T. "Treatment of Effluent by Flotation Processes," *Effluent Water Treatment J.* (November 1976): 563–567.
15. Winans, T.O. and D.E. Dreier. "Innovative Acid Pickling Wastes Treatment at Universal-Cyclops Stainless Steel Plant," paper presented at the 34th Annual Purdue Industrial Waste Conference, West Lafayette, IN (May 1979).
16. Kennedy, I.F.T. and S.D. Gupta. "Electrochemical Removal of Trace Metals for Metal Wastes with Simultaneous Cyanide Destruction," *Fin. Manag.* (July 1978): 7–18.
17. Jashnani, I.L. and R. Green. "Pretreatment System for a Circuit Board Plant," paper presented at the 8th National Conference on Municipal Sludge Management, Municipal and Industrial Sludge Management Disposal and Utilization, Miami Beach, FL (March 19–21, 1979).
18. "Aluminum Manufacturer Removes Chromium," *Ind. Wastes* (January/February 1978): 24, 34.
19. Olver, J., W. Kreye, R. Roberts, and H. Sutton. "Anodizing Wastewater Treatment and Sludge Disposal: A Total Systems Design," *Proceedings of the 3rd Annual Conference on Treatment and Disposal of Industrial Wastewaters and Residues*, Houston, TX (April 18–20, 1978), pp. 94–101.
20. Rouse, J.V. "Removal of Heavy Metals from Industrial Effluents," *ASCE J. Environ. Eng. Div.* (October 1976): EE5 929–936.
21. Soboroff, D.M., J.D. Troyer, and A.A. Cochran. "A One-Step Method for Recycling Waste Chromic Acid-Sulfuric Acid Etching Solutions," paper presented at the 33rd Annual Purdue Industrial Waste Conference, West Lafayette, IN (May 1978).
22. Carter, J., M. Jones, and R. Tiefenbrunn. "Treatment for Removal of Heavy Metals and Oil from a Piston Ring Manufacturing Waste," paper presented at the 37th Annual Purdue Industrial Waste Conference, West Lafayette, IN (May 1982).
23. Wangen, L. and J. Williams. "Control by Alkaline Neutralization of Trace Elements in Acidic Coal Cleaning Waste Leachates," *J. Water Pollution Control. Fed.* 54 (1982): 1302–1310.
24. *Treatability Studies for the Inorganic Chemicals Manufacturing Point Source Category*, USEPA 440/1-80/103 (July 1980).
25. Eldridge, J.D. and T.L. Judell. "Recovery of Materials from Process Effluents," *Solid Waste Treatment and Disposal*, N.Y. Kirov, ed. (Ann Arbor: Ann Arbor Science Publishers, 1971).

26. Besselievcr, E.B. *The Treatment of Industrial Wastes* (New York: McGraw-Hill, 1969).
27. Parsons, W.A. *Chemical Treatment of Sewage and Industrial Wastes* (Washington, D.C.: National Lime Association, 1965).
28. Wang, L.K., R.P. Leonard, J.G. Michalouci, and D.W. Goupil. "Glue Treatment—Pick a Way," *Water Wastes Eng.* 10 (1973): 20–24.
29. Netzer, A., A. Bowers, and J. Norman. "Removal of Heavy Metals from Wastewater by Lime," *Proceedings of the 16th Conference on Great Lakes Research*, Huron, OH (1973), pp 260–265.
30. Yeats, A.R. "Ion Exchange Selectively Removes Heavy Metals from Mixed Plating Wastes," paper presented at the 32nd Annual Purdue Industrial Waste Conference, West Lafayette, IN (May 1977).
31. Zievers, J.F., R.W. Crain, and F.G. Barclay. "Waste Treatment in Metal Finishings, U.S. and European Practices," *Plating* 55 (1968): 1171–1179.
32. "Liquid SO_2 in New Reduction Role," *Ind. Eng. Chem.* (July 1961): 29A–30A.
33. Aurutskii, P.I. "Control of Chromium (VI) Concentration in Waste Waters," *Chem. Abstr.* 70 (1969): 206–207.
34. Hansen, N.H. and W. Zabban. "Design and Operation Problems of a Continuous Automatic Plating Waste Treatment Plant at the Data Processing Division, IBM, Rochester, Minnesota," *Proceedings of the 14th Industrial Waste Conference, Purdue University* (May 1959), pp 227–249.
35. O'Connor, S.F., B.W. Montjoy, Jr., and N.S. Chamberlin. "Western Electric Builds Modern Plant for Treating Metal Finishing Wastes," *Water Wastes Eng./Ind.* 9 (1972): D16–D19.
36. "Development Document for Effluent Limitations Guidelines and New Source Performance Standards for the Major Inorganic Products Segment of the Inorganic Chemicals Manufacturing Point Source Category," USEPA 440/1-74-007-a (March 1974).
37. Stone, E.H.F. "Treatment of Non-Ferrous Metal Process Waste at Kynoch Works Birmingham, England," *Proceedings of the 22nd Industrial Waste Conference, Purdue University* (May 1967), pp 848–865.
38. Werner, H.W. "Treatment of Metal Finishing Wastes by Robertshaw Control Company," paper presented at the 27th Annual Purdue Industrial Waste Conference, West Lafayette, IN (May 1972).
39. Curry, N.A. "Philosophy and Methodology of Metallic Waste Treatment," *Proceedings of the 27th Industrial Waste Conference, Purdue University* (May 1972).
40. Teer, E.H. and L.V. Russell. "Heavy Metals Removal from Wood Preserving Wastewater," paper presented at the 27th Annual Purdue Industrial Waste Conference, West Lafayette, IN (May 1972).
41. Kraljik, J. "Practical Guide to Treatment of Chromium Waste Liquors," *Metal Fin. J.* 73 (1975): 49–55.
42. Anderson, J.W., J.M. Phillips, and C.B. Schriver. "Water Contamination by Certain Heavy Metals," paper presented at the 44th Annual Water Pollution Control Federation Conference, San Francisco, CA (1971).
43. Kelsey, G.D. and W.E. Seeds. "Fully Automated Effluent Treatment Plant," *Metal Fin. J.* 18 (1972): 246–248.
44. Chalmers, R.K. "Treatment of Wastes from Metal Finishing and Engineering Industries," in *Progress in Water Technology*, vol. I, *Applications of New Concepts*

of Physical-Chemical Wastewater Treatment, W.W. Eckenfelder and L.K. Cecil, eds. (New York: Pergamon Press, 1974).
45. Ross, R.D., ed. *Industrial Waste Disposal* (New York: Reinhold, 1968).
46. Thompson, J. and V.J. Miller. "Role of Ion Exchange in Treatment of Metal Finishing Wastes," *Plating* 58 (1971): 809–812.
47. McGarvey, F.X. and S.A. Fisher. "Effluent Control in Zinc Bonderizing Process, II. The Recycle of Dilute Chromic Acid Baths via an Ion Exchange Process," *Plating* 62 (1975): 335–339.
48. Lacy, W.J. and A. Cywin. "The Federal Water Pollution Control Administration's Research and Development Program: Industrial Pollution Control," *Plating* 55 (1968): 1299–1301.
49. McVaugh, J. and W.T. Wall, Jr. "Optimization of Heavy Metals Wastewater Treatment. Effluent Quality Versus Sludge Treatment," *Proceedings of the 31st Industrial Waste Conference, Purdue University* (May 1976), pp 17–25.
50. Yost, K.J. and D.R. Masarik. "A Study of Chemical-Destruct Waste Treatment Systems in the Electroplating Industry," *Plating Surface Fin.* (January 1977): 35–40.
51. Hannah, S.A., M. Jelus, and J.M. Cohen. "Removal of Uncommon Trace Metals by Physical and Chemical Treatment Processes," *J. Water Pollution Control. Fed.* 49 (1977): 2297–2309.
52. "Development Document for Effluent Limitations Guidelines and New Source Performance Standards for the Copper, Nickel, Chromium and Zinc Segment of the Electroplating Point Source Category," USEPA 440/1-74-003-a (March 1974).

7
Copper

A principal source of copper in industrial waste streams is metal-cleaning and -plating baths and rinses. Brass and copper metal working requires periodic oxide removal by immersing the metals in strong acid baths. Solution adhering to the cleaned metal surface, referred to as *dragout*, is rinsed from the metal and contaminates the waste rinsewater. Similarly, metal parts undergoing copper or brass plating drag out some of the concentrated plating solution. Alkaline cyanide-based baths are most popular for copper plating, although noncyanide solutions containing proprietary chelators, ammonium salts, alkanolamines, alkaline pyrophosphate, acid sulfate, or acid fluoborate, have come into recent use. Copper concentration in the plating bath depends upon the bath type, and may range from 3 to 50 g/l. For a given bath, the rinsewater copper concentration will be a function of many factors including drainage time over the bath, shape of the part and total surface area, method of rinsing and the rate of rinse water flow. Rinsewater copper concentration may range from 0.02% to 1.0% of the process bath concentration [1, 2].

Jewelry manufacturers employ copper plating either directly or as a base metal for silver and other precious metals. Copper is used extensively in the manufacture of printed circuit boards. Copper is also employed in the alkaline Bemberg rayon process as cupro-ammonium salt. Copper-bearing mining wastes and acid mine drainage contribute significant quantities of dissolved copper to waste streams. Additional potential sources of copper-bearing waste include pulp, paper and paper board mills, chlorosilane (silicone) synthesis, wood preserving, fertilizer manufacturing, petroleum refining, paint and pigment manufacturing, basic steel works and foundries, nonferrous metal works and foundries, and motor vehicle and aircraft plating and finishing [3]. A copper concentration of 23 mg/l has been reported for one municipal refuse incinerator scrubber water [4], and copper levels exceeding 700 mg/l have been found in power-plant boiler-cleaning wastes [5, 6]. When anhydrous ammonia is used as one of the boiler-cleaning chemicals, the ammonia acts as a copper-complexing agent in the wastewater and interferes with copper treatment by conventional technologies [6]. Industrial process wastewater concentrations are summarized in Table 7.1.

Table 7.2 presents wastewater data for the primary and secondary copper-production segments of the nonferrous metals industry. Primary copper production involves raw ore smelting and refining. Secondary copper is produced

Table 7.1 Concentrations of Copper in Industrial Process Wastewaters

Process	*Copper Concentration (mg/l)*	*Reference*
Plating rinse	20–120	7
	0–7.9	8
	20 (avg)	9
	5.2–41	10
	6.2–88	11
	2.0–36.0	12
	3–30	13
	11.4	14
	2.8–7.8 (4.5 avg)	15
	21	16
	24	16
Copper plating bath rinse		
First rinse	183	17
Second rinse	2.2	
Welding wire copper plating		
Plating bath	3,640	18
Spent acid	34	
Rinsewater	2–10	
Appliance manufacturing		
Spent acids	0.06–11.0	14
Alkaline wastes	1.0	
Integrated circuit manufacture	0.23	19
Circuit board manufacture	16.5–77	20
	2.3	16
Automobile heater production	24–33 (28 avg)	21
Metal finishing	0.5–5	22
Silver plating		
Silver-bearing	3–900 (12 avg)	15
Acid wastes	30–590 (135 avg)	
Alkaline wastes	3.2–19 (6.1 avg)	
Brass plating		
Pickling bath wastes	4.0–23	15
Bright dip wastes	7.0–44	
Aluminum anodizing	0.2–2.0	23
Anodizing and plating	1.3	16
	4.7	16
Brass mill rinse	4.4–8.5	24
	74–888	25
Brass mill bichromate pickle	4.5–74	25
Brass and copper wire mill	75–124	13
	60–90	13
	20–35	13

Table 7.1 (*continued*)

Process	*Copper Concentration (mg/l)*	*Reference*
Copper mill rinse	19–74	24
Copper tube mill	70 (avg)	26
	800 (avg)	26
Copper ore extraction	0.28–0.33	27
Gold ore extraction	20	28
	3.2	29
Acid mine drainage	0.12–3.9	30
	51.6–128.0	31
	3.6–76	32
Paint formulation	0.04–0.40	33
Ink formulation	0.01–6.4	33
Paint and ink formulation	0–100	34
Wood preservation	0.05–1.1	35
Copper sulfate manufacture	433	36
	221	37
Power plant boiler cleaning wastes	0–334	5
	100–790 (440 avg)	6
Petroleum refining	0.0–1.4	34
Gum and wood chemicals	0.0–6.0	34
Porcelain enameling	0.0–12.0	34

Table 7.2 Wastewater Characteristics from Primary and Secondary Copper Manufacture [38, 39]

		Copper Concentration (mg/l)	
Category	*pH*	*Range*	*Avg*
Primary copper			
Slag granulation	6.4–8.1	0.05–0.60	0.26
Scrubber blowdown	1.8–2.5	0.06–7,350	2,450
Cast metal cooling	7.0–8.4	0.20–3.5	1.6
Arsenic recovery	3.8–4.4	—	88.4
Electrolytic refining	6.4–7.6	0.20–25.3	5.7
Secondary copper			
Slag granulation	8.3–9.8	0.07–1,250	317
Scrubber water	7.1–7.6	0.04–280	140
Cast metal cooling	7.9–8.9	0.07–3.38	1.3

from scrap metal or metallurgical slag. There is a strong correlation between the pH and copper content of wastewaters of the industry, with higher metal levels occurring at pH below 7 [39].

TREATMENT TECHNOLOGY

As with most heavy-metal wastes, treatment processes employed for reduction of soluble copper in wastewaters may involve precipitation and disposal of resulting sludge solids, or recovery processes such as ion exchange, evaporation, and electrolysis. The value of copper metal frequently makes recovery processes attractive. In precipitation treatment, a copper-rich sludge may have significant value for copper recovery by an outside processor.

The importance of good housekeeping and process modification in reducing overall treatment requirements, particularly in metal processing and plating operations, has been shown in many instances [40–44]. Principally, modification of rinsing techniques by installing cascading or countercurrent rinse systems increases the contaminant concentration while reducing the total wastewater volume. Recovery processes may become economically and technologically feasible through such modifications. Ion exchange is an appropriate treatment method for wastewaters containing copper at less than 200 mg/l; precipitation is applicable for copper levels of 1.0–1,000 mg/l, and electrolytic recovery is advantageous for treatment at wastewater copper concentrations above 10,000 mg/l [3].

Precipitation

The standard treatment method for copper and most heavy metals is precipitation as the relatively insoluble hydroxide, at alkaline pH, or occasionally as the sulfide. Coprecipitation by addition of ferrous sulfate has also been suggested for strongly complexed copper wastewaters such as originate from electroless plating baths and rinses [45, 46]. However, the process is somewhat complicated, requiring pH adjustment in two steps to break the copper complex, and relying upon ferrous iron reduction of the copper to the cuprous form to prevent recomplexation [47].

Generally, hydroxide precipitation is accomplished by alkali addition to an acidic wastewater. Lime is usually preferred to the more expensive sodium hydroxide (caustic). In the presence of high sulfate, considerable calcium sulfate precipitate may form, and recovery value of the copper sludge may dictate use of a more expensive base to obtain a purer sludge.

Cupric oxide, which forms from the hydroxide, has minimal solubility between pH 9.0 and 10.3 [48]. A minimal solubility of 0.01 mg/l has been reported in that pH range, from laboratory studies [49]. This value corresponds closely to the theoretical minimum effluent level which could be achieved by

Table 7.3 Effect of Treatment pH on Copper Precipitation

Lime Treatment [50][a]		*Lime Treatment [37]*[b]		*Caustic Treatment [37]*[b]		*Caustic Treatment [18]*[c]	
pH	*Copper*	*pH*	*Copper*	*pH*	*Copper*	*pH*	*Copper*
5.2[d]	93[d]	3.1[d]	221[d]	3.1[d]	221[d]	2.5[d]	10[d]
6.6	35	—	—	—	—	—	—
7.0	5.3	7.0	2.6	7.0	7.6	—	—
8.0	<0.1	8.0	2.5	8.0	9.7	—	—
—	—	9.0	3.5	9.0	9.6	9.0	<0.1
9.5	<0.1	—	—	—	—	9.5	<0.1
—	—	10.0	4.7	10.0	13.4	10.0	<0.1
—	—	—	—	—	—	10.5	<0.1
11.0	<0.1	—	—	—	—	—	

[a] Synthetic copper chloride solution.
[b] Copper sulfate-manufacturing wastewater.
[c] Steel wire copper-plating rinse.
[d] Initial conditions.

precipitation. Theoretical levels are seldom attained in actual practice, due to poor separation of colloidal precipitates, slow reaction rates, pH fluctuations, and the influence of other ions or complexing agents in solution.

The influence of such factors is demonstrated by Table 7.3, which compares the results of treatability studies performed on a synthetic wastewater [50], a copper sulfate manufacturing wastewater [37], and a plating rinsewater [18]. Results are presented, for both lime and caustic precipitation, over a range of treatment pH values. For lime treatment, minimum residual copper is achieved by pH 8.0. However, the level achieved is much lower in the synthetic wastewater than in the real (copper sulfate-manufacturing) wastewater. Caustic treatment of the latter waste yielded very poor results, which may however not be representative (see discussion regarding Table 7.5). Caustic treatment of the plating rinsewater yielded results comparable to that observed for lime treatment of the synthetic wastewater.

The data of Table 7.4 demonstrate the impact of solids separation on effluent copper levels. Two plants with similar wastes and employing identical

Table 7.4 Comparison of Plating Shop Effluent Copper Concentrations [51]

Plant	*Treatment pH*	*Effluent Total Copper (mg/l)*		*Effluent Soluble Copper (mg/l)*	
		Range	*Avg*	*Range*	*Avg*
A	8.8–10.1	0.05–0.25	0.12	0.01–0.04	0.02
B	9.4–10.2	0.10–1.06	0.49	0.06–0.08	0.07

Table 7.5 Treatability Results for a Copper Sulfate-Manufacturing Wastewater [after 37]

Parameter	*Lime Treatment*	*Caustic Treatment*
Initial pH	3.0–3.3	3.0–3.4
Treatment pH	9.7–10.4	9.6–10.5
Initial copper, mg/l	240	250
Clarifier effluent copper, mg/l	1.10	1.30
Filter effluent copper, mg/l	0.18	0.24

treatment processes were monitored for treatment pH, and total soluble effluent copper [51]. As shown in Table 7.4, effluent soluble copper was equivalent for the two plants, despite a fourfold greater effluent total copper level in plant B as a result of poorer solids separation.

Table 7.5 compares the average performance for 12 treatability runs, for the same copper sulfate-manufacturing wastewater described in Table 7.3. Lime and caustic treatability results for both clarifier effluent and filter effluent are comparable. Filtration significantly reduced the residual copper levels.

Reported treatment levels achieved by precipitation in full-scale industrial treatment operations are presented in Tables 7.6 and 7.7. These data also demonstrate the impact of effective solids separation. Pilot treatment studies on acid mine drainage, involving lime addition, gravity clarification and sand filtration, resulted in reduction of copper initially at 3.6–30.4 mg/l to a final effluent level of 0.1 mg/l or less [32].

Maruyama et al [64] and Hannah et al [65] studied copper removal from municipal wastewater enriched with copper and other metals. Alum and various combinations of lime and ferrous sulfate addition were examined for treatment effectiveness in a 4 gpm pilot plant. After chemical addition, flocculation, and settling, the effluent was sand filtered and passed through activated carbon columns. Treatment levels achieved are summarized in Table 7.8. It is noteworthy that at the higher initial copper concentrations, best copper removal occurred at pH 6.0, with the addition of ferrous sulfate only, rather than at higher pH with lime neutralization. Similar improved treatment, as in the presence of iron, has also been reported in another instance for alum flocculation at pH 6.8–7.0 [66], and reflects the phenomenon of coprecipitation, wherein a secondary heavy metal is coprecipitated with a primary metal (iron or aluminum) hydroxide.

A significant problem in achieving low residual concentrations of copper is the presence of complexing agents, such as cyanide and ammonia. The presence of high levels of ammonia in one power-plant boiler cleaning waste, for example, resulted in a postprecipitation copper effluent level of 15 mg/l [5]. Removal of the complexing agent is essential for good copper treatment. Therefore, effective copper treatment may depend strongly upon preceding pro-

Table 7.6 Summary of Effluent Copper Concentrations After Precipitation Treatment

	Copper Concentration (mg/l)		
Source (Treatment)	*Initial*	*Final*	*Reference*
Metal processing (lime)	204–385	0.5	52
Nonferrous metal processing (lime)	—	0.2–2.3	53,54
Metal processing (lime)	—	1.4–7.8 (prior to sand filtration)	55
Electroplating (caustic, soda ash + hydrazine)	6.0–15.5	0.09–0.24 (sol.) 0.30–0.45 (tot.)	56
Machine plating (lime + coagulant)	—	2.2	57
Metal finishing (lime)	—	0–1.2 (0.19 avg)	58
Brass mill (lime)	10–20	1–2	26
Plating	—	0.02–0.2	59
Plating (CN oxidation, Cr reduction, neutralization)	11.4	2.0	60
Wood preserving (lime)	0.25–1.1	0.1–0.35	61
Brass mill (hydrazine + caustic)	75–124	0.25–0.85	62
Silver plating (CN oxidation, lime + $FeCl_3$)	30 (avg)	0.16–0.3 (with sand filtration)	15
Copper sulfate manufacture (lime)	433	0.14–1.25 (0.48 avg)	36
Integrated circuit manufacture (lime)	0.23	0.05	19

Table 7.7 Precipitation Treatment Results in the Electroplating Industry [after 63]

	Copper Concentration (mg/l)	
Treatment pH[a]	*Influent*	*Effluent*
6.6	—	0.14–0.21
6.8–8.8[b]	—	0.3–1.2
6.9–7.5	—	0.05–0.90
7.0–8.0[c]	—	0.5
7.7	4.0	2.0
8.8	1.64	0.11–0.16

[a] Treatment by lime addition and gravity clarification, except as specified.

[b] Hydrazine, plus soda ash and caustic.

[c] Clarifier effluent sand filtered.

Table 7.8 Treatment of Copper Wastewater [64, 65]

Process	*Chemical Addition*	*pH*	*Copper Concentration (mg/l)*	
			Influent	*Effluent*
Alum only	Alum at 220 mg/l	6.4	0.7	0.022
Iron only	$FeSO_4$ at 45 mg/l Fe	6.0	5.0	0.155
Low-lime	Lime at 260 mg/l $FeSO_4$ at 20 mg/l Fe	10.0	5.0	0.170
High-lime	Lime at 600 mg/l	11.5	5.0	0.352

cesses. Copper cyanide-complexed wastewaters can be treated by first destroying the cyanide, followed by conventional copper treatment [67]. Typically, cyanide destruction is by the alkaline chlorination process (see Chapter 8). Excess chlorine may be required in the presence of copper, or other metal ions which catalyze the spontaneous destruction of hypochlorite.

Recent studies on copper-plating wastes have produced evidence that, for copper-cyanide wastes, interfering ions may remain even after total cyanide destruction. Yost and Scarfi [68, 69] performed solubility vs pH experiments on freshly prepared and working-bath acidic copper sulfate and alkaline copper-cyanide solutions, to determine the pH of minimum solubility and its associated residual soluble copper concentrations. Their data are summarized in Table 7.9. For the initial acidic copper sulfate samples, optimum treatment was achieved at pH 10–10.5, yielding residual soluble copper levels of less than 0.1 mg/l. No pH effect and only slight residual solubility effect resulted due to initial copper concentration, and the freshly prepared and working-bath copper sulfate samples responded in a similar manner to treatment.

A much different treatment response was observed for the copper cyanide solutions, however. These samples were first treated by two-stage alkaline chlorination to destroy the cyanide. Optimum treatment for the freshly prepared copper cyanide solution was obtained at pH values of 11.5–12.0, indicating a treatment requirement approximately 1.5 pH units higher than from the copper sulfate waste. As with the two copper sulfate wastes, some slight influence of initial waste concentration is evident, perhaps involving an ionic strength effect.

For the working copper cyanide bath solution, even higher optimum treatment pH of 12.0–13.0 was indicated, plus a major interfering effect on copper solubility at higher initial waste concentrations (see Table 7.9). This interfering effect is apparently due to the presence of one or more complexing agents (other than cyanide) which were diluted out in the weaker initial waste solutions. The principal interfering agent appears to be carbonate, which accumulates to extremely high levels in the alkaline copper cyanide working baths through uptake of atmospheric carbon dioxide.

These results suggest that with copper, a metal for which the carbonate ion is a particularly strong ligand, treatment of highly alkaline working solutions

Table 7.9 Minimum Solubility and Optimum pH of Precipitation for Copper-Plating Wastes [68, 69]

Initial Copper Concentration (mg/l)	*Freshly Prepared $CuSO_4$ Solution*[a]		*Diluted $CuSO_4$ Plating Bath*	
	Opt. pH	*Min. Cu*	*Opt. pH*	*Min. Cu*
1,000	10.5	0.008	10.0	0.06
500	10.5	0.008	10.0	0.04
200	10.5	0.008	10.0	0.02
50	10.5	0.003	10.0	0.01
Initial Copper Concentration (mg/l)	*Freshly Prepared $Cu(CN)_2$ Solution*[a]		*Diluted $Cu(CN)_2$ Plating Bath*	
	Opt. pH	*Min. Cu*	*Opt. pH*	*Min. Cu*
1,000	11.7	0.05	13.0	7.0
500	12.0	0.02	12.5	1.2
200	12.0	0.012	12.0	0.06
50	11.5	0.001	12.0	0.02

[a] Cyanide not initially present for $CuSO_4$ solutions, and destroyed by two-stage alkaline chlorination for $Cu(CN)_2$ solutions.

in their derived rinsewaters may represent an additional complication due to the presence of the carbonate. Recent treatability studies by Yeats [70] confirm this carbonate influence. As shown in Table 7.10, treatability studies were performed for copper in rinsewaters associated with initial acidic, initial near-neutral, and initial alkaline plating baths. Effective copper treatment for the former two rinsewaters was achieved at pH below 10, yielding residual soluble copper levels below 0.05 mg/l. For the alkaline waste, however, even at pH 11.5, poor treatment results were obtained. A similar interfering effect of carbonate on copper precipitation is suggested by filtered effluent results reported by Osantowski and Ruppersberger [71], who compared a lime treatment system with a parallel system utilizing lime plus soda ash (sodium carbonate).

As with other complexing agents such as cyanide or ammonia, effective copper treatment for initial alkaline wastewaters containing excess carbonate would require prior removal of the interfering ion. One possible approach would

Table 7.10 Treatment Results for Copper in Electroplating Rinse Waters [70]

Wastewater Source	*pH*		*Copper (mg/l)*	
	Initial	*Final*	*Initial*	*Final*
Chrome-plating rinse	2.0	9.5	33.6	<0.05
Nickel-plating rinse	7.4	8.0	2.1	<0.05
Zinc-plating rinse	9.4	11.5	16.8	2.97

be pH reduction to a near-neutral value plus air stripping, followed by pH readjustment for copper precipitation. An alternate approach might involve carbonate depletion by precipitation of calcium carbonate, upon addition of lime or some other calcium salt such as the chloride.

One method of avoiding cyanide treatment for electroplating wastes is to use the pyrophosphate copper-plating process [72]. Conventional lime treatment of copper pyrophosphate waste is not effective, since the bath is already at pH 8.3–8.8. Elevation of pH to 12.0 with lime will precipitate calcium pyrophosphate, and thus allow simultaneous precipitation of copper hydroxide [46]. High-lime treatment has reduced copper pyrophosphate rinsewater copper concentration from 33.4 mg/l to less than 1 mg/l [7]. High-lime treatment, at pH 11.5–12 with supplemental calcium added as calcium chloride, has also been reported successful in treating an electroless copper tartrate-plating waste to effluent copper levels of 0.04–0.3 mg/l. Initial copper concentration ranged from 50 to 300 mg/l [46]. In both the pyrophosphate and tartrate applications, the calcium serves to precipitate the copper-complexing agent, and thus frees the copper for hydroxide precipitation. The major disadvantages are high lime demand, need for pH readjustment, high-volume solids production, and low copper content in the sludge. The sludge produced in a typical hydroxide precipitation system is acceptable for copper recovery only if the sludge copper concentration is in the range of 6–10% or greater copper, on a dry-weight basis. Typical lime-precipitated sludge copper content has been reported in the range of 0.13–6.0% (dry weight) in sludges of 21–40% total solids [73].

Cuprous oxide, on the other hand, forms a very dense, easy-to-settle, easy-to-handle sludge of 40–60% solids [52, 72, 73]. Treatment of cupric wastewater with hydrazine or some other reducing agent plus sodium hydroxide reduces copper to the cuprous form for cuprous oxide precipitation, and is recommended by several authors [62, 72–74]. Such a system may be operated at pH 7–8. A full-scale system has been reported in operation (see Table 7.6), using the integrated treatment method [62]. The resulting dense, high-copper sludge is readily marketable. In another application of this technique, copper is reduced from 6–15.5 mg/l down to 0.09–0.25 mg/l soluble and 0.30–0.45 total copper. The system is operated with caustic and soda ash at pH 9.5–10.5, and hydrazine at 300–700 mg/l [56].

Limited information is available on sulfide precipitation of copper, principally from mining and ore extraction wastes [28, 31]. Copper was reduced from 20 mg/l to below detectable limits by sulfide precipitation from a gold ore-extraction wastewater. Treatment by coprecipitation with alum yielded equivalent results [28]. In field studies on an acid mine drainage waste, the waste was first lime neutralized to pH 5.0 to precipitate iron and aluminum, and then in a second stage treated at pH 6.5 with a sulfide salt for removal of other heavy metals. Copper was reduced from initial levels of 50–115 mg/l to below 0.5 mg/l [31].

Table 7.11 compares treatability results for wastewaters from several industries. Lime was used for one treatment and lime plus sodium sulfide for the

Table 7.11 Comparison of Lime versus Lime plus Sulfide Precipitation Treatment for Copper [after 16]

Treatment pH	*Initial Concentration*	*Lime Treatment*		*Lime plus Sulfide Treatment*	
		Clarified	*Filtered*	*Clarified*	*Filtered*
8.5[a]	21.0	1.30	0.37	2.25	0.17
8.5[b]	7.0	0.04	<0.01	0.04	<0.01
8.5[c]	4.7	0.14	<0.01	0.08	0.02
8.5[d]	2.3	1.80	0.20	1.90	<0.01
9.3[c]	1.3	0.24	0.24	0.21	0.17
10.0[a]	2.0	0.91	0.94	0.06	0.16

Wastewater source: [a] electroplating; [b] smelter scrubbing; [c] electroplating plus anodizing; [d] printed circuit board.

other. Sodium sulfide was metered into the reactor to maintain a residual 0.5 mg/l excess sulfide level. The data of Table 7.11 indicate no consistent effect of sulfide addition on the clarifier effluent copper, but some slight improvement in the filter effluent. Sulfide demand ranged from 1 to 400 mg/l of wastewater treated, and process economics may not justify the slight improvement in filter effluent quality.

The proprietary Sulfex process (Permutit Co.) utilizes solid ferrous sulfide addition to precipitate copper sulfide. The premise of the technique is slight solubilization of the ferrous sulfide, thereby releasing sulfide ion to react with the copper. The sludge produced in the Sulfex process is a mixture of iron hydroxide and metallic sulfides. Since insoluble ferrous sulfide is added in the process, the weight of sludge produced is greater than that in the conventional hydroxide precipitation process [75]. Scott [76] has reported that the sludge disposal cost for this sulfide process is twice the cost of sludge disposal for conventional precipitation treatment.

Costs associated with treatment systems involving hydroxide precipitation of copper-bearing wastes are summarized in Table 7.12. Where possible, costs pertaining directly to copper wastes have been isolated. Figure 7.1 presents general capital investment cost curves for precipitation systems. A typical process will incorporate a reactor with lime addition, and a clarifier. Filters may also be required in order to meet stringent effluent standards or to provide consistent effluent copper levels. Capital and operating costs for conventional precipitation treatment systems, including filtration, were also presented in Table 1.1. When copper wastewaters include noncyanide chelating agents, greater treatment costs may be incurred. A treatment plant for one 56,300 gpd printed circuit board wastewater, which included pretreatment for a chelated/ammoniated copper rinsewater, was estimated to cost $13.95/gpd treatment capacity [20]. Operating costs were $10.66/1,000 gal treated, not including sludge disposal. A substantial proportion of capital costs may be related to

Table 7.12 Capital Costs of Precipitation Treatment of Copper-Bearing Wastewaters

Process	*Flow (1,000 gpd)*	*Capital Costs[a] ($1,000 gpd)*	*Reference*
Lime precipitation, rapid sand filtration	2,000	960	53
Integrated hydrazine-NaOH (copper recovery)	220	1,260	62
Lime precipitation, chromium reduction	480	910	77
Integrated, lime (no recovery)	240	245	77
Integrated, lime, CN destruction	480	290	77
Continuous, lime, Cr reduction	72	1,295	61

[a] Costs are for plants installed since 1967. All costs updated to 1979 dollars based on the Marshall & Swift Equipment Cost Index.

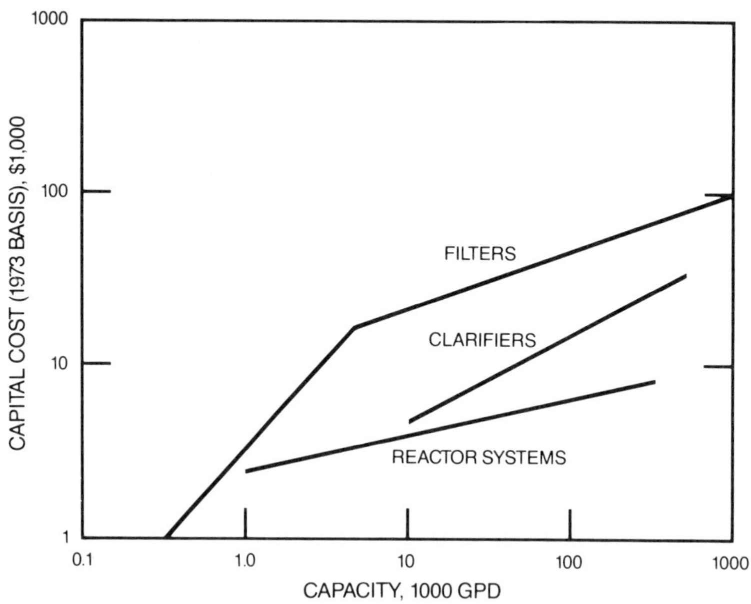

Figure 7.1 Capital costs (in 1973) of reactor systems, clarifiers, and filters [78].

Table 7.13 General Treatment Costs (in 1974) for the Electroplating Industry [63]

Wastewater Volume (gpm)	*Precipitation Treatment* Capital ($/1,000 gpd)	*Precipitation Treatment* Operating ($/1,000 gal)	*Precipitation plus Cyanide Treatment* Capital ($/1,000 gpd)	*Precipitation plus Cyanide Treatment* Operating ($/1,000 gal)
26	74.63	1.07	97.97	1.36
60	48.70	0.92	59.65	1.24
102	37.79	0.87	46.14	1.12

collection of the waste. In one reported case, 47% of the capital costs for a metal-finishing waste-treatment system was attributed to installation of collection lines [57].

Frequently, copper removal is associated with cyanide destruction, chromium reduction, and nonmetal waste treatment procedures. Table 7.13 presents generalized cost data (1974 basis) for precipitation treatment, with and without cyanide destruction. Addition of cyanide treatment increments precipitation treatment capital investments by 22–31%, and operating costs by 27–35%. A comparison of these costs with recently reported actual costs for copper treatment (see Table 7.12) indicates generally higher reported capital costs.

Evaporative Recovery

Use of evaporative recovery has been practiced for over 20 years [79]. Recovery of copper is often more attractive than is that of some other metals such as zinc, due to copper's higher value. Smith and Roy [80] have reported that makeup solution for copper cyanide baths costs $1.49/gal, and for acid copper sulfate baths, $1.76/gal. A recent study [81] indicated that the recovery value for copper solutions ranging from 100 to 500 mg/l is $0.60–3.00/1,000 gal. In brass rinsewaters containing 40–250 mg/l copper, the recovery values are given as $0.20–1.20/1,000 gal. Unless a recovery process is a closed-loop operation, however, the concentrated wastewater may still require final treatment for residual metal removal.

Process modification for waste stream segregation and flow reduction is usually an integral part of the recovery process. In fact, direct recovery can result from such modifications. In one case, use of countercurrent rinse after copper-plating baths allowed direct return of the rinse flow into the plating baths as evaporative makeup [10]. Rinsewater flow was reduced from 8 to 0.3 gpm for the copper-plating lines. The final rinse tank concentration of copper was reduced to 0.9 mg/l, from a previous value of 40 mg/l. In the event that water evaporation from the plating bath is not sufficient to allow direct return of the dilute rinsewater, preliminary evaporation of rinsewater may be necessary in order to increase its copper concentration.

Domey and Steifel [43] report that copper-plating bath rinsewater volume could be reduced by 85%, from 4,100 to 600 gal/day, by rinsing modifications. Such modification is generally essential to the economic feasibility of evaporative recovery systems, since the major cost item is the steam [82]. Energy requirements may be reduced by using multiple effect evaporation, at a greater capital cost.

One disadvantage of evaporative recovery is the buildup of impurities in the plating bath, since both the impurities which are added with plating chemicals and those formed by decomposition of additives are returned quantitatively to the plating bath. As these impurities build up, defective plating may result.

Ion Exchange

Ion exchange is capable of achieving very high levels of copper removal, particularly from dilute wastes. Copper removal by ion exchange from 1.02 mg/l to less than 0.03 mg/l has been reported [83]. The initial costs of an ion exchange unit depend upon the required resin volume. For dilute wastes, flow rate will be the limiting factor, while the resin exchange capacity will dictate the volume for more concentrated systems. Costs of various units and types of resins are given in Figure 7.2. These costs are for complete units (valves, resin, etc.). Capital investments of $400–1,200/1,000 gpd capacity have been cited, with operating costs of $0.31–0.81/1,000 gal [77, 85, 86]. When compared to other treatment methods, for cases where metal recovery is not feasible, ion exchange does not appear to be practical from a cost standpoint. In a direct comparison between two nearly identical plating lines producing equal waste volumes (0.48 MGD) of equal composition and concentration, one system employed precipitation treatment and 30% water recycle, the other ion exchange treatment and 90% water recycle [77]. Costs were $461 vs $1,075/1,000 gpd treatment capacity, and $0.12 vs $0.18/1,000 gal treated, respectively, by precipitation and by ion exchange.

Treatment of copper rinse solutions containing high ammonia concentrations by ion exchange has been reported [87]. A selective ion exchange resin reduced copper from 45 mg/l to nondetectable levels. The initial ammonia content of 70 mg/l was reduced to 1 mg/l. Pelosi and McCarthy [88] warn that copper complexed by high molecular-weight organic chelates will foul ion exchange resin, as will cuprous copper, which converts to the cupric form after exchange. Special chelating resins are available to remove most chelated copper complexes [89].

When copper recovery is feasible, ion exchange economics are greatly improved. Schore [90] reported that ion exchange reclamation of copper from pickle liquor allowed electrolytic copper recovery and regenerated the acid pickle solution. He also reported that ion exchange was used to treat 330 gpm (0.43 MGD) of solution from a printed-circuit etching machine. In both cases,

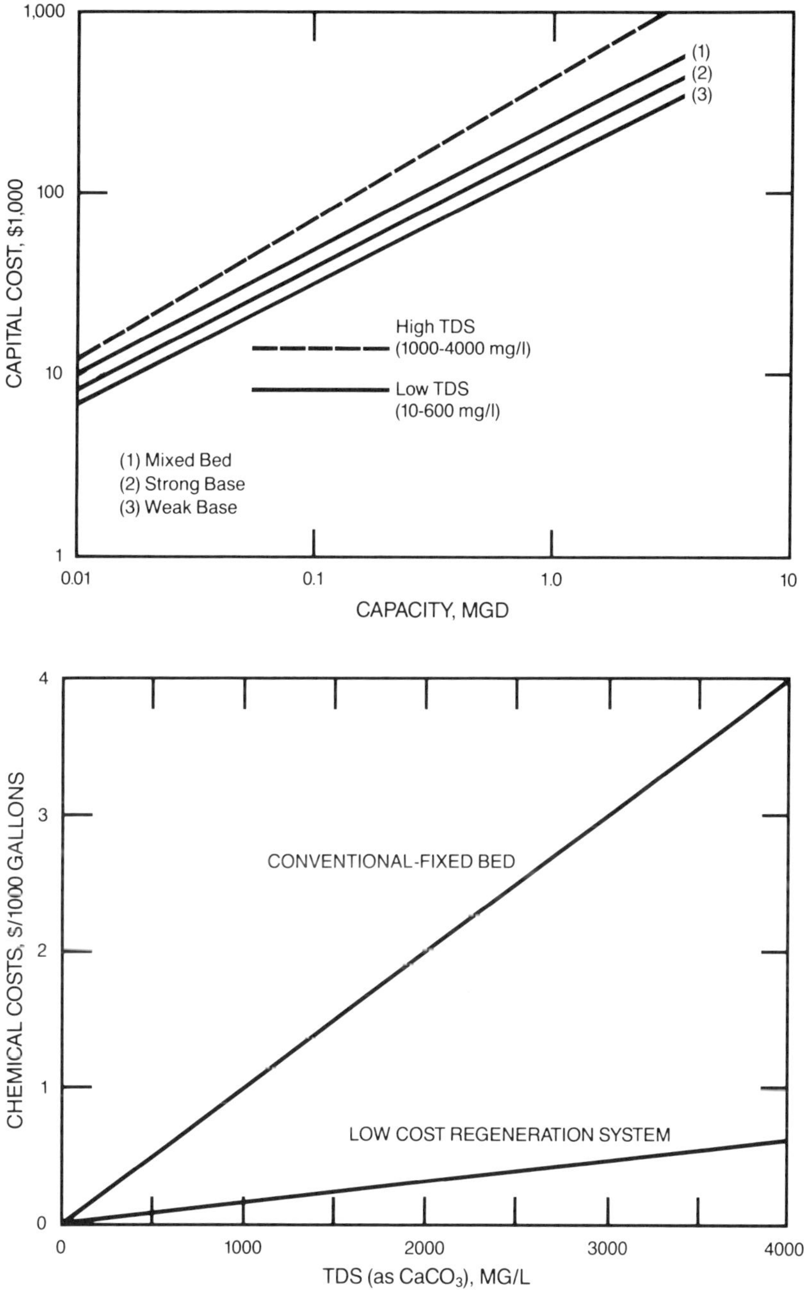

Figure 7.2 Costs (in 1974) for ion exchange demineralization [84].

it was claimed that the treatment systems were paid for in 3–4 years by the reclaimed copper.

The major difficulty in copper recovery by ion exchange treatment is to utilize sufficiently small regenerant volume to recover a concentrated copper solution, while adequately regenerating the resin. A countercurrent, continuous-regeneration system, such as the patented Asahi process [91], accomplishes this, reportedly yielding regenerant containing 3–4 N copper. In this type of system, of which there are several available, the resin moves intermittently down and out of the bed, whereas the wastewater moves up the resin column. Regeneration typically occurs in a tall thin column, to provide efficient resin-regenerant contact. Compared with conventional ion-exchange systems, the continuous countercurrent system is usually more expensive, complex, and inflexible in operation.

Electrolytic Recovery

Direct electrolytic recovery of copper metal is possible from relatively concentrated waste solutions. Treatment of dilute solutions (less than 2 g/l) using conventional electrolysis is impractical, because of the high power consumption [73]. The electrolytic process generally requires a preconcentration step such as ion exchange or evaporation. Acid leach solution from copper ore is electrolyzed directly in one application to recover copper. The resultant regenerated sulfuric acid is also recycled [8]. The electrolytic acid recovery process has also been utilized to purify a chromic-sulfuric acid bath plating brass parts, to plate out copper and zinc impurities while converting trivalent chromium to chromate for recycle to the bath [92]. Copper removal efficiency is reported to be 41%.

Electrolytic recovery of copper from copper pickling baths presents no problem, but brass pickle solutions may cause problems from tin sludge formation [73]. Proper control of electrolysis current can reduce or eliminate problems of other metals plating out [62].

For copper cyanide baths, copper recovery by electrolysis can be accomplished in conjunction with cyanide destruction [9, 93, 94]. In the past, the process has not been feasible for dilute solutions but was appropriate for treatment of spent plating-bath solutions or ion exchange concentrates. The process economics compare favorably with other copper-cyanide treatment processes. Operating data are presented in Table 7.14 and an economic analysis of the process is given in Table 7.15.

A new process has been reported [42, 73, 96], which purports to make electrolytic treatment of dilute copper-cyanide solutions economically attractive. The process consists of a fluidized semiconducting carbon bed across which a potential is applied. The advantage of this type of system is that it has a very large surface area per unit of volume, which provides a low current density on the electrode surfaces via a high current density across the cell.

Table 7.14 Reduction of Copper and Cyanide by Batch Electrolysis [93]

Days of Treatment	*Copper (g/l)*	*Cyanide (mg/l)*	*pH*
Start	26	75,000	12.2
1	23	50,000	11.7
4	11	12,500	10.4
6	6	5,980	10.0
8	2	2,200	9.8
11	0	750	9.7
18	—	0.2	9.5

The system is commercially available in 5 and 10 gpm units. It is claimed that 20 mg/l cyanide is reduced to 0.5 mg/l, and 40 mg/l copper is reduced to below 1.0 mg/l [96].

Other Processes

One recently developed process with significant potential application for both concentrated and dilute copper wastewaters is cementation [3, 30, 41, 73, 97–100]. The process involves percolating the copper wastewater through a bed of scrap iron. By oxidation-reduction reaction, copper plates (cements) onto the iron, with sacrificial iron going into solution. In one pilot-scale application, influent copper averaging 80 mg/l (range 34–220 mg/l) was reduced to an average effluent of 0.33 mg/l (range 0.2–0.84 mg/l). The copper can be recovered as essentially pure metal, representing a significant offsetting savings to treatment costs.

Table 7.15 Electrolysis Costs per 1,000 Gallons of Plating Wastes [93]

Item	*Cost ($) (1967)*
Power consumption	11.66
Water consumption	0.14
Steam consumption	3.50
Plate usage	14.80
Labor	21.00
Total	51.10
Copper salvaged	24.00
Net Cost	27.10

The cementation process is reported to proceed most effectively at pH below 2.5. The scrap iron must be fully degreased before use, to avoid interference with the cementation reaction [100].

Electrodialysis has been cited as economically feasible for treatment of process solutions and rinsewaters from plating and metal-finishing operations, and for treatment of acid mine drainage. Anderson et al [101] report operating energy requirements of 10 KWH/1,000 gal for electrodialysis, and indicate that membrane fouling can be a serious problem.

Granular activated-carbon treatment has been reported to reduce copper in a gum and wood chemical wastewater from 1.30 to 0.36 mg/l [102]. Carbon filtration has also been mentioned in conjunction with pilot plant studies on municipal wastes [64, 103]. Some copper reduction by activated carbon beds was indicated. Pilot plant studies involving the removal of copper cyanide on activated carbon have also been reported [12]. For copper cyanide concentrations of 100–112 mg/l copper, copper removals were reported greater than 98% in all cases. Flows were 25–50 gpm.

Successful reverse osmosis treatment (99+% rejection) has been reported for copper pyrophosphate, acid copper, and copper-cyanide plating-bath rinsewaters [61, 104–106]. Advantages claimed for reverse osmosis include low capital and energy costs. However, available membranes are suitable only over a pH range of 2.5–11.0 for extended use, and the process is incapable of achieving high solution concentrations. There is also a possibility of decomposition of the pyrophosphate, with consequent adverse effects when the recovered brine is returned to the plating bath [107]. Maximum operating pressure is 400–1,500 psig, depending upon membrane type and configuration, and capital costs (1974) are reported as $1,250–$2,650/1,000 gpd [107]. Recent field studies on treatment of copper-cyanide rinsewater were of limited success. At one plant, rapid membrane deterioration made reverse osmosis treatment unattractive on the basis of membrane replacement costs. Performance was better at a second plant, with salt rejection about 85% over a 100-day operating period [108].

SUMMARY

Copper treatment is predominantly accomplished by precipitation with lime. Effluent concentrations below 0.5 mg/l are achieved on a consistent basis only with proper pH control and either proper clarifier design accompanied by good sludge settling characteristics or tertiary effluent treatment by filtration. Precipitation as the cuprous oxide seems to offer distinct advantages with respect to solids handling, especially when the sludge can be sold for copper reclamation. Final copper concentrations below 0.1 mg/l are reported but do not seem readily attainable on a consistent basis with current precipitation methods.

Process modification and waste stream segregation provide means of enhanced treatment of copper-bearing wastes. When such changes are implemented, recovery processes become more attractive. In fact, to be economically feasible, ion exchange treatment would seem to require recovery, unless available space is the limiting factor. Recovery, of one type or another, is practiced in many industrial operations, and appears economically attractive, primarily due to the high value of recovered copper.

REFERENCES

1. Golomb, A. "Application of Reverse Osmosis to Electroplating Waste Water Treatment. Part II. The Potential of Reverse Osmosis in the Treatment of Some Plating Wastes," *Plating* 59 (1972): 316–319.
2. Rudolfs, W., ed. *Industrial Waste Treatment* (New York: Reinhold, 1953).
3. Dean, J.G., F.L. Bosqui, and K.H. Lanouette. "Removing Heavy Metals from Waste Water," *Environ. Sci. Technol.* 6 (1972): 518–522.
4. Iammartino, N.R. "Mercury Cleanup Routes—II," *Chem. Eng.* 82 (1975): 36–37.
5. Kuppusamy, N. "Copper Removal from Power Plant Boiler Cleaning Waste," *Ind. Wastes* (March/April 1977): 43–45.
6. Chu, J., G. Steiner, and C. McEntyre. "Removal of Complex Copper-Ammonia Ions from Aqueous Wastes with Fly Ash," paper presented at the 32nd Annual Purdue Industrial Waste Conference, West Lafayette, IN, (May 1977).
7. Parsons, W.A. and W. Rudolfs. "Lime Treatment of Copper Pyrophosphate Plating Wastes," *Sew. Ind. Wastes Eng.* 22 (1951): 313–315.
8. Simpson, R.W. and K. Thompson. "Chlorine Treatment of Cyanide Wastes," *Sew. Ind. Wastes Eng.* 21 (1950): 302–304.
9. Pinkerton, H.L. "Waste Disposal. Inorganic Wastes," in *Electroplating Engineering Handbook*, 2nd ed., A. Kenneth Graham, ed. (New York: Reinhold, 1962)
10. Barnes, G.E. "Disposal and Recovery of Electroplating Wastes," *J. Water Poll. Control. Fed.* 40 (1968): 1459–1470.
11. Wise, W.S. "The Industrial Waste Problem, IV. Brass and Copper Electroplating and Textile Wastes," *Sew. Ind. Wastes* 20 (1948): 96–102.
12. Battelle Columbus Laboratories. "An Investigation of Techniques for Removal of Cyanide from Electroplating Wastes," USEPA 12010 EIE 11/71 (1971).
13. Lowe, W. "The Origin and Characteristics of Toxic Wastes with Particular Reference to the Metals Industries," *Water Poll. Cont.* (London, 1970): 270–280.
14. Anderson, J.S. and E.H. Iobst, Jr. "Case History of Wastewater Treatment in a General Electric Appliance Plant," *J. Water Poll. Control. Fed.* 10 (1968): 1786–1795.
15. Nemerow, N.L. *Theories and Practices of Industrial Waste Treatment* (Reading, MA: Addison-Wesley, 1963).
16. Robinson, A. and J. Sum. "Sulfide Precipitation of Heavy Metals," USEPA 600/2-80-139 (June 1980).
17. Kennedy, I.F.T. and S.D. Gupta. "Electrochemical Removal of Trace Metals for Metal Wastes with Simultaneous Cyanide Destruction," *Fin. Manag.* (July 7, 1978): 7–18.

18. Reece. C. "Steel Welding Wire Manufacturer Meets Stringent Effluent and Sludge Disposal Criteria," paper presented at the 37th Annual Purdue Industrial Waste Conference, West Lafayette, IN (May 1982).
19. Hango, R.A., W.J. Jewell, and E.D. Driscoll. "Wastewater Treatment for Integrated Circuit Manufacturing," paper presented at the 28th Annual Purdue Industrial Waste Conference, West Lafayette, IN (May 1973).
20. Jashnani, I.L., and R. Green. "Pretreatment System for a Circuit Board Plant," paper presented at the 8th National Conference on Municipal Sludge Management, Municipal and Industrial Sludge Management, Disposal and Utilization, Miami Beach, FL (March 19–21, 1979).
21. Gard, S.M., C.A. Snavely, and D.J. Lemon. "Design and Operation of a Metal Wastes Treatment Plant," *Sew. Ind. Wastes* 23 (1951): 1429–1438.
22. McDonald, C.W. and R.S. Bajwa. "Removal of Toxic Metal Ions from Metal-Finishing Wastewater by Solvent Extraction," *Sep. Science* 12 (1977): 435–445.
23. Olver, J., W. Kreye, R. Roberts, and H. Sutton. "Anodizing Wastewater Treatment and Sludge Disposal: A Total Systems Design" *Proceedings of the 3rd Annual Conference on Treatment and Disposal of Industrial Wastewaters and Residues*, Houston, TX (April 18–20, 1978), pp 94–101.
24. McGarvey, F.X. "The Application of Ion Exchange Resins to Metallurgical Waste Problems," *Proceedings of the 7th Industrial Waste Conference, Purdue University* (May 1952), pp 289–304.
25. McGarvey, F.X., R.E. Tenhoor, and R.P. Nevers. "Brass and Copper Industry: Cation Exchangers for Metal Concentration from Pickle Rinse Waters," *Ind. Eng. Chem.* 44 (1952): 534–541.
26. Tallmange, J.A. "Nonferrous Metals," *Chemical Technology*, vol. 2. *Industrial Waste Water Control*, C. Fred Gurnham, ed. (New York: Academic Press, 1965).
27. Hallowell, J.B., J.F. Shea, G.R. Smithson, Jr., A.B. Tripler, and B.W. Gonser. *Water Pollution Control in the Primary Non-Ferrous Metals Industry*, vol. I. *Copper, Zinc, and Lead Industries*, USEPA R2-73-247-a (September 1973).
28. Rosehart, R. and J. Lee. "Effective Methods of Arsenic Removal from Gold Mine Wastes," *Can. Min. J.* (June 1972): 53–57.
29. Schmidt, J., L. Simovic, and E. Shannon. "Development Studies for Suitable Technologies for the Removal of Cyanide and Heavy Metals from Gold Mining Effluents," paper presented at the 36th Annual Purdue Industrial Waste Conference, West Lafayette, IN (May 1981).
30. Hill, R.D. "Control and Prevention of Mine Drainage," paper presented at the Environmental Resources Conference on Cycling and Control of Metals, Battelle Columbus Laboratories, Columbus, OH (1972).
31. Larsen, H.P., J.K.P. Shou, and L.W. Ross. "Chemical Treatment of Metal-Bearing Mine Drainage," *J. Water Poll. Control. Fed.* 45 (1973): 1682–1695.
32. Huck, P.M. and B.P. LeClair. "Treatment of Base Metal Mine Drainage," paper presented at the 30th Annual Purdue Industrial Waste Conference, West Lafayette, IN (May 1975).
33. "Development Document for Effluent Limitations Guidelines and New Source Performance Standards for the Oil Base Solvent Wash Subcategories of the Paint Formulating and the Ink Formulating Point Source Category," USEPA Publication 440/1-75/050-a, Group II (July 1975).
34. Cawley, W., ed. *Treatability Manual*, vol. I. *Treatability Data*, USEPA Report 600/8-80-042-a (July 1980).

35. "Development Document for Effluent Limitations Guidelines and New Source Performance Standards for the Plywood, Hardboard, and Wood Preserving Segment of the Timber Products Processing Point Source Category," USEPA 440/1-74-023-a (April 1974).
36. "Development Document for Interim Final Effluent Limitations Guidelines and Proposed New Source Performance Standards for the Significant Inorganic Products Segment of the Inorganic Chemicals Manufacturing Point Source Category," USEPA 440/1-75/037 Group I, Phase II (May 1975).
37. *Treatability Studies for the Inorganic Chemicals Manufacturing Point Source Category*, USEPA 440/1-80/103 (July 1980).
38. "Development Document for Interim Final Effluent Limitations Guidelines and Proposed New Source Performance Standards for the Primary Copper Smelting Subcategory and the Primary Copper Refining Subcategory of the Copper Segment of the Nonferrous Metals Manufacturing Point Source Category," USEPA 440/1-75/032-b, Group I, Phase II (February 1975).
39. "Development Document for Interim Final Effluent Limitations Guidelines and Proposed New Source Performance Standards for the Secondary Copper Subcategory of the Copper Segment of the Nonferrous Metals Manufacturing Point Source Category," USEPA 440/1-75/032-c, Group I, Phase II (February 1975).
40. Chalmers, R.K. "Pretreatment of Toxic Wastes," *Water Poll. Control* (London, 1970): 281–291.
41. Ciancia, J. "Pollution Abatement in the Metal Finishing Industry," paper presented at the Environmental Resources Conference on Cycling and Control of Metals, Battelle-Columbus Laboratories, Columbus, OH (1972).
42. Silman, H. "Treatment of Rinse Water from Electrochemical Processes," *Metal Fin.* 69 (1971): 62–66.
43. Domey, W.R. and R.C. Stiefel. "Wastewater Reduction in Metal Finishing Operations," paper presented at the 43rd Annual Conference of the Water Pollution Control Federation (1970).
44. Foulk, D.G. "Metal Finishing Products," in *Chemical Technology*. vol. 2. *Industrial Waste Water Control*, C. Fred Gurnham, ed. (New York: Academic Press, 1965).
45. Wing, R.E., W.M. Doane, and W.E. Rayford. "Treatment Process for Some Rinse Waters from Electroless Plating of Copper," *Plating Surface Fin.* (October 1977): 39–43.
46. Wing, R.E. and W.E. Rayford. "Heavy Metal Removal Processes for Plating Rinse Water," *Proceedings of the 32nd Industrial Waste Conference, Purdue University* (May 1977), pp 838–852.
47. Wing, R.E., W.E. Rayford, and W.M. Doane. "Ferrous Sulfate Treatment for Rinse Waters from the Electroless Plating of Copper," *Plating Surface Fin.* (October 1977): 39–43.
48. Stumm, W. and J. J. Morgan. *Aquatic Chemistry* (New York: Wiley-Interscience, 1970).
49. Jenkins, S.H., D.G. Knight, and R.E. Humphreys. "The Solubility of Heavy Metal Hydroxides in Water, Sewage and Sewage Sludge. I. Solubility of Some Metal Hydroxides," *Int. J. Air Water Poll.* 8 (1964): 537–556.
50. Netzer, A., A. Bowers, and J. Norman. "Removal of Heavy Metals from Wastewaters by Lime," *Proceedings of the 16th Conference on Great Lakes Research* (1973), pp 260–265.

51. Yost, K.J. and D.R. Masarik. "A Study of Chemical-Destruct Waste Treatment Systems in the Electroplating Industry," *Plating Surface Fin.* (January 1977): 35–40.
52. Nyquist, O.W. and H.R. Carroll, "Design and Treatment of Metal Processing Wastewaters," *Sew. Ind. Wastes* 31 (1959): 941–948.
53. Stone, E.H.F. "Treatment of Non-Ferrous Metal Process Waste of Kynoch Works, Birmingham, England," *Proceedings of the 25th Industrial Waste Conference, Purdue University* (May 1967), pp 848–865.
54. Stone, E.H.F. "Treatment of Non-Ferrous Metal Process Waste," *Metal Fin. J.* 18 (1972): 280–290.
55. Ogawa, H., Y. Makoto, and K. Hurikawa. "Industrial Waste-Water Treatment. Treatment of Waste Water Containing Copper Ions," *Mizu Shari Gijustu* (Japan) 13 (1972): 15–20.
56. Martin, J.J. Jr. "Chemical Treatment of Plating Waste for Removal of Heavy Metals," USEPA R2-73-044 (May 1973).
57. Hansen, N.H. and W. Zabban. "Design and Operation Problems of a Continuous Automatic Plating Waste Treatment Plant at Data Processing Division, IBM, Rochester, Minnesota," *Proceedings of the 14th Industrial Waste Conference, Purdue University* (May 1959), pp 227–249.
58. Watson, K.S. "Treatment of Complex Metal-Finishing Wastes," *Sew. Ind. Wastes* 26 (1954): 182–194.
59. "Effluent Treatment at B.S.R.," *Metal Fin. J.* 17 (1971): 248.
60. Chalmers, R.K. "Trade Effluent Treatment at the Cannock Factory of Joseph Lucas, Ltd.," *J. Proc. Inst. Sew. Purif.* (1965): 357–359.
61. Teer, E.H. and L.V. Russell. "Heavy Metals Removal from Wood Preserving Wastewater," paper presented at the 27th Annual Purdue Industrial Waste Conference, West Lafayette, IN (May 1972).
62. Volco Brass and Copper Co. "Brass Wire Mill Process Changes and Waste Abatement, Recovery and Reuse," USEPA 12010 DPF 11/71 (1971).
63. "Development Document for Interim Final Effluent Limitations Guidelines and Proposed New Source Performance Standards for the Common and Precious Metals Segment of the Electroplating Point Source Category," USEPA 440/1-75/040, Group I, Phase II (April 1975).
64. Maruyama, T., S.A. Hannah, and J.M. Cohen. "Removal of Heavy Metals by Physical and Chemical Treatment Processes," *J. Water Poll. Control. Fed.* 47 (1975): 962–975.
65. Hannah, S.A., M. Jelus, and J.M. Cohen. "Removal of Uncommon Trace Metals by Physical and Chemical Treatment Processes," *J. Water Poll. Control Fed.* 49 (1977): 2297–2309.
66. Culp, G.L. and R.L. Culp. *New Concepts in Water Purification* (New York: Van Nostrand Reinhold, 1974).
67. Wagner, R. "Behavior of Noncyandic Metal Complexes in Detoxification of Electroplating Waste Waters," *Von Wasser* (Germany) 39 (1972): 263–286.
68. Yost, K.J. and A. Scarfi. "Effects of Alkaline Cleaner on Metal Solubility in Treated Cyanide Plating Solutions," *Fin. Manag.* (July 19–24, 1978).
69. Yost, K.J. and A. Scarfi. "Factors Affecting Copper Solubility in Electroplating Waste," *J. Water Poll. Control Fed.* 51 (1979): 7.

70. Yeats, A.R. "Ion Exchange Selectively Removes Heavy Metals from Mixed Plating Wastes," paper presented at the 32nd Annual Purdue Industrial Waste Conference, West Lafayette, IN (May 1977).
71. Osantowski, R. and J. Ruppersberger. "Upgrading Foundry Wastewater Treatment," paper presented at the 39th Annual Purdue Industrial Waste Conference, West Lafayette, IN (May 1984).
72. Crowle, V. "Effluent Problems as They Affect the Zinc Die-Casting and Plating-on-Plastics Industries," *Metal Fin. J.* 17 (1971): 51–54.
73. Jackson, D.V. "Metal Recovery from Effluents and Sludges," *Metal Fin. J.* 18 (1972): 235–242.
74. Curry, N.A. "Philosophy and Methodology of Metallic Waste Treatment," *Proceedings of the 27th Industrial Waste Conference, Purdue University* (May 1972).
75. Feigenbaum, H.N. "Removing Heavy Metals in Textile Waste," *Ind. Wastes* (March/April 1977): 31–33.
76. Scott, M.C. "Sulfide Process Removes Metals, Produces Disposable Sludge," *Ind. Wastes* (July/August 1979): 34–38.
77. Pinner, R. and V. Crowle. "Cost Factors for Effluent Treatment and Recovery of Materials in the Metal Finishing Department," *Electroplat. Metal Fin.* 3 (1971): 13–31.
78. "Development Document for Effluent Limitations Guidelines and New Source Performance Standards for the Steam Electric Power Generating Point Source Category," USEPA 440/1-74-029-a (October 1974).
79. Culotta, J.M. and W.F. Swanton. "Case Histories of Plating Waste Recovery Systems," *Plating* 57 (1970): 251–255.
80. Smith, D. and C. Roy. "Disposal: What Does it Really Cost," *Proceedings of the Conference on Advanced Pollution Control for the Metal Finishing Industry*, USEPA 600/9-82-022 (December 1982).
81. *State of the Art: Review of Product Recovery*, Water Pollution Control Research Services 17-7 ODJW 11/69, USDI, Washington, DC (1969).
82. Culotta, J.M. and W.F. Swanton. "Controls for Plating Recovery System," *Plating* 58 (1971): 783–785.
83. Botham, G.H. and W.R. Bryson. "Note on the Corrosion of Aluminum Dairy Equipment by Water," *J. Dairy Res.* 20 (1953): 154–155.
84. "Development Document for Effluent Limitations Guidelines and New Source Performance Standards for the Major Inorganic Products Segment of the Inorganic Chemicals Manufacturing Point Source Category," USEPA 440/1-74-007-a (March 1974).
85. Lancy, L., W. Nohse, and D. Wystroch. "Practical and Economic Comparison of the Most Common Metal Finishing Waste Treatment Systems," *Plating* 59 (1972): 126–130.
86. Zievers, J.F. and C.J. Novotny. "Recovery of Mixed Rinse Water by Means of Ion Exchange," *Plating* 58 (1971): 482–485.
87. "Winning Heavy Metals from Waste Streams," *Chem. Eng.* 78 (1971): 62–64.
88. Pelosi, P. and J. McCarthy. "Preventing Fouling of Ion-Exchange Resins—I," *Chem. Eng.* 89 (1982): 75–78.
89. Courduvelis, C., G. Gallager, and B. Whalen. "New Developments for the Treatment of Wastewater Containing Metal Complexes," *Proceedings of the 4th Conference on Advanced Pollution Control for the Metal Finishing Industry*, USEPA 600/9-82-022 (December 1982).

90. Schore, G. "Electronic Equipment and Ion Exchange for Use in Automated Treatment Systems," paper presented at the 27th Annual Purdue Industrial Waste Conference, West Lafayette, IN (May 1972).
91. Solt, G.S. "Waste Treatment by Ion Exchange," *Effluent Water Treatment J.* 13 (1973): 768–773.
92. Herdrich, W. "Recovery of Acid Etchants at Imperial Clevite, Inc.," *Proceedings of the 4th Conference on Advanced Pollution Control for the Metal Finishing Industry*, USEPA 600/9-82-022 (December 1982).
93. Easton, J.K. "Electrolytic Decomposition of Concentrated Cyanide Plating Wastes," *J. Water Poll. Control. Fed.* 39 (1967): 1621–1625.
94. American Electroplating Society, "A Report on the Control of Cyanides in Plating Shop Effluents," *Plating* 56 (1969): 1107–1112.
95. Hollis, M. "Treatment of Cyanide Wastes by Electrolysis," *Trans. Ins. Metal Fin.* 55 (1975): 65–73.
96. "Electrolysis Speeds Up Waste Treatment," *Environ. Sci. Technol.* 4 (1970): 201.
97. Joster, T.L. and T.H. Taylor. "Industrial Waste Treatment at Scoville Manufacturing Company," paper presented at the 28th Annual Purdue Industrial Waste Conference, West Lafayette, IN (May 1973).
98. Case, O.P. "Metallic Recovery from Waste Waters Utilizing Cementation," USEPA 670/2-74-008 (January 1974).
99. Anderson, J.R. and R.P. Hollingworth. "Removal of Copper from Aqueous Solutions with Highly Stressed Activated Ferrous Metals," U.S. Patent 3,674,466 (July 4, 1972).
100. Joster, T.L. and T.H. Taylor. "Cementation Plant Reduces Chromium, Recovers Copper," *Ind. Wastes* 21 (1975): 26–28.
101. Anderson, J.S., J.M. Philips, and C.B. Schriver. "Water Contamination by Certain Heavy Metals," paper presented at the 44th Annual Meeting, Water Pollution Control Federation, San Francisco, CA (1971).
102. Cawley, W., ed. *Treatability Manual*, Vol. III. *Technologies for Control/Removal of Pollutants*, USEPA 600-8-80-0420 (July 1980).
103. Nilsson, R. "Removal of Metals by Chemical Treatment of Municipal Waste Water," *Water Res.* 5 (1971): 51–60.
104. Donnelly, R.G., R.L. Goldsmith, K.J. McNulty, and M. Tan. "Reverse Osmosis Treatment of Electroplating Wastes," *Plating* 61 (1974): 432–442.
105. Klein, E.W. and J. Del Pico. "RO Comes of Age for Plating Pollution Control," *Prod. Fin.* 38 (1974): 74–78.
106. Crampton, P. and R. Wilmoth. "Reverse Osmosis in the Metal Finishing Industry," *Metal Fin.* (March 21–27, 1982).
107. Jakobsen, K. and R. Laska. "Advanced Treatment Methods for Electroplating Wastes," *Poll. Eng.* (October 1977): 42–46.
108. McNulty, K.J., D.C. Grant, A. Gollan, and R.L. Goldsmith. "Field Demonstration of Reverse Osmosis Treatment of Cyanide Rinse Water," paper presented at the 30th Annual Purdue Industrial Waste Conference, West Lafayette, IN (May 1975).

8

Cyanide

Cyanide, as sodium cyanide (NaCN) or hydrocyanic acid (HCN), is a widely used industrial material. Cyanide waste streams result from ore extracting and mining, photographic processing, coke furnaces, synthetic fiber manufacturing, case hardening of steel, and industrial gas scrubbing. Cyanide is produced as a side reactant in the catalytic cracking of petroleum feed and in coking of distillation residues [1]. A major source of waste cyanide is the electroplating industry. Electroplaters use concentrated cyanide baths to hold metallic ions such as cadmium, copper, and zinc in solution. Dragout of the plating solution, containing cyanide ions and metal-cyanide complexes, contaminates rinsewaters.

Table 8.1 presents typical cyanide levels reported for plating wastewaters. Cyanide concentrations in other industrial wastes are presented in Table 8.2. Of these, the byproduct coke oven wastes are among the most complex waste streams, due to the presence of high ammonia and thiocyanate, plus considerable concentrations of a wide variety of organic compounds ([15, 18]. Spent case-hardening salt baths, used for surface hardening of ferrous materials, represent a particularly concentrated (10–15%) waste cyanide source [29, 30].

TREATMENT TECHNOLOGY

Several methods of treating cyanide wastes are in current use. The process most frequently employed is cyanide destruction by chlorination. Cyanide may be either partially oxidized to cyanate (CNO^-), or completely oxidized to carbon dioxide (CO_2) and nitrogen (N_2). Another cyanide destruction process, electrolytic decomposition, is applicable where waste cyanide concentrations are very high, although new technology shows promise of application to more dilute solutions. Ozone oxidation of cyanide has been employed with some success, and is claimed to be equivalent to destruction by chlorination, on the basis of effectiveness and costs. Evaporative recovery has also been successfully employed to treat cyanide wastewaters. Other cyanide treatment processes, which are still primarily in the development state, include reverse osmosis, ion exchange, and catalytic and thermal oxidation. Treatment of cyanide-bearing coke plant wastes is considered in a separate section of this chapter, in view of the complexity of these wastes.

Table 8.1 Concentrations of Cyanide in Plating Wastewaters

Process	*Average (mg/l)*	*Range (mg/l)*	*Reference*
Plating rinse	2	0.3–4	2
	700		3
		10–25	4
	32.5		5
	25		6
		60–80	7
		30–50	8
	3		9
	55.6	1.4–256	2
Bright dip		15–20	8
Alkaline cleaning bath		4,000–8,000	8
Plating Bath	30,000		10
Plating Bath		45,000–100,000	11
Plating Bath			
Brass		16,000–48,000	8,12–14
Bronze		40,000–50,000	
Cadmium		20,000–67,000	
Copper		15,000–67,000	
Silver		12,000–60,000	
Tin-zinc		40,000–50,000	
Zinc		4,000–64,000	

Chlorination

Destruction of cyanide by chlorination may be accomplished by direct addition of sodium hypochlorite or by addition of chlorine gas plus sodium hydroxide to the waste. Sodium hydroxide reacts with the chlorine to form sodium hypochlorite. Selection between the two methods is on the basis of economics and safety. Chlorine gas treatment is about half as expensive as direct hypochlorite addition, but handling is more dangerous and equipment costs are higher. The hypochlorite added or produced oxidizes cyanide to cyanate. Oxidation of cyanide to cyanate is accomplished most completely and rapidly under alkaline conditions at pH 10 or higher, and this cyanide treatment process is often termed *alkaline chlorination*. An oxidation period of 30 min–2 hr is usually allowed [7]. To avoid solid cyanide precipitates which may resist chlorination, the wastewater should be continuously agitated during treatment. Close process control, utilizing pH and ORP sensors and automated chemical feed techniques, is required to assure maximum reaction rate, minimum detention time, and complete reaction [31].

The resulting cyanate is much less toxic than cyanide [32], but can if necessary be further oxidized to carbon dioxide and nitrogen. Oxidation of cyanate to carbon dioxide and nitrogen can be accomplished by additional

Table 8.2 Cyanide Levels in Wastewaters Other Than from the Plating Industry

Industrial Source	*Concentration*	*Reference*
Blast furnace scrubber water	0.2–1.4	15
	2.4	15
	48.5	16
Ferroalloy scrubber waters	0.7–5.4	17
Coke plant waste streams		18
Coke oven liquor	0–8	
Decantation tank	8	
Final cooler condensate	196	
Benzole separator	2736	
Oil generation plant separator	104	
Spent limed liquor	4	
Coke plant ammonia liquor	2–44.5	19
	20–60	20
	7.5–39.6	21
Coke plant waste	10.0–38.1	22
	100	23
	91–110	15
Color film bleaching process	71	24
Hydrogen cyanide manufacturing	14–42 (28 avg)	25
Coal conversion wastes	2–30	26
Coal conversion (synthane)	1–6	23
Gold ore extraction	18.2–22.3	27
Explosives manufacture	0.0–2.6	28
Petroleum refining	0.0–1.5	28
Paint and ink formulation	0.0–2.0	28

chlorination, or to carbon dioxide and ammonia by acid hydrolysis. Acid hydrolysis must take place at pH 2–3, however, usually obtained by addition of sulfuric acid. Thus, acid hydrolysis requires addition of acid to achieve the required pH, and subsequent neutralization of the acidic wastewater before the effluent can be discharged. These stages of pH adjustment increase the treatment costs as well as the total dissolved solids of the wastewater. Cyanate will be destroyed by acid hydrolysis within about 5 minutes of contact time.

Complete cyanide oxidation to CO_2 and N_2 can be accomplished solely through chlorination, if close pH control is maintained. After initial oxidation to cyanate, further oxidation to yield CO_2 and N_2 will occur slowly over several hours at pH 10+. However, if the waste pH is reduced to 8–8.5, cyanate oxidation by chlorine can be completed within 1 hour. At the lower pH, sufficient chlorine must be added to ensure an excess beyond that needed to oxidize the cyanide to cyanate, in order to avoid liberation of highly toxic cyanogen chloride gas. Cyanogen chloride is the intermediate product of the oxidation of cyanide to cyanate. It breaks down very rapidly at pH 10+ and temperatures above 20°C [33]. However, at lower pH or temperature, excess chlorine is needed to speed the breakdown [32]. Complete cyanide treatment at high pH, although requiring long reaction times, may be advantageous if the treated wastewater can be utilized to neutralize acidic waste streams in the plant.

Cyanide baths and rinses are frequently treated together, with the concentrated bath solutions being blended into the more dilute rinsewaters. It is common practice to use batch treatment for alkaline chlorination when average flows do not exceed 15 gpm. Batch processes have been used for waste flows up to 30 gpm, however, to provide close process control in the absence of automated treatment [34].

One modification of the batch process is the patented Integrated Process [12, 13, 35]. In plating shops, when articles leave any solution which contains cyanide, they first enter a rinse which consists of a concentrated solution of sodium hypochlorite. Dragout cyanide is rapidly oxidized to cyanate at pH 10–12.5 and the free chlorine content of 0.3–1.0 mg/l of this static rinse bath. The cyanate slowly decomposes to carbon dioxide and nitrogen. It is claimed that any metal in the dragout which was previously complexed by the cyanide is released by the oxidation of the cyanide and, at the pH maintained, is precipitated [35]. After the work piece leaves the static treatment bath, it is rinsed with water.

Several manufacturers have developed small "package" cyanide treatment plants which employ the chlorination process. Figure 8.1 shows basic types of package plants. The tank-type plant shown at the top of Figure 8.1 provides cyanide treatment to the level of cyanate only. If complete destruction is required, a second tank is employed for secondary pH and ORP control and second-stage chlorination. The two-stage reaction tower-type plant shown at the bottom of Figure 8.1 provides complete cyanide treatment to the level of CO_2 and N_2. Package plants such as these range from treatment capacities of 15 gpm of waste containing a maximum of 450 mg/l of cyanide, to units of 30 gpm and a maximum cyanide content of 350 mg/l. A unit of 15 gpm capacity costs about $30,000 (1979 basis), complete [37]. Hardware costs of $30,000–35,000 (1978 basis) have been reported for two-stage continuous alkaline chlorination systems treating 15–40 gpm, with installation representing an additional 25% increment [38]. Batch treatment systems are reported to range from $12,000 at 2 gpm capacity to $30,000 at 30 gpm capacity, complete [38].

O'Neill [10] has described a 15-gpm two-stage cyanide oxidation plant. Chlorine gas and caustic are used to treat a cyanide-cleaning bath rinsewater to less than 0.4 mg/l effluent cyanide. The treatment plant, including a chromate reduction and precipitation process, cost (in 1970) $16,000 for equipment plus

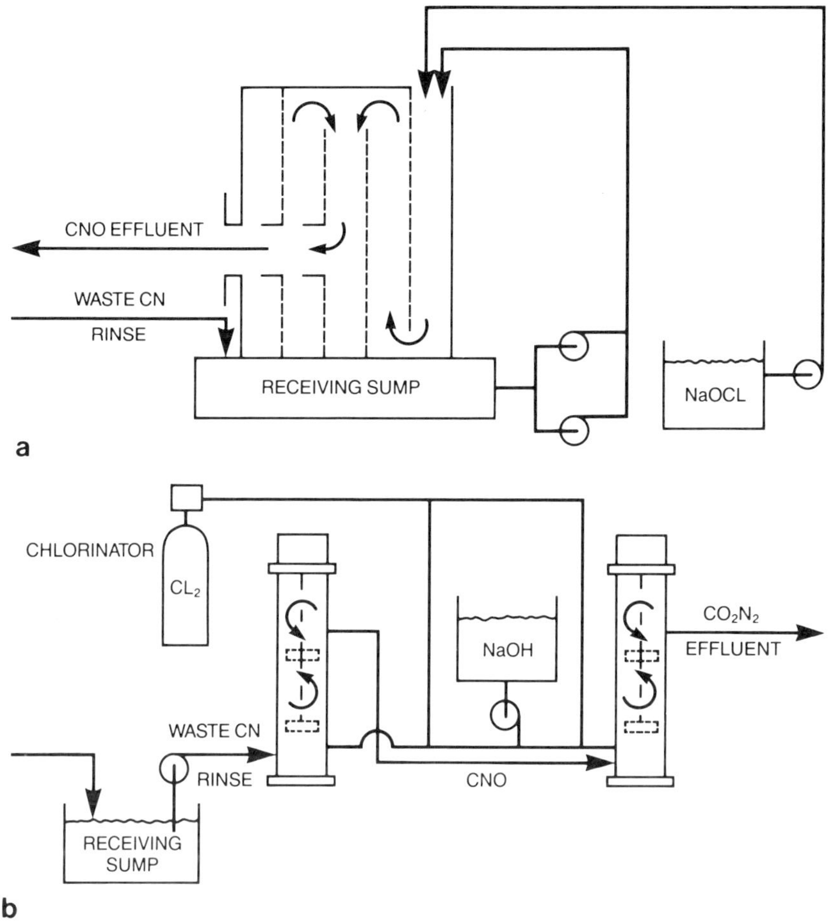

Figure 8.1 Commercial package plants for cyanide waste treatment. (From Zievers et al [36], courtesy of *Plating*.) *a*. Retention-tank-type package cyanide treatment plant. *b*. Reaction-tower-type package cyanide treatment plant.

another $10,000 for the building and piping. The cyanide treatment segment was continuous, but not automated.

Larger volumes of cyanide waste require specifically designed and constructed treatment plants, tailored to the individual industrial process and flow configuration producing the waste. Fisco [39] has described such a treatment plant, which oxidizes the cyanide component of a waste stream totaling 1.5 million gpd. Treatment involves total cyanide destruction by a two-step process of chlorination at high and low alkaline pH as discussed above.

Treatment of a cyanide waste flow of 40–50 gpm by alkaline chlorination to the level of cyanate has been reported [40]. An effluent cyanide content of

0.1 mg/l was achieved by the process. Complete oxidation of cyanide to CO_2 was achieved in treating another waste by alkaline chlorination plus acid hydrolysis. Cyanide content of the rinsewater was reduced from 700 mg/l to zero cyanide plus cyanate [3].

The cyanide content of electroplating rinsewaters is quite variable (see Table 8.1). Watson [5] has reported the treatment of a dilute cyanide waste for a metal-plating plant. The waste contained an average cyanide content of 32.5 mg/l, and was treated to zero cyanide on a batch basis at the rate of 200,000 gal/month. Effluent cyanate concentration was not reported. Cyanide treatment was to the level of cyanate by alkaline chlorination, 3 lb chlorine being fed/lb cyanide treated.

It has been reported [32] that oxidation of cyanide to cyanate requires approximately 1.75 lb chlorine/lb cyanide, and complete oxidation to CO_2 plus N_2 requires about 4.3 lb chlorine/lb cyanide. Others, however, disagree with this dosage, reporting that 3.4–4.0, and 8 lb chlorine/lb cyanide are respectively required for the two levels of oxidation [34, 41]. Overall waste composition will affect the actual requirement. Unless metal-bearing wastes are segregated from cyanide wastes, an extra chlorine demand will be exerted as a result of metal oxidation by the chlorine (e.g., Fe^{2+} to Fe^{3+}, Cr^{3+} to Cr^{6+}) and/or catalytic decomposition of chlorine and hypochlorite. Any extra chlorine demand exerted by a mixed-metal cyanide waste would increase cyanide treatment costs significantly. Furthermore, aqueous chlorine is not stable, and will decompose with time, a process catalyzed by light.

Cost curves for single-stage cyanide treatment (to cyanate) and for complete cyanide oxidation have been published. Costs, which are reproduced in Figures 8.2 and 8.3, indicate single-stage treatment costs of \$0.28–1.28/lb cyanide destroyed, and two-stage costs of \$0.50–\$2.45/lb cyanide. Capital costs are also shown in both figures, expressed as thousands of dollars capital investment versus waste flow in gallons per minute.

One major problem associated with alkaline chlorination of cyanide wastes occurs if soluble iron is present with the cyanide. Iron, in the presence of cyanide, forms extremely stable ferrocyanide complexes, which prevent the cyanide from being oxidized [42]. Similar difficulties result from formation of nickelocyanides [43]. Successful treatment of cyanide in ferri- and ferrocyanide wastes has been reported, however, by alkaline chlorination at pH 10+ and temperature 90°C [44]. Cyanide was reduced from 5.1 to 0.1 mg/l. Curry [45], in discussing treatment of ferrocyanides, suggests precipitation with ferrous salts as the ferro-ferrocyanide, or alkaline chlorination at a temperature of 70°C and pH of 12.0. The ferro-ferrocyanide precipitation process does not appear to be very effective, as effluent cyanide levels of 5–10 mg/l have been reported [46]. Furthermore, stability of the sludge upon disposal is of concern.

Chlorination of cyanide in process waste streams other than from plating operations has been reported. Combination of iron blast-furnace scrubber water with sinter plant wastewater and subsequent settling, chlorination, and filtration

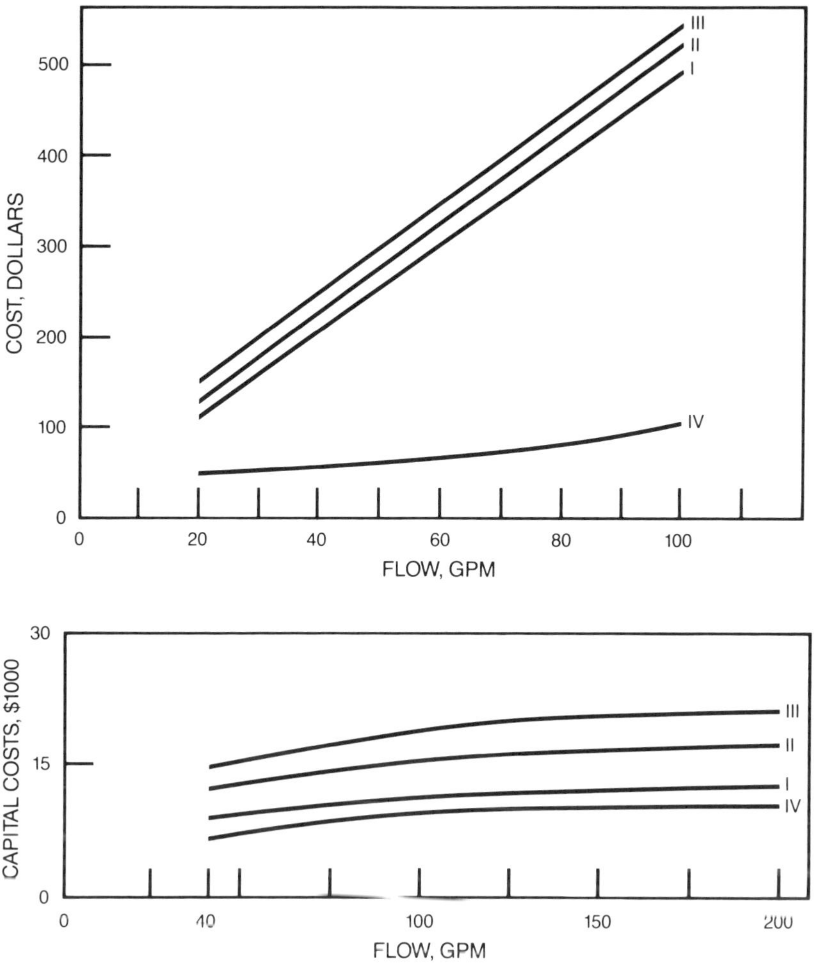

Figure 8.2 Costs (in 1975) of single-stage cyanide treatment to the level of cyanate. (From Zievers et al [36], courtesy of *Plating*.) (*Top*) Monthly operating costs (plus depreciation) at 50 mg/l cyanide, 172 hr/month. I, no instrumentation; II, medium instrumentation; III, full instrumentation; IV, using "low cyanide" type bath. (*Bottom*) Initial capital costs, including installation. I, manual controls; II, medium instrumentation; III, full instrumentation; IV, package plant.

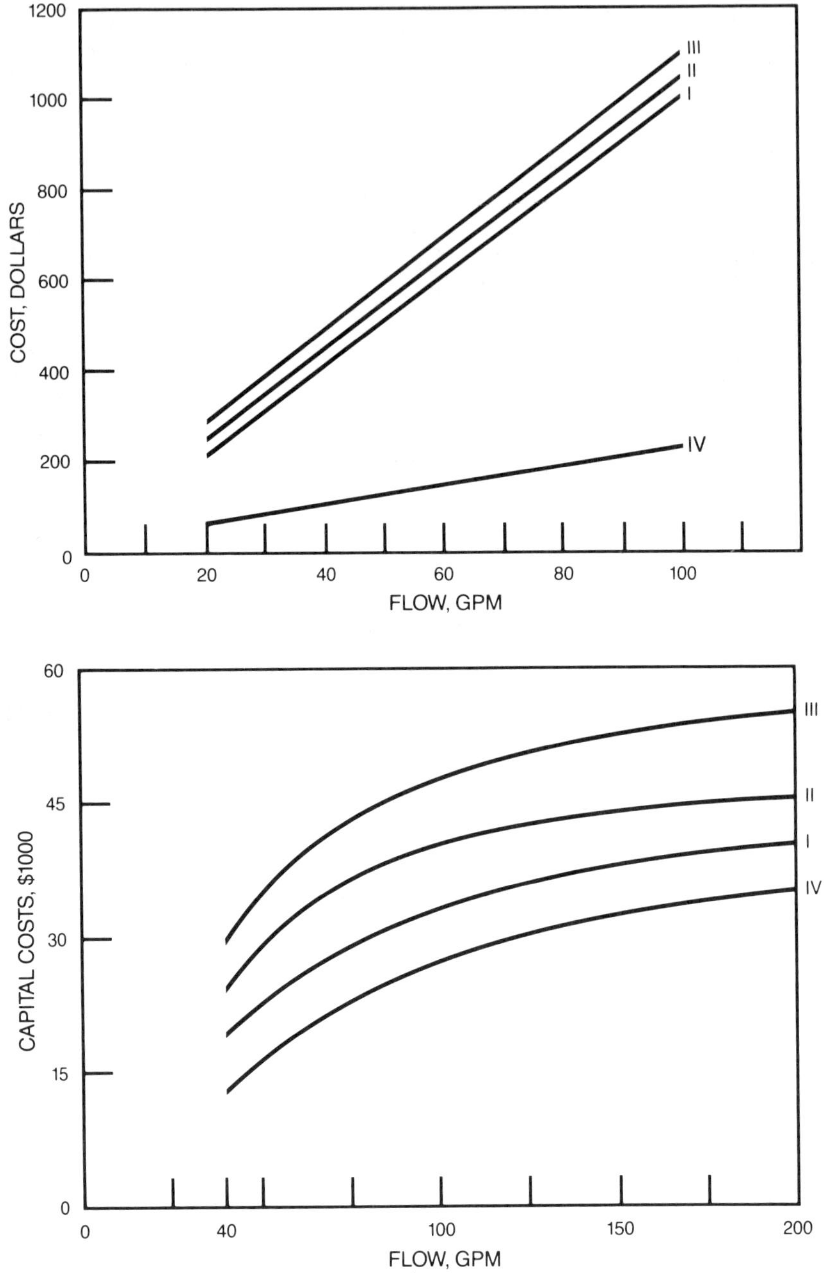

Figure 8.3 Costs (in 1975) of complete cyanide destruction to the level of carbon dioxide. (From Zievers et al [36], courtesy of *Plating*.) (*Top*) Monthly operating costs (plus depreciation) at 50 mg/l cyanide, 172 hr/month. I, no instrumentation; II, medium instrumentation; III, full instrumentation; IV, using "low cyanide" type bath. (*Bottom*) Initial capital costs, including installation. I, manual controls; II, medium instrumentation; III, full instrumentation; IV, package plant.

for water recycle resulted in 0.005 mg/l cyanide in the recycle blowdown stream [15]. Alkaline chlorination of raw process wastes from hydrogen cyanide manufacture reduced cyanide of 14–42 mg/l (28 avg) to 2.0 mg/l in the effluent [25].

Ozonation

Ozonation shows some promise as a substitute for chlorination of cyanide waste, although it has had limited full-scale use. Further, there is some question as to the ability of the process to effectively oxidize cyanate to end products. One advantage of ozonation is that no dissolved solids are added in the treatment step. Kandzas and Mokina [47] have reported complete ozone oxidation of cyanide (via cyanate intermediate) to end products. Complex zinc, nickel, and copper cyanides were easily destroyed, but complex cobalt cyanide was resistant to treatment ozonation. The use of ultraviolet light and heat to assist decomposition of the more stable cyanides has been suggested [13].

Ozonation has been employed to treat a cyanide waste at a metal-working plant [6]. Waste flow through the plant is 500 gpm, and cyanide concentrations to 25 mg/l are encountered. Oxidation of cyanide to cyanate was reported, with partial oxidation of cyanate to final end products. Capital costs (in 1958) for the process, installed, totaled $70,000. Ozone production costs (in 1972) of 14–15¢/lb [41] compare with chlorine costs (in 1968) of 13¢/lb [36].

In ozone treatment, cyanide oxidation to cyanate is very rapid (10–15 min) at pH 9–12, and is virtually instantaneous in the presence of traces of copper, which acts as a catalyst [41]. Copper-catalyzed reduction of cyanide to less than 1 mg/l with ozone at a cost of $2.35/1,000 gal has been reported for initial cyanide concentration of 20 mg/l [48]. Oxidation of cyanate to final end products is much slower, especially in the case of complex cyanides other than copper and nickel, unless catalysts such as iron, copper, or manganese are present. However, for iron and nickel cyanide complexes themselves, additional treatment, such as elevated temperature or ultraviolet light, seems to be required [48].

Electrolytic Decomposition

Wastes containing high concentrations of cyanide are most successfully treated by electrolytic decomposition, a well-established technique primarily used by industry for the destruction of cyanide in concentrated spent stripping and dip solutions for copper and nickel, alkaline descalers, and derusters. With the simple types of electrolytic cells normally used, it is usually not economical to carry the electrolytic decomposition process below a few hundred mg/l residual cyanide. Final destruction is typically accomplished by chlorination [30]. It has not generally been practical to treat rinsewaters by electrolysis, due to their initial low cyanide concentrations [4, 29, 30]. Hillis has described sharply

Table 8.3 Electrolytic Decomposition of Cyanide Waste [11]

Run	*Initial Cyanide Concentration (mg/l)*	*Time to Decompose (Days)*	*Final Cyanide Concentration (mg/l)*
1	95,000	16	0.1
2	75,000	17	0.2
3	50,000	10	0.4
4	75,000	18	0.2
5	65,000	12	0.2
6	100,000	17	0.3
7	55,000	14	0.4
8	45,000	7	0.1
9	50,000	14	0.1
10	55,000	8	0.2
11	48,000	12	0.4

escalating energy demand per unit of cyanide destroyed, for increasingly higher degrees of treatment [30]. A total of 15 kWh/kg cyanide destroyed is required to reduce an initial 400–100 mg/l residual cyanide. This energy consumption doubles to achieve a residual 10 mg/l cyanide.

The concentrated cyanide waste is subjected to anodic electrolysis at high temperature (125–200°F) for periods of several hours to days. Initially, cyanide is completely broken down to the gaseous products carbon dioxide, nitrogen, and ammonia, with cyanate as an intermediate [41]. Urea may also be produced [30]. As the process continues, the waste electrolyte becomes less capable of conducting electricity, and the reaction may not go to final completion. Anodic current densities below 200 A/m^2 are typical, this value representing a practical limit on the use of graphite electrodes.

Low levels of residual cyanide can be achieved by electrolytic decomposition, providing that sufficient treatment time is allowed [11]. Results of such treatment of several batches of plating wastes are presented in Table 8.3. Wastes containing several tens of thousands of mg/l of cyanide were batch treated, over 7- to 18-day periods, down to cyanide levels of less than 0.5 mg/l. Electrolytic decomposition is less effective on cyanide wastes containing sulfate. Cyanide reductions to only 695–750 mg/l have been reported, before heavy scaling at the anode prevented further electrolysis [11, 13]. Initial cyanide concentrations were equivalent to those presented in Table 8.3. Sulfate concentrations in the cyanide wastes were not reported.

One plating plant has been reported to use electrolytic decomposition on a continuous basis for cyanide destruction, achieving an effluent cyanide level of 0.35 mg/l [13]. Improved effective surface area is achieved by filling the space between the flat electrodes with carbonaceous particles. Flow control is critical to effective operation.

Operating costs reported for electrolytic decomposition were 8.2¢/lb of cyanide destroyed [11]. At cyanide waste concentrations shown in Table 8.3, these costs convert to $55.00/1,000 gal. Hillis has demonstrated that at high initial cyanide concentrations, economics strongly favor electrolysis over chlorination [30]. Utilizing electrolysis to reduce initial cyanide to approximately 1,000 mg/l, followed by chlorination for final polishing treatment, yields costs which are approximately 10% of those required for direct chlorination. Depending upon the initial cyanide concentrations and waste volume to be treated, the capital cost of the electrolysis system may be recovered from savings in operating costs within 6–18 months [30].

If sodium chloride is added to the cyanide solution, the oxidation rate, power consumption, and efficiency of electrolytic decomposition at lower cyanide concentrations (below 200 mg/l) is greatly enhanced [30, 35, 41, 49]. In addition to improving solution conductivity, the anodic oxidation of the chloride ion to chlorine gas yields chlorination oxidation concurrent with electrolytic destruction [30]. One particular advantage for electrolytic decomposition is its effectiveness in destroying nickel, copper, and ferro- and ferricyanide complexes, which are difficult to treat by alkaline chlorination [41, 45].

A package electrolytic decomposition treatment plant has been developed specifically to treat rinsewaters containing low cyanide content, and is claimed to purify a waste stream with 20 mg/l to less than 0.5 mg/l cyanide [50]. Unit capacity is 5–10 gpm and capital cost is less than $20,000 (1979 basis) for the typical plating shop [50].

Coke Plant Wastes

For the specific case of by-product coke-oven waste streams, the presence of high ammonia and organic constituents severely suppresses the effectiveness of conventional oxidative treatment of cyanide prior to ammonia removal and biological treatment. The presence of cyanide and thiocyanate also affects the design and performance of biological treatment systems for coke wastes [51]. One full-scale activated-sludge system adds 0.3 MGD cooling water to 0.112 MGD flow of coke wastes, to assist function of the biological unit [20]. Even so, cyanide and thiocyanate removals are inconsistent and incomplete. It was observed that removal of these constituents was greatly enhanced when influent ammonia levels were low, yielding effluent cyanide of 1–3 mg/l.

Activated sludge treatment of waste ammonia liquor in one instance yielded an effluent cyanide concentration of 9.5 mg/l [15]. Influent gross plant-waste loading was reported at 33.7 mg/l (reduced from 110 mg/l in the raw waste ammonia liquor), and ammonia stripping was practiced at the coking plant.

Table 8.4 Effectiveness and Cost (1969) for Chemical Removal of Residual Cyanide in Effluent Resulting from Biological Treatment of Coke Plant Wastes [19]

Treatment	*Influent Cyanide (mg/l)*	*Effluent Cyanide (mg/l)*	*Chemical Cost ($/1,000 gal)*
Oxidation—Cl_2	10	<1.0	9.50
Oxidation—ClO_2	4.0–5.2	1.8–3.6	18.00
Activated carbon	2.3–2.8	0.6–1.4	12.00

The high cost of postbiological treatment for further reduction of cyanide was reported in one study [19] and is presented in Table 8.4. Treatment results are based on laboratory studies performed on a full-scale activated-sludge effluent, and costs are for treatment chemicals only.

Other Processes

Recently, a patented process (Kastone, E.I. du Pont de Nemours & Company) has become commercially available. The process utilizes hydrogen peroxide plus formaldehyde to treat cyanide wastes at 120–130°F. End products include cyanate, ammonia, and various organic acids [50]. The process simultaneously precipitates zinc and cadmium, if present. At one commercial installation, the Kastone process reduced cyanide from initial values of 350–400 mg/l to less than 0.3 mg/l [52]. The process is reported effective for cyanide levels of 50–500 mg/l. Treatment costs of $0.41/lb cyanide were cited, which were compared to alkaline chlorination costs of $0.18/lb [50].

Economic benefits may accrue to the Kastone process, due to the simultaneous precipitation of zinc or cadmium. Savings on land disposal of sludge have been suggested for zinc systems by recycle of zinc oxide, which can be obtained in high purity [48]. Sludge haulage distance and associated cost structure would obviously be important variables. One major disadvantage of the process is its inability to oxidize cyanide below the level of cyanate. Another is the production of glycolic acid amide, which renders the waste unsuitable for direct stream discharge [13]. The cyanide present is converted on a molar basis to roughly 33% cyanate, 32% ammonia (a portion of which is volatilized), and 35% combined ammonia as glycolic acid amide [12].

Evaporative recovery of cyanide wastes has been shown to be practical and economical under proper circumstances, with the advantage of recovery frequently outweighing the disadvantages of high capital and operating costs and individual units required for each plating line [13]. Evaporative units have been in service for more than 10 years, and several types are available. The present 100 units operating represent only a small fraction of the plating industry

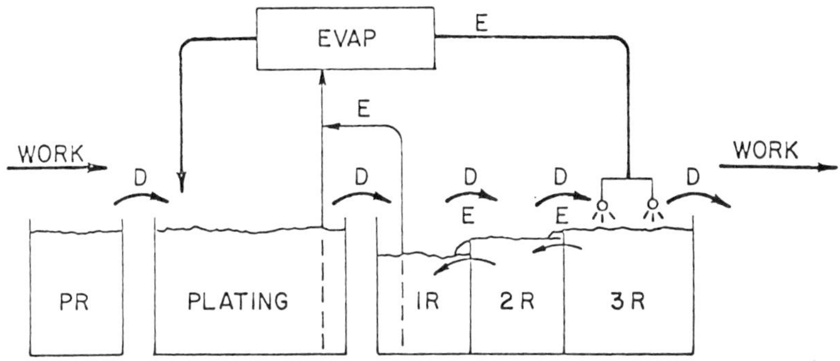

Figure 8.4 Schematic of a closed-loop evaporative recovery system. PR, prerinse tank; D, dragout; R, rinse tank; P, returned plating solution; E, rinse or distillate flow. (From Culotta and Swanton [53], courtesy of *Plating*.)

treatment [13]. The evaporation process is used primarily for concentrating rinsewaters for process chemical recovery. It is best applied for segregated plating lines with countercurrent rinse tanks and reduced dragout [13]. Under these conditions the system may operate as a closed loop, with recondensed water vapor returned to the head of the rinse line, and concentrated plating solution returned to the plating tank. A schematic of a typical closed-loop evaporative-recovery system is shown in Figure 8.4.

Evaporative recovery is perhaps best suited to the treatment of cyanide plating solution above all other solutions, due to the noncorrosive nature of the distillate. This results in lower equipment and maintenance costs. Operating costs are primarily influenced by cost of steam, cooling water, and power. Utilities costs become a direct function of evaporative rate. Countercurrent rinsing, wherein only the most contaminated rinsewater is treated, thus improves the economics of the process. Reduced utilities cost of multiple-effect evaporation results from improved efficiency of steam usage. This is offset by increased capital cost for each additional evaporation effect [54]. As a general rule, at evaporation rates below 2.5 gpm, additional investment for multiple-effect evaporators is not justified [38].

The installed cost of a 20 gal/hr (0.33 gpm) single-effect evaporator has been estimated at $35,700 [38]. This represents a capital investment of in excess of $74,000/1,000 gpd treatment capacity. Associated operating costs (1978 basis) were $52.28/1,000 gal treated. Operating costs would be partially offset by the value of the recovered chemicals and water.

Other cyanide treatment processes include ion exchange [48, 55, 56], reverse osmosis [14, 48, 57–59], copper-catalyzed activated-carbon adsorption and oxidation [2, 22, 46, 48, 56], and thermal hydrolysis [60]. Ion exchange recovery of cyanide generally has not appeared feasible because, among other reasons, cyanide deteriorates the exchange resins [34]. Metal cyanide fouling has also been reported [61]. Zievers and Novotny [55] report ion exchange effluents containing 6–15 mg/l of cyanide, and point out the danger of evolution of highly toxic hydrogen-cyanide gas upon resin regeneration. One potential application of ion exchange is the use of weak base anion exchange resin to remove ferri- and ferrocyanide [24, 33, 62]. Reductions of 200 mg/l $Fe(CN)6^{+3}$ to less than 1 mg/l are reported [24, 62]. Beyond this limited scope, however, ion exchange is not generally feasible for cyanide wastewaters.

Cellulose acetate and polymide reverse osmosis membranes deteriorate rapidly by alkaline hydrolysis at the normal pH of cyanide wastes (pH 11+). Reduction of pH precipitates metals in the waste and clogs the membrane. More resistant membranes have shown poor performance, rejecting only 28% of the cyanide [57]. Further, most effective membrane rejection of cyanide occurs at pH above 10 [63]. With recent developments in membrane technology resulting in improved pH and temperature resistance, the applicability of reverse osmosis for copper, cadmium, and zinc cyanide wastes has been suggested [13, 48, 64] with good performance reported for zinc cyanide [65]. However, near-neutral pH (moderately alkaline) baths are required for all three applications with current membranes [59]. For high-pH cyanide baths, no currently commercially available membranes can withstand the high pH [48, 59].

A recently developed process involving cyanide adsorption plus copper-catalyzed oxidation on granular activated carbon has proved effective in preliminary pilot plant tests [2, 22, 41, 65]. Under controlled conditions the process yielded effluent cyanide levels below 0.1 mg/l for a wide range of influent cyanide concentrations on both plating wastes and coking wastes [22]. The process has not yet been tested on a full-scale, long-term basis, however.

Thermal hydrolysis of concentrated cyanide-plating bath solutions has been tested at laboratory scale and, on the basis of performance and projected economics, reported to warrant full-scale testing [60]. The process, involving high temperature (240–280°C) and pressure (550–600 psi), converts cyanide to ammonia and formic acid. The process, operating in batch mode, is reported to reduce an initial 50,000 mg/l cyanide to 0.02 mg/l within 2 hours of achieving the desired operating temperature [60].

SUMMARY

Processes are available to treat cyanide wastewater to almost any effluent level required. Treatment of strong cyanide wastes appears to be most feasible by electrolytic decomposition. Low-to-intermediate effluent cyanide content is reported by this process, and either partial treatment to cyanate or complete destruction can be achieved as required. Chemical oxidation by chlorination or ozonation is appropriate for relatively dilute cyanide wastes, whether carried over from the electrolytic decomposition process or as dilute rinsewater wastes. If proper rinsing procedures are used, yielding a concentrated cyanide rinsewater, evaporative recovery is feasible and economical. This results in zero cyanide discharge, plus reuse of distillate and the concentrated cyanide solu-

Table 8.5 Treatment Levels for Cyanide Wastewaters

Treatment Process	*Cyanide Concentration (mg/l)* Initial	*Final*	*Percent Removal*	*Reference*
Alkaline chlorination[a]	28 (avg)	2.0	92.9	25
	—	1.7	—	66
	—	0.1	—	40
Alkaline chlorination[b]	—	0.4	—	10
	700	0.0	100	3
	32.5	0.0	100	5
	1.3	0.25	80.7	67
	—	0.6	—	67
Alkaline chlorination[c]	5.1	0.1	98	44
Ferro-ferrocyanide precipitation	—	5–10	—	46
Electrolytic decomposition	45,000–100,000	0.1–0.5	99.99+	11
Electrolytic decomposition	20	0.5	97.5	50
Ozonation	25	0.0	100	7
Kastone process	350–400	0.3	99.9+	52

[a] Single-stage chlorination.
[b] Two-stage chlorination.
[c] High-pH, high-temperature process.

tion. Table 8.5 presents a summary of cyanide treatment processes and reported efficiencies. Reported effluent cyanide concentrations are summarized in Table 8.6 for 37 individual plating operations which generate cyanide-bearing wastes. For the alkaline chlorination process, improved treatment efficiencies result from proper pH control and increase chlorine dosages.

Capital investments for alkaline chlorination plants, as reported by several authors [10, 36, 37, 68] appear to be accurately described by the cost curves of Figures 8.2 and 8.3. Green and Smith [41] have reported investments for

Table 8.6 Summary of Plating Industry Cyanide Treatment [13]

Treatment Process	*No. Plants*	*Range of Flows (1,000 gpd)*	*Effluent Cyanide (mg/l)* Range	*Avg*
Batch chlorination	9	1–624	<0.01–0.2	<0.04
Continuous chlorination	16	23–1,000	0.0–<1.0	<0.10
Integrated process	5	18–192	0.02–<0.10	<0.05
Electrolytic decomposition	2	60–630	0.35–1.0	0.68
Evaporation	5	8–100	<0.3–1.1	<0.55

Table 8.7 Comparison of Capital Costs for Cyanide Treatment Processes [after 41]

Treatment Process	*Plant Capacity (1,000 gpd)*	*Cost Ratio per 1,000 gpd (1972 $)*
Alkaline chlorination	14.27	1.00
	11.36	1.05
	12.68	1.50
Electrolysis	2.38	2.67
Ozonation	4.44	3.59
Ozonation	18.49	3.43
Evaporation	2.39	9.33

Table 8.8 Operating Costs for Cyanide Waste Treatment

Treatment Process	*Treatment Cost ($/lb cyanide destroyed) (various years)*	*Reference*
Alkaline chlorination[a]	0.28–1.28	36
	4.50	5
Alkaline chlorination[b]	0.50–2.45	36
Kastone process[a]	0.41	50
Electrolysis	0.082	11
Ozonation	0.14–0.64	41

[a] Oxidation to level of cyanate.
[b] Complete oxidation to CO_2.

common cyanide treatment processes by seven English industries. Their data, converted to cost ratios, are presented in Table 8.7. The least expensive alkaline chlorination plant has been taken as a base cost of one unit, and all other capital investments compared to that. Increased capital investments are often offset by lower operating costs, however. Table 8.8 summarizes reported operating costs.

REFERENCES

1. Kunz, R., J. Casey, and J. Huff. "A Review of Cyanide in Refinery Wastewaters," *Proceedings of the 3rd Annual Conference on Treatment and Disposal of Industrial Wastewaters and Residues*, Houston, TX (April 18–20, 1978).
2. Bernardin, F.E. "Cyanide Detoxification Using Adsorption and Catalytic Oxidation on Granular Activated Carbon," *J. Water Poll. Control. Fed.* 45 (1973): 221–231.
3. Hansen, N.H., and W. Zabban. "Design and Operation Problems of a Continuous Automatic Plating Waste Treatment Plant at the Data Processing Division, IBM,

Rochester, Minnesota," *Proceedings of the 14th Industrial Waste Conference, Purdue University* (May 1959), pp 227–249.

4. Palla, L.T. and R.G. Spicher. "Cyanide Treatment in Profit and Cure," paper presented at the 26th Annual Purdue Industrial Waste Conference, West Lafayette, IN (May 1971).
5. Watson, K.S. "Treatment of Complex Metal-Finishing Wastes," *Sew. Ind. Wastes* 26 (1954): 182–194.
6. "Ozone Counters Waste Cyanide's Lethal Punch," *Chem. Eng.* 24 (1958): 63–64.
7. Crowle, V.A. "Effluent Problems as They Affect the Zinc Die-Casting and Plating-on-Plastics Industries," *Metal Fin. J.* 17 (194) (1971): 51–54.
8. Lowe, W. "The Origin and Characteristics of Toxic Wastes, with Particular Reference to the Metal Industry," *Water Poll. Control* (London, 1970): 270–280.
9. Jashnani, I.L. and R. Green. "Pretreatment System for a Circuit Board Plant," paper presented at the 8th National Conference on Municipal Sludge Management, Municipal and Industrial Sludge Management, Disposal and Utilization, Miami Beach, FL (March 19–21, 1979).
10. O'Neill, F. "Facing up to Pollution," *Plating* 57 (1970): 1211–1213.
11. Easton, J.K. "Electrolytic Decomposition of Concentrated Cyanide Plating Wastes," *J. Water Poll. Control. Fed.* 39 (1967): 1621–1625.
12. "Development Document for Effluent Limitations Guidelines and New Source Performance Standards for the Copper, Nickel, Chromium and Zinc Segment of the Electroplating Point Source Category," USEPA 440/1-74-003a (March 1974).
13. "Development Document for Interim Final Effluent Limitations Guidelines and Proposed New Source Performance Standards for the Common and Precious Metals Segment of the Electroplating Point Source Category," USEPA 440/1-75/040, Group I, Phase II (April 1975).
14. McNulty, K.J., D.C. Grant, A. Gollan, and R.L. Goldsmith. "Field Demonstration of Reverse Osmosis Treatment of Cyanide Rinse Water," paper presented at the 30th Annual Purdue Industrial Waste Conference, West Lafayette, IN (May 1975).
15. "Development Document for Effluent Limitations Guidelines and New Source Performance Standards for the Steel Making Segment of the Iron and Steel Manufacturing Point Source Category," USEPA 440/1-74-024-a (June 1974).
16. Brower, G.R., S.J. Ryckman, and P.A. Luther. "Control of Water Problems in a Steel Mill's Blast Furnace Gas Wash Recirculation System," paper presented at the 32nd Annual Purdue Industrial Waste Conference, West Lafayette, IN (May 1977).
17. "Development Document for Effluent Limitations Guidelines and New Source Performance Standards for the Smelting and Slag Processing Segments of the Ferroalloy Manufacturing Point Source Category," USEPA 440/1-74-008-a (February 1974).
18. Pearce, A.S. and S.E. Punt. "Biological Treatment of Liquid Toxic Wastes, Part 1," *Effluent Water Treatment J.* 115 (1975): 32–39.
19. Kostenbader, P.D. and J.W. Flecksteiner. "Biological Oxidation of Coke Plant Weak Ammonia Liquor," *J. Water Poll. Control. Fed.* 41 (1969): 199–207.
20. Cousins, W.G. and A.B. Mindler. "Tertiary Treatment of Weak Ammonia Liquor from Coke By-Products Plant," Report, International Hydronics Corporation, P.O. Box 910, R-4, Princeton, NJ 08540.

21. Eklund, C.W. and J.W. Irvin. "Alternate Methods of Treating Steel Mill Coke Plant Wastewaters," paper presented at the 33rd Annual Purdue Industrial Waste Conference, West Lafayette, IN (May 1978).
22. Bernadin, F.E. "Detoxification of Cyanide by Adsorption and Catalytic Oxidation on Granular Activated Carbon," paper presented at the 48th Annual Meeting, Water Pollution Controllers Federation, Miami Beach, FL (October 7, 1975).
23. Valiknac, T. and R.D. Neufeld. "Thiocyanate Toxic Inhibition to Phenol Bio-oxidation," paper presented at the 33rd Annual Purdue Industrial Waste Conference, West Lafayette IN (May 1978).
24. Ohmi, T. and Y. Moritsu. "Removal of Ferricyanide and Ferrocyanide from Waste Water with Weakly Basic Ion Exchange Resin," *Japan Kokai* 74 49,800 (May 15, 1974); *Chem. Abstr.* 81 (1974): 158–159.
25. "Development Document for Interim Final Effluent Limitations Guidelines and Proposed New Source Performance Standards for the Significant Inorganic Products Segment of the Inorganic Chemicals Manufacturing Point Source Category," USEPA 440/1-75/037, Group I, Phase II (May 1975).
26. Sack, W.A. and W.R. Bokey. "Biological Treatment of Coal Gasification Wastewater," paper presented at the 33rd Annual Purdue Industrial Waste Conference, West Lafayette, IN (May 1978).
27. Schmidt, J., S. Simovic, and E. Shannon. "Development Studies for Suitable Technologies for the Removal of Cyanide and Heavy Metals from Gold Milling Effluents," paper presented at the 36th Annual Purdue Industrial Waste Conference, West Lafayette, IN (May 1981).
28. Cawley, W., ed. *Treatability Manual*, vol. I. *Treatability Data* USEPA 600-8-800-42-a (July 1980).
29. Entwistle, J.E. "The Electrolytic Processing of Cyanide Wastes," *Effluent Water Treatment J.* (March 1976): 123–128.
30. Hillis, M.R. "Treatment of Cyanide Wastes by Electrolysis," *Trans. Inst. Metal Fin.* 53 (1975): 65–73.
31. Metzner, A.W. "Removing Soluble Metals from Wastewater," *Water Sew. Works* (April 1977): 98–101.
32. American Electroplaters Society, "A Report on the Control of Cyanides in Plating Shop Effluents," *Plating* 56 (1969): 1107–1112.
33. Beckenn, W.E. "Treatment of Cyanide Waste," *Electroplating Metal Fin.* 25 (1974): 20–21.
34. Dvorin, R. "Water and Waste Treatment—A Review of Methods for the Metal Finishing Industry," *Metal Finishing Guidebook-Directory*, 28th edition (1960).
35. Crowle, V.A. "Effluent Treatment Materials Recovery from the Metal Finishing Industry Using the Integrated Method of Treatment," *Water Poll. Control* (London, 1972): 636–645.
36. Zievers, J.F., R.W. Crain, and F.G. Barclay. "Waste Treatment in Metal Finishing: U.S. and European Practice," *Plating* 55 (1968): 1171–1179.
37. Besselievre, E.B. *The Treatment of Industrial Wastes* (New York: McGraw-Hill, 1969).
38. "Environmental Pollution Control Alternatives: Economics of Wastewater Treatment Alternatives for the Electroplating Industry," USEPA 625/5-79-016 (June 1979).

39. Fisco, B. "Plating and Industrial Waste Treatment at the Expanded Plant of Fisher Body, Elyria, OH," *Proceedings of the 25th Industrial Waste Conference, Purdue University* (May 1970).
40. Schink, C.A. "Plating Wastes: A Simplified Approach to Treatment," *Plating* 55 (1968): 1302–1305.
41. Green, J. and D.H. Smith. "Processes for the Detoxification of Waste Cyanides," *Metal Fin. J.* 18 (1972): 229–232.
42. O'Connor, S.F., B.W. Mountjoy, Jr., and N.S. Chamberlin. "Western Electric Builds Modern Plant for Treating Metal Finishing Wastes," *Water Waste Eng./Ind.* 9 (1972): D16–D19.
43. Beevers, M. "Chlorine and Sulfur Dioxide in the Treatment of Cyanide and Chromium Wastes," *Metal Fin. J.* 18 (1972): 232–235.
44. Yagishita, A. "Removal of Ferri- and Ferrocyanides from Industrial Wastes," *Chem. Abstr.* 78 (1973): 163794m.
45. Curry, N.A. "Philosophy and Methodology of Metallic Waste Treatment," *Proceedings of the 27th Industrial Waste Conference, Purdue University* (May 1972).
46. "An Investigation of Techniques for Removal of Cyanide from Electroplating Wastes," USEPA 12010 EIE 11/71 (1971).
47. Kandzas, P.F. and A.A. Mokina. "Use of Ozone for Purifying Industrial Waste Waters," *Chem. Abstr.* 71 (1969): 237.
48. Skovronek, H.S. and M.K. Stinson. "Advanced Treatment Approaches for Metal Finishing Wastewaters, Part I and Part II," *Plating Surface Fin.* (October 1977): 30–38; (November 1977): 24–31.
49. Tanji, H., T. Saski, Y. Horita, N. Tarasaki, and S. Fuchigami. "Removal of Cyanides from Waste Water by Electrolysis," *Japan Kokai* 75 (III): 117, 253; *Chem. Abstr.* 84 (1976): 21865k.
50. "New Process Detoxifies Cyanide Wastes," *Environ. Sci. Technol.* 5 (1971): 496–497.
51. Pearce, A.S. and S.E. Punt. "Biological Treatment of Liquid Toxic Wastes, Conclusion," *Effluent Water Treatment J.* 15 (1975): 87–95.
52. Kibbel, W.H. Jr., C.W. Raleigh, and J.A. Shepherd. "Hydrogen Peroxide for Industrial Pollution Control," paper presented at the 27th Annual Purdue Industrial Waste Conference, West Lafayette, IN (May 1972).
53. Culotta, J.M. and W.F. Swanton. "Recovery of Plating Wastes: Selection of Lowest Cost Evaporator," *Plating* 57 (1970): 1221–1223.
54. Gallo, B.R. and J.M. Culotta. "Save on Plating Waste System," *Water Wastes Eng.* 9 (1972): A18–A19.
55. Zievers, J.F. and C.N. Novotny. "Recovery of Mixed Rinse Waters by Means of Ion Exchange," *Plating* 58 (1971): 482–485.
56. Moore, F.L. "An Improved Ion Exchange Resin Method for Removal and Recovery of Zinc Cyanide and Cyanide from Electroplating Wastes," *J. Environ. Sci. Health* 411 (1976): 459–467.
57. Golomb, A. "Application of Reverse Osmosis to Electroplating Waste Water Treatment, Part II: The Potential Role of Reverse Osmosis in the Treatment of Some Plating Wastes," *Plating* 59 (1972): 316–319.
58. "Ultrathin Membranes for Treating Metal Finishing Effluents by Reverse Osmosis," USEPA 12010 DRH 11/71 (1971).
59. Jakobsen, K. and R. Laska. "Advanced Treatment Methods for Electroplating Wastes," *Poll. Eng.* (October 1977): 42–46.

60. Laughlin, R., H. Roboy, and P. Gooderham, "Batch Hydrolysis System for the Destruction of Cyanides in Electroplating Effluents," *Proceedings of the 4th Conference on Advanced Pollution Control for the Metal Finishing Industry*, USEPA Publication EPA-600/9-82-022 (December 1982).
61. Phasey, N.W. "Experiences with Ion Exchange Resins for Effluent Treatment," *Prod. Fin.* 25 (1972): 28–34.
62. Saito, H., T. Hashino, and N. Nishida. "Treatment of Wastewater Containing Ferricyanide Ions," *Japan Kokai* 74 (11): 790 (February 1, 1974); *Chem. Abstr.* 81 (1974): 5072k.
63. Spatz, D.D. "Electroplating Waste Water Processing with Reverse Osmosis," *Prod. Fin.* 36 (1972): 70–98.
64. Klein, E.W. and J. DelPico. "RO Comes of Age for Plating Pollution Control," *Prod. Fin.* 38 (1974): 74–78.
65. Cheremisinoff, P.H. and Y.H. Habib. "Cyanide—An Assessment of Alternatives for Water Pollution Control," *Water Sew. Works* (1973): R95–R107.
66. Coulter, K.R. "Pollution Control and the Plating Industry," *Plating* 57 (1970): 1197–1202.
67. Yost, K.J. and D.R. Masarik. "A Study of Chemical-Destruct Waste Treatment Systems in the Electroplating Industry," *Plating Surface Fin.* (January 1977): 35–40.
68. Lavin, D. and C. Lavin. "A Practical Approach to Water Pollution Control for the Plating Company," paper presented at the 27th Annual Purdue Industrial Waste Conference, West Lafayette, IN (May 1972).

9
Fluoride

Industries which discharge significant quantities of fluorides in process waste streams include fluorosilicone and fluorocarbon polymer synthesis, coke manufacture, glass and ceramic manufacture, electronics manufacture, electroplating operations, steel and aluminum manufacture, metal etching (with hydrofluoric acid), wood preservatives, and pesticide and fertilizer manufacture. Fluoride discharged from fertilizer-manufacturing processes is typically in the form of silica tetrafluoride (SiF_4), as a result of processing of phosphate rock [1, 2]. Power-plant boiler-cleaning wastes may contain fluoride, due to its presence in the cleaning formulation.

Fluoride wastes originate in the steel-manufacturing industry from the sintering and steel-making processes. In the sintering plant, fluoride occurs as an impurity of limestone, plus dust and fines returned from the basic oxygen furnace (BOF). Both fluorspar (CaF_2) and limestone are among the basic materials used in steel making. Air pollution scrubber water associated with steel manufacture in the basic oxygen, open hearth, and electric arc furnaces, as well as in the sintering plant, is the principal source for this industry of fluoride-containing wastewater [3].

The primary aluminum processing industry utilizes the fluoride compound cryolite (Na_3AlF_6) as a catalyst in bauxite ore reduction. Previously, gaseous fluorides resulting from this process were discharged directly into the atmosphere. Wet scrubbing of the process fumes, an air pollution abatement procedure, now results in transfer of the fluorides to aqueous waste streams. Average fluoride values for aluminum reduction plants are reported as 107–145 mg/l in wastewater streams [4]. Concentrations an order of magnitude greater have been reported for glass manufacturing, ranging from 1,000 to 3,000 mg/l of fluoride [1].

Glass and plating wastes typically contain fluoride in the form of hydrogen fluoride (HF) or fluoride ion (F^-), depending upon the pH of the waste. Hydrofluoric acid is used for acid polishing of television picture-tube screens and funnels, to provide smooth edges when the two pieces are sealed together. Hydrofluoric acid is also used to frost incandescent lamp bulb envelopes, and to acid-polish various pressed and blown glass products. Fluoride wastes originate from fume control, rinsewaters, and periodic dumping of the concentrated acid [5].

Table 9.1 Reported Fluoride Levels in Industrial Wastewaters

Source	*Fluoride Concentration (mg/l)*		*Reference*
	Range	*Avg*	
Computer circuits	—	57.8	7
Printed circuit boards	—	47.5	8
Aluminum ore smelting	10.2–1,400	147.7	9
Coke plant ammonia			
Recovery still	10–100	—	10
Steel manufacture			3
Sintering plants	—	8.5	
Blast furnace	0.49–23.0	14.0	
Basic oxygen furnace	3.75–14.5	9.1	
Open hearth furnace	65–148	106.5	
Electric arc furnace	0–20.7	8.2	
Aluminum production (gas scrubber waste)	—	<1,000	11
Phosphate ore			
Furnace slag quench	73–270	—	12
Phosphoric acid production	30–150	—	13
	4,000–12,000	—	14
	—	11,100	14
	—	1,460	14
Phosphate fertilizer plant waste	—	308	15
	—	1,050	16
Hydrogen fluoride manufacture	—	13.0	17
	—	193	18
Glass manufacture			5
TV picture tube	—	143	
Incandescent bulb frost	—	2,800	
Pressed and blown glass	194–1,980	—	
Fluoborate plating bath	—	134	6
Titanium descaling bath	60,000–99,500	—	19
Aluminum deoxidizer bath	—	2,250	19
Steel alloy descaling bath	16,000–39,600	—	19
Acidic coal-cleaning waste	—	81	20

In the electroplating industry, fluoborate baths are used for several metals, including lead, tin, and their alloys. Upon dilution in rinsewater, the fluoborate ion (BF_4^-) hydrolyzes to the very stable boron trifluoride (BF_3) and the fluoride ion. Treatment is thus directed toward removal of the fluoride resulting from hydrolysis upon dilution of fluoborate [6]. Table 9.1 presents data on fluoride levels in industrial wastewaters.

TREATMENT TECHNOLOGY

A variety of treatment methods are available for fluoride-bearing waste streams. A summary of the more important of these treatment processes, and their removal efficiencies, is presented in Table 9.2. Current treatment methods can be divided into two categories: precipitation methods and adsorption methods.

Precipitation methods involve addition of treatment chemicals, and formation of fluoride precipitates or simultaneous coprecipitation of fluoride with a resulting precipitate. Coprecipitation incorporates sorptive aspects. Removal is accomplished by solids separation of the precipitate. Treatment efficiency is therefore in part dependent upon the effectiveness of liquid-solid separation. Chemicals employed include lime, magnesium compounds (e.g., dolomite), and aluminum sulfate (alum).

Lime

Direct addition of lime is the standard technique for precipitation of high concentrations of fluoride ion. The lime reacts with fluoride in the wastewater to produce calcium fluoride. Calcium fluoride has a theoretical maximum solubility of approximately 8 mg/l fluoride at stoichiometric concentrations of calcium. Thus, concentrations of calcium fluoride above this solubility limit will form a precipitate. Reported fluoride removals down to residuals of only 10–20 mg/l reflect the slow rates of precipitate formation [1, 18, 21]. However, 24-hour contact with lime has been reported to reduce fluoride to near the theoretical limit of 8 mg/l [1]. An additional difficulty with lime precipitation is the poor settling characteristics of the precipitate [14, 18]. Laboratory treatability studies at an electronic equipment-manufacturing plant resulted in improved residual fluoride levels (below 10 mg/l) when lime treatment was followed by the addition of a coagulant and coagulant aid [35].

At very high lime dosages, improved fluoride removal has been reported (Figure 9.1; see Table 9.2). Successful lime treatment may require addition of 4–8 g lime/l wastewater, resulting in an effluent pH above 12 [23, 37]. Lime dosing at the 4 g/l level has been reported to reduce fluoride from 200 to 3 mg/l, in 19 hr reaction time [23]. Two-stage lime treatment of a phosphoric acid waste by application of 6.8 g/l of lime (equivalent to 4.7 mg lime per mg fluoride) reduces an initial 1,460 mg/l fluoride level to 9 mg/l [14].

Recent laboratory studies on synthetic and industrial wastes indicate that both pH and calcium concentration may influence the lime treatment process (Figure 9.2). Minimum soluble fluoride levels were observed at pH 8 and above pH 12. Less effective treatment resulted at intermediate pH, however [38]. A similar pattern has been reported for lime treatment of fluoride in phosphoric acid wastewaters, although the curve can shift with respect to both pH and residual fluoride level [14].

Table 9.2 Summary of Fluoride Treatment Processes and Levels of Treatment Achieved

Treatment Process	*Fluoride Concentration (mg/l)*		*Current Application*	*Reference*
	Initial	*Final*		
Lime	—	10	Industrial	21
	1,000–3,000	20		1
	500–1,000	20–40		22
	200–700	6 (16-hr settling)		23
	45	8		24
	4–20	5.9 (avg)		25
	590	80		25
	57.8	29.1 (avg) 14–16 (best)		7
	93,000	0.8–8.8	Industrial (pilot scale)	19
	—	10.6 (Clarified) 10.4 (Filtered)	Industrial	26
Lime, two-stage	1,460	9		14
Lime + calcium chloride	—	12		27
Lime + alum	—	1.5		28
Lime + alum	2,020	2.4	Industrial (pilot scale)	19
Calcium carbonate + lime, two-stage	11,100	6	Industrial	14
Alum	3.6	0.6–1.5	Municipal	29
Alum	60	2	Lab scale	27
Hydroxylapatite beds				
Synthetic	12–13	0.5–0.7	Municipal	30
Synthetic	10	1.6		31
Bone char	6.5	1.5		32
Bone char	9–12	0.6		33
Alumina contact bed	4.5–7.5	0.1–1	Municipal	34
	8	1	Municipal	32
	9	1.3	Industrial (lab scale)	1
	20–40	2–3	Industrial (pilot scale)	22

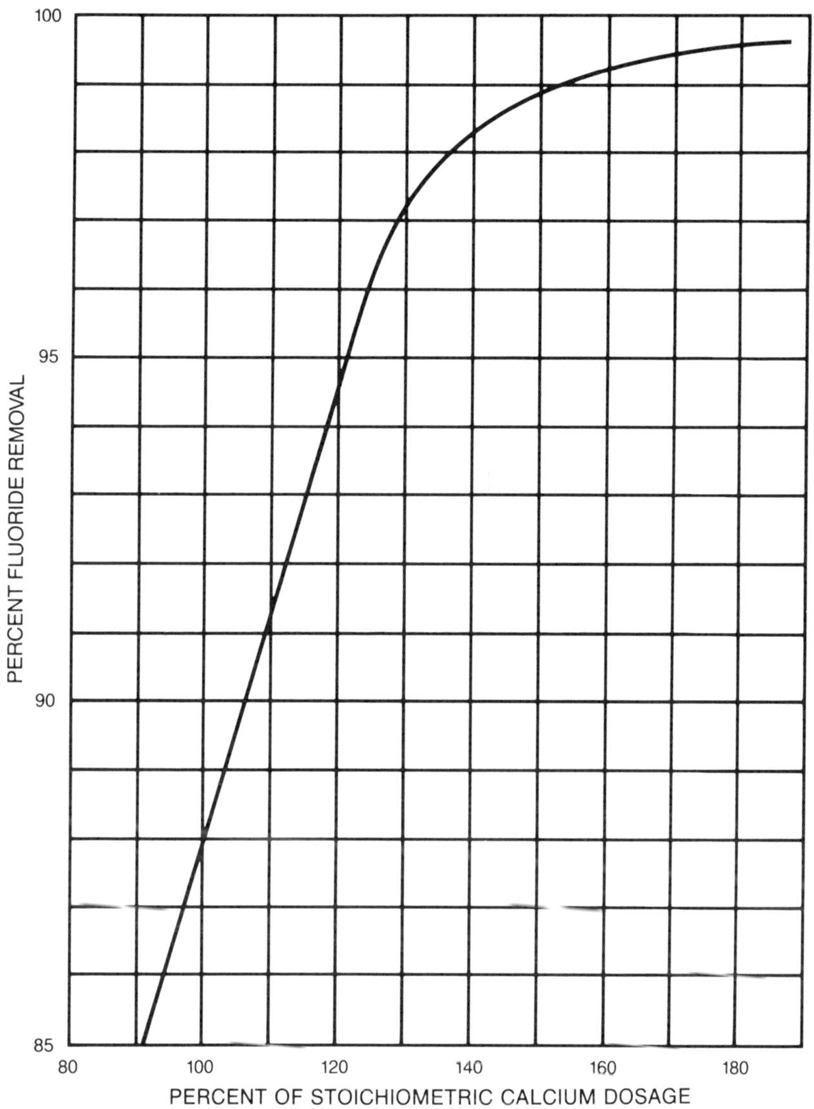

Figure 9.1 Calcium requirement for fluoride removal. (From Arnold and Wolfram [36], courtesy Purdue University Industrial Waste Conference.)

Extensive lime precipitation treatability studies performed on a hydrofluoric acid-manufacturing waste, all at pH near 10.0, confirm the poor performance indicated for lime treatment at that pH indicated by Figure 9.2. An initial raw wastewater fluoride level of about 200 mg/l was reduced on average to 110 mg/l by lime precipitation plus clarification and to 90 mg/l by effluent filtration [18].

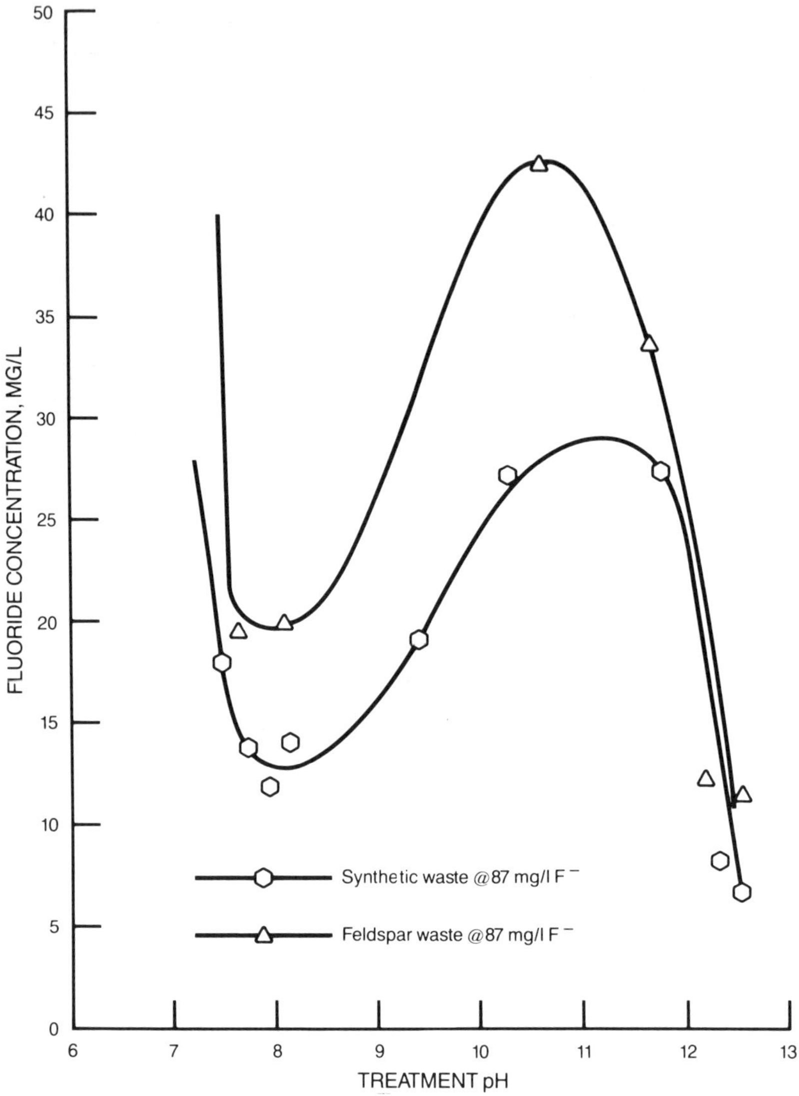

Figure 9.2 Effect of pH on residual fluoride after lime treatment. (After Parker and Fong [38], courtesy of *Industrial Wastes*.)

Mooney et al [14] report that if lime precipitation is performed in two stages, with intermediate clarification, the maxima shown in Figure 9.2 at intermediate pH is not observed. First-stage treatment would be performed at a pH near the first minima (about pH 5.0 for their wastewater), followed by second-stage treatment at higher pH. They reported results for one plant, operating at a second stage pH of 9.3, which reduced fluoride from 1,460 to 9

Table 9.3 Performance of Two-Stage Precipitation Treatment at Two Plants [14]

Parameter	*Plant A* *Lime + $CaCO_3$*	*Plant B* *Lime Only*
pH		
Initial	1.5	2.6
First stage	5.5	5.5–6.0
Second stage	9.4	9.3
Fluoride		
Initial	11,100	1,460
Effluent	6	9
Chemical Demand		
$CaCO_3$, g/l	10	0
mg/mgF	0.9	0
CaO, g/l	24	6.8
mg/mgF	2.2	4.7

mg/l. Higher second-stage operating pH would achieve even lower effluent fluoride levels. This experience suggests that in two-stage lime precipitation different fluoride salts are precipitated at each stage, with the first precipitating at the first pH minima (see Figure 9.2), then resolubilizing before the second salt begins to precipitate.

In view of the effectiveness of the two-stage process, Mooney et al [14] recommend the use of calcium carbonate plus lime in the first stage, to reduce overall lime demand. Bulk calcium carbonate is about one-third the cost of lime. Full-scale results for two phosphoric acid plants presented in Table 9.3 demonstrate that the lime demand per mg fluoride can be reduced by more than 50% through the application of calcium carbonate in the first stage. Some lime may also be required in the first stage, to achieve the desired first-stage treatment pH.

A mixture of calcium chloride and hydrated lime has been used successfully at one plant, to reduce total fluoride levels to approximately 12 mg/l after neutralization, clarification, and filtration, at pH 7–8 [27]. The addition of calcium chloride increments the soluble calcium concentration, without concurrent increase in pH.

At one plant, coprecipitation effects may have contributed to reduction of fluoride [23]. A television tube-manufacturing plant combined lead precipitation with fluoride precipitation in a sequential treatment step, followed by combined sedimentation. Lead wastes were treated with trisodium phosphate, followed by introduction of the fluoride waste stream, lime, and polyelectrolyte addition. Sixteen hours of sedimentation produced a 6 mg/l fluoride effluent. Addition of excess lime further reduced fluoride concentration, at the expense of additional treatment chemical costs, and increased pH and calcium concentration in the waste effluent. Recarbonation of the effluent provided a means of excess calcium removal and pH adjustment. An alternative explanation of the improved treatment observed in this instance may be the presence of phosphate, added in the lead treatment step. In two separate studies, the addition of sodium hexametaphosphate with alum has been reported to enhance fluoride treatment above that observed for alum alone [27, 39]. In one application of alum plus sodium hexametaphosphate (4:1 dosage ratio), an initial fluoride level of 20 mg/l was reduced to below 2 mg/l at pH 7 [27].

Lime-treated effluent fluoride levels in the plating industry have been reported for four plants as ranging from 2.2 to 65 mg/l, with an average of 34.3 mg/l. The lowest reported fluoride level of 2.2 mg/l resulted upon lime treatment plus sand filtration [6].

High-lime treatment of concentrated fluoride wastewaters from hydrofluoric acid frosting of lamp bulbs has resulted in a high effluent level of 29 mg/l fluoride. The waste is high in ammonia (650 mg/l), and it has been postulated that the ammonia or some other waste constituent may interfere with the fluoride precipitation reaction [5, 35]. Lime precipitation of television-tube wastewaters has yielded effluent fluoride of 15 or less mg/l [5].

One very concentrated fluoride waste (40,000 mg/l fluoride in 30% sulfuric acid) was treated by pumping the 240 gal batch of waste into a 20% lime slurry and then discharging the resultant 15% solids sludge to a landfill [40]. This operation was conducted 3–5 times/week and employed 400 g lime/1 of waste treated.

In another instance of lime treatment of concentrated (99,500 mg/l) fluoride wastes, lime slurry added to the wastewater on a 1:1 volume basis at a lime concentration of 178 g/l yielded effluent fluoride levels of 6.2–6.8 mg/l. Treatment was at pH 11.7–12.2 [19]. In these pilot studies, extended reaction times to 24 hours enhanced treatment efficiency. The resultant treated slurry required centrifugation for solids separation, producing about 1–2 gal dewatered sludge/5 gal wastewater centrifuged [19].

Treatment of individual fluoride wastewaters prior to mixing of waste streams can greatly enhance fluoride discharge levels. Two South African steel mills with total waste flows of 1 and 2.5 MGD have reported effluent levels of 2–2.5 and 1.0 mg/l fluoride respectively [41]. Fluoride, originating from electrolytic tinning units, is first precipitated with lime, and the resulting effluent is then diluted upon combination with the total plant effluent.

Treatment costs specifically for fluoride removal are frequently difficult to isolate, as fluoride is often not the only wastewater contaminant being removed and individual unit costs are normally not reported. The lime-treatment process for fluoride waste is composed of standard unit operations of the sort employed in precipitation treatment. These units include a rapid-mix basin for chemical addition, a flocculating unit, and a settling basin for solids separation. General cost curves are presented in Figures 9.3 and 9.4. Necessity for improved precipitate solids removal may require effluent filtration.

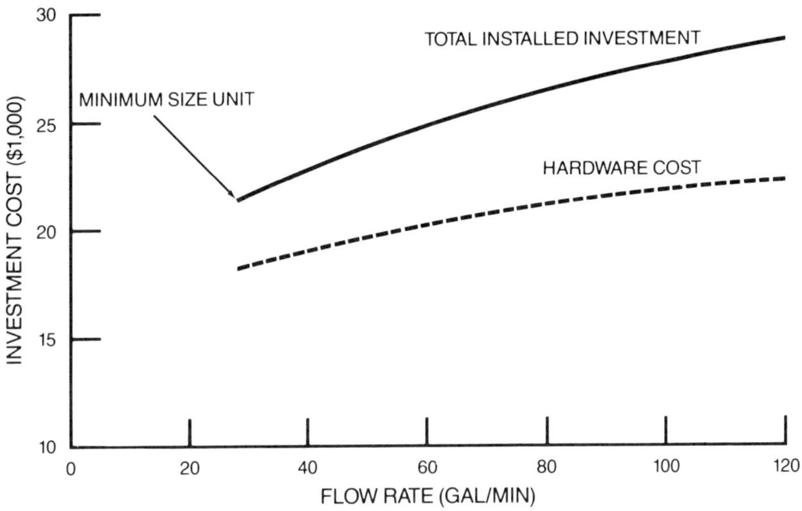

Figure 9.3 Capital cost (1978 basis) of a continuous single-stage neutralization/precipitation system [42].

Magnesium

Early work in the treatment of drinking waters containing excessive fluoride demonstrated that fluoride ion was removed during the standard lime-soda ash softening process. Removal in both pilot and full-scale systems was determined to be a function of initial fluoride level and proportional to the square root of

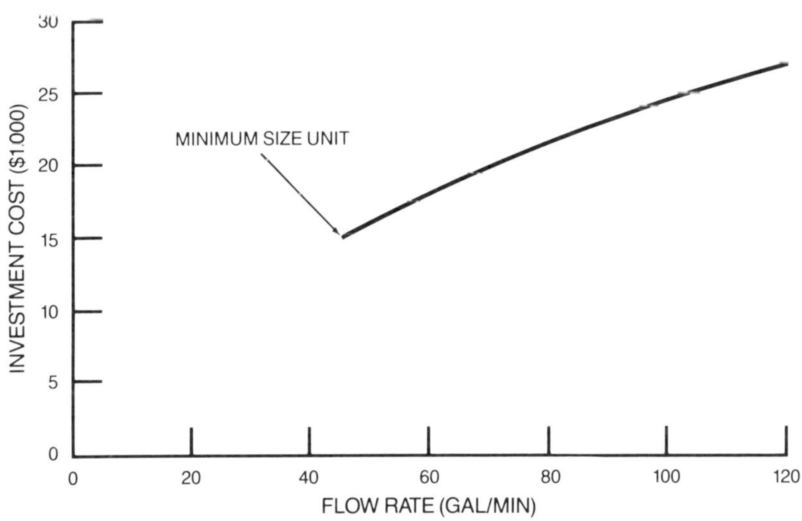

Figure 9.4 Capital cost (1978 basis) of flocculation/clarification system [42].

the amount of magnesium hardness removed; it was attributed to adsorption of fluoride ion onto the magnesium hydroxide floc formed in the softening process. Fluoride concentrations of 3–4 mg/l were reported reduced to 0.8–1.1 mg/l [29]. However, in the absence of appreciable magnesium occurring naturally in the water or wastewater to be treated, the requirement for large supplementary additions of magnesium may make this process uneconomical. In addition, a recent study has revealed that in the absence of calcium hardness, the precipitate has poor settling and sludge-handling qualities, although the calcium has little effect on the fluoride removal efficiency [43]. No fluoride removal plant has been specifically designed to use this process. Fluoride reduction is, however, accomplished in conjunction with magnesium hardness removal at several municipal water-treatment plants in the United States where fluoride is present in the raw water at concentrations of 2–3 mg/l [44].

Aluminum Sulfate (Alum)

Fluoride removal by alum addition is essentially a coprecipitation phenomenon whereby fluoride is removed with aluminum hydroxide precipitate. Best removal appears to occur at pH 6–7, with initial fluoride of about 100 mg/l reported reduced first to 5–5.2 mg/l by lime precipitation (pH 12.9) and to 1.7 mg/l by alum coagulation at pH 5.9–7 [37]. For a water of low hardness containing 3.6 mg/l fluoride ion, simple alum addition plus small quantities of lime for pH control resulted in fluoride reduction to 1 mg/l [29]. For hard waters, other methods of removal were more effective.

In one recent study, final fluoride level was found to be independent of initial fluoride concentration up to an initial fluoride level of at least 100 mg/l, and was dependent only upon alum dosage and treatment pH [27]. Best treatment was reported to occur at pH 7, with fluoride reduced to 4.8 mg/l at an alum dosage of 40 mg/mg fluoride. At alum dosage above 200 mg/mg fluoride, effluent fluoride levels below 2 mg/l were achieved. Even at moderate alum dosages, however, sludge volumes near 40% of the original wastewater volume treated resulted [27].

A second recent study, at low initial fluoride levels of about 5 mg/l, confirms that the two parameters discussed above are significant in determining fluoride removal efficiency [43]. These parameters are ratio of alum dosage to initial fluoride level (Figure 9.5) and treatment pH (optimum results at pH 6.2–6.4). Calcium concentration also has some influence, although slight, as compared to the other two (Table 9.4).

Many industrial wastewaters originating from hydrofluoric acid etching of aluminum parts already contain significant background aluminum concentration. For these wastewaters, simple neutralization to precipitate aluminum effectively constitutes the type of treatment described herein.

Two-stage lime plus alum addition has been pilot tested on one concentrated aluminum conversion coating spent bath. Lime, at 255 g/l wastewater,

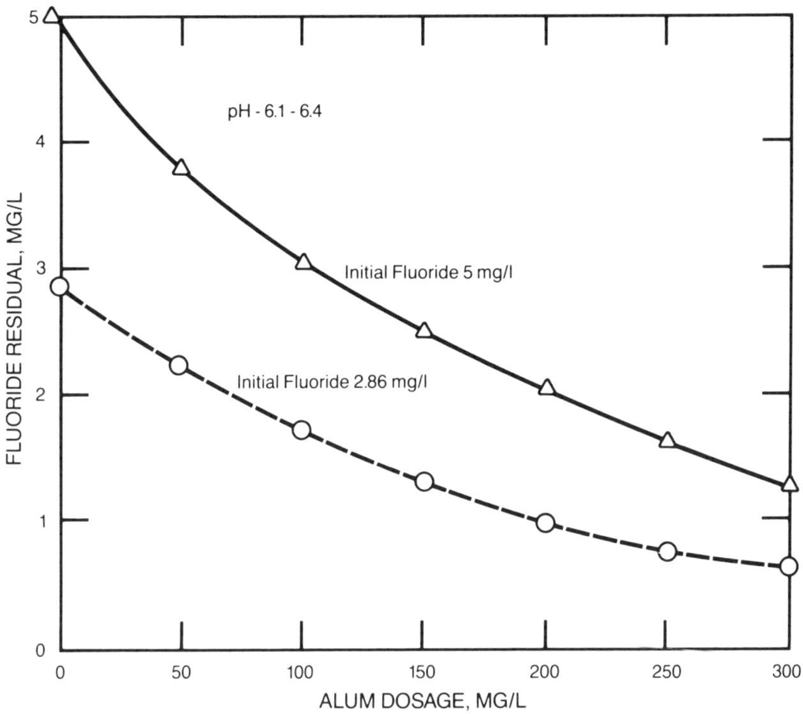

Figure 9.5 Fluoride residuals obtained after alum coagulation for initial fluoride concentrations of 2.86 mg/l and 5.0 mg/l [43].

Table 9.4 Effect of Alum Dosage and Calcium Concentration on Fluoride Removal [43]

Treatment pH	*Alum Dosage (mg/l)*	*Initial Fluoride (mg/l)*	*Alum: Fluoride Ratio (mg/mg)*	*Calcium Added (mg/l)*	*Residual Fluoride (mg/l)*	*% Fluoride Removal*
6.4	100	5.0	20.0	0	3.3	35
				50	3.2	37
				200	3.1	39
6.3	200	4.7	42.6	0	2.1	56
				50	2.0	58
				200	1.9	59
6.2	300	4.8	62.5	0	1.3	73
				50	1.3	74
				200	1.2	75

was added to a waste initially containing fluoride at 2,020 mg/l [19]. Lime treatment reduced the fluoride level to 16 mg/l at 0.5 hr reaction time, and 10.5 mg/l at 21 hr reaction time. This effluent was then treated with alum at 5.7 g alum/l wastewater, yielding a final effluent of 2.4 mg/l fluoride [19].

Fluoride removal in conjunction with arsenic treatment of a rare-earth industrial waste was reported to produce an effluent fluoride concentration of 1.5 mg/l [28]. Initial lime precipitation was followed by further addition of lime plus superphosphate and aluminum sulfate. Laboratory treatability studies on alum treatment of a fluorspar-process wastewater initially containing 8 mg/l fluoride resulted in effluent fluoride levels of 6.4, 4.3, and 1.8 mg/l for respective alum dosages of 100, 300, and 600 mg/l [38]. For the same wastewater, activated alumina adsorption reduced fluoride to below 1 mg/l.

Adsorption

Treatment involves the passage of the wastewater through a contact bed, with fluoride being removed by general or specific ion exchange or chemical reaction with the solid bed matrix. Although solids removal is not required for these operations, bed regeneration and subsequent treatment of the concentrated regenerant are an integral part of the overall treatment process.

Since the basic mechanism of a contact process is one of ion exchange or surface reaction, these methods are usually appropriate only for low-level fluoride wastes, or polishing processes after fluoride reduction by previous treatment to the 10–20 mg/l level. Otherwise, the requirement for frequent bed regeneration makes the process economically unfeasible. Bed media that have been used are hydroxyapatite (synthetic, processed bone, and bone char), ion-exchange resins, and activated alumina.

Hydroxyapatite

Reduction of fluoride from 12–13 mg/l to 0.5–0.7 mg/l was reported, upon passage through a full-scale contact bed of synthetic hydroxyapatite [30]. Later reports [44] indicate that this unit and others employing the same process were ultimately abandoned due to high chemical costs and bed attrition (up to 42% loss/yr). For a similar pilot plant-scale operation in South Africa, 10 mg/l fluoride could be reduced to 1.5 mg/l [31]. The presence of chlorides in the water reduced filter-bed capacity and increased regeneration chemical requirements.

Natural hydroxyapatite in the form of bone char was reported capable of reducing 10 mg/l fluoride to 0.6 mg/l, but regeneration with phosphoric acid and sodium hydroxide resulted in excessive chemical costs [33].

Ion Exchange

Natural and synthetic zeolite ion exchangers have been examined and reported to be effective for removal of fluoride. The exchangers were both pretreated and regenerated with aluminum salts. Further investigation indicated that fluoride removal was attributable to aluminum oxide, which precipitated in the column bed [32, 45].

In a separate pilot study, strong-base anion exchange resin, regenerated on the hydroxide cycle with caustic, was reported effective in treatment of sodium fluoride solutions [19]. A two-stage system of ion exchange, followed by an activated-alumina column for effluent polishing, was recommended.

Alumina

Contact beds and columns of activated alumina have been used for many years in municipal water treatment plants for removal of fluoride ion. Regenerating the activated alumina can be accomplished with caustic, sulfuric acid, hydrochloric acid, or alum. In one industrial application, sulfuric acid regenerant was recommended [19]. Caustic regeneration is practiced at most water treatment plants, with water or dilute acid being used to remove residual caustic from the bed following regeneration. One unit, in Bartlett, Texas, has operated successfully since 1952 using a 500 ft^3 bed regenerated by sodium hydroxide and neutralized with sulfuric acid. Fluoride is reduced from 8 mg/l to 1 mg/l [32]. Water is treated at the rate of 400 gpm. Bellock [46] reported operating costs for fluoride and arsenic removal with activated alumina columns to range from $0.015 to 0.05/1,000 gal. A 1977–1978 equipment and construction cost of $270/1,000 gpd treatment capacity was reported from the Gila Bend, Arizona activated-alumina fluoride control facility [34].

Wu and Nitya [47] have evaluated the fluoride adsorption of activated alumina as a function of alumina:fluoride ratio and treatment pH. Adsorption results were described by the Langmuir isotherm, which indicated the maximum adsorptive capacity of the alumina (Alcoa Type F-1) for fluoride to be 12 mg/g. Maximum adsorption rate occurs at pH 5, with reduced rate at higher or lower pH. Rubel and Woosley [34], however, recommend an operating pH of 5.5, to reduce interference by competing ions. They report operating costs of $0.11–0.15/1,000 gal, for three full-scale municipal water supply plants treating fluoride at 4.5–7.5 mg/l.

Sorg [48] has compared the capital (Table 9.5) and operating (Table 9.6) costs of ion exchange versus activated alumina treatment for fluoride removal from drinking water supplies. Activated alumina involves lower capital investment, but higher operating cost except for treatment capacities of 500,000 gpd or higher.

Use of an alumina bed as a polishing unit to follow lime precipitation of high-fluoride (1,000–3,000 mg/l) wastes resulted in 30 mg/l residual fluoride carryover from the precipitation process being reduced to approximately 2 mg/l upon passage through the contact bed [1]. At a pH of 11.0–11.5, a concentration of 9 mg/l fluoride was reduced to 1.3 mg/l. In regenerating, only 4% of the alumina was lost for every 100 regeneration cycles.

Table 9.5 Capital Cost (in 1980) of Fluoride Removal for Drinking-Water Supplies [48]

	Capital, \$/1,000 gpd Capacity	
Capacity (gpd)	*Ion Exchange*	*Activated Alumina*
50,000	1,210	1,000
100,000	680	580
250,000	340	288
500,000	212	184
1,000,000	150	130

One laboratory study of activated alumina has produced evidence that alkalinity in the influent competes successfully with fluoride for the alumina exchange sites, resulting in over 60% of the exchange (on an equivalence basis) attributed to alkalinity removal [43]. These data indicate that high background alkalinity may, through competitive exchange, greatly reduce the fluoride removal capacity of an alumina bed. Fluoride exchange capacities of about 4,600 g/m^3 of activated alumina have been reported [34, 42].

Other Processes

A proprietary package electrochemical process has been developed which is claimed to reduce fluoride to 1.5 mg/l or less [49]. The system operates as continuous flow and utilizes consumable electrodes. No pH adjustment or chemical additions are required, and economics are reported comparable to precipitation and adsorption treatment methods.

Both use of iron salts as coagulants, and activated carbon adsorption have been evaluated at bench scale for fluoride removal. Both processes are pH dependent (iron, optimum pH near 6; carbon, optimum pH near 3), but neither was capable of fluoride removals in excess of 20% [43].

Table 9.6 Operating Cost (in 1980) of Fluoride Removal for Drinking-Water Supplies [48]

	Operating Cost, \$/1,000 gal	
Capacity (gpd)	*Ion Exchange*	*Activated Alumina*
50,000	0.63	0.71
100,000	0.33	0.41
250,000	0.17	0.18
500,000	0.11	0.10
1,000,000	0.07	0.06

Table 9.7 Fluoride Pollution Abatement Costs (in 1974) for the Glass Industry [5]

Process	*Flow (gpm)*	*Treatment Level*[a]	*Influent Fluoride (mg/l)*	*Effluent Fluoride (mg/l)*	*Capital ($/1,000 gpd)*	*Operating*[b] *($/1,000 gal)*
Television tube[c] envelope manufacture	525	A	335	15	0	0
		B		13	132.94	0.06
		C		2	1,111.11	0.45
Pressed and blown glass manufacture	10	A	442	20	30,537.63	6.52
		B		13	34,946.24	7.63
		C		2	44,086.02	8.75

[a] A = Lime precipitation; B = A + sand filtration; C = B + activated alumina polishing.
[b] Based upon 16 hr/day, 300 day/yr operation.
[c] Treatment level A currently in place for example plant, treatment levels B and C costs are increments above current capital investment and operating costs.

SUMMARY

Industrial wastes containing high fluoride levels often require two-stage treatment. Lime precipitation removes fluorides to 10mg/l or less. Further reductions can be accomplished to the 1 mg/l level with other existing techniques. The preferred method seems to be activated alumina. The units involved are essentially those used for ion-exchange water softening, and the regenerant can be treated in the high-lime removal unit by recycling. Although treatment levels obtainable by the various fluoride removal techniques are well defined in the literature, there is much less information available on costs. Published data, however, indicate that industrial operating costs are generally in the range of \$0.10–0.50/1,000 gal. Much higher costs may result for treatment of low-volume, high-strength wastes, however (Table 9.7).

REFERENCES

1. Zabban, W. and H.W. Jewett. "The Treatment of Fluoride Wastes," *Proceedings of the 22nd Industrial Waste Conference, Purdue University* (May 1967), pp 706–716.
2. Cherry, J.M. "Fluorine Recovery in Phosphate Manufacture," *Water Wastes Eng.* 7 (1970): D5.
3. "Development Document for Effluent Limitations Guidelines and New Source Performance Standards for the Steel Making Segment of the Iron and Steel Manufacturing Point Source Category," USEPA 440/1-74-024-a (June 1974).
4. Sylvester, R.O., R.T. Oglesby, D.A. Carlson, and R.F. Christman. "Factors Involved in the Location and Operation of an Aluminum Reduction Plant," *Proceedings of the 22nd Industrial Waste Conference, Purdue University* (May 1967), p 441.
5. "Development Document for Effluent Limitations Guidelines and New Source Performance Standards for the Pressed and Blown Glass Segment of the Glass Manufacturing Point Source Category," USEPA 440/1-75/034-a, Group I, Phase II (January 1975).
6. "Development Document for Interim Final Effluent Limitations Guidelines and Proposed New Source Performance Standards for the Common and Precious Metals Segment of the Electroplating Point Source Category," USEPA 440/1-75/040 (April 1975).
7. Hango, R.A., W.J. Jewell, and E.D. Driscoll. "Wastewater Treatment for Integrated Circuit Manufacturing," paper presented at the 28th Annual Purdue Industrial Waste Conference, West Lafayette, IN (May 1973).
8. Jashnani, I.L. and R. Green. "Pretreatment System for a Circuit Board Plant," paper presented at the 8th National Conference on Municipal Sludge Management, Municipal and Industrial Sludge Management, Disposal and Utilization, Miami Beach, FL (March 19–21, 1979).
9. "Development Document for Effluent Limitations Guidelines and New Source Performance Standards for the Primary Aluminum Smelting Subcategory of the

Aluminum Segment of the Nonferrous Metals Manufacturing Point Source Category," USEPA 440/1-74-019-d (March 1974).

10. Adams, C.E. Jr., R.M. Stein, and W.W. Eckenfelder, Jr. "Treatment of Two Coke Plant Wastewaters to Meet EPA Effluent Criteria," paper presented at the 29th Annual Purdue Industrial Waste Conference, West Lafayette, IN (May 1974).
11. "Development Document for Interim Final Effluent Limitations Guidelines and Proposed New Source Performance Standards for the Significant Inorganic Products Segment of the Inorganic Chemicals Manufacturing Point Source Category," USEPA 440/1-75/037 (May 1975).
12. "Development Document for Effluent Limitations Guidelines and New Source Performance Standards for the Phosphorus Derived Chemicals Segment of the Phosphate Manufacturing Point Source Category," USEPA 440/1-74-006-a (January 1974).
13. Khripunov, N.F., A.Y. Abramovich, and V.A. Koyylov. "Ways of Lowering the Amount of Fluorine-Containing Waste Waters in the Production of Wet-Process Phosphoric Acid," *Chem. Abstr.* 82 (1975): 248.
14. Mooney, G., A. Nogueira, and D. Cope. "Two-Stage Lime Treatment in Practice," *Environ. Prog.* 1 (1982): 274–280.
15. Yasutomi, H., N. Takeshima, T. Matsuno, and H. Momo. "Fluoride Removal from Waste Water," *Chem. Abstr.* 87 (1977): 341.
16. Etoh, Y. and T. Takadoi. "Treatment of Waste Water Containing Fluoride and Boron Compounds," *Chem. Abstr.* 87 (1977): 330.
17. "Development Document for Effluent Limitations Guidelines and New Source Performance Standards for the Major Inorganic Products Segment of the Inorganic Chemicals Manufacturing Point Source Category," USEPA 440/1-74-007-a (March 1974).
18. "Treatability Studies for the Inorganic Chemicals Manufacturing Point Source Category," USEPA 440/1-80/103 (July 1980).
19. Staebler, C.J., Jr. "Treatment and Recovery of Fluoride and Nitrate Industrial Wastes," *Proceedings of the National Conference on Management and Disposal of Residues from the Treatment of Industrial Wastewaters*, Washington, DC (February 3–5, 1979).
20. Wangen, L. and J. Williams. "Control of Alkaline Neutralization of Trace Elements in Acidic Coal Cleaning Waste Leachates," *J. Water Poll. Control. Fed.* 54 (1982): 1302–1310.
21. Schink, C.A. "Plating Wastes: A Simplified Approach to Treatment," *Plating* 55 (1968): 1302–1305.
22. Petrova, T.N. and U.G. Bakhurov. "Purification of Waste Solutions from Fluorine," *Atom. Eng.* 26 (1969): 552–553.
23. Rohrer, K.L. "An Integrated Facility for the Treatment of Lead and Fluoride Wastes," *Ind. Wastes* 17 (1971): 36–39.
24. "Pulling Effluents Into Line," *Chem. Eng.* 78 (1971): 40.
25. Teer, E.H. and L.V. Russell. "Heavy Metals Removal from Wood Preserving Wastewater," paper presented at the 27th Annual Purdue Industrial Waste Conference, West Lafayette, IN (May 1972).
26. Knapp, R. and E. Paulson. "Gravity Filtration Reduces Suspended Metals in a Lime Precipitation Treatment System," paper presented at the 37th Annual Purdue Industrial Waste Conference, West Lafayette, IN (May 1982).

27. Zabban, W. and R. Helwick. "Defluoridation of Wastewater," paper presented at the 30th Annual Purdue Industrial Waste Conference, West Lafayette, IN (May 1975).
28. Skripach, T., V. Kagan, M. Ramanov, L. Kamin, and A. Semina. "Removal of Fluorine and Arsenic from the Waste Water of the Rare-Earth Industry," *Proceedings of the 5th International Conference on Water Pollution Research* 1970, 2:Pap-No. III-34 (1971); *Water Poll. Abstr.* 564 (1971): 2770.
29. Culp, R.L. and H.A. Stoltenburg. "Fluoride Reduction at LaCrosse, Kansas," *J. Am. Water Works Assoc.* 50 (1958): 423–431.
30. Wamsley, R. and W.F. Jones. "Fluoride Removal," *Water Sew. Works* 94 (1947): 372–376.
31. Cillie, G.G., O.O. Hart, and G.J. Standy. "Defluoridation of Water Supplies," *J. Inst. Water Eng.* 12 (1958): 203–210.
32. Maier, F.J. "Defluoridation of Municipal Water Supplies," *J. Am. Water Works Assoc.* 45 (1953): 879–888.
33. "Treating a High Fluoride Water," *Public Works* 86 (1955): 67.
34. Rubel, F., Jr. and R.D. Woosley. "The Removal of Excess Fluoride from Drinking Water by Activated Alumina," *J. Am. Water Works Assoc.* (January 1979).
35. Fraust, C.L. "Modifying a Conventional Chemical Waste Treatment Plant to Handle Fluoride and Ammonia Wastes," paper presented at the American Electroplating Society 62nd Annual Meeting, Toronto (1975).
36. Arnold, D.W. and W.E. Wolfram. "Ammonia Removal and Recovery from Fertilizer Complex Wastewaters," paper presented at the 30th Annual Purdue Industrial Waste Conference, West Lafayette, IN (May 1975).
37. Rabosky, J.G. and J.P. Miller, "Fluoride Removal by Lime Precipitation and Alum and Polyelectrolyte Coagulation," *Proceedings of the 29th Industrial Waste Conference, Purdue University* (May 1974).
38. Parker, C.L. and C.C. Fong. "Fluoride Removal: Technology and Cost Estimates," *Ind. Wastes* 21 (1975): 23–27.
39. Miller, D.G. "Fluoride Precipitation in Metal Finishing Waste Effluent, Water—1974, I. Industrial Wastewater Treatment," AIChE Symposium Series 144, vol. 70 (1974).
40. Lewder, L.R. "Fluoride Waste Puzzle Solved," *Water Wastes Eng.* 8 (1971): B6.
41. Heynike, J.J.C. and F.V.K. von Reiche. "Water and Pollution Control in the Iron and Steel Industry with Special Reference to the South African Iron and Steel Industrial Corporation," *Water Poll. Control* (London, 1969): pp 569–673.
42. "Environmental Pollution Control Alternatives: Economics of Wastewater Treatment Alternatives for the Electroplating Industry," USEPA 625/5-79-016 (June 1979).
43. Sollo, F.W., T.E. Larson, and H.F. Mueller. "Fluoride Removal from Potable Water Supplies," Illinois Water Resources Center Report 136, Urbana (September 1978).
44. Savinelli, E.A. and A.P. Black. "Defluoridation of Water and Activated Alumina," *J. Am. Water Works Assoc.* 50 (1958): 33–44.
45. Maier, F.J. "Water Defluoridation at Britton: End of an Era," *Public Works* 102 (1971): 70.
46. Bellock, E. "Methods of Removing Fluoride from Water," *Am. J. Public Health* 37 (1947): 1559–1968.

47. Wu, Y.C. and A. Nitya. "Water Defluoridation with Activated Alumina," *ASCE J. Environ. Eng. Div.* (April 1979): EE2 357.
48. Sorg, T. "Compare Nitrate Removal Methods," *Water Waste Eng.* (December 1980): 26–31.
49. Duffey, J.G. "Andco Fluoride Removal System," Technical Bulletin, Andco Environmental Processes, Inc., Buffalo, NY.

10
Iron

Iron exists in the ferric or ferrous form, depending upon conditions of pH and dissolved oxygen concentration. At neutral pH and in the presence of oxygen, soluble ferrous iron (Fe^{+2}) rapidly oxidizes to ferric iron (Fe^{+3}), which readily hydrolyzes to form the insoluble precipitate, ferric hydroxide, $Fe(OH)_3$. At pH values above 12, ferric hydroxide will resolubilize. At pH below 6, the rate of ferrous iron oxidation to the ferric form is extremely slow. Therefore, acidic, or extreme basic, and/or anaerobic conditions are necessary in order for appreciable concentrations of soluble iron to exist.

Industrial sources of iron-containing wastewaters include mining operations, ore milling, chemical industries (organic, inorganic, petrochemical), dye and pigment formulators, metal-processing industries, textile mills, food canneries, tanneries, titanium dioxide production, petroleum refining, and fertilizers [1–3]. Acid coal-mine drainage is a major source of iron wastewater (Table 10.1). In some industrial wastes, ferrous and ferric iron may be solubilized in the presence of cyanide, due to the formation of stable ferro- and ferricyanide complexes. Such species present considerable difficulties for both iron and cyanide treatment. A color film-bleaching waste has been reported to contain ferro- plus ferricyanide at a total iron concentration of 71 mg/l [5]. Pigments known under the general name of "iron blues" are based upon ferrocyanide formulations [6].

Iron in some wastewaters may be principally in the particulate form. Ferric oxides are used as pigments in certain paints. In addition, iron oxide is used as a polishing powder for glass, metal, and ceramic materials [1]. Such operations may represent a source of particulate iron pollution, due to the colloidal nature of polishing compounds and the fact that water is usually involved as a coolant-lubricant and as a rinsing agent. In the processing of phosphoric acid, iron phosphate precipitates are formed and recovered as a sludge, representing a potential waste source in the absence of efficient solids separation [2].

Iron and steel production represents a major source of fine iron particles from the production furnaces, by virtue of air pollution control measures involving flue-gas scrubbing [7, 8]. Steel mill washwaters may contain 1,000–2,000 mg/l suspended solids, with the solids metal composition roughly reflecting the furnace charge. Of blast furnace suspended solids, 50% may be less than 50μ in diameter, while open-hearth steel furnace solids may be even

Table 10.1 Characteristics of Coal Mine Drainage [4]

	Acid Mine Drainage				*Alkaline Mine Drainage*			
	Surface Mines		*Subsurface*		*Surface Mines*		*Subsurface*	
Parameter[a]	*Range*	*Avg*	*Range*	*Avg*	*Range*	*Avg*	*Range*	*Avg*
pH	2.67–7.7	3.6	2.04–8.02	4.0	6.02–8.02	7.7	6.06–8.05	7.9
Alkalinity	0–184	5	0–720	59	30–860	313	22–1,840	469
Total iron	0.08–440	52	0.24–9,300	352	0.07–6.7	0.78	0.03–9.10	1.54
Soluble iron	0.01–440	50	0.05–5,000	268	0.01–2.7	0.15	0.01–0.95	0.25
Manganese	0.29–127	45	0.04–92	7.3	0.01–6.8	0.61	0.01–0.41	0.08
Aluminum	0.10–271	71	0.10–533	43	0.01–0.85	0.20	0.01–0.60	0.13
Zinc	0.06–7.7	1.7	0.02–12.7	1.5	0.01–0.59	0.14	0.01–0.30	0.06
Nickel	0.01–5.0	0.71	0.01–5.6	0.72	0.01–0.18	0.02	0.01–0.02	0.01
Sulfate	22–3,860	1,842	300–9,711	2,370	42–3,700	1,297	10–1,370	495
Ammonia-N	0.53–22.0	6.5	0.00–57	12	0.04–36	4.2	0.02–4.0	0.94
TDS[b]	120–8,870	4,060	12–15,572	4,749	152–8,358	2,867	418–22,658	2,702
TSS[c]	4–15,878	549	1–1,740	228	1–684	96	1–76	26

[a] All parameters in mg/l, except pH.
[b] Total Dissolved Solids.
[c] Total Suspended Solids.

smaller [7]. Ferromanganese steel processing produces the highest percentage of semicolloidal particles, with 50% being less than 10μ in diameter [8].

Milling of iron or steel products yields solids in the form of rust scale. The total iron loss during milling is roughly 2.5% of the finished product weight [8]. Much of this scale is of large, readily settleable size, but as metal product thickness decreases so does the scale size, with finishing-mill scale approaching the micron level. Quench water [7] and casting equipment-cleaning water [9] in iron foundries represent another source of finely divided iron and iron oxide particles. Much of the particulate iron resulting from iron and steel industry milling and air pollution abatement procedures is recovered and reused [10].

Concentrated sulfuric or hydrochloric acid baths are used to remove surface oxides from iron or steel parts before further processing. These pickling baths accumulate large concentrations of iron. The spent pickle liquors are extremely acidic, and iron concentrations can approach 10% [11]. Spent iron pickle liquors are sometimes used beneficially for phosphorus removal in domestic sewage treatment plants. Table 10.2 presents a summary of industrial sources and iron concentrations.

TREATMENT TECHNOLOGY

Oxidation–Precipitation

The predominant treatment process for iron is conversion of ferrous iron to the ferric state, and precipitation of ferric hydroxide. Minimum ferric hydroxide solubility is achieved by pH 5.5 [37, 38], and resolubilization does not occur until pH 12. Conversion of ferrous to ferric iron occurs rapidly upon aeration at pH 7.0–7.5. Spontaneous formation of ferric hydroxide then results in iron removal by precipitation. In concentrated wastewaters, the hydrolysis effect associated with formation of ferric hydroxide may cause a drift in pH toward the acidic side, thereby interfering with effective precipitation.

Precipitation of ferrous hydroxide is also effective at pH above 8 [37, 39]. However, Kelly and Olthof [39] have reported three reasons why oxidation and precipitation of ferric hydroxide are preferable to direct precipitation of ferrous hydroxide: (a) At neutral treatment pH, lower residual ferric than ferrous iron levels are achievable; (b) ferrous sludges are more difficult to thicken and dewater than are ferric sludges; and (c) ferrous sludges will undergo slow oxidation and consequent acidification, releasing soluble iron within the sludge disposal site.

The rate of oxidation of ferrous to ferric iron is dependent upon pH, buffer intensity in bicarbonate systems, availability of dissolved oxygen, and the presence of dissolved organic matter [40]. These organic compounds, which contain hydroxyl and carbonyl functional groups, may greatly reduce oxidation rates and can result in high soluble-iron levels if the organic substances do not readily break down.

Table 10.2 Iron Concentrations Reported for Industrial Wastewaters

Source	*Iron Concentration (mg/l)*	*Reference*
Heavy farm equipment manufacture	56	12
Motor vehicle assembly	3–4	13
Metal forge shop rinsewaters	25	14
Steel manufacture		
Waste pickle liquor	96,800	15
Waste pickle liquor	70,000	16
Pickle bath rinse	200–5,000	17
Pickle bath rinse	60–1,300 (210 avg)	18
Pickle bath rinse	175	19
Steel cold-finishing mills	60–150	20
Steel mill plant wastes	25–75	21
Metal processing and plating		
Appliance manufacture	0.09–1.9	22
Automobile heating controls	1.5–31	23
Appliances		24
Mixed wastes	0.2–20	
Spent acids	25–60	
Chrome plating	40	16
Chrome plating	64	25
Zinc plating	3.4	25
Copper plating	11.6–120	26
Plating wastes	2–4	27
Plating wastes	1.9–7.8	28
Plating wastes	6–25	29
Hydrochloric pickle acid	98,000	26
Copper plating bath	23,000	26
Printed circuit board manufacture	12.5	30
Titanium dioxide manufacture	0.02–31,000	31
	136	32
Aluminum hot rolling	3,210	33
Paint manufacture	3.8–37.3 (15.2 avg)	34
Ink formulation	134	34
Dye house wastes	31.5	35
Power plant operations		
Boiler tube cleaning	1,125	36
Air preheater cleaning	1,860	
Acidic coal-cleaning waste	1,680 (ferrous)	
	1,630 (ferric)	37

Table 10.3 Summary of Treatment Performance for Municipal Iron-Treatment Plants

Process	*Iron Concentration (mg/l)* *Influent*	*Effluent*	*Reference*
1. Chlorination, alum-lime-sodium silicate precipitation, sand filtration	1.5	0.05	41
2. Aeration, lime, sand filtration	2.4	0.0	42
3. Aeration, sand filtration	2.5	0.13	43
4. Aeration, coke bed filtration, sedimentation, sand filtration	5.7	0.13	44
5. Lime, aeration, diatomite filtration	10	0.1	45

Kelly and Olthof [39] have studied the oxidation of ferrous iron in concentrated industrial wastes. They report that oxidation rates increase with increasing temperature and that neutralization with lime or caustic yielded equivalent, and greater, oxidation rates than did neutralization with soda ash. For concentrated wastes, the rate-limiting aspect may be the oxygen availability. Oxidation of ferrous to ferric iron requires about 0.14 mg oxygen/mg ferrous iron oxidized. The maximum practical oxygen transfer rate, using conventional aeration equipment, is about 14 mg oxygen/liter-minute. For concentrated wastes (2,500 mg/l), the oxidation rate is zero order and, at pH above 6.0 oxygen transfer would limit the oxidation rate.

Insoluble iron oxide may exist in suspension as a stable colloidal dispersion, which resists sedimentation due to the small particle size. In addition, the particles tend not to agglomerate into larger particles due to their characteristic surface charge, which causes them to repel each other. Ferric hydroxide ultimately becomes ferric oxide in solution through chemical reaction. However, freshly precipitated ferric hydroxide has a characteristic low specific gravity, which makes settling somewhat difficult without long detention times or additional treatment.

Iron removal from municipal water supplies has been successfully accomplished by the water treatment industry for many years. Iron levels in groundwater supplies are characteristically in the 1–10 mg/l range, as ferrous iron. Groundwaters are typically anoxic and weakly acidic, usually high in CO_2. Aeration for removal of CO_2 (and consequently increase in pH) and to oxidize ferrous to ferric iron comprises the standard municipal water treatment. Chlorination may be substituted for aeration, to achieve iron oxidation. Lime is frequently employed for pH adjustment. Table 10.3 summarizes treatment efficiencies obtained from full-scale municipal water treatment plants.

Table 10.4 Comparison of Costs (in 1973–1974) of Neutralizing Chemicals for Acid Wastes [31, 47]

Chemical	*Cost ($/ton)*	*lb Required/100-lb Waste Acid, as* H_2SO_4	*HCl*	HNO_3
Pulverized limestone, high Ca	24.00	110	148	86
Pulverized limestone, dolomitic	18.00	94	127	73
Hydrated lime, high Ca	34.00	79	107	62
Hydrated lime, dolomitic	27.00	65	87	51
Pebble lime, high Ca	32.00	60	80	46
Pebble lime, dolomitic		54	73	42
Pulverized quicklime, high Ca	32.00	60	80	46
Sodium bicarbonate	200.00	173	233	135
Soda ash (58%)	60.00	205	275	160
Caustic soda (98%)	160.00	82	110	114

Iron treatment processes for most industrial wastes differ from that of the water industry, in that for iron removal it is often necessary to neutralize fairly high concentrations of acid contained in iron-bearing wastewaters. Neutralization increases treatment costs considerably, and since lime is the preferred base and sulfuric acid is often present in the waste, large quantities of insoluble gypsum (calcium sulfate) are also precipitated, which complicates the sludge disposal problem. A lime demand of 0.95–1.30 (wt/wt) has been cited for neutralization of acid mine drainage acidity, expressed as equivalent calcium carbonate [46]. Table 10.4 is a cost comparison of neutralizing chemicals for acid wastewaters. The use of limestone slurries may offer advantages with respect to solids removal and handling. These advantages may not necessarily always offset the cost differences (Table 10.5).

Iron-bearing wastes in the steel and plating industries frequently are the result of acid pickling of iron and steel to remove surface oxides. Two types of waste are generated by this cleaning process. Rinsewaters may typically contain 50–5,000 mg/l of ferrous iron and considerable acidity (see Table 10.2). The pickling baths themselves are extremely concentrated in acid and, ultimately, soluble iron. This may necessitate periodic disposal of the spent bath,

Table 10.5 Treatment of Acid Mine Drainage [48]

Treatment Chemical	*Iron (mg/l) Initial*	*Final*	*% Iron Removal*	*Chemical Cost (¢/1,000 gal)*	*Chemical Cost/ mg Acidity (¢/MG)*
Limestone slurry	145	5	96.2	5.9	10
Lime	157	1.6	99.0	3.2	5.2
Soda ash	126	2.3	98.2	26.8	4.9

Table 10.6 Cost of Waste Pickle Liquor Treatment

Treatment	*Plant Capacity (MGD)*	*Capital Cost ($/1,000 gpd)*	*Operating Cost ($/1,000 gal)*	*Reference*
Neutralization, precipitation, and sludge disposal	na[a]	na	$30–40	15 (1972)
Neutralization and precipitation	na	na	$20	49 (1967)
Neutralization, precipitation, and sludge lagooning	na	na	$10–40	50 (1970)
Neutralization, precipitation, and sludge disposal	0.06	$1,667	$20	51 (1958)

[a] Not available.

which can involve a volume of several tens of thousands of gallons. Treatment of waste pickle liquor is normally best handled separately from rinsewaters [20]. Cost of rinsewater treatment is a function of many variables, including plant location, extent of waste stream segregation, total flow, particular waste characteristics, auxiliary facilities required, and the effluent standards in force.

In general, one of the most costly processes involving iron removal is treatment of the concentrated waste pickle liquor. In addition to high iron levels and ferrous-to-ferric ratios, the waste is strongly acidic. Treatment cost is more directly related to neutralization of acidity than to iron removal. Neutralization and subsequent precipitation with lime have been most commonly used by steel mills to treat these wastes. Treatment costs are summarized in Table 10.6.

Rinsewaters from pickle lines are treated by neutralization and aeration. The extent of conversion of ferrous iron to ferric iron will influence the final soluble iron concentration. An additional factor pertaining to effluent iron concentration is solids removal efficiency.

In the case of ferric hydroxide carryover from a sedimentation clarifier or settling basin which does not provide adequate settling, chemical conditioning of the floc to obtain better settling characteristics may be implemented at increased operating costs, particularly in the case of polyelectrolyte use. Treatment of one stainless-steel pickle rinsewater involves lime neutralization followed by clarification with polymer addition [19]. Treatment cost for this 1.5 MGD system averages $0.32/1,000 gal treated. Polymer consumption, at $3.00/lb, constitutes 20% of the total operating cost. The facility reduces influent iron averaging 175 mg/l to 0.24 mg/l in the effluent [19].

Armco Steel [18] reported synthetic polymer flocculant aid costs to be 17% of total operating costs for limestone neutralization and precipitation of acid pickle rinsewaters. Effluent total iron averaged 4 mg/l, with a range of 0.32–23.0 mg/l. Cationic polymers have been reported to range from $1.25 to $3.00/lb dry weight, and to be used at dosages of 2–5 mg/l. Anionic and some nonanionic polymers cost $1.50–3.00/lb dry weight, and are used at dosages

Table 10.7 Suspended Solids Removal from a Foundry Waste [17]

Treatment	*Suspended Solids (mg/l)*		*Cumulative % Removal*
	Influent	*Effluent*	
Primary sedimentation	3,760	363	90.3
Coagulation and secondary sedimentation	363	30	99.2
Secondary coagulation and flotation	30	10	99.7

of 0.25–1.0 mg/l [52]. Installed cost of polymer feed facilities is reported as \$1.00–2.00/1,000 gpd capacity [47].

Milling operations utilize a rough settling basin referred to as a scale pit, which collects the larger iron scale for recycle [7]. Older practice was to discharge the scale pit overflow, containing approximately 200 mg/l of suspended solids, without further treatment. Modern practice is to follow the scale pit with a clarifier or pressure filter to remove residual particles, and recycle the effluent water [8].

Fine suspended solids in an iron foundry waste have been reduced from 3,760 to 10 mg/l by a sequence of primary sedimentation, addition of lime and alum coagulant, secondary sedimentation, further addition of lime and alum, and flotation. Table 10.7 presents the reported suspended solids levels after each unit step of treatment.

Standard operations for particulate iron removal include sedimentation with or without coagulation, and possible filtration through sand, gravel, crushed anthracite, or other media. Use of polyelectrolyte, if required to improve settling, increases operating costs substantially [53].

Nebolsine and Sanday [54] have reported the use of ultrahigh-rate gravity and pressure filtration for the treatment of hot-strip steel-mill process waters. These mixed-media filters are capable of reducing 100–300 mg/l suspended solids to less than 5 mg/l at loadings of 15 gpm/ft^2. Suspended solids reductions of 95–99% are claimed for steel plant discharges. A 4.3-MGD unit is reported operating at the Fukuyama plant of Nippon Kokum Steel Co., Japan, since 1966, and a 72-MGD unit at U.S. Steel in Chicago. Dissolved air flotation (microflotation) has been used after automatic lime feed to treat 3.5 gpm steel pickle rinsewater to which small volumes of spent pickle liquor were added [21]. Iron was reduced from 25–75 mg/l to 1 mg/l.

Treatment of mine drainage is similar to steel industry wastes. Long-detention sedimentation ponds are commonly used for treating alkaline mine drainage wastewater. The holding ponds serve to remove suspended solids and to equalize the discharge volumes, and these wastes are normally discharged

Table 10.8 Treatment Results for Acid Coal Mine Drainage [4]

	Raw Wastewater[b]			*Treated Effluent*[b]		
Treatment Sequence[a]	*pH*	*Total Iron*	*Soluble Iron*	*pH*	*Total Iron*	*Soluble Iron*
1, 2, 4	2.9	212	185	7.2	1.44	0.28
1, 2, 4	3.1	276	276	8.2	1.68	0.04
1, 5	3.0	164	139	8.9	0.35	0.01
1, 3, 4	5.8	187	63.7	8.0	0.48	0.01
1, 2, 4, 5	2.7	412	95	6.9	0.15	0.06
1, 2, 5	5.9	123	55	7.8	1.77	0.03
1, 3, 5	2.6	980	970	6.8	1.60	0.08
1, 2, 4	2.7	760	760	8.2	1.34	0.26
1, 3, 4	2.5	380	370	7.9	1.00	0.02
2, 1, 4	2.9	87.8	82.8	7.9	1.70	0.05
1, 4	2.5	802	32	8.8	1.80	0.03
6, 2, 5, 5	6.35	140	140	7.7	2.50	0.02

[a] 1, Lime neutralization, 2, mechanical aeration, 3, natural aeration, 4, gravity clarifier, 5, settling pond, 6, caustic neutralization.

[b] All values in mg/l, except pH.

without additional treatment. Since mine drainage contains both particulate and soluble iron (see Table 10.1), and natural aeration can result in ferrous iron oxidation with resultant precipitate formation, reductions in both total and soluble iron are normally observed in these ponds [4]. Depending upon land availability and land value, use of large holding ponds may provide an economical alternative to either clarifiers or filtration.

Treatment of acid mine drainage typically involves neutralization, natural or mechanical aeration, and use of either a gravity clarifier or a settling pond. Treatment results for acid coal-mining wastes are summarized in Table 10.8. Results for other industrial wastewaters are given in Table 10.9. Typical effluent iron levels of about 0.5 mg/l or less are reported. The somewhat high effluent iron level for iron-blue pigment manufacture (Table 10.9) likely reflects the presence of ferrocyanide. Cost data for various industrial and mine water treatments are presented in Tables 10.10 and 10.11.

Frequently two waste streams can be combined to provide treatment for one or both. The most obvious example is neutralization by combination of acidic and basic wastes. Occasionally, chromate wastes may be present in the same plant with waste pickle liquor or rinsewater containing appreciable ferrous iron. In place of separate treatment, the pickle liquor can be employed as a reductant for chromate (see Chapter 5), with simultaneous oxidation of ferrous to ferric iron, followed by coprecipitation of chromium and iron hydroxide by

Table 10.9 Precipitation Treatment Results for Iron Wastewaters

Source	*Iron Concentration (mg/l)*		*Treatment*	*Comments*	*Reference*
	Initial	*Final*	*pH*		
Base-metal acid mine drainage[a]	718	0.54			46
	1,202	0.58		Without filtration	
		0.25		With filtration	
	1,138	0.42		Without filtration	
		0.16		With filtration	
	93	0.53		Without filtration	
		0.22		With filtration	
Boiler and air preheater cleaning wastes	1,125–1,860	<0.1	8.5		36
Titanium dioxide manufacture		0.20		Range 0.1–3.0	31
	159	0.48	10.0	Without filtration	32
		0.19	10.0	With filtration	
Steel mill wastes (rinse and spent pickle)	25–75	1.0	6.9	Microflotation for solids separation	21
Graphite mine drainage	—	0.10			55
Plating rinse	7.8	0.11			28
Chrome-plating rinse	64	<0.05	11.0	With filtration	25
Zinc-plating rinse	3.4	<0.05	11.6	With filtration	25
Iron blue pigment manufacture	—	1.0			6
Heavy farm equipment manufacture	56	1.6–3.1	8.5–10.5		11

[a] Pilot plant results.

lime addition. One such process has been reported to reduce iron and chromium in the final effluent to 0.5 and 0.05 mg/l respectively [57].

Deep-Well Disposal

Deep-well injection is popular as a disposal technique, particularly for steel mill-waste pickle liquor [50]. Costs are widely variable depending upon well depth and injection pressure requirements. Wells may cost \$50,000–150,000 to drill and case, and operating costs may be of the order of \$2.50–10.00/1,000 gal [15]. Capital costs may vary from \$30,000 for a 1,800 ft deep well where no surface equipment is required, to \$1,400,000 for another 12,000 ft deep, plus a clarifier, dual filters, and four displacement-type pumps [58]. The filter requirement applies to any waste containing suspended solids, as they clog the

Table 10.10 Cost Summary for Treatment of Industrial and Acid-Mine Iron-Bearing Waters by Neutralization and Precipitation

Waste	*Flow (MGD)*	*Capital Costs ($/1,000 gpd)*	*Operating Costs (¢/1,000 gal)*	*Reference*
Pickling rinsewater	2.16	630	24	18 (1971)
Steel cold-finishing mill wastes[a]	14.4–21.6	208–625	7–25	20 (1970)
Acid mine drainage[a] (limestone)				56 (1971)
90 mg/l Fe^{+2}, 190 mg/l acidity	4.0	165	9.7	
500 mg/l Fe^{+2}, 1,000 mg/l acidity	0.1	1,303	66	
	1.0	352	27	
	7.0	194	22	
5,000 mg/l Fe^{+2}, 8,000 mg/l acidity	0.1	1,732	213	
	1.0	765	171	
	7.0	556	166	

[a] Includes sludge disposal costs.

well. Figures 10.1 and 10.2 present capital and operating costs for deep-well injection systems. Deep-well injection is not feasible for pickling rinsewater disposal, due to the large waste volume involved.

Other Methods

Neutralization of waste pickle liquor is expensive, and often impractical, due to the quantities of sludge produced. Recovery processes for sulfuric acid pickle liquor and for hydrochloric acid pickle liquor are available [59]. Various pickle-liquor recovery processes are in operation, and their nature and some cost data

Table 10.11 Costs (in 1975) for Lime Treatment of Acid Mine Drainage [4]

Type Aeration	*Design Capacity (MGD)*	*Capital Cost[a] ($/1,000 gpd)*	*Volume Treated (MGD)*	*Operating Cost[b] ($/1,000 gal)*
Natural aeration	1.441	119	0.264	0.75
	0.865	413	0.824	0.56
Mechanical aeration	1.202	377	1.067	0.49
	6.861	69	5.328	0.16
	7.598	71	7.586	0.11

[a] Based upon design capacity.
[b] Based upon actual volume treated, 300 days/yr.

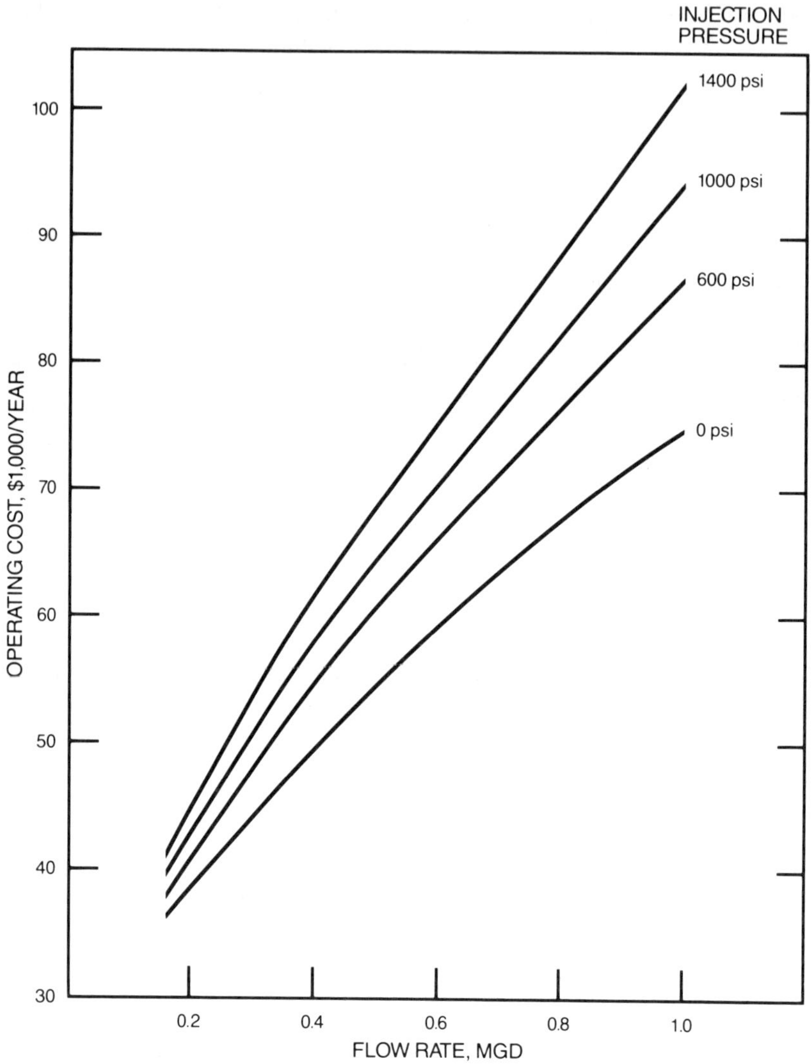

Figure 10.1 Annual operating costs (in 1970) of waste disposal by deep-well injection [2].

have been summarized [60]. For each process available, however, the primary emphasis is on recovery of acid rather than removal of iron. In brief, the processes evaporate the waste liquor and remove iron as a ferrous or ferric salt, and acid as either concentrated hydrochloric or sulfuric acid.

Jackson [61] has described a recovery process for hydrochloric acid pickle liquor, which can convert the ferrous chloride to free iron or ferric oxide while

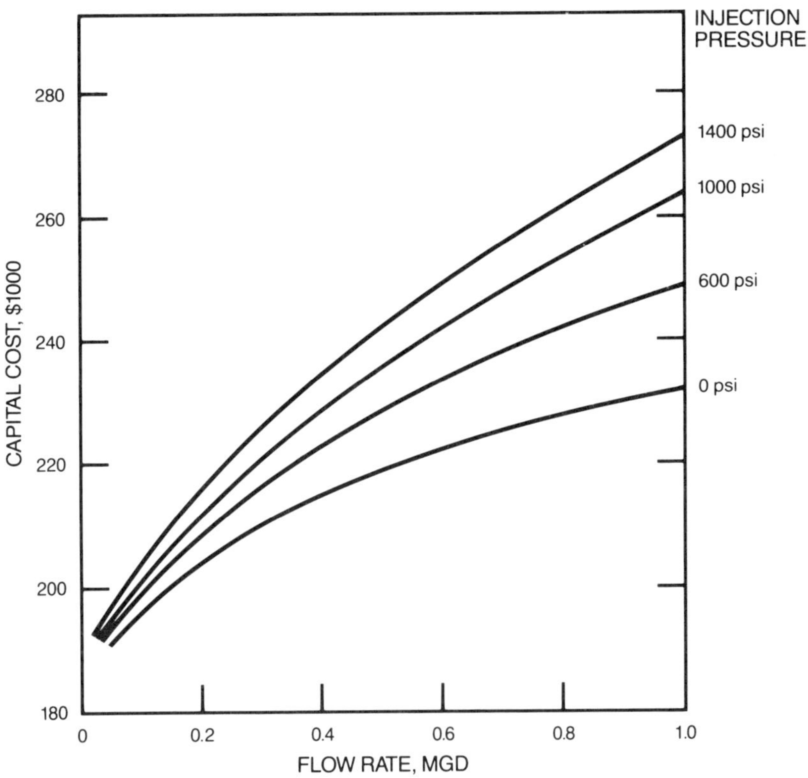

Figure 10.2 Capital costs (in 1970) for deep-well injection systems [2].

regenerating the hydrochloric acid. A flowchart of a commercial unit in operation in Canada is presented in Figure 10.3. This unit employs reduction of the iron by hydrogen gas. Although the original unit was designed for use with scrap, the application to pickle liquor is direct. The system is claimed to be capable of successful and economical operation.

Ion exchange treatment, utilizing a two-column cation-anion exchange system, has been investigated for acid mine drainage. The process is effective (Table 10.12) but produces a regenerant liquor requiring additional treatment [62]. Further, some ferrous iron leaks through the cation exchange column and apparently oxidizes to ferric iron and precipitates in the anion exchange column. This behavior could result in column clogging or resin fouling, and could present a serious operating difficulty [63].

Reverse osmosis has been suggested for the treatment of acid mine drainage. A pilot-plant reverse osmosis unit was operated on-site for 4 months, and achieved iron reduction from 67 to 0.60 mg/l total iron and 64 to 0.57 mg/l ferrous iron [64]. The concentrated brine was subjected to neutralization and precipitation, with reported effluent iron concentrations from this process of

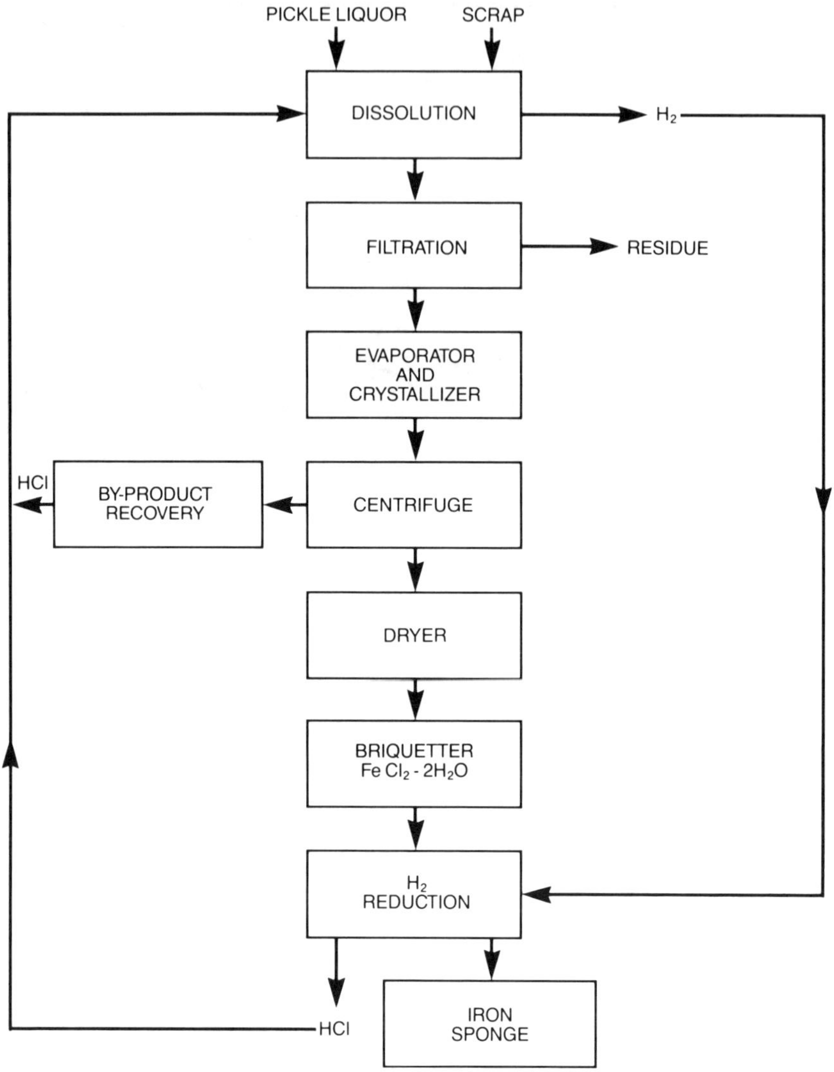

Figure 10.3 Outline flowchart of the Peace River Mining and Smelting Company process for treatment of ferrous scrap. (After Jackson [61], courtesy of *Metal Finishing Journal*.)

0.9–610 mg/l. Projected costs for a 0.50 MGD system were given as \$607/1,000 gpd capital and \$0.49/1,000 gal operating.

While designed primarily for treatment of oil wastes, specifically spent soluble-oil solutions used in the hot rolling of aluminum plate and sheet, ultrafiltration has been reported to reduce an iron concentration of 3,210 mg/l to 4.2 mg/l [33]. Further treatment of the ultrafiltration permeate by reverse osmosis

Table 10.12 Summary of Ion-Exchange Treatment of Acid Mine Drainage [62]

Sample	*pH*	*Acidity (Alkalinity)*[a]	*Iron (mg/l)* Total	*Iron (mg/l)* Ferrous
Raw feed	4.9	410	210	190
Cation effluent	1.5	1,780	23	22
Anion effluent	9.4	(280)	3.1	0.3

[a] Expressed as equivalent calcium carbonate.

yielded a final effluent iron concentration of 0.3 mg/l. Treatment was preceded by pH adjustment.

SUMMARY

Reduction of the solids fraction of total iron can be readily accomplished by existing technology and is essentially a matter of suspended solids removal. Two distinct situations exist: (a) the iron originates in a particle form, and (b) the particulate iron is formed by precipitation in a wastewater treatment process. Particulate iron removal is then through sedimentation, with or without coagulant aids, or by filtration. The use of treatment chemicals, if necessary, will have a marked effect upon treatment costs.

Existing technology is capable of iron removal to levels below 0.5 mg/l. Failure to achieve these levels appears to be the result of improper pH control, resulting in increased solubility of ferric hydroxide during coagulation, incomplete oxidation of the more soluble ferrous iron to the ferric state, or poor solids removal. Apparent soluble iron may actually be present in effluent waters in a finely divided colloidal state, due to inefficient removal of the ferric hydroxide. Precipitation treatment systems which consistently produce low-iron waters employ some polishing treatment, such as filters.

Oxidation and precipitation are capable of providing effective iron removal from most wastes, providing that prior treatment has removed iron-complexing agents (e.g. cyanides), and that the treatment pH is suitable. However, treatment costs may be extremely high for chemicals and for sludge dewatering and disposal, particularly when wastes contain high concentrations of sulfuric acid, due to precipitation of calcium sulfate sludge.

Recovery processes appear to be of economical value currently only for large industries such as steel mills, although certain process modifications may make recovery more feasible and enhance treatment efficiencies at reduced costs. Other than precipitation and recovery, only deep-well disposal seems economically competitive for industries with high-acid, high-metal waste solutions. The costs, if geological conditions are favorable and regulations allow this disposal technique, are often more attractive to smaller industries than are those of other treatment alternatives.

REFERENCES

1. McKee, J.E. and H.W. Wolf. *Water Quality Criteria*, 2nd ed., California State Water Quality Board Publication 3-A (1963).
2. *The Economics of Clean Water,* Vol. III, *Inorganic Chemicals Industry Profile* (1970) USDI, Washington, DC.
3. Dean, J.C., F.L. Bosqui, and K.H. Lanouette. "Removing Heavy Metals from Waste Water," *Environ. Sci. Technol.* 6 (1972): 518–522.
4. "Development Document for Interim Final Effluent Limitations Guidelines and New Source Performance Standards for the Coal Mining Point Source Category," USEPA 440/1-75/057 (October 1975).
5. Ohmi, T. and Y. Moritsu. "Removal of Ferricyanide and Ferrocyanide from Wastewater with Weakly Basic Ion Exchange Resin," *Chem. Abstr.* 81 (1974): 346.
6. "Development Document for Interim Final Effluent Limitations Guidelines and Proposed New Source Performance Standards for the Significant Inorganic Products Segment of the Inorganic Chemicals Manufacturing Point Source Category," USEPA 440/1-75/037, Group I, Phase II (May 1975).
7. Kemmer, F.N. "Pollution Control in the Steel Industry," *Industrial Pollution Control Handbook*, Herbert F. Lund, ed. (New York: McGraw-Hill, 1971).
8. Bramer, H.C. "Iron and Steel," *Chemical Technology*, vol. 2, *Industrial Wastewater Control*, C.F. Gurnham, ed. (New York: Academic Press, 1965).
9. Barzler, R.P. and D.J. Giffels. "Pollution Control in Foundry Operations," *Industrial Pollution Control Handbook,* Herbert F. Lund, ed. (New York: McGraw-Hill, 1971).
10. "Steel Industry Sludge is Being Reused," *Environ. Sci. Technol.* 9 (1975): 624–625.
11. Howell, J. "Beneficial Disposal of Pickle Liquor," *Ind. Wastes*, (May/June: 21–22, 1982).
12. McVaugh, J. and W.T. Wall, Jr. "Optimization of Heavy Metals Wastewater Treatment Effluent Quality Versus Sludge Treatment," *Proceedings of the 31st Industrial Waste Conference, Purdue University* (May 1976), pp 17–25.
13. *The Cost of Clean Water*, vol. III. *Industrial Waste Profiles, No. 2, Motor Vehicles and Parts* USDI, (Washington, DC: USDI, 1967).
14. Hockenbury, M.R. and A.W. Loven. "Treating Metal Forging and Processing Water," *Ind. Wastes* (May/June 1977): 45–49.
15. Southern Research Institute. "An Electromembrane Process for Regenerating Acid from Spent Pickle Liquor," USEPA 12010 EQF 3/72 (1972).
16. Parsons, W.A. *Chemical Treatment of Sewage and Industrial Wastes*, (Washington, DC: National Lime Association, 1965).
17. Nemerow, N.L. *Theories and Practice of Industrial Waste Treatment* (Reading, MA: Addison-Wesley, 1963).
18. Armco Steel Corporation, "Limestone Treatment of Rinse Waters from Hydrochloric Acid Pickling of Steel," USEPA 12010 DUL 02/71 (1971).
19. Winans, T.O. and D.E. Dreier. "Innovative Acid Pickling Wastes Treatment at Universal-Cyclops Stainless Steel Plant," paper presented at the 34th Annual Purdue Industrial Waste Conference, West Lafayette, IN (May 1979).

20. Donovan, E.J. Jr. "Treatment of Wastewater for Steel Cold Finishing Mills," *Water Wastes Eng.* 7 (November 1970): F22–F25.
21. Cottrell, W.R.T. "Treatment of Effluent by Flotation Processes," *Effluent Water Treatment J.* (November 1976): 563–567.
22. Watson, K.S. "Treatment of Metal Finishing Wastes," *Sew. Ind. Wastes* 26 (1954): 182–194.
23. Gard, C.M., C.A. Snavley, and D.J. Lemon. "Design and Operation of a Metal Works Waste Treatment Plant," *Sew. Ind. Wastes* 23 (1951): 1429–1438.
24. Anderson, J.S. and E.H. Iobst, Jr. "Case History of Wastewater Treatment in a General Electric Appliance Plant," *J. Water Poll. Control. Fed.* 40 (1968): 1786–1795.
25. Yeats, A. "Ion Exchange Selectively Removes Heavy Metals from Mixed Plating Wastes," paper presented at the 32nd Annual Purdue Industrial Waste Conference, West Lafayette, IN (May 1977).
26. Reese, C. "Steel Welding Wire Manufacturer Meets Stringent Effluent and Sludge Disposal Criteria," paper presented at the 37th Annual Purdue Industrial Waste Conference, West Lafayette, IN (May 1982).
27. Battelle-Columbus Laboratories, "An Investigation of Technique for Removal of Cyanide from Electroplating Wastes," USEPA 12010 EIE 11/71 (1971).
28. "Development Document for Interim Final Effluent Limitations Guidelines and Proposed New Source Performance Standards for the Common and Precious Metals Segment of the Electroplating Point Source Category," USEPA 440/1-75/040, Group I, Phase II (April 1975).
29. Yost, K.J. and D.R. Masarik. "A Study of Chemical-Destruct Waste Treatment Systems in the Electroplating Industry," *Plating Surface Fin.* (January 1977): 35–40.
30. Jashnani, I. and R. Green. "Pretreatment System for a Circuit Board Plant," paper presented at the 8th National Conference on Municipal Sludge Management, Municipal and Industrial Sludge Management, Disposal and Utilization, Miami Beach, FL (March 19–21, 1979).
31. "Development Document for Effluent Limitations Guidelines and New Source Performance Standards for the Major Inorganic Products Segment of the Inorganic Chemicals Manufacturing Point Source Category," USEPA 440/1-74-007-a (March 1974).
32. Treatability Studies of the Inorganic Chemicals Manufacturing Point Source Category, USEPA 440/1-80/103 (July 1980).
33. Sonksen, M.K., F.M. Sittig, and E.F. Maziarz. "Treatment of Oily Wastes by Ultrafiltration/Reverse Osmosis. A Case History," paper presented at the 33rd Annual Purdue Industrial Waste Conference, West Lafayette, IN (May 1978).
34. "Development Document for Effluent Limitations Guidelines and New Source Performance Standards for the Oil Base Solvent Wash Subcategories of the Paint Formulating and the Ink Formulating Point Source Category," USEPA 440/1-75/050-a, Group II (July 1975).
35. Rouse, J.V. "Removal of Heavy Metals from Industrial Effluents," *ASCE J. Environ. Eng. Div.* (October 1977) EE5-929-936.
36. Kuppusamy, N. "Copper Removal from Power Plant Boiler Cleaning Waste," *Ind. Wastes* (March/April 1977): 43–45.

37. Wangen, L. and J. Williams. "Control by Alkaline Neutralization of Trace Elements in Acidic Coal Cleaning Waste Leachates," *J. Water Poll. Control. Fed.* 54 (1982): 1302–1310.
38. Netzer, A., A. Bowers, and J. Norman. "Removal of Heavy Metals from Wastewaters by Lime," *Proceedings of the 16th Conference on Great Lakes Research*, Huron, OH (1973).
39. Kelly, W. and M. Olthof. "Iron Oxidation in Industrial Waste Streams," paper presented at the 37th Annual Purdue Industrial Waste Conference, West Lafayette, IN (May 1982).
40. Jobin, R. and M.M. Ghosh. "Effect of Buffer Intensity and Organic Matter on the Oxygenation of Ferrous Iron," *J. Am. Water Works Assoc.* 64 (1972): 590–595.
41. Haney, C.G. "Potassium Permanganate Solves a Problem at Charlottesville, VA," *Water Sew. Works* 111 (1964): R353–R354.
42. Owens, L.V. "Iron and Manganese Removal by Split Flow Treatment," *Water Sew. Works* 110 (1963): R76–R86.
43. Gardner, J.R. and A.J. Stika. "Valveless Filter Installed for Iron Removal," *Public Works* 95 (1964): 109–110.
44. McCracken, R.A. "Study of Color and Iron Removal by Means of Pilot Plant at Amesbury," *J. New Eng. Water Works Assoc.* 75 (1961): 102–114.
45. Kollin, W. "Iron Removal Methods by Diatomite Filtration," *Water Sew. Works* 109 (1962): R182.
46. Huck, P.M. and B.P. LeClair. "Treatment of Base Metal Mine Drainage," paper presented at the 30th Annual Purdue Industrial Waste Conference, West Lafayette, IN (May 1975).
47. Edwards, A.W., W.W. Eckenfelder, Jr., and J.A. Roth. *Industrial Water Cost Studies*, vol. I, Office of Research and Development, USEPA Contract 68-01-2485 (1975).
48. Wilmoth, R.C. and R.D. Hill. "Neutralization of High Ferric Iron Acid Mine Drainage," USEPA 14010 ETV 8/70 (1970).
49. *The Cost of Clean Water*, vol. III, *Industrial Waste Profiles, No. 1, Blast Furnace and Steel Mills*, FWPCA Publication (Washington, DC: USDI 1967).
50. "Water Pollutant or Reusable Resources?" *Environ. Sci. Technol.* 4 (1970): 380–382.
51. Heise, L.W. and M. Johnson. "Practical Development Aspects of Waste Pickle Liquor Disposal, *Proceedings of the 13th Industrial Waste Conference, Purdue University* (May 1958), pp 140–150.
52. Tofflemire, T.J. and L.J. Hetling. "Chemicals and Clarifiers: Which are Best?" *Water Wastes Eng.* 10 (1973): F24–F26.
53. Smith, R. "Cost of Conventional and Advanced Treatment of Wastewater," *J. Water Poll. Control. Fed.* 40 (1968): 1546–1574.
54. Nebolsine, R. and R. Sanday. "Ultra-High Rate Filtration, A New Technique for Purification and Reuse of Water," *Iron Steel Eng.* (December 1967): 1–8.
55. "Development Document for Interim Final Effluent Limitations Guidelines and New Source Performance Standards for the Clay Ceramic, Refractory and Miscellaneous Minerals Segment of the Mining and Processing Industry Point Source Category," USEPA 440/1-75-059-d (October 1975).
56. Bituminous Coal Research, Inc. "Studies of Limestone Treatment of Acid Mine Drainage. Part II," USEPA EIZ 12/71 (1971).

57. Cupps, C.C. "Treatment of Wastes for Automobile Bumper Finishing," *Ind. Water Wastes* 6 (1961): 111–114.
58. Spencer, E.F. Jr. "Pollution Control in the Chemical Industry," in *Industrial Pollution Control Handbook*, Herbert F. Lund, ed. (New York: McGraw-Hill, 1971).
59. Delaine, J. "Management to Achieve Water Economy in the Iron and Steel Industry—Conclusion," *Effluent Water Treatment J.* 6 (1966): 543–551.
60. Besselievre, E.B. *The Treatment of Industrial Wastes* (New York: McGraw-Hill, 1969).
61. Jackson, D.V. "Metal Recovery from Effluents and Sludges," *Metal Fin. J.* 18 (1972): 235–242.
62. Wilmoth, R.C., R.B. Scott, and E.F. Harris. "Application of Ion Exchange to Acid Mine Drainage Treatment," paper presented at the 32nd Annual Purdue Industrial Waste Conference, West Lafayette, IN (May 1977).
63. Pelosi, P. and J. McCarthy. "Preventing Fouling of Ion-Exchange Resins—I," *Chem. Eng.* (August 9, 1982).
64. Rex Chainbelt, Inc. "Reverse Osmosis Demineralization of Acid Mine Drainage," USEPA 14010 FQR 03/72 (1972).

11
Lead

Lead is used as an industrial raw material in the manufacture of storage batteries, pigments, leaded glass, fuels, photographic materials, matches, and explosives. It is one of the most widely used nonferrous metals in industry.

The storage battery industry is the largest consumer of lead, followed by the petroleum industry in producing gasoline additives [1]. In a survey of eight battery manufacturing plants, it was reported that lead losses/battery manufactured ranged from 4.54 to 6,810 mg, and that water usage ranged from 11 to 77 gal/battery [2]. Highest lead concentrations originated from the plate-forming area. This water has a pH below 3.0, which results in high concentrations of dissolved lead [3]. A wastewater lead level of 11.7 mg/l has been reported at one secondary lead-battery plant reclaiming lead from old batteries [4].

The use of lead (as sodium plumbite) in petroleum refining produces a lead sludge, from which lead is normally recovered [5]. Lead is also a constituent of many plating wastes, although not so frequently encountered as are copper, zinc, cadmium, or chromium. Lead is plated as a protective coating on nuts, bolts, and bearings, and on storage battery parts. Fluorosilicate and fluoroborate plating baths are most widely used. Plating rinse wastewater lead concentrations as high as 98 mg/l have been cited [6, 7]. The lead content of wastewater from onc engine parts plating plant has been reported as ranging from 2 to 140 mg/l [8]. The high concentrations resulted from dumping of spent plating-bath solutions.

Chrome-yellow pigments, used primarily as paint pigment, are manufactured by reacting sodium chromate or dichromate with lead nitrate or acetate. Chrome-yellow pigments are the most widely used pigments in the United States [9]. A similar, but less widely used pigment is molybdate chrome orange, which contains lead sulfate, chromate, and molybdate. Effluent lead levels for chrome-yellow pigment production of 0.2–172 mg/l have been reported [9]. A molybdate chrome-orange pigment production waste-lead content of 843 mg/l was measured at one plant [9]. It is probable that this lead is primarily in an insoluble form.

A lead-solder glass frit mixture is used to fuse the front glass panel to the funnel of television tubes. Quality control of color picture tubes may require

Table 11.1 Lead Content of Ore-Processing Wastewaters

Industry	*Process*	*pH*	*Lead (mg/l)*	*Reference*
Lead	Smelting	—	0.07–0.16	15
	Slag granulation	7.6–8.3	0.30–0.85	16
	Scrubber water	6.5–6.8	0.50–0.52	16
Copper	Slag granulation	6.4–8.1	0.04–7.4	17
	Scrubber water	1.8–2.5	0.91–9.77	17
	Arsenic recovery	3.8–4.4	7.7	17

the salvage and reuse of the glass envelopes. Separation of the two components is accomplished by dissolving the frit in dilute acid, which yields an acidic wastewater containing lead and fluoride. A lead level of 400 mg/l has been measured in the effluent from one such salvage operation [10].

Pollution from lead mines, mining, and smelting has also been reported in the literature [11–14]. Characteristics of ore-processing wastewaters are given in Table 11.1. Significant lead levels have been found in acid mine drainage, as shown in Table 11.2. At the indicated pH values, the lead would be predominantly in the soluble form.

Most lead wastes are in the inorganic form. However, wastewaters of the tetraethyl lead industry contain significant concentrations of organic lead compounds. Because these compounds do not respond to the standard precipitation treatment procedures employed for inorganic lead, wastewaters of the tetraethyl lead industry present particular difficulty.

Other industrial sources of lead-containing wastewaters are the vacuum-degassing process in steel manufacture [20], leakage through resin seals around lead mounts in the diaphragm cell process for chlorine manufacture [21], and (primarily as particulate metal) in the lead sheath process for rubber hose manufacture [22]. Table 11.3 summarizes reported lead levels in industrial wastewaters.

Table 11.2 Lead Content of Acid Mine Drainage

Source	*pH*	*Lead (mg/l)*	*Reference*
Tailing pond	2.0	0.67	18
Mine drainage	2.6	0.3	18
	3.2	2.5	18
	—	0.02	18
	2.5–2.9	1.6–4.5	18
	2.0–3.4	1.4–4.6	19
	2.9–3.5	0.9–3.1	19
	—	0.04–0.5	14

Table 11.3 Reported Lead Levels in Industrial Wastewaters

Industry	*Lead (mg/l)*	*Reference*
Battery manufacture		
Particulate lead	5–48	2
Soluble lead	0.5–25	
Battery manufacture		
Particulate lead	0.4–66.5	3
Soluble lead	2.6–5.1	
Battery manufacture	40.3–319.4	23
	11.7	4
Plating	2–140	8
	0–30	6
	0.2–2.0	24
Plating pickle liquor	10	25
Television tube manufacture	380–400	26
Printed circuit board manufacture	1.65	27
Glass manufacture	0.43–100	28
Porcelain enameling	2.9	29
Chlor-alkali plant	1,160	30
Mining process water	0.018–0.098	31
Ammunition plant	6.5	32
Tetraethyl lead manufacture		
Organic	126.7–144.8	33
Inorganic	66.1–84.9	
Tetraethyl lead manufacture	45	34
Spent ink	94	35
Paint manufacture	1.1–10.0	36
Paint and ink formulation	86	29
Pigment manufacture	1–200	37
	0.2–843	9
Textile dyeing	8.4	38,39
Steel manufacture, vacuum degassing process	0.47–1.39	20
Rubber hose manufacture, lead sheath process	63	22
Foundry	7.7	29
	29–170	40
Piston ring manufacture	94.6	41

TREATMENT TECHNOLOGY

Treatment of soluble lead involves for the most part reaction to form a lead precipitate, and solids removal. Precipitation chemicals used include lime or caustic to precipitate lead hydroxide, soda ash to precipitate lead carbonate, and phosphate to precipitate lead phosphate. Alum and ferrous and ferric sulfate coagulation have also been reported for lead treatment, as has ion exchange.

Precipitation

In the precipitation process, lead is normally precipitated as the carbonate, $PbCO_3$ or hydroxide, $Pb(OH)_2$. The lead form precipitated depends upon the amount of carbonate in or added to the wastewater, and the treatment pH. Initial acidic wastes are typically low in carbonate, and precipitation treatment of these wastewaters would normally yield lead hydroxide unless supplemental carbonate was added. Lead carbonate precipitate is more crystalline than is lead hydroxide, resulting in desirable settling and sludge-dewatering characteristics. In addition, at intermediate pH, lead carbonate is normally more insoluble than is lead hydroxide. Optimum carbonate level for lead carbonate precipitation is reported at 200 mg/l as equivalent calcium carbonate [42]. A large excess of carbonate, or treatment pH above 9.0, may yield less effective precipitation treatment, however. Optimum pH range for lead carbonate precipitation is reported to be pH 7.5–9.0 [26, 42–45].

The literature contains conflicting values of pH optima for precipitation of lead hydroxide. For example, a lead precipitation pH of 6.0 has been suggested in one instance [46], and a pH range of 6.0–7.0 has been suggested in another [18]. Effective lead treatment of a metal-finishing waste has been reported at pH 8.7–9.3 [47], and even higher pH (up to 10.0) has also been recommended [48]. Both theoretical solubility calculations [43] and plant data [44, 47] indicate that most efficient lead hydroxide precipitation occurs at pH near 10. Effluent lead of 0.01–0.03 has been reported for treatment in this pH range [42, 44]. The pH optima for lead hydroxide precipitation is apparently narrow, and efficiency of hydroxide precipitate formation decreases rapidly at more extreme pH values. Lead is reported as being recovered from the sludge yielded in the precipitation process [5, 49].

In most precipitation treatment systems, two factors influence lead removal. These are (1) lead solubility, and (2) lead precipitate settleability, where effluent filtration is not applied. Table 11.4 presents data from two separate studies which relate to these factors. In one study [50], lead was precipitated at various pH values, and the supernatant sampled. These results (Table 11.4) indicate that treatment pH 10.8 yielded lowest supernatant lead levels. The second study [51] measured soluble lead resulting at various treatment pH values, in the absence of detectable carbonate and with carbonate present at 3–5 mg/l. Without carbonate, where lead hydroxide precipitation is probable, minimum soluble lead occurred at pH 8.5, and increased with higher pH. In the presence of low levels of carbon, minimum soluble lead occurred near pH 7.5, at a concentration of one-half that observed without carbon. At higher pH, lead was much more soluble in the carbonate system.

Table 11.4 Effect of pH on Lead Removal

Settled Supernatant [50]		*Soluble Concentration [51]*		
pH	*Lead (mg/l)*	*pH*	*Lead[a] (mg/l)*	*Lead[b] (mg/l)*
5.2	107	—	—	—
—	—	6.3	24.6	1.30
—	—	6.6	1.10	—
7.1	37	7.1	—	0.035
—	—	7.4	0.131	0.025
—	—	7.6	—	0.040
8.0	11.9	8.5	0.055	0.075
9.2	10.7	9.4	0.215	4.10
10.5	2.9	10.5	0.150	—
10.8	1.5	10.8	—	8.36
11.0	4.2	—	—	—
11.6	8.9	—	—	—

[a] Inorganic carbon less than 2 mg/l.
[b] Inorganic carbon is 3–5 mg/l.

These experimental results suggest that actual treatment plant performance involves a complex interaction of treatment pH and wastewater carbonate concentration, plus solids removal efficiency. In a plant utilizing only clarification to treat a wastewater of low carbonate, best results would likely be achieved at pH 10.5–11.0 (see Table 11.4). In the same plant, but employing filtration, a treatment pH of about 8.5 would yield lowest soluble lead levels. However, if the wastewater contained even small amounts of carbonate, the data of Table 11.4 suggest an optimum treatment pH of 7.0–7.5.

Narrow and possible high pH optima such as is required for lead hydroxide precipitation represent a particular problem when a wastewater contains several metals each having a different pH optima. The problem is demonstrated in Table 11.5, which is a summary of data from one full-scale treatment system operated over a range of pH levels. As demonstrated in Table 11.5, best treatment for lead, and best suspended solids clarification, occurred above pH 10. These results are consistent with the data of Table 11.4. However, at this high pH, the efficiency of zinc precipitation declined significantly. Noncompatible pH optima among metals contained in a multimetal waste often requires use of a compromise, nonoptimal treatment pH, or operation of the system at the pH for which best results are achieved for the metal representing the most stringent effluent limitation.

The impact of lead hydroxide precipitation treatment at nonoptimal pH is evidenced by the results of one plating rinsewater treatment system. Treatment was at pH 7.7 by batch reaction, followed by 8 hr sedimentation. Influent lead of 98 mg/l was reduced by only 40% to 39 mg/l, at the treatment pH used [7]. In another instance, lime treatment at pH 5.5–8.0 resulted in an average

Table 11.5 Effect of Treatment pH on Effluent Lead, Zinc and Suspended Solids [52]

	Effluent Concentration (mg/l)[a]		
Treatment pH	*Lead*	*Zinc*	*Suspended Solids*
8.5	1.67	2.93	87
9.2	1.05	1.91	43
9.8	1.00	2.80	27
10.5	0.62	2.88	28

[a] Results are averages of 4–9 sets of effluent samples.

effluent lead level of 3 mg/l, but peak discharge of 100 mg/l [9]. These treatment results typify the performance of processes operating at nonoptimal pH, often aggravated by ineffective precipitate solids removal. As a partial solution, some plants incorporate effluent filtration to reduce effluent suspended solids [9, 19]. During times of erratic clarifier performance, a filter provides overall improvement and more consistent effluent quality. Filter effluent lead levels of 0.03–0.08 mg/l have been reported from one direct-filtration demonstration study of a battery-manufacturing wastewater [23]. Filtration in a full-scale radiator-manufacturing wastewater treatment plant reduced clarifier effluent lead from 0.32 to 0.08 mg/l [53].

In forming insoluble lead hydroxide, lime is the treatment chemical of choice, although caustic has been used. Treatment results from one study, which compared caustic and lime, were equivalent [42]. However, that same study indicated an interference with lead hydroxide precipitation as calcium ion concentration increased. At treatment pH 10.5, residual lead increased from 0.4 to 0.8 mg/l as calcium concentration increased from 25 to 375 mg/l, expressed as equivalent calcium carbonate.

A survey of lead wastewater treatment practices at eight battery-manufacturing plants indicated that simple sedimentation was relatively ineffective in removing particulate lead and, of course, ineffective in reducing dissolved lead [2]. Data reported for five plants which used pH adjustment, followed by (for four plants) sedimentation, are presented in Table 11.6. Improved effluent levels were observed at more alkaline pH and for long settling times. The benefit of effective settling is emphasized by the performance of Plant E, which had no settling and a high particulate effluent lead content. Plant E also had the highest soluble effluent lead, probably as a result of using a complexing agent (ammonium hydroxide) for pH adjustment. Complexation of lead with ammonia would prevent effective precipitation. Among the plants surveyed, the best effluent was obtained by addition of caustic (NaOH) to pH 7.0, followed by 24 hr settling.

In pilot studies on carbonate precipitation of lead battery wastes, sodium carbonate was employed as the treatment chemical [3]. Precipitation plus sand filtration yielded effluent total lead of 0.2–3.8 mg/l, and soluble lead of 0.1 mg/l.

Table 11.6 Treatment Practice and Effluent Lead Levels from Battery Plants [2, 4]

Plant	*Treatment Chemical*	*pH*	*Settling Time (hr)*	*Effluent Lead (mg/l)*		
				Dissolved	*Suspended*	*Total*
A	Caustic	5.5	1	0.6	1.0	1.6
B	Lime	6.8	30	0.30	0.18	0.48
C	Caustic	7.0	24	0.02	0.02	0.04
D	Lime	8.3	—[b]	0.13	0.14	0.27
E	Ammonia[a]	7.8	0	1.9	22.0	23.9

[a] Anhydrous ammonia used for pH adjustment.
[b] Clarifier plus polishing lagoon.

Treatment was at pH 6.4–8.7. One bench-scale study of lead carbonate precipitation indicated best treatment at a total carbonate alkalinity (as equivalent $CaCO_3$) of 200 mg/l, with a soluble lead level of 0.05 mg/l achieved at treatment pH 8.5 [42]. However, comparison treatability studies for a foundry wastewater wherein lime was compared with lime plus soda ash (50–500 mg/l Na_2CO_3) at treatment pH 10.0–10.5 resulted in higher lead levels in the clarifier effluent of the lime plus soda ash system, but equal lead levels in the filter effluents of the two systems [40]. The performance of the two systems is compared in Table 11.7.

Precipitation of dissolved lead by dolomite ($CaCO_3 \cdot MgCO_3$) to yield lead carbonate has been reported [54]. The process involved filtration of lead wastewater through a calcined dolomite bed, which reacted with and precipitated the dissolved lead. The dolomite filters were regenerated by backwashing with water, which removed the lead carbonate formed in the process. The impact of magnesium ion, derived from the dolomite, on lead precipitation is somewhat complex. One study indicated maximum interference with lead hydroxide precipitation at a magnesium level of 150 mg/l expressed as equivalent calcium carbonate, with less interference at both higher and lower magnesium concentrations [42].

The reaction of lead with dolomite, yielding insoluble lead carbonate, results in very little lead entering receiving streams from southeastern Missouri lead mines [11, 12, 31]. The lead veins are located in dolomitic deposits, and

Table 11.7 Comparison of Lime with Lime plus Soda Ash for Lead Removal from a Foundry Wastewater [40]

Lead (mg/l) Waste	*Lime System*	*Lime + Soda Ash System*
Influent	60.5	58.1
Clarifier effluent	1.8	2.4
Filter effluent	0.6	0.6

Table 11.8 Sulfide Precipitation Treatment for Lead [30]

	Treatment pH	
Treatment Sequence	*8.4–9.1*[a]	*9.9–10.0*[b]
Raw wastewater	578	468
Lime, clarified		
Total	12.1	35.5
Soluble	6.1	5.1
plus Sulfide, clarified		
Total	8.5	11.4
Soluble	5.0	5.1

[a] Average of four runs.
[b] Average of nine runs.

the mine tailings thus contain crushed dolomite. As the mine tailings are channeled through sedimentation basins prior to release of the waste effluent, dolomite plus reacted lead carbonate are readily removed. Under existing conditions, any heavy metals in the wastewater are precipitated in the settling lagoons at the slightly basic pH (7.5–8.2) of the waste.

The use of lime plus ferrous sulfide has been proposed to treat wastewater from the chlor-alkali industry [30]. Averaged results for several treatability runs are presented in Table 11.8. The raw wastewater was at initial pH of 11.4–11.7, and high in carbonate alkalinity. After acidification, the wastewaters were neutralized with lime and clarified. Residual soluble lead levels (Table 11.8) are consistent with the carbonate system results presented in Table 11.4. Lime clarified supernatant was then treated with ferrous sulfide, and further clarified. The sulfide treatment did not enhance soluble lead removal. Total lead removal was improved, apparently due to the increment in settling time associated with the sulfide treatment.

The use of sodium phosphate (Na_3PO_4) has been reported for lead treatment, to precipitate lead phosphate [3, 10]. The solubility of lead phosphate is determined by treatment pH and phosphate dosage. Treatment conditions generally specified are pH above 7.0, with up to three times excess sodium phosphate added (3.7 lb sodium phosphate/1b lead). This may however result in unacceptable effluent phosphate levels. Treatment of a battery waste with sodium phosphate at pH 7.2–7.4 has been reported to achieve 0.2–0.6 mg/l total lead and 0.1 mg/l soluble lead [3]. Equivalent lead treatment has been reported for phosphate treatment of an electronic tube-manufacturing waste [10]. Table 11.9 presents a cost comparison (in 1975) for lead precipitation treatment, based upon laboratory studies. Most effective single-chemical treatment was obtained at pH 7.0 with trisodium phosphate, to yield a residual soluble lead level of 0.1 mg/l. Projected treatment costs were extremely high, however.

Conventional treatment for lead involves lime or caustic precipitation, followed by clarification and perhaps filtration. Operating costs are $1.00–2.00/

Table 11.9 Comparison of Lead Wastewater Treatment Chemicals and Costs (1975) [26]

Treatment Chemical[a,b]	*Optimal Dosage (g/l)*	*Treatment pH*	*Residual Soluble Lead*[c] *(mg/l)*	*Treatment Cost ($/1,000 gal)*
Lime	2.80	10.2	0.20	0.52
Caustic	2.91	8.9	0.50	1.93
Soda ash	6.40	7.1	2.50	3.84
Trisodium phosphate (TSP)	12.0	7.0	0.10	14.22
Caustic + TSP	2.5 + 1.0	8.5	0.02	2.84

[a] Initial wastewater pH = 1.6; initial lead = 380 mg/l.
[b] Lime = $0.0221/lb; caustic = $0.0798/lb; soda ash = $0.0720/lb; TSP = $0.1420/lb.
[c] Analytical detection limit = 0.02 mg/l lead.

1,000 gal treated. Capital costs for systems of this type were presented in Table 1.5.

Coagulation

Treatment of a wastewater from a tetraethyl lead plant has been described [49]. The effluent was alkaline, and contained inorganic lead salts in solution and suspension, plus organic lead salts. The pH was adjusted to 8–9, and inorganic lead precipitated by addition of lime. The precipitation was assisted by the addition of ferrous sulfate, which acted as a coagulating agent. The treated effluent flowed to a sedimentation basin, from which the lead sludge was collected, dewatered on a filter, and reclaimed by a lead slag refinery.

Nozaki and Hatotani [33] also have described treatment of wastewaters from a tetraethyl lead-manufacturing process. The two major categories of waste were inorganic lead wastewaters and organic lead wastewaters. After sedimentation in a holding basin to recover solid lead and lead oxide, the inorganic lead waste fraction (66.1 mg/l) was effectively treated by coagulation with ferric and ferrous sulfate. Treatment of combined organic and inorganic lead waste streams containing respective concentrations of 13.3 mg/l and 84.9 mg/l was not as effective, however. Effluent contained "trace" to 1.4 mg/l inorganic lead, and most of the influent organic lead. A separate bench-scale treatment process was therefore evaluated for the organic lead waste stream. The organic lead compounds were removed effectively on a strongly acid cation exchange resin. Reduction of organic lead from 126.7 to 144.8 mg/l to 0.020–0.53 mg/l (as lead) were reported. The organic lead compounds adsorbed on the ion exchange resin were eluted by caustic soda. This eluate was collected in a reaction basin into which chlorine gas was injected at 95°C for 45 min. The organic lead compounds were thereby almost completely converted to

Table 11.10 Lead Removal by Precipitation and Sedimentation plus Filtration [56]

Treatment Chemical	*Treatment pH*	*Effluent Lead Concentration (mg/l)*	
		Sedimentation	*Filtration*
Ferric sulfate at 45 mg/l Fe	6.0	0.25	0.030
Lime at 260 mg/l + ferric sulfate at 20 mg/l Fe	10.0	0.25	0.029
Lime at 600 mg/l	11.5	0.20	0.019

inorganic lead compounds, which could be treated as described on preceding page. Overall lead removals to less than 1 mg/l residual were projected for the final plant effluent, after installation of the organic-lead wastewater treatment sequence [33].

A third report on treatment of lead-bearing waste from a tetraethyl plant described a similar treatment sequence for inorganic lead wastewaters [34]. Lead occurring in the manufacturing process effluent at a concentration of 45 mg/l was reduced to 1.7 mg/l by precipitation at pH 10.4–10.8, aided by addition of ferrous sulfate. The lead sludge was recovered, and lead was extracted and reclaimed. Waste-treatment capital costs represented 4.2% of the total plant costs. Waste treatment included settleable solids and chloride control, in addition to lead removal.

Similar treatment has been reported for lead removal from a municipal water supply [55]. Lime and ferrous sulfate are added to the water to precipitate and coagulate lead leached into the water supply from a nearby abandoned mine. Pilot plant results, presented in Table 11.10, indicate that coagulation at pH 6 by ferric sulfate treatment provides near-equivalent lead removal results to lime precipitation. The process involved coagulation, sedimentation, and filtration. These pilot plant data emphasize the importance of effective solids removal. Initial lead concentration was 5 mg/l, and filtration was through dual media. As the data of Table 11.10 reveal, effluent levels almost an order of magnitude lower were achieved by sedimentation plus filtration, as opposed to sedimentation only.

Alum coagulation at pH 6.8–7.0 has been reported to reduce lead from 17 to 1.3 mg/l [48]. A second study on alum coagulation revealed that lead removal efficiency increases with alum dosage. Alum at 50 mg/l yielded residual lead of 1.3 mg/l, while alum at 200 mg/l resulted in a lead level of 0.7 mg/l [42].

Other Processes

Successful ion exchange treatment has been reported for both organic and inorganic lead [32, 33, 57]. In one study on treatment of an ammunition industry

Table 11.11 Summary of Reported Lead Treatment Processes and Efficiencies

Treatment Process	Treatment pH	Lead Concentration (mg/l) Initial	Lead Concentration (mg/l) Final	Removal %	Reference
Ion exchange	5.0–5.2	0.1	0.01	90	32
Ion exchange[a]	—	0.055	0.0015	97.3	57
	—	126.7–144.8	0.020–0.053	99.9+	33
Lime + sedimentation	8.3	11.7	0.27	97.7	4
	7.1	0.91	0.19	79.1	17
	8.2	1.2	0.15	87.5	17
	—	30	1	96+	28
	—	6.5	0.1	98.5	32
	—	0.31	0.1	97+	13
Lime + 8-hr sedimentation	7.7	98	39	40	7
Lime + 30-hr sedimentation	6.8	—	0.48	—	2
Caustic + 1-hr sedimentation	5.5	—	1.6		
Caustic + 24-hr sedimentation	7.0	—	0.04		
Ammonium hydroxide	7.8	—	23.9		
Lime + ferric sulfate +	10.0	5.0	0.25	95	56
sedimentation + filtration[a]			0.029	99.4	
Lime + sedimentation +	11.5	5.0	0.20	96	56
filtration[a]			0.019	99.6	
	9.5	0.9–4.6	0.11–0.21	—	19
	9.5	0.9–4.6	0.10–0.17	—	
Caustic + soda ash[a]	9.0–9.5	5.0	0.01–0.03	99+	44
Sodium carbonate + filtration[a]	6.4–8.7	10.2–70.0	0.2–3.6	82–99+	3
Sodium phosphate + filtration[a]	7.2–7.5	3.0–5.0	0.2–0.6	83–93	3
Ferric sulfate + sedimentation +	6.0	5.0	0.25	95	56
filtration[a]	6.0	5.0	0.03	99.4	
Ferrous sulfate + sedimentation	10.4–10.8	45	1.7	96.2	34

[a] Pilot-plant results.

waste, lead was initially reduced from 6.5 to 0.1 mg/l by precipitation. Ion exchange treatment of the effluent at pH 5.0–5.2 with a phosphoric acid resin further reduced the lead to 0.01 mg/l [32]. In another application, an effluent level of 0.002 mg/l was achieved [57]. Organic lead reductions from above 100 mg/l to below 0.5 mg/l were reported, using a strong-acid cation-exchange resin [33].

Activated carbon treatment has been suggested for organic lead [58, 59], although there are no reports of full-scale applications. In one laboratory study, organic lead initially at 6 mg/l was reduced to 0.4 mg/l by carbon treatment [58]. A separate laboratory study yielded organic lead removal of 58–96% by carbon adsorption [59].

SUMMARY

Information on the treatment of lead-bearing wastewaters is summarized in Table 11.11. Precipitation by lime or dolomite and sedimentation, with sludge disposal or lead recovery, have been shown to be effective, as has ion exchange.

Treatment efficiencies by precipitation plus sedimentation exceeding 99% have been observed in both pilot plant and full-scale operations, and results indicate that filtration can improve effluent levels an order of magnitude. There is also an indication that long settling-tank detention times (one day or greater) may match filter performance.

REFERENCES

1. Hall, S.K. "Lead Pollution and Poisoning," *Environ. Sci. Technol.* 6 (1972): 30–35.
2. Haas, W.R. and S. Miller. "Evaluations of Lead Plant Wastewater Treatment Methods," IIT Research Institute C8213-2 (1972).
3. Crandall, C.J. and J.R. Rondenberg. "Waste Lead Oxide Treatment of Lead Acid Battery Manufacturing Wastewater," paper presented at the 29th Annual Purdue Industrial Waste Conference, West Lafayette, IN (May 1974).
4. Mezey, E.J. "Characterization of Priority Pollutants from a Secondary Lead and Battery Manufacturing Facility," USEPA 600/2-79-039 (January 1979).
5. Hill, J.B. "Waste Problems in the Petroleum Industry," *Ind. Eng. Chem.* 31 (1939): 1361–1363.
6. Pinkerton, H.L. "Waste Disposal," *Electroplating Engineering Handbook, 2nd ed.*, A. Kenneth Graham, ed. (New York: Reinhold, 1962).
7. "Development Document for Interim Final Effluent Limitations Guidelines and Proposed New Source Performance Standards for the Common and Precious Metals Segment of the Electroplating Point Source Category," USEPA 440/1-75/040, Group I, Phase II (April 1975).
8. Fales, A.L. "A Plating Waste Disposal Problem," *Sew. Works J.* 20 (1948): 857–860.

9. "Development Document for Interim Final Effluent Limitations Guidelines and Proposed New Source Performance Standards for the Significant Inorganic Products Segment of the Inorganic Chemicals Manufacturing Point Source Category," USEPA 440/1-75/037, Group I, Phase II (May 1975).
10. Rohrer, K.L. "An Integrated Facility for the Treatment of Lead and Fluoride Wastes," *Ind. Water* 17 (1971): 36–39.
11. Wixson, B.G., E.A. Bolter, N.H. Tibbs, and A.R. Handler. "Pollution from Mines in the New Lead Belt of Southeastern Missouri," *Proceedings of the 24th Industrial Waste Conference, Purdue University* (May 1969), pp 632–643.
12. Jennett, J.C. and B.G. Wixson. "Treatment and Control of Lead Mining Wastes, in S.E. Missouri," *Proceedings of the 26th Industrial Waste Conference, Purdue University* (May 1971).
13. "Annual Report Western Australian Government Chemical Laboratories," Department of Mines, Australia (1962).
14. Larsen, H.P., J.K.P. Shou, and L.W. Ross. "Chemical Treatment of Metal-Bearing Mine Drainage," *J. Water Poll. Control. Fed.* 45 (1973): 1682–1695.
15. Hallowell, J.G., J.F. Shea, G.R. Smithson, Jr., A.B. Tripler, and B.W. Gonser. *Water Pollution Control in the Primary Nonferrous Metals Industry*, vol. I. *Copper, Zinc and Lead Industries*, USEPA R2-73-247-a (September 1973).
16. "Development Document for Interim Final Effluent Limitations Guidelines and Proposed New Source Performance Standards for the Lead Segment of the Nonferrous Metals Manufacturing Point Source Category," USEPA 440/1-75/032-a (February 1975).
17. "Development Document for Interim Final Effluent Limitations Guidelines and Proposed New Source Performance Standards for the Primary Copper Smelting Subcategory and the Primary Copper Refining Subcategory of the Copper Segment of the Nonferrous Metals Manufacturing Point Source Category," USEPA 440/1-75/032-b (February 1975).
18. Hill, R.D. "Control and Prevention of Mine Drainage," paper presented at the Environmental Resources Conference on Cycling and Control of Metals, Battelle-Columbus Laboratories, Columbus, OH (1972).
19. Huck, P.M. and B.P. LeClair. "Treatment of Base Metal Mine Drainage," paper presented at the 30th Annual Purdue Industrial Waste Conference, West Lafayette, IN (May 1975).
20. "Development Document for Effluent Limitations Guidelines and New Source Performance Standards for the Steel Making Segment of the Iron and Steel Manufacturing Point Source Category," USEPA 440/1-74-024-a (June 1974).
21. "Development Document for Effluent Limitations Guidelines and New Source Performance Standards for the Major Inorganic Products Segment of the Inorganic Chemicals Manufacturing Point Source Category," USEPA 440/1-74-007-a (March 1974).
22. "Development Document for Effluent Limitations Guidelines and New Source Performance Standards for the Fabricated and Reclaimed Rubber Segment of the Rubber Processing Point Source Category," USEPA 440/1-74/030-a Group I, Phase II (December 1974).
23. Liu, H. and J. Blacklidge. "Cross Flow Filtration Technology for Metal Finishers," *Proceedings of the 4th Conference on Advanced Pollution Control for the Metal Finishing Industry*, USEPA 600/9-82-022 (December 1982).

24. Yeats, A.R. "Ion Exchange Selectively Removes Heavy Metals from Mixed Plating Wastes," paper presented at the 32nd Annual Purdue Industrial Waste Conference, West Lafayette, IN (May 1977).
25. Reece, C. "Steel Welding Wire Manufacturer Meets Stringent Effluent and Sludge Disposal Criteria," paper presented at the 37th Annual Purdue Industrial Waste Conference, West Lafayette, IN (May 1982).
26. Rohrer, K.L. "Chemical Precipitants for Lead-Bearing Wastewaters," *Ind. Water Eng*. (1975): 13–17.
27. Jashnani, I.L. and R. Green, "Pretreatment System for a Circuit Board Plant," paper presented at the 8th National Conference on Municipal Sludge Management, Municipal and Industrial Sludge Management, Disposal and Utilization, Miami Beach, FL (March 19–21, 1979).
28. "Development Document for Effluent Limitations Guidelines and New Source Performance Standards for the Pressed and Blown Glass Segment of the Glass Manufacturing Point Source Category," USEPA 440/1-75/034-a, Group I, Phase II (January 1975).
29. Cawley, W., ed. *Treatability Manual*, vol. I. *Treatability Data*, USEPA 600-8-80-042-a (July 1980).
30. *Treatability Studies for the Inorganic Chemicals Manufacturing Point Source Category*, USEPA 440/1-80-103 (July 1980).
31. Bolter, E., J.C. Jennett, and B.G. Wixson. "Geochemical Impact of Lead Mining Wastewaters on Streams in Southwestern Missouri," paper presented at the 27th Annual Purdue Industrial Waste Conference, West Lafayette, IN (May 1972).
32. Shannon, R.E. "Some Examples of the Concentration of Trace Heavy Metals with Ion Exchange Resins," *Traces of Heavy Metals in Water, Removal Processes and Monitoring*. Center for Environmental Studies, Princeton University (1973).
33. Nozaki, M. and H. Hatotani. "Treatment of Tetraethyl Lead Manufacturing Wastes," *Water Res*. 1 (1967): 167–177.
34. Robb, L.A. "Waste Water Treating Facilities at Ethyl Corporation of Canada, Ltd.," *Proceedings of the 8th Ontario Industrial Waste Conference* (1961), pp 90–96.
35. Cottrell, W.R.T. "Treatment of Effluent by Flotation Process," *Effluent Water Treatment J.* (November 1976): 563–567.
36. "Development Document for Effluent Limitations Guidelines and New Source Performance Standards for the Oil Base Solvent Wash Subcategories of the Paint Formulating and the Ink Formulating Point Source Category," USEPA 440/1-75/050-a, Group II (July 1975).
37. Campbell, H.J. Jr., N.C. Scrivner, K. Batzar, and R.F. White. "Evaluation of Chromium Removal from a Highly Variable Wastewater Stream," *Proceedings of the 32nd Annual Industrial Waste Conference, Purdue University* (May 1977), pp 102–115.
38. Rouse, J.V. "Removal of Heavy Metals from Industrial Effluents," *ASCE J. Environ. Eng. Div*. (October 1976) EE5 929–930.
39. Arthur D. Little, Inc. *Physical, Chemical and Biological Treatment Techniques for Industrial Wastes*, vol. II, USEPA Report (1977).
40. Osantowski, R. and J. Ruppersberger. "Upgrading Foundry Wastewater Treatment," paper presented at the 39th Annual Purdue Industrial Waste Conference, West Layfayette, IN (May 1984).

41. Carter, J., M. Jones, and R. Tiefenbrunn. "Treatment for Removal of Heavy Metals and Oil from a Piston Ring Manufacturing Waste," paper presented at the 37th Annual Purdue Industrial Waste Conference, West Lafayette, IN (May 1982).
42. Naylor, L.M. and R.R. Dague. "Simulation of Lead Removal by Chemical Treatment," *J. Am. Water Works Assoc.* 67 (1975): 560–565.
43. Hem, J.D. and W.H. Durum. "Solubility and Occurrence of Lead in Surface Water," *J. Am. Water Works Assoc.* 651 (1973): 562–568.
44. Day, R.V., E.T. Lee, and E.S. Hochuli. "Bell Systems' Metals Recovery Plant," *Ind. Wastes* 20 (1974): 26–29.
45. Patterson, J.W., H.E. Allen, and J.J. Scala. "Carbonate Precipitation of Heavy Metals Pollutants," *J. Water Poll. Control. Fed.* 49 (December 1977): 2397–2410.
46. Dean, J.G., F.L. Bosqui, and K.L. Canouette. "Removing Heavy Metals from Waste Water," *Environ. Sci. Technol.* 6 (1972): 518–522.
47. "Effluent Treatment at B.S.R.," *Metal Fin. J.* 17 (1971): 201–248.
48. Nilsson, R. "Removal of Metals by Chemical Treatment of Municipal Waste Water," *Water Res.* 5 (1971): 51–60.
49. "An Integrated Plant for Tetraethyl Lead," *Ind. Chemist.* 30 (1954): 429–436.
50. Netzer, A., A. Bowers, and J. Norman. "Removal of Heavy Metals from Wastewater by Lime," *Proceedings of the 16th Conference on Great Lakes Research*, Huron, OH (1973).
51. McAndrew, P. "Corrosion, Control through Carbonate Supplementation," M.S. thesis, Illinois Institute of Technology (December 1979).
52. McVaugh, J. and W.T. Wall, Jr. "Optimization of Heavy Metals Wastewater Treatment Effluent Quality versus Sludge Treatment," *Proceedings of the 31st Industrial Waste Conference, Purdue University* (May 1976), pp 17–25.
53. Knapp, R. and E. Paulson. "Gravity Filtration Reduces Suspended Solids in a Lime Precipitation Treatment System," paper presented at the 37th Annual Purdue Industrial Waste Conference, West Lafayette, IN (May 1982).
54. Voznesenskic, S.A., A.B. Evalonova, and R. V. Suvorova. "Physico-Chemical Methods of Purification of Industrial Waste Waters," *Water Poll. Abstr.* 13 (1940): 135.
55. "East Midland Districting Meeting," *J. Inst. Munic. Eng.* 84 (1957): XXV–XXVCC.
56. Maruyama, T., S.A. Hannah, and J.M. Cohen. "Removal of Heavy Metals by Physical and Chemical Treatment Processes," *J. Water Poll. Control. Fed.* 47 (1975): 962–975.
57. Licbig, C.F., Jr., A.P. Vanselow and H.D. Chapman. "The Suitability of Water Purified by Synthetic Ion-Exchange Resins for the Growing of Plants in Controlled Nutrient Cultures," *Soil. Sci.* 55 (1943): 371–376.
58. Otto, J.M. "Removing Dissolved Organo-Lead Compounds from Water," *Chem. Abstr.* 88 (1978): 372.
59. Edmondson, T. "Removal of Organo-Lead Compounds from Aqueous Media by Reaction and Adsorption," *Chem. Abstr.* 87 (1977): 307.

12
Manganese

Manganese and its salts are used in the steel alloy, dry-cell battery, glass and ceramics, paint and varnish, ink and dye, fertilizer, and match and fireworks industries [1].

Among the many forms and compounds of manganese, only the manganous salts and the highly oxidized permanganate anion are appreciably soluble. The latter is a strong oxidant which is reduced under normal circumstances to insoluble manganese dioxide. The highly soluble manganous chloride is used in dyeing operations, linseed oil dryers, and electric batteries, while the equally soluble manganous sulfate is used in porcelain glazing, varnishes, and production of specialized fertilizers.

Manganese concentrations of 1.2–8.0 mg/l have been reported for mine waters [2], with a value of 50 mg/l for one acid mine water [3] and 30 mg/l for another [4]. Reported ranges for coal mine waters by subcategory are presented in Table 12.1. Manganese concentrations have been reported to be 60 mg/l in a municipal refuse-incinerator scrubber water [6], 833 mg/l in venturi scrubber water from a ferromanganese blast furnace [7], and 5.7–13.3 mg/l in steel industry degassing-plant wastes [7] and to range up to 1,322 in various wet-air pollution-control scrubber waters from the smelting of ferroalloys [8]. In the latter industry, slag pond discharges were reported to contain 0.5–10.8 mg/l of manganese [8]. One manganese value of 10 mg/l in paint-formulating plant wastes was reported [9]. Concentrated wastewaters from electrolytic manganese production may contain 1,000–6,700 mg/l manganese, while dilute wastewaters may range from 22 to 128 mg/l [10].

Since the chemistry of manganese is similar to that of iron [11], it would be expected that any pickling operation involving manganese steel alloy results in dissolution of manganous ion and the presence of this ion in pickling and rinse solutions. In contrast to iron, the divalent (manganous) form is not readily oxidized to the insoluble manganic form, other than at elevated pH [11].

TREATMENT TECHNOLOGY

Removal of soluble manganese has been practiced for many years in the treatment of municipal water supplies and industrial process waters. Low-manganese waters are essential for many industries, including food and beverage.

Table 12.1 Manganese Concentrations in Coal Mine Waters [5]

	Manganese Concentration (mg/l)	
Mine Category	*Range*	*Avg*
Subsurface—alkaline drainage	0.01–0.41	0.08
Subsurface—acidic drainage	0.04–92	7.3
Surface—alkaline drainage	0.01–6.8	0.61
Surface—acidic drainage	0.29–440	50.1
Coal-preparation plant process water	0.30–4.21	1.67

Treatment technology basically involves conversion of the soluble manganous ion to an insoluble precipitant. In water treatment, this generally means simultaneous removal of iron, since conditions under which high iron levels occur in water supplies are essentially the same as those for soluble manganese. Removal is effected by oxidation of the manganous ion, and separation of the resulting insoluble oxides and hydroxides of manganese. Manganese removal processes are listed in Table 12.2.

Aeration

Because of the low reactivity of manganous ion with oxygen, simple aeration is not an effective oxidation technique below pH 9, and even above that pH the oxidation reaction is slow [12]. Organic matter in solution can combine with manganese and prevent its oxidation by simple aeration, even at high pH [13]. Figure 12.1 illustrates the effect of pH on manganese removal by lime. A reaction pH above 9.4 is required to achieve significant manganese removal.

Hill [2] indicates that manganese can be reduced to less than 0.5 mg/l by simple neutralization and precipitation, but no pH values are reported. He further states that the use of calcium carbonate plus lime for iron and manganese removal in Japanese acid mine waters has proved successful. Treatment of deep coal-mining drainage waters by addition of lime slurry or sodium hydroxide has been practiced in Pennsylvania and West Virginia [5]. Generally, mechanical or natural aeration devices are employed, primarily for iron oxidation. Manganese removal is also accomplished, even though pH is not optimal for manganese oxidation. It is probable that the iron acts as a coagulant to enhance manganese removal. Performance data for 13 mines are presented in Table 12.3. Generally, holding tanks and lagoon discharge were employed. Removals range from 53.9 to 99.3%, with an average of 87.3%.

A combined waste from an electrolytic manganese facility was reduced from 3,158 mg/l to 91 mg/l manganese by liming and discharging into a lagoon [10]. Sulfide precipitation, used to remove heavy metals from acid mine drain-

Table 12.2 Manganese Removal Processes [12]

Process	*Comments*
1. Aeration	Slow and ineffective below pH 9
2. Chlorine and hypochlorite oxidation	Not effective for organically bound manganese.
3. Adjustment of pH	Lime-soda type treatment gives removal at pH 9.5.
4. Catalysis	Copper ion enhances air oxidation.
5. Ion exchange	Effective for small quantities of iron and manganese. Resins quickly fouled by iron and manganese oxides.
6. Chlorine dioxide	Rapidly oxidizes manganese to the insoluble form, but expensive.
7. Manganese dioxide or potassium permanganate	Regeneration, greensand filter process. Economic disadvantage of requiring excess permanganate.
8. Direct potassium permanganate addition	Requires filtration (sand, diatomite, or mixed media).

age at pilot scale, achieved reduction of manganese from 2 mg/l to 0.4 mg/l [15].

Chemical Oxidation

The use of chemical oxidants to convert manganous ion to insoluble manganese dioxide, in conjunction with coagulation and filtration, has been recommended, although this technique may cause filtration problems. The character of manganese flocs produced when strong oxidants are used has been reported to rapidly clog slow sand filters [13]. Despite reports that manganese oxidation by chlorination may not be effective [12], successful manganese removals have been reported. Owens [14] reports removal of manganese at concentrations as high as 5.7 mg/l, with chlorination followed by lime coagulation and rapid sand filtration. Split-flow treatment was employed, to achieve intermediate manganese levels in the final effluent. Chemical costs (in 1963), adjusted to the treated stream only, were about 2.1¢/1,000 gal. Sodium silicate was used as a coagulant aid. Effective manganese oxidation by chlorination requires a free chlorine residual of 0.5 mg/l, and a dosage of 1.29 mg chlorine/mg manganese

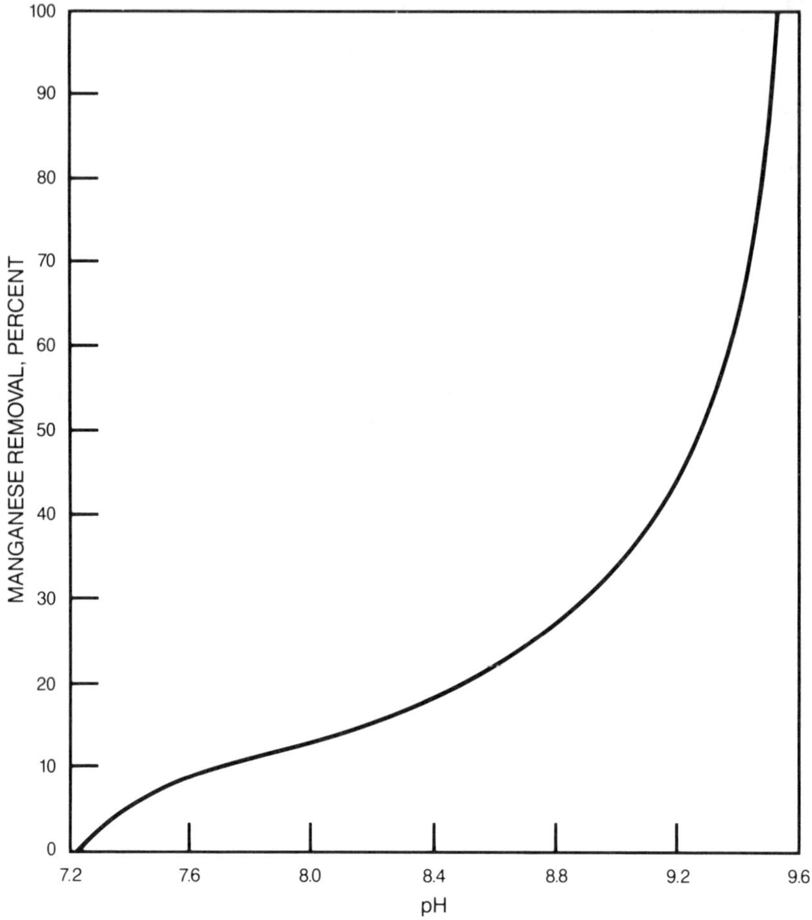

Figure 12.1 Effect of pH on efficiency of manganese removal using lime. (From Owens [14], courtesy of *Water and Sewage Works*.)

oxidized [16]. Figures 12.2–12.4 summarize costs associated with chlorine oxidation systems.

The successful use of potassium permanganate as a chemical oxidant has been reported. Addition of 1.5–2.0 mg/l potassium permanganate reduced 1.0 mg/l manganese to less than 0.05 mg/l [18]. Addition is 30 minutes upstream from the prechlorination point, and is followed by lime-alum-sodium silicate coagulation (4–10 hr detention), and rapid sand filtration. Costs attributed to permanganate addition are less than 0.3¢/1,000 gal. Since iron is generally present, the quantity of permanganate used must include that required for oxidation of the iron content. The stoichiometric relationships between permanganate required and the quantities of iron and manganese present are shown in Figures 12.5 and 12.6, respectively. The ability to remove manganese at lower pH (an

Table 12.3 Manganese Removal Performance for Treatment of Subsurface Acid Coal-Mine Drainage Waters [5]

		pH		*Manganese (mg/l)*	
Identification	*Treatment System*[a]	*Influent*	*Effluent*	*Influent*	*Effluent*
A	HLS	2.9	7.2	9.17	1.09
B	HLS, NA	2.6	6.8	21.0	1.1
C	HLS, NA	2.5	7.9	4.3	0.12
D	HLS, NA	3.0	8.9	3.83	0.07
E	DHL, NA	5.8	8.0	8.13	2.46
F	HLS, MA	2.7	6.9	8.8	0.47
G	HLS, MA	5.9	7.8	3.2	0.66
H	HLS, MA	2.9	7.9	3.2	0.25
I	HLS, MA	2.7	8.2	7.0	0.55
J	HLS, MA	3.1	8.4	11.5	0.78
K	HLS, MA	2.5	8.8	4.25	0.03
L	NaOH, MA	6.4	7.7	16.0	2.8
M	Aeration + Coagulant Aid	7.7	8.0	4.19	1.93

[a] HLS, hydrated lime slurry; DHL, dry hydrated lime; NA, natural aeration; MA, mechanical aeration; NaOH, sodium hydroxide.

advantage when simultaneous iron removal is desired) by permanganate oxidation is demonstrated by the data of Holland [20]. The addition of 0.04 mg/l sodium alginate reduced 0.35 mg/l manganese to less than 0.02 mg/l in the filtered supernatant. The absence of residual taste or odor has been cited as an additional advantage of permanganate use over chlorination in water supply applications [13].

The application of permanganate oxidation in conjunction with vacuum diatomite filtration has been used to treat a municipal water supply [21]. Manganese is reduced from 1.23 mg/l to 0.05 mg/l or less. Capital cost for the 2.33 MGD unit was given as \$107/1,000 gpd. No operating costs were provided, but the unit requires a total of 250 lb diatomite/million gallons of water filtered.

Permanganate may be used indirectly for manganese removal in the manganese greensand filter method. The filter is a combination of New Jersey glauconite (greensand), manganese sulfate, and potassium sulfate, which results in a filter bed of sand grains containing high oxides of manganese [19, 22]. Upon contact with manganous ion, an oxidation-reduction reaction occurs, resulting in the formation of insoluble manganese dioxide, which remains in the filter bed. The bed is periodically reoxidized with permanganate, and solids are backwashed from the bed. With this system, manganese was reduced from 1.7 mg/l to below 0.1 mg/l at pH 7.2, using a pressure greensand system [23]. The plant, of 700 gpm capacity, cost (in 1963) \$26/1,000 gpd completely installed and operating, compared with an estimated \$74/1,000 gpd for a conventional

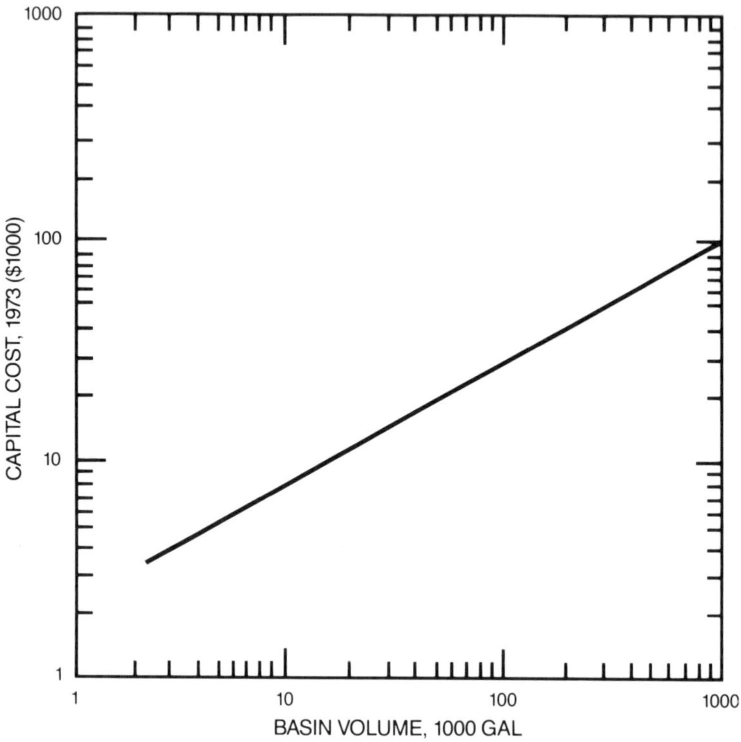

Figure 12.2 Capital cost (in 1973) of chlorine contact basins [17].

aeration, lime coagulation-clarifier-filter-recarbonation system. Chemical costs were 3.5¢/1,000 gal. The water had a manganese concentration of 13 mg/l, which required excess permanganate. Treatment costs were roughly equivalent to lime chemical costs required for a similar treatment level. Regeneration for this plant was on a continuous basis with periodic backwashing of the greensand filter. Generalized removal capacities for greensand filters are given as 0.09 lb manganese/ft^3 of filter volume, with regeneration requirements of 0.18 lb/ft^3 potassium permanganate. A flow rate of 3 gpm/ft^2 is typical [24].

Ozone

The use of ozone in conjunction with lime has been investigated as a manganese removal technique in laboratory studies. The data of Table 12.4 are results for solutions of manganous chloride initially containing 94 mg/l manganese.

Ion Exchange

The ion exchange process has been utilized for iron and manganese removal. Many advantages are cited, including smaller capital investment than coagu-

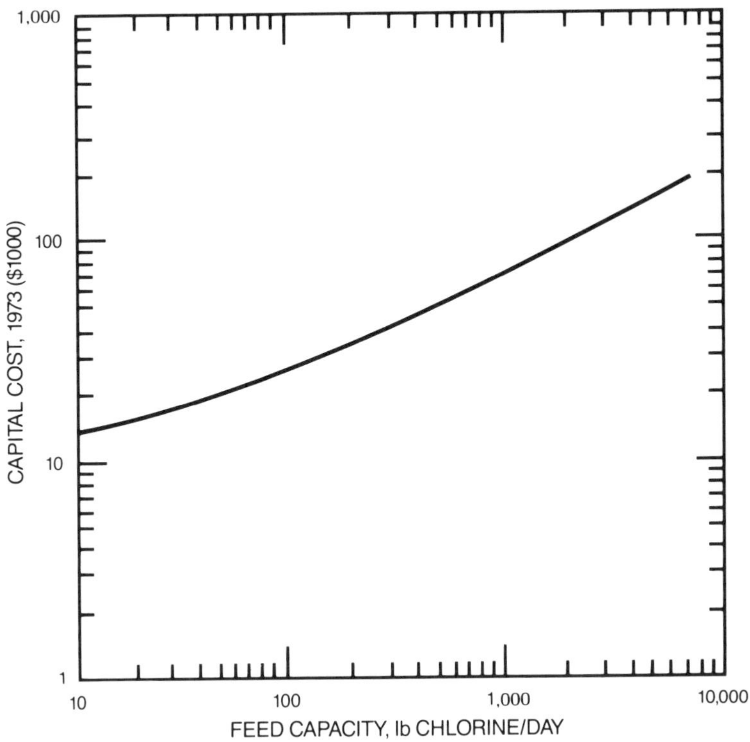

Figure 12.3 Capital cost (in 1973) of chlorination feed system [17].

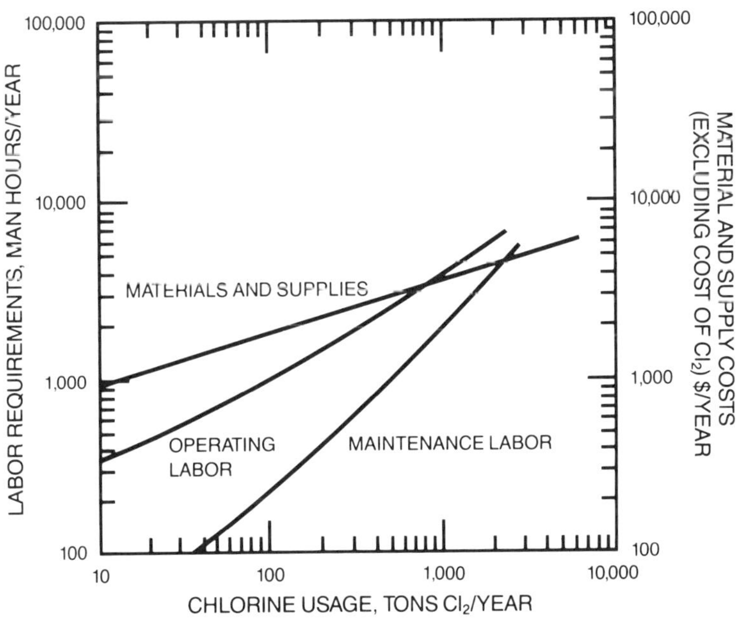

Figure 12.4 Operation and maintenance relationships for chlorination [17].

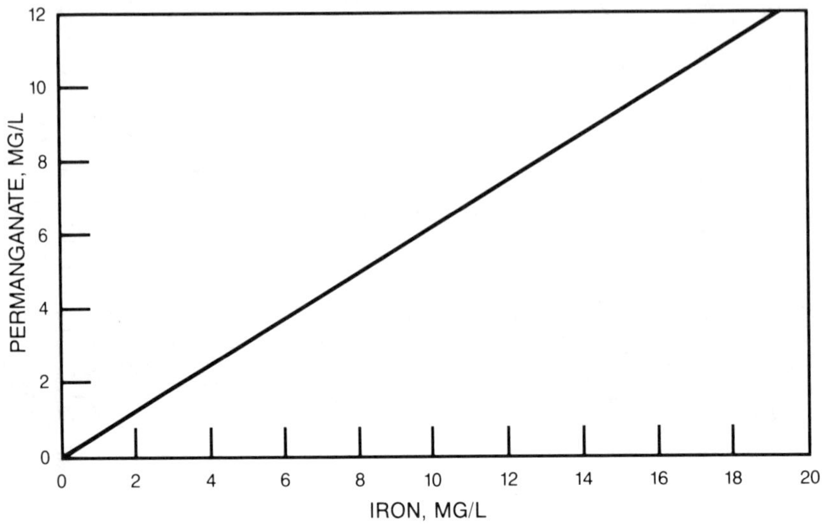

Figure 12.5 Permanganate requirement for oxidation of ferrous iron. (From Willey and Jennings [19], courtesy of American Water Works Association.)

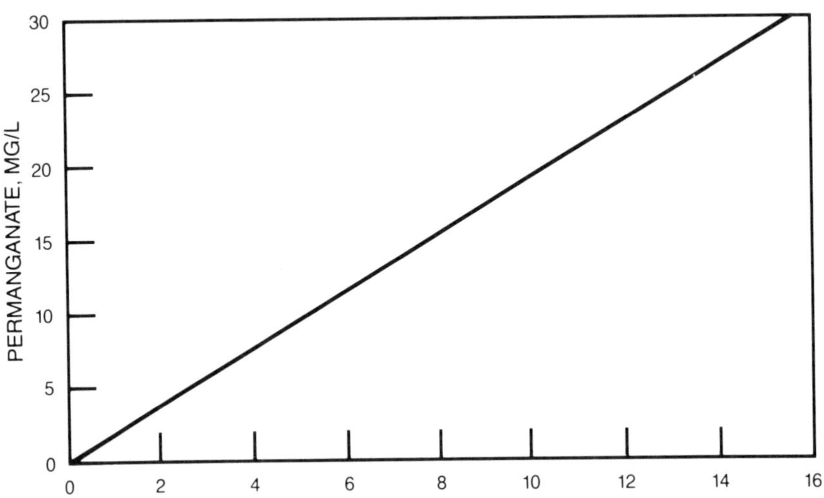

Figure 12.6 Permanganate requirement for oxidation of manganous manganese. (From Willey and Jennings [19], courtesy of American Water Works Association.)

Table 12.4 Ozone Treatment for Manganese [25]

pH	*Final Manganese Concentration (mg/l)*	
	Lime Only	*Lime Followed by Ozone*
7.0	—	0.8
8.0	90	0.3
9.0	17	<0.1
10.0	0.15	<0.1
11.0	<0.1	—

lation-filtration, nearly four times greater flow rates, smaller plant requirements, and simple operation [24]. The obvious drawback to ion exchange is the nonselective removal of other ions, which increases operation costs. Regeneration brine disposal poses an additional treatment problem since the manganese remains in soluble form in the brine. An anaerobic or acidic waste stream is necessary for the ion-exchange system, to avoid air oxidation of the manganese with resultant precipitation and clogging of the ion exchange bed. Treatment data for a full-scale ion exchange unit treating well water are summarized in Table 12.5.

SUMMARY

Manganese removal to levels of 0.05 mg/l or less has been accomplished for many years in the treatment of municipal drinking and industrial process waters. Several methods of removal are available, each of which may have distinct advantages depending upon the overall nature of the waste, other required waste treatment operations, and the total treatment goals for a particular waste. The most common general approach involves oxidation of the soluble

Table 12.5 Performance of Ion Exchange in Manganese Removal [24]

Parameter	*Influent*[a]		*Effluent*[a]	
	Range	*Avg*	*Range*	*Avg*
pH	5.6–6.1	—	6.4–6.7	—
Hardness, as $CaCO_3$	76–96	86	2	2
Bicarbonate, as $CaCO_3$	15–62	42	130–138	133
Free CO_2, as $CaCO_3$	80–95	88	50	50
Iron	37–53	47	0.1–0.5	0.3
Manganese	1.0–1.7	1.3	0.0–0.1	<0.1
Sodium	49–74	62	224–280	252

[a] All concentrations except pH in mg/l.

manganous form to insoluble manganese hydroxide or oxide, with subsequent precipitation. Ion exchange treatment has also proved effective, however.

REFERENCES

1. McKee, J.E. and H.W. Wolf. *Water Quality Criteria,* 2nd ed., California State Water Quality Control Board Publication 3-A (1963).
2. Hill, R.D. "Control and Prevention of Mine Drainage," paper presented at the Environmental Resources Conference on Cycling and Control of Metals, Battelle-Columbus Laboratories, Columbus, OH (1972).
3. Kunin, R. and D.G. Downing. "Ion Exchange System Boasts More Pulling Power," *Chem. Eng.* 78: 67–69.
4. Huck, P.M., K.L. Murphy, C. Reed, and B.P. LeClair. "Optimization of Polymer Flocculation of Heavy Metal Hydroxides," *J. Water Poll. Control. Fed.* 49 (1977): 2411–2418.
5. "Development Document for Interim Final Effluent Guidelines and New Source Performance Standards for the Coal Mining Point Source Category," USEPA 440/1-75/057, Group II (October 1975).
6. Iammartino, N.R. "Mercury Cleanup Routes—II," *Chem. Eng.* 82 (1975): 36–37.
7. "Development Document for Effluent Limitations Guidelines and New Source Performance Standards for the Steel Making Segment of the Iron and Steel Manufacturing Point Source Category," USEPA 440/1-74-024-a (June 1974).
8. "Development Document for Effluent Limitations Guidelines and New Source Performance Standards for the Smelting and Slag Processing Segments of the Ferroalloy Manufacturing Point Source Category,"USEPA 440/1-74-008-a (February 1974).
9. "Development Document for Effluent Limitations Guidelines and New Source Performance Standards for the Oil Base Solvent Wash Subcategories of the Paint Formulating and Ink Formulating Point Source Category," USEPA 440/1-75-050-a, Group II (July 1975).
10. "Development Document for Interim Final Effluent Limitations Guidelines and Proposed New Source Performance Standards for the Electrolytic Ferroalloys Segment of the Ferroalloy Manufacturing Point Source Category," USEPA 440/1-75-038-a, Group I, Phase II (February 1975).
11. Cotton, F.A. and G. Wilkinson. *Advanced Inorganic Chemistry,* (New York: Interscience, 1962).
12. Welch, A. "Potassium Permanganate in Water Treatment," *J. Am. Water Works Assoc.* 55 (1963): 734–741.
13. Borgren, G.R. "Removal of Iron and Manganese from Ground Waters," *J. New Eng. Water Works Assoc.* 55 (1963): 734–741.
14. Owens, L.V. "Iron and Manganese Removal by Split Flow Treatment," *Water Sew. Works* 110 (1963): R76–R87.
15. Rouse, J.V. "Removal of Heavy Metals from Industrial Effluents," *ASCE J. Environ. Eng. Div.* (October 1976) EE5 929–936.
16. Harris, W.C. "Iron and Manganese Removal," Engineering Experimental Station Bulletin 68, Louisiana State University (1962).

17. Edwards, A.W., W.W. Eckenfelder, Jr., and J.A. Roth. *Industrial Water Cost Studies,* vol. 1, chaps. I, II, III and IV, Office of Research and Development, USEPA Contract 68-01-248S (1975).
18. Haney, C.G. "Potassium Permanganate Solves a Problem of Charlottesville, VA," *Water Sew. Works* 112 (1964): R353–R354.
19. Willey, B.F. and H. Jennings. "Iron and Manganese Removal with Potassium Permanganate," *J. Am. Water Works Assoc.* 55 (1963): 729–734.
20. Holland, G.J. "Removal of Dissolved Aluminum, Iron and Manganese," *Chem. Ind.* (London, (1964): 2016.
21. Costabile, F.J. and C.H. Perron. "Diatomite Filter Ends Manganese Problem," *J. Am. Water Works Assoc.* 63 (1971): 230–232.
22. Sayell, K.M. and R.R. Davis. "Removal of Iron and Manganese from Raw Water Supplies Using Manganese Greensand Zeolite," *Ind. Water Eng.* (October/November 1975): 20–23.
23. Alling, S.F. "Continuously Regenerated Greensand for Iron and Manganese Removal," *J. Am. Water Works Assoc.* 55 (1963): 749–752.
24. Alzentzer, H.A. "Ion Exchange in Water Treatment," *J. Am. Water Works Assoc.* 55 (1963): 742–748.
25. Netzer, A., A. Bowers, and J.D. Norman. "Removal of Heavy Metals from Wastewater by Lime," *Proceedings of the 16th Conference on Great Lakes Research,* Huron, OH (1973).

13
Mercury

The major consumptive user of mercury is the chlor-alkali industry, using the Downs, DeNora, or similar electrolytic cells [1–6]. The second largest user is the electrical and electronics industry [2, 3, 5]. Other users of various mercury forms include explosives manufacturing, the photographic industry, and the pesticide and preservative industry [1, 2, 4–6]. Within the chemical and petrochemical industry, mercury compounds have found use as catalysts in plastics production and in hydrogenation and dehydrogenation, sulfonation, oxidation, chlorination, and acidolysis reactions [1, 4].

Both the paint and the pharmaceutical-cosmetic industries have used mercury compounds, as preservatives and as antifouling or mildew-proofing agents [1, 2, 4]. A mercury level of 300 μg/l has been reported for the wastewater of one water-base paint plant [7]. Pulp and paper operations have employed mercurials as fungicides, slimicides, and biocides [1, 2, 4], but recently use has been curtailed [2]. Other sources of mercury wastes are seed preservatives; research and hospital laboratories; sealants used in machines, thermometers and manometers; and film production and processing [1, 4].

Limited data are available regarding actual wastewater mercury levels. Wastewater emanating from an acetaldehyde production unit was reported to contain 20 mg/l mercury, and the combined wastewater, prior to treatment, contained mercury at 18 mg/l [8]. Effluent mercury from one chlor-alkali plant was estimated at 1.4–2.8 mg/l, based on 600 gpm waste discharge and 10–20 lb/day mercury loss [5]. Mercury effluent data are summarized in Table 13.1.

It has been reported that commercial bleach preparations may contain up to 0.2 mg/l mercury [9]. Caustic, one product of the chlor-alkali process, may also contain high levels of mercury (Table 13.2).

TREATMENT TECHNOLOGY

Many types of mercury treatment technology have been reported. Most treatment results described in the technical literature are based upon laboratory or pilot scale studies. The effectiveness, and economics, of treatment by each type of technology depends upon the chemical nature and initial concentration of mercury, the presence of other constituents in the wastewater which may interfere with that specific treatment, and the degree of mercury removal which

Table 13.1 Levels of Mercury in Industrial Wastewaters

Waste	*Mercury (ug/l)*	*Reference*
Paper mill	20–34	9
Fertilizer mill	0.26–40	9
Smelting plant	20–40	9
Chlor-alkali plant	80–2,000	9
	4,600–5,100	10
	1,400–2,800	5
	3,000–8,000	11
	300–6,000	12
	21,500	13
Water-base paint	300	7
Paint and ink formulation	0–120,000	14
Acetaldehyde production	20,000	8
Fluorescent lamp production	2	15
Coal fly-ash pond effluent	2–3	16
Textile dyeing waste	15,000	17
Textile mill waste	11	18
Secondary lead battery recovery	0.66 (Total) <0.20 (Soluble)	19
Rubber processing	0–720	14

must be achieved. Among the more common types of treatment are sulfide precipitation, ion exchange, adsorption and coagulation, and reduction of ionic mercury to elemental form with removal by filtration. Hydroxide precipitation of mercury is ineffective, with soluble levels exceeding 75 mg/l mercury over a pH range of 3.5 to 11.5 [22].

Precipitation

Sulfide addition, in order to precipitate highly insoluble mercury sulfide, is the most common precipitation treatment reported. Precipitation may be combined

Table 13.2 Mercury Content of Caustic (NaOH) from Chlor-Alkali Plants

Mercury (mg/l)	*Reference*
0.07–18.8	9
125–150	20
7.0–42.0	21

Table 13.3 Sulfide Precipitation Treatment for Mercury

Treatment Chemical	*Mercury Concentration (μg/l)* Initial	Final	Treatment pH	Additional Treatment	Reference
Sodium sulfide	na[a]	<3	na	Vacuum filter	23
	300–6,000	10–125	na	Pressure filter	23
	1,000–50,000	10	na	Flocculation + activated carbon	24
Sodium hydrosulfide	131,500	20	3.0	"Filter"	25
Magnesium sulfide	5,000–10,000	10–50	10–11	—	26
Sulfide	300–6,000	10–125 (50 avg)	5.1–8.2	Filtration	12
	na	<1,000	<9	—	27
	na	100–300	na	—	28
	na	100	na	1. None	29
		10–20	na	2. Activated carbon	

[a] Not available.

with flocculation and separation by gravity settling, filtration, or dissolved air flotation [9]. These latter steps improve the removal of the precipitated mercury sulfide, but do not enhance the efficiency of precipitation of the soluble mercury itself. Table 13.3 presents a summary of available treatment data for sulfide precipitation. For high initial mercury levels, sulfide precipitation will achieve 99.9+% removal, but even with filtration or activated carbon-polishing treatment, minimum effluent mercury achievable appears to be 10–20 μg/l. Most effective precipitation, with regard to minimizing sulfide dosage, appears to occur in the near-neutral pH range. Precipitation efficiency declines significantly at pH above 9 [12]. In addition to lack of capability to reduce mercury below 10 μg/l, other drawbacks of this method include (1) the formation of soluble mercury sulfide complexes at high levels of excess sulfide, (2) the difficulty of monitoring excess sulfide levels, and (3) the problem of toxic sulfide residual in the treated effluent [12].

Sulfide precipitation appears to be the standard practice for mercury control in many U.S. and foreign chlor-alkali plants. Levels of mercury removal of 95–99.9% are reported for well-designed and -managed treatment systems [12, 30]. A mercury effluent level of about 65 μg/l has been cited for sodium sulfide treatment of wastewaters from the chlor-alkali industry [30]. Treatment costs for one chlor-alkali plant utilizing sodium sulfide addition plus diatomaceous earth filtration for a 100 gpm flow were $0.50/1,000 gal (1973 basis). Capital cost for the treatment system (1973 basis) was $959.33/1,000 gpd capacity [12]. One consequence of sulfide precipitation is increasing stockpiles of mercury-rich process sludges, which must be either disposed of in an environmentally acceptable manner, or further processed for mercury recovery [31].

Ion Exchange

Most ion exchange treatment of inorganic mercury involves formation of a negatively charged mercuric chloride complex by addition of chlorine or hypochlorite (to oxidize metallic mercury), or chloride salts, and removal of the mercuric chloride complex on an anion exchange resin. Most experience with ion exchange treatment has been with chlor-alkali plant wastes, which contain high background chloride levels. There are some limited data [12, 32] to indicate that the best ion exchange treatment results from an initial, roughing stage followed by a second polishing ion exchange (Table 13.4). Ion exchange has also been suggested as a polishing step after precipitation [12].

In wastes where chloride is not high, cation exchange resins are effective. Although information on the cation exchange behavior of mercury is scarce, Amberlite IR-120 and Dowex-50W-X8 are both reported to be effective [40]. Resins which contain mercapto groups (R-S-H) such as polythiostyrene are highly specific for mercuric ion [40]. Thiol resin (AKZO Chemie, Holland) is widely used for cationic mercury removal in Europe [41]. Other high-affinity cation resins available are an isothiouronium resin (Ionac Chemical Co. & Ayalon, Israel) and a carbamate type resin (Nippon Soda, Japan). The isothiouronium resin is reported to be effective for both inorganic and methylmercury, and the carbamate resin to have an extremely high affinity or selectivity for mercury [41].

Regardless of whether anionic resins are used for mercuric chloride complex removal, or cationic resins for mercuric ion removal, the best ion exchange treatment for inorganic mercury appears to yield an effluent of 1–5 μg/l. Most effective treatment results from two-stage treatment, at neutral to slightly acidic pH.

Most available data relate to treatment of inorganic mercury, and there is relatively little information on organic mercurials. However, efficient treatment of methylmercury acetate has been reported with Dowex A-1 Chelate [42] and Ionac SRXL resins [43]. Ionac SRXL is no longer commercially available although a substitute, Srafion NMRR, is reported to have similar properties [43].

Costs of the Osaka ion exchange process (see Table 13.4) have been reported (in 1971) as $400,000–575,000 for chlor-alkali plants of 300–600 tons chlorine/day capacity, including licensing fee. Applying these costs to plants of 0.018–0.32 MGD, capital costs would be $18,000–22,000/1,000 gpd. Operating costs are given as $1.00/1,000 gal. Credit for recovered mercury may not be included in this cost analysis, however [44].

Table 13.4 Ion-Exchange Treatment for Mercury

Resin Type	Mercury	Treatment pH	Mercury (μg/l) Initial	Mercury (μg/l) Final	Additional Treatment	Reference
Mtylon-T	Inorganic	5–6	5,000–25,000	1	—	33
Lewatit	Inorganic	na	na	15	—	34
Anion[a]	Inorganic	7	850	2.5	—	35
Macroreticular	Inorganic	na	10,000	<10	—	36
Anion[a]	Inorganic	na	470	30	—	37
Osaka IE[a]	Inorganic	"acidic"	3,000–10,000	100–150	Prefilter	32
Osaka MR[a]	Inorganic	na	100–150	2–5	Osaka IE resin	32
Imac-TMR	Inorganic	3	na	5	Carbon prefilter	38
Anion[a]	Inorganic	5–7	na	100	—	28
Activex	Inorganic	na	10	<5	—	39
Ajinomoto	Inorganic	1.5	60	5	—	12
		6.0	87	3	—	
		11.0	1,800	990	—	
Billingsfors-Langed Q13	Inorganic	6.5	35	1	—	12
		na	2,000–5,000	10–20	Absorption filter	29

[a] Mercury removed as mercuric chloride complex, $HgCl_x^{-(x-2)}$.

A Swedish company (Aktiebolaget Billingsfors-Langed) has developed an ion-exchange system which is similar in concept to the Osaka process [45, 46]. The strong-base first-stage exchanger is regenerable, while the second-stage polisher is not. Removal is as the mercuric chloride complex anion, thus requiring pretreatment of mercurous and organic mercury. Capital costs are reported (in 1971) as $1,500/1,000 gpd for a 0.040 MGD unit. Treatment levels are equivalent to the Osaka process and the unit has been on the market since 1969.

Coagulation Treatment

Removal of mercury by coagulation has been reported for a variety of mercury-containing wastewaters. Coagulants employed include aluminum sulfate (alum), iron salts, and lime. The process has been applied with some success to both organic and inorganic mercury. In studies on coagulation treatment for removal of inorganic mercury dosed to domestic sewage, both iron and alum coagulation, followed by filtration, reduced initial mercury levels of 50–60 μg/l by 94–98% [47]. Lime coagulation treatment, applied at a higher mercury level of 500 μg/l, provided 70% removal with filtration.

In pilot plant tests of alum and iron treatment of inorganic and methylmercury, inorganic mercury was removed effectively only by iron, while neither the iron nor the aluminum coagulant provided effective methylmercury removal [48]. Similar results have been reported in another study, which also found that increasing coagulant dosage to 100–150 mg/l did not improve mercury removal [49]. Treatment data for coagulation are summarized in Table 13.5. Effluent levels of mercury achieved by alum treatment range from 1.5 to 102 μg/l, and by iron treatment from 0.5 to 12.8 μg/l. However, at comparable initial mercury levels, both alum and iron coagulation provide equivalent treatment results, with a lower effluent limit for mercury of 0.5–5.0 μg/l.

Activated Carbon

The effectiveness of carbon treatment is dependent upon several factors, including initial form and concentration of mercury, dosage and type of activated carbon, treatment pH, and contact period between the carbon and mercury containing wastewater. Increasing carbon dosages and increasing contact times improve removal of both inorganic and organic mercury [49]. Organic mercury is more effectively removed than is inorganic mercury.

A review of available treatment data (Table 13.6) indicates that the highest percentage removals (85–99%) result from carbon treatment of more concentrated mercury solutions. However, lowest effluent mercury results from treatment of lower initial mercury concentrations, although the relative efficiency is less. Thus carbon treatment of initial mercury below 1 μg/l yields an effi-

Table 13.5 Coagulation Treatment Results for Mercury

Coagulation Salt	*Dosage (mg/l)*	*Treatment pH*	*Mercury, µg/l*		*Additional Treatment*	*Reference*
			Initial	*Final*		
Alum	1,000	3	11,300	102	Filtration	50
	100	na	90	11	—	51
	100	na	na[a]	10	—	51
	21–41	6.7–7.2	5.9–8.0	5.3–7.4	Filtration	39
	na	7.0	50	26.5	Filtration	48
	220	6.4	60	3.6	Filtration	47
	20–30	na	3–8	1.5–6.4	—	49
	20–30	na	3–16[a]	2.3–21.3	—	49
Iron	34–72	6.9–7.4	4.0–5.0	2.5–2.5	Filtration	39
	na	8.0	50	3.5	Filtration	48
	40	6.2	50	1.0	Filtration	47
	20–30	na	1–17	0.5–6.8	—	49
	20–30	na	2–17[a]	1.2–12.8	—	49
Lime	415	11.5	500	150	Filtration	47
	na	8.3	0.66	<0.2	—	19

[a] Organic mercury.

Table 13.6 Activated Carbon Treatment for Mercury

Mercury (μg/l)		*Additional Treatment*	*Reference*
Initial	*Final*		
na[a]	<100	Presettling	52
12–46	<0.5	—	
5.8	0.9	Iron coagulation	48
9.3	1.6	Alum coagulation	39
na	0.8	—	49
na	0.2	—	49
100	10–20	Precipitation	29
60–87	4.5–6.0	—	12
0.40–0.81	0.13–0.19	Polymer added	53
0.5	0.21	Polymer added	53
150	40	Lime pretreatment to pH 11.5	47

[a] Not available.

ciency of removal of less than 70%, but effluent mercury below 0.25 μg/l. Carbon treatment of initial mercury between 10 and 100 μg/l yields 90% or greater efficiency, but highest effluent mercury levels of 20 μg/l or less. There is some evidence that efficiency of mercury removal is increased at lower pH of 2–4 [54, 55]. A pH effect was also observed in another study, with removal efficiency increasing from 50 to 80% as treatment pH was reduced from pH 9 to 7 [56]. Treatment was with powdered carbon dosed at 100 mg/l to a water initially containing 10 μg/l mercury.

Greatly enhanced mercury removal has been reported when activated carbon is pretreated with carbon disulfide solution before use [54]. Laboratory batch tests with CS_2-soaked carbon resulted in mercury removal from an initial level of 10 μg/l to below 0.2 μg/l. The best treatment with the CS_2 system occurred at pH 10, with reduced performance at pH 4 [54].

Reduction Processes

Inorganic ionic mercury can be converted to the metallic form by reduction and separated by filtration or other solids separation techniques. A variety of reducing agents are available, including aluminum, zinc, hydrazine, stannous chloride, and sodium borohydride. A summary of available treatment data on these processes is presented in Table 13.7. Although there is much discussion of reduction processes in the literature, there are little actual treatment data presented. The main advantage claimed for reduction is that mercury can be recovered in the metallic state. However, the data of Table 13.7 indicate that reduction processes cannot effectively achieve mercury levels below 100 μg/l,

Table 13.7 Performance of Reduction Processes for Mercury Treatment

Reductant Used	*Mercury (μg/l)*		*Treatment pH*	*Reference*
	Initial	*Final*		
Zinc	5,000–10,000	5–10	—	57
	1,800	140	11.5	12
	12,500	830	10.0	
	12,500	750	6.0	
	12,500	470	2.5	
	na[a]	600	—	29
Stannous chloride	2,800	500	—	12
Sodium borohydride	10,000	220	—	12
	4,000	420	—	
	26,000	820	—	
	4,700	200	—	58
	na	<10	0–11	28

[a] Not available.

and that their use would require second-stage polishing by one of the more rigorous treatment methods discussed in this chapter.

A full-scale commercial process, based on reduction of mercury by sodium borohydride, is marketed as the Ventron process [46, 47]. The process claims to reduce effluent mercury to 10 μg/l. A 15,000 gal/day plant is in operation at a Wood Ridge, New Jersey mercury chemicals plant. Capital costs are reported (in 1971) as $667/1,000 gpd, with operating costs of 6.7–26.7¢/

Table 13.8 Other Processes for Mercury Treatment

Process	*Mercury (μg/l)*		*Comments*	*Reference*
	Initial	*Final*		
Solvent extraction with high molecular-weight amines	na[a]	na	99.9% removal	59, 60
Silicon alloys	9,000	0.8–2.6	Mg-Fe-Si alloy	61
	9,900	3.6–75.6	$CaSi_2$ alloy	
Starch xanthate-cationic polymer complex	100,000	3.8	pH 7.0	62, 63
Ground vulcanized rubber	1,800–11,000	10	170 bed volumes throughput	64
Polyethylenimine-modified wool	1,100	20	—	65
Cationic polymer flocculation	na	2.5–15	—	29
Proprietary—Terraneers, Ltd.	na	10	—	29

[a] Not available.

Table 13.9 Summary of Treatment Technology for Mercury

Technology	*Lower Limit of Treatment Capability (Hg, μg/l)*
Sulfide precipitation	10–20
Ion exchange	1–5
Alum coagulation	1–10
Iron coagulation	0.5–5
Activated carbon	
High initial Hg	20
Moderate initial Hg	2.0
Low initial Hg	0.25

1,000 gal treated [46]. The unit recovers 1–4 lb mercury per day, at a treatment cost of $1.00/lb mercury recovered.

Other Processes

A variety of other treatment processes are mentioned in the technical literature, although treatment efficiency data are limited. These processes include solvent extraction with high molecular-weight amines [23, 59, 60], use of silicon alloys for reduction of elemental mercury [61], removal by insoluble starch xanthate-cationic polymer complexation [62, 63], adsorption onto ground rubber [64] and wool [65] at sulfhydral sites, and various proprietary processes [29]. Although several of the processes appear promising (Table 13.8), they are all in the experimental stages and most have not been proved in either pilot or full-scale application.

SUMMARY

There are many types of treatment technology available for mercury control. The more effective are summarized in Table 13.9, with lower limits of treatment capability indicated. Among the processes shown, ion exchange, iron or alum coagulation, and activated carbon adsorption have the capability to reduce mercury below 3 μg/l. Conventional precipitation systems employing sulfide and flocculating aids seem capable of yielding mercury levels of 10–20 μg/l, based upon the information available. Other methods, particularly reduction for small-scale use, are capable of producing very low effluent concentrations of mercury. Many of these methods have not been tested in full scale application, however.

REFERENCES

1. "Mercury Pollution: Interest in It Now on the Upswing," *Wat. Wastes Eng.* 8 (January 1971): A13–A16.
2. National Industrial Pollution Control Council, *Mercury,* Staff Report, U.S. Government Printing Office 0-409-779 (October 1970).
3. Zugger, P.D. and M.W. Ghosh. "The Effect of Mercury on the Activated Sludge Process," paper presented at the 27th Annual Purdue Industrial Waste Conference, West Lafayette, IN (May 1972).
4. Cheremisinoff, P.N. and Y.H. Habib. "Cadmium, Chromium, Lead, Mercury: A Plenary Account for Water Pollution, Part 1, Occurrence, Toxicity and Detection," *Water Sew. Works* 119 (1972): 73–85.
5. Gurney, W.G. "Mercury Pollution: Michigan's Action Program," *J. Water Poll. Control. Fed.* 43 (1971): 1427–1438.
6. D'Itri, F.M. "Mercury in the Aquatic Ecosystem," Technical Report 23, Institute of Water Research, Michigan State University, East Lansing (1972).
7. "Development Document for Effluent Limitations Guidelines and New Source Performance Standards for the Oil Base Solvent Wash Subcategories of the Paint Formulating and the Ink Formulating Point Source Category," USEPA 440/1-75/050-a, Group II (July 1975).
8. Irukayama, K. "The Pollution of Minamata Bay and Minamata Disease," *Proceedings of the 3rd International Conference on Water Pollution Research, Munich* (1966), pp 153–180.
9. Bouveng, H.O. "Control of Mercury in Effluents from Chlorine Plants," *J. Pure Applied Chem.* 29 (1979):75–91.
10. Smith, H. and D.G. Ward. "Removal of Mercury from Aqueous Systems," *Chem. Abstr.* 84 (1976): 381.
11. Bieling, H., H.J. Koch, and K. Thormann. "Separation of Mercury from Mercury-Containing Solutions of Waste Waters," *Chem. Abstr.* 83 (1975): 349.
12. Perry, R. "Mercury Recovery from Contaminated Waste Water and Sludges," USEPA 660/2-74-086 (December 1974).
13. Diaz, N.E., A.L. Redondo Abad, M. Baile, and V.J.M. Regife. "Separation of Mercury from Industrial Waste Waters," *Chem. Abstr.* 87 (1977): 306.
14. Cawley, W., ed. *Treatability Manual,* vol. I. *Treatability Data*, USEPA 600-8-80-042-a (July 1980).
15. Ohtsubo, K. "Mercury-Containing Wastewater Treatment," *Chem. Abstr.* 87 (1977): 306.
16. Chu, T.J. and R.J. Ruane. "Wastewater Treatment for Coal-Fired Electric Generating Stations," paper presented at WWEMA Industrial Pollution Conference, St. Louis (April 11–13, 1978).
17. Rouse, J.V. "Removal of Heavy Metals from Industrial Effluents," *ASCE J. Environ. Eng. Div.* (October 1976) EE5 929–936.
18. Branden, C.A., J.J. Porter, and T.N. Sargent. "Reuse of Total Composite Wastewater Renovated by Hyperfiltration in Textile Dyeing Operations," *Proceedings of the National Conference on Management and Disposal of Residues from the Treatment of Industrial Wastewaters, Washington, DC* (February 3–5, 1975).
19. Mezey, E.J. "Characterization of Priority Pollutants from a Secondary Lead and Battery Manufacturing Facility," USEPA 600/2-79-039 (January 1979).

20. Newell, I.L. "Mercury and Other Heavy Metals in Water Supplies," *J. New England Water Works Assoc.* 85 (1971): 289.
21. Bouveng, H.P. "The Chlorine Industry and the Mercury Problem," *Modern Kemi* (Sweden) 3 (1968): 45.
22. Netzer, A., A. Bowers, and J. Norman. "Removal of Heavy Metals from Wastewater by Lime," *Proceedings of the 16th Conference on Great Lakes Research*, Huron, OH (1973).
23. Immartino, N.R. "Mercury Cleanup Routes—II," *Chem. Eng.* 82 (1975): 36–37.
24. Fujii, M. "Treatment of Mercury in Waste Water," *Chem. Abstr.* 78 (1973): 256.
25. Sato, Y. et al. "Recovery Method for Mercury," *Chem. Abstr.* 77 (1972): 295.
26. Ishida, T. and K. Aiba. "Recovery Method for Mercury," *Chem. Abstr.* 77 (1972): 295.
27. Dean, W.E. "Mercury Removal from a Brine Solution," *Chem. Abstr.* 77 (1972): 242.
28. Cheremisinoff, P.N. and Y.H. Habib. "Cadmium, Lead, Mercury: A Plenary Account for Water Pollution, Part 2—Removal Techniques," *Water Sew. Works* 119 (August 1972): 46–51.
29. Industrial Waste Study—Mercury Using Industries," USEPA 805/25-18,000 HIP 07/71 (1971).
30. "Development Document for Effluent Limitations Guidelines and New Source Performance Standards for the Major Inorganic Products Segment of the Inorganic Chemicals Manufacturing Point Source Category," USEPA 440/1-74-007-a (March 1974).
31. Perry, R.A. "Mercury Recovery from Process Sludges," *Chem. Eng. Prog.* 70 (1974): 73–80.
32. Gardiner, W.C. and F. Munoz. "Mercury Removed from Waste Effluent via Ion Exchange," *Chem. Eng.* 78 (1971): 57–59.
33. Fedonina, V.F., M.O. Lisheuskaya, N.S. Torocheshnikov, and V.V. Voronenko. "Removal of Mercury from Industrial Waste Waters," *Chem. Abstr.* 76 (1972): 267.
34. Oehme, C. "Purification of Mercury-Containing Water and Waste Waters by Lewatit Ion Exchangers," *Von Wasser* (Germany) 38 (1971): 345–56.
35. Yokota, N. "Removal of Mercury from Waste Water," *Chem. Abstr.* 76 (1972): 222.
36. Egawa, H. "Selective Absorption Resins. 3. Recovery and Removal of Mercuric Ion with Macroreticular Chelate Resins Containing Polyethylenepolyamino Groups," *Chem. Abstr.* 75 (1971): 252–253.
37. Oehme, C., H.R. Brost and H. Hoffman. "Removal of Mercury from Industrial Wastes," *Chem. Abstr.* 77 (1972): 271.
38. Rosenzweig, M.D. "Mercury Cleanup Routes—1," *Chem. Eng.* 82 (1975): 60–61.
39. Logsdon, G.S. and J.M. Symons. "Removal of Heavy Metals by Conventional Treatment," *Proceedings of the 16th Water Quality Conference,* University of Illinois, Urbana (February 12–13, 1974).
40. Roesmer, J. "Radiochemistry of Mercury," NAS-NS-3026 (rev.), National Academy of Sciences-National Research Council (September 1970).
41. Calmon, C., J. Casana, and H. Gold. "Ion Exchange and Resin Adsorption Data for the Removal of Toxic Contaminants from Wastewaters," report by Water Purification Association, Cambridge, MA (September 1980).

42. Muzzarelli, R. and A. Isolati. "Methyl Mercury Acetate Removal from Water by Chromatography on Chelating Polymers," *Water, Air and Soil Poll.* 1 (1971): 65–71.
43. Law, S.L. "Methyl Mercury and Inorganic Mercury Collection by a Selective Chelating Resin," *Science* 174 (1971): 285–287.
44. Gardiner, W.G. and F. Munoz. "Mercury Removed from Waste Effluent Via Ion Exchange," *Chem. Eng.* 78 (1971): 57–59.
45. Cheremisinoff, P.N. and Y.H. Habib. "Cadmium, Chromium, Lead, Mercury: A Plenary Account for Water Pollution. Part 2. Removal Techniques," *Water Sew. Works* 119 (1972): 46–51.
46. Rosenzweig, M.D. "Parring Mercury Pollution," *Chem. Eng.* 78 (1971): 70–71.
47. Hannah, S.A., M. Jelus, and J.M. Cohen. "Removal of Uncommon Trace Metals by Physical and Chemical Treatment Processes," *J. Water Poll. Control. Fed.* 49 (1977): 2297–2309.
48. O'Connor, J.T. "Trace Metals vs. Conventional Water Treatment Systems," *Water Sew. Works* 12 (1974): 72–75.
49. Logsdon, G.S. and J.M. Symons. "Mercury Removal by Conventional Water Treatment Techniques," *J. Am. Water Works Assoc.* 651 (1973): 554–562.
50. Honda, A. "Mercury Treatment in the Wastes and Waste Water from the Fluorescent Lamp Industry," *Chem. Abstr.* 78 (1973): 220.
51. Hellestan, C.J.S. "Compositions and Method for Precipitating or Binding Various Forms of Mercury Occurring in Contaminated Water, Such as Lakes," *Chem. Abstr.* 77 (1972): 296.
52. Suzuki, G. "Mercury Pollution Control at Mercury-Process Sodium Chloride Electrolysis Plants. 2. Removal of Mercury from the Solution," *Chem. Abstr.* 78 (1973): 216.
53. "Progress Report—420 GPH Laundry Wastewater Treatment Unit." Sanitary Science Division Military Technical Department, US Army Mobility Equipment R&D Center, Fort Belvoir, VA (January 1974).
54. Humenick, M.J. Jr. and J.L. Schnoor. "Improving Mercury (II) Removal by Activated Carbon," *ASCE J. Environ. Eng. Div.* (1974): EE6 1249–1262.
55. Fife, K.W., J.D. Williams, and H.G. Folster. "Mercury (II) Removal from Aqueous Solutions Using Ion Exchange and Carbon Adsorption," paper presented at the 32nd Annual Purdue Industrial Waste Conference, West Lafayette, IN (May 1977).
56. Thien, L., D. Badorek, and J.T. O'Connor. "Removal of Mercury from Drinking Water Using Activated Carbon," *J. Am. Water Works Assoc.* 68 (1976): 447–451.
57. Waltrich, P.F. "Apparatus for the Removal of Mercury from Effluent Streams," *Chem. Abstr.* 78 (1973): 235.
58. Coulter, M.O. "Removal of Mercury from Spent Brine," *Chem. Abstr.* 78 (1973): 226.
59. Moore, F.L. "Removal of Mercury and Other Toxic Metals from Plant Effluent Solutions by Solvent Extraction," *Ecology and Analysis of Trace Contaminants* Progress Report ORFNL-NSF-EATC-1, Oak Ridge National Laboratory, Oak Ridge, TN (March 1973).
60. Moore, F.L. "Solvent Extraction of Mercury from Brine Solution with High-Molecular-Weight Amines," *Environ. Science Technol.* 6 (1972): 525–529.

61. McKaveney, J.P. "Removal of Heavy Metals from Water and Brine Using Silicon Alloys," *Environ. Science Technol.* 6 (1972): 1109–1113.
62. Swanson, C.L. "Mercury Removal from Wastewater with Starch-Xanthate Cationic Polymer Complex," *Environ. Science Technol.* 7 (1973): 614–618.
63. Wing, R.E. "Heavy Metal Removal with Starch Xanthate-Cationic Polymer Complex," *J. Water Poll. Control. Fed.* 46 (1974): 2043–2047.
64. "Mercury Removal from Aqueous Process Streams," *Nuclear Science Abstr.* 29 (1974): 2316–2317.
65. Freeland, G.N. "Adsorption of Mercury from Aqueous Solutions by Polyethylenimine-Modified Wool Fibers," *Environ. Science Technol.* 8 (1974): 943–946.

14
Nickel

Wastewaters containing nickel originate from many industrial activities, although plating operations are the most common source. High levels of nickel have also been reported in wastes from silver refineries [1]. In addition, basic steel works and foundries, motor vehicle and aircraft industries, and printing are potential sources of nickel [2]. Nickel concentrations reported for a variety of wastewater sources are summarized in Table 14.1.

Wastewaters resulting from pickling and plating of nickel and nickel alloys pose far more problems in treatment than is the case for other plating metals such as copper. These difficulties result from the wide spectrum of nickel alloys, and the variety of pickling liquors used. Acids used in pickling include sulfuric, nitric, hydrochloric, and hydrofluoric [33]. Alkaline liquors are also used for surface preparation of some nickel alloys. Acid and sulfamate nickel plating baths generally contain nickel at 40,000–90,000 mg/kg solution, while electroless process baths are of weaker concentration, containing about 7,500 mg/kg solution [8]. Nickel from the electroless bath deposits spontaneously by catalysis on metals such as iron and cobalt after alkaline cleaning and acid dipping. Catalysis deposition can be induced on copper and silver by first applying a strike coat in a dilute palladium chloride bath. There are two distinct types of electroless nickel baths: (1) the more common acid bath (pH 7) based upon nickel chloride or sulfate, sodium hypophosphite, and an organic complexing agent; and (2) alkaline ammoniacal baths containing nickel chloride, sodium hypophosphite, sodium citrate, and ammonium chloride and hydroxide [34]. Waste rinsewater constituents reflect the bath characteristics. Generally, plating rinsewaters contain about 1% of the plating bath concentration [4].

Industries concerned primarily with copper or brass plating or processing may generate nickel containing wastewaters, but generally at low levels of 5 mg/l or less. Nickel has been reported in acid mine drainage. Additional reported sources of nickel wastewater are from the use of spent nickel plating bath solution to manufacture nickel sulfate, and the use of impure copper metal to produce copper sulfate [35]. In the former case, a wastewater nickel level of 12.2 mg/l was reported, while nickel in wastewater from the latter process was present at 159 mg/l.

Nickel exists in waste streams predominantly as the soluble ion. In the presence of complexing agents such as ammonia, EDTA, or cyanide, nickel may exist in an extremely stable soluble complexed form. The formation of

Table 14.1 Summary of Nickel Concentrations Reported in Wastewaters

	Nickel Concentration (mg/l)		
Source	*Range*	*Avg*	*Reference*
Plating plants			
Four plants	2–205	—	3
Five plants	5–58	24	4
Rinsewaters	2–900	—	5
Large plating plant			6
Segregated flow	—	88	
Combined flow	—	46	
Large job shop	—	5.7	7
Plating of zinc castings	45–55	—	8
Plating of plastics	30–40	—	8
Manual barrel and rack	15–25	—	8
Nickel plate rinse	—	132	9
	—	134	10
	—	110	11
	—	119	12
	—	99	12
Plating and anodizing rinse	—	3.2	12
Mixed plating rinse	—	35	12
	0.93–2.2	—	13
Electroless copper plating	0–150	—	14
Tableware plating			
Silver-bearing waste	0–30	5	15
Acid waste	10–130	33	15
Alkaline waste	0.4–3.2	1.9	15
Metal finishing			
Mixed wastes	17–51	—	16
Acid wastes	12–48	—	16
Alkaline wastes	2–21	—	16
Small parts fabrication	179–184	181	17
Brass pickling	—	3	18
Metal-forging rinse	0.77–1.06	—	19
Steel pickling	5–10	—	20
Stainless steel pickling	0.2–400	—	21
	—	250	22
Business machine manufacture			
Plating wastes	5–35	11	15
Pickling wastes	6–32	17	15

Table 14.1 *(continued)*

Source	Nickel Concentration (mg/l) Range	Avg	Reference
Other			
Mine drainage	0.19–0.51	—	23
Acid mine drainage	0.46–3.4	—	24
	0.01–5.6	0.72	25
Alkaline mine drainage	0.01–0.18	0.02	25
Coal storage pile drainage	1–10	—	26
Fly ash ponds	0.06–0.15	—	26
Acidic coal-cleaning waste	—	7.5	27
Gold-ore extraction	—	1.4	28
	—	6.5	29
Boiler tube cleaning	0.5–15.8	—	30
Boiler cleaning			
Acid waste	80–400		
Alkaline waste	1–10		
Dye house effluent	67.5	—	31
Paint and ink formulation	0–40	0.5	32
Porcelain enameling	0.25–67	14	32
Nickel sulfate manufacture	470–890	—	22
Copper sulfate manufacture	—	22	22
Titanium dioxide manufacture	—	1.3	22
Sodium dichromate manufacture	0.6–7.8	—	22

nickel-cyanide complexes interferes with both cyanide and nickel treatment, and the presence of this species may be responsible for high levels of both cyanide and nickel in treated wastewater effluents.

TREATMENT TECHNOLOGY

Regardless of the type of treatment employed for nickel, the level of treatment required and associated costs can often be reduced by good housekeeping practices to reduce accidental loss, spillage, or leaks of plating or pickling solutions containing nickel. Process modifications to reduce the major contributor, dragout loss of plating-bath solution into rinsewaters, have been successful. As one example, a nickel-plating line which initially used 4.5 gpm rinsewater, reduced this volume over 75% by process changes [6].

Usually, total flow reduction is an integral part of recovery systems, in order to reduce capital costs. Countercurrent-flow multiple rinse tanks have been employed to allow direct recovery of nickel-plating solution dragout. The first (and therefore most concentrated) rinse bath solution may serve as plating-

Table 14.2 Effect of pH on Nickel Solubility

	Nickel Concentration (mg/l)					
Treatment pH	*I*	*II*	*III*	*IV*	*V*	*VI*
6.0	—	93.0	—	—	—	—
7.0	—	77.0	7.4	16.2	—	0.85
7.3–7.5	83.0	—	—	—	—	0.65
8.0	—	47.7	6.7	10.0	4.13	0.47
8.4–8.5	75.0	—	—	—	3.74	0.25
9.0	69.0	—	—	—	2.48	0.13
10.0	—	0.20	1.60	0.70	2.58	0.12
11.0	0.40	0.12	—	—	4.13	0.12
11.6–12.0	<0.10	—	—	—	—	0.10

I. Synthetic wastewater [37], initial concentration = 96 mg/l.
II. Plating rinsewater [11], initial concentration = 110 mg/l.
III. Copper sulfate manufacturing waste, treated with caustic [22], initial concentration = 22 mg/l.
IV. Same as III, treated with lime [22].
V. Chlor-alkali wastewater [22], initial concentration not reported.
VI. Titanium dioxide wastewater [22], initial concentration = 1.3 mg/l.

bath solution makeup, to compensate for evaporation from the hot plating bath. In one application, use of four sequential rinse tanks resulted in a nickel concentration of only 0.9 mg/l in the final rinse tank effluent [36].

Precipitation

The formation and precipitation of nickel hydroxide is the primary method for treatment of nickel-bearing wastes. Precipitation with carbonate is commonly associated wtih nickel recovery systems. Nickel forms insoluble nickel hydroxide upon addition of lime. Table 14.2 presents treatability data, for five different wastewaters, as a function of treatment pH. A comparison of the results indicates (a) that most effective treatment occurs at pH above 11 (e.g. wastes I, II and VI), and (b) that other wastewater characteristics may prevent the achievement of nickel solubility below 1 mg/l (e.g., wastes III, IV and V). Such interference has also been noted in the treatment of a nickel sulfate-manufacturing wastewater [22], wherein minimum solubility (2.5 mg/l) was achieved at pH 10.0, with increased solubility thereafter. Table 14.3 summarizes full-scale results of lime precipitation of nickel in various industrial wastewaters. Effluent nickel values greater than 1 mg/l appear to result from improper pH control, poor solids separation [15, 40], or interference in precipitate formation.

Kantawala and Tomlinson [42] have reported the reduction of 100 mg/l nickel to 1.5 mg/l at pH 9.9, by addition of 250 mg/l lime and resultant precipitation of nickel hydroxide. Figure 14.1 shows the relationship between nickel

Table 14.3 Summary of Precipitation Treatment Results for Nickel Wastewaters

Source	Treatment pH	Nickel Concentration (mg/l) Initial	Final	Comments	Reference
Plating	—	46	0.8		6
	8.8	5.6	0.6		34
	—	7.7	2.5		9
Mixed plating	5.5–10.1	16.3	1.0		13
	9.4–10.2	5.7	2.5	Poor solids removal	13
	8.5	35	1.00	Clarification only	12
			0.75	Filtration	
	10.0	35	0.04	Clarification only	
			0.02	Filtration	
Tableware plating	—	21	0.09–1.9	$FeCl_3$ + sand filtration	15
Electroless plating	12.0–12.5	85–140	<0.5	High pH, lime, polymer	14
	11.6–12.0	51.9	0.1		14
Steel pickle rinse	—	5–10	2.0		20
Stainless steel pickle rinse	8.5–9.2	9.5	<0.1		21
Nonferrous metal works	8.5–9.0	—	0.05–0.13		38
	8.1	—	15		39
Appliance manufacturing	—	35	0.4		16
Office machine manufacturing	—	39	0.17	Polymer	40
Record changer manufacturing	—	—	0.1–0.2		41
Mixed boiler wastes	8.2–8.5	15–30	<0.1		30
	11.2–11.8	15–30	<0.1		30
Copper sulfate manufacture	—	159	<0.5	Sand filtration	35
Nickel sulfate manufacture	10.2	12.2	3.0	Sand filtration	35

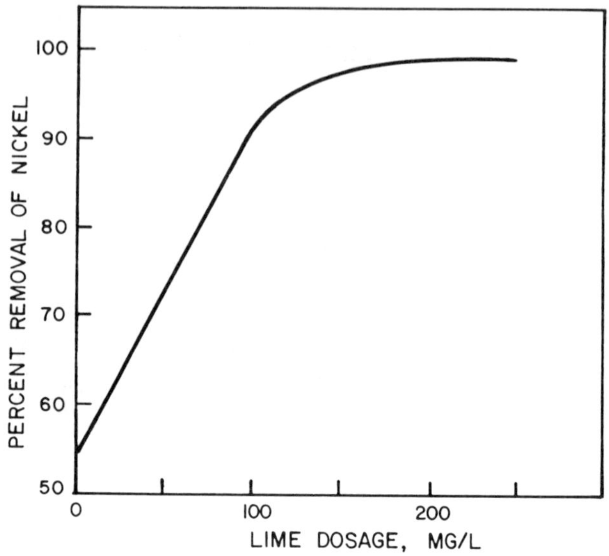

Figure 14.1 Treatment efficiency for nickel removal by chemical precipitation with lime. (From Kantawala and Tomlinson [42], courtesy of *Water and Sewage Works*.)

removal and lime dose for their laboratory studies. Maruyama et al [43], in pilot plant studies (15 gpm) added heavy metals to municipal waste at concentrations of 5 mg/l. Lime addition at 260 mg/l (pH 10.0) plus 20 mg/l ferrous sulfate yielded 0.35 mg/l nickel after sedimentation plus mixed media filtration. Lime addition alone at 600 mg/l (pH 11.5) yielded residual nickel of 0.15 mg/l, after sedimentation and filtration.

Control of pH for precipitation treatment is normally accomplished by addition of either lime or sodium hydroxide (caustic). In smaller installations, liquid sodium hydroxide is frequently chosen, despite its higher chemical cost, as a result of ease of storing and feeding. However, recent monitoring studies on seven plating plants using lime and 18 plants using caustic for nickel treatment revealed that the lime-treated effluents contained less than half the nickel present in the caustic-treated effluents [44]. Similar results are seen in Table 14.4, for treatability studies on wastewaters from one copper sulfate-manufacturing and two nickel sulfate-manufacturing plants. Both clarifier and filter effluent nickel is higher when caustic is used as the treatment chemical. The reasons for this dissimilarity in treatment results are not clear. One possible factor is that coprecipitation of calcium carbonate may enhance nickel precipitation treatment upon calcium hydroxide addition. Some nickel wastes are somewhat alkaline, and contain background carbonate due to atmospheric contribution. However, in the instance of the copper sulfate wastewater (Table

Table 14.4 Treatability Results on Copper and Nickel Sulfate-Manufacturing Wastewaters [after 22]

	Copper Sulfate		Nickel Sulfate		Nickel Sulfate	
Parameter	*Lime*	*Caustic*	*Lime*	*Caustic*	*Lime*	*Caustic*
Treatment pH	9.7–10.4	9.6–10.5	10.0	10.0	10.0–10.5	10.5–10.7
Initial nickel (mg/l)	34.5	35.3	1,676	1,860	15.0	14.0
Clarifier effluent	0.17	0.25	2.11	2.92	1.60	2.27
Filter effluent	<0.11	<0.14	1.84	2.79	<0.05	0.13

14.4), the initial waste pH was acidic (pH 3.0–3.3) and low carbonate levels would be expected.

Effective treatment may depend to a large extent on the precipitate solids separation efficiency. Patel and Pearson [45] report that rapid batch precipitation of nickel hydroxide yields gelatinous, poor settling and filtering solids. Continuous flow precipitation results in much better clarifier performance, due to enhanced solids settling rates. Sludge recycle to the nickel precipitation reactor yields further improvement in settleability, and in sludge-dewatering characteristics [45]. The data of Table 14.5 show the distribution between dissolved and suspended nickel in the hydroxide precipitation treatment effluent from a plating operation [13]. These data illustrate the importance of solids separation. Continuous treatment with gravity clarification has been claimed to yield residual total effluent nickel concentrations below 2 mg/l, with lower values possible by polymer addition [46]. The data of Table 14.6 compare the results of gravity clarification with those obtained by either direct filtration without presettling, or polishing filtration of clarifier effluent. General precipitation-treatment cost data (in 1975) are presented in Table 14.7 (see also Table 4.11), and costs of filtration are shown in Table 14.8 and Figures 14.2 and 14.3.

Sulfide precipitation treatment has been recommended for nickel, with one reported advantage of effective treatment at more near-neutral pH than is required for nickel hydroxide precipitation. One gold-ore extraction waste-

Table 14.5 Distribution of Effluent Nickel in a Precipitation Treated Plating Wastewater [13]

	Nickel Concentration (mg/l)		
Treatment pH	*Soluble*	*Particulate*	*Total*
8.8	0.08	0.17	0.25
9.1	0.16	0.27	0.43
9.6	0.13	1.10	1.23
10.1	0.06	3.4	3.45

Table 14.6 Average Effluent Nickel Concentrations from Alternative Solids Removal Processes [after 44]

Process	*Number of Plants*	*Effluent Nickel (mg/l)*	
		Avg	*Range*
Gravity clarification	21	1.4	0.1–12.5
Direct filtration	4	0.75	0.3–1.3
Clarification plus filtration	4	0.63	0.1–1.1

water, initially at 6.5 mg/l nickel, was reduced to 2.1 mg/l after treatment with sodium sulfide at pH 7 [29]. Use of alum at pH 7–8 with the same waste reduced nickel to 0.5 mg/l, however, and of lime or caustic at pH 10+ to below the detectable limit. In a second application of sulfide precipitation, this time to acid mine drainage, nickel levels of 0.19–0.51 mg/l were reduced to 0.13–0.29 mg/l at pH 5.5–6.5. The best treatment occurred at the higher pH [23].

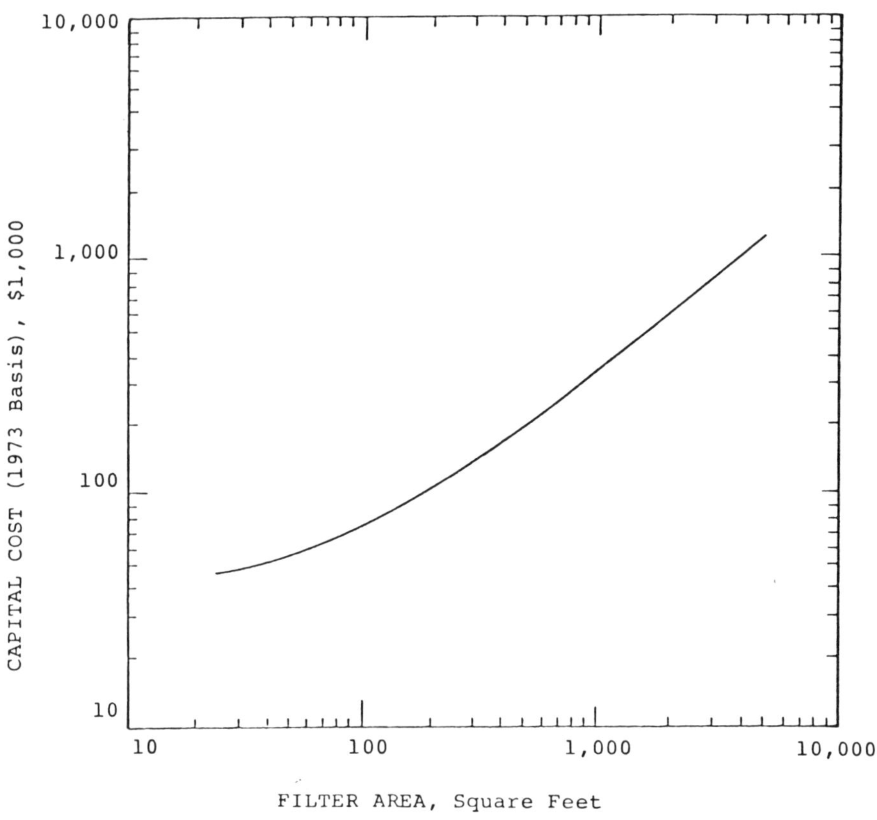

Figure 14.2 Capital cost (in 1973) of granular media filtration [47].

Table 14.7 General Precipitation Treatment Costs (in 1975) for the Electroplating Industry [34]

Wastewater Volume (gpm)	*Capital Cost ($/1,000 gpd)*	*Operating Cost ($/1,000 gal)*
26	74.63	1.07
60	48.70	0.92
102	37.79	0.87

In one study, three electroplating wastewaters were treated by lime (at pH below the optimal for hydroxide precipitation) and by lime plus sulfide, with residual soluble sulfide maintained at 0.5 mg/l [12]. The results are presented in Table 14.9. At treatment pH below 9.0, sulfide addition improved treatment, although the effluent levels were far higher than those achievable by hydroxide precipitation alone at pH 9 or above. These results suggest that,

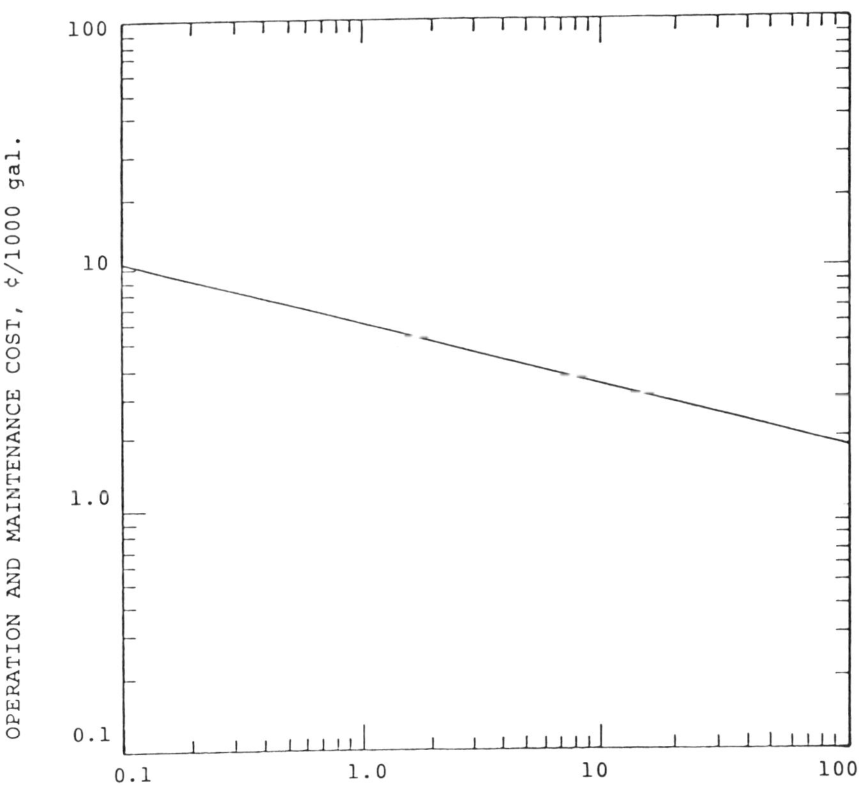

Figure 14.3 Operation and maintenance cost (in 1973) for granular media filtration [47].

Table 14.8 Diatomaceous Earth Filtration Treatment Costs (1979) [44]

Wastewater Volume (gpm)	*Capital Cost ($/1,000 gpd)*	*Operating Cost ($/1,000 gal)*
1	7,400	10.04
20	1,000	0.70
200	230	0.35

at moderate pH, sulfide addition is not capable of significant improvement in treatment efficiency.

McAnally et al [11] compared the effectiveness and economics of nickel carbonate versus nickel sulfide precipitation at pH 9.5 and above. Below that pH, neither process was able to reduce a nickel-plating wastewater, originally at 110 mg/l nickel, below 0.5 mg/l nickel. They reported that both carbonate and sulfide precipitation were effective, although carbonate precipitation was most cost effective (Table 14.10). At high sulfide dosage (S:Ni = 2.0 mg/mg), and pH 10, residual soluble nickel levels above 3 mg/l were observed [11].

Ion Exchange

Recovery of nickel can be accomplished by ion exchange. Recovery of nickel from process waste streams is more attractive than for some other metals, due to the high value of nickel. One report [5] estimates rinsewater recovery value to be $0.80–7.00/1,000 gal, for nickel concentrations ranging from 150 to 900

Table 14.9 Comparison of Lime Versus Lime Plus Sulfide Precipitation of Nickel in Electroplating Wastewaters [after 12]

	Wastewater		
Parameter	*A*	*B*	*C*
Treatment pH	8.5	8.75	9.0
Initial nickel, mg/l	119.0	99.0	3.2
Lime treatment			
Clarifier effluent	12.0	16.0	0.47
Filter effluent	9.4	12.0	0.07
Lime plus sulfide			
Clarifier effluent	11.0	7.0	0.35
Filter effluent	3.5	4.2	0.20

Table 14.10 Ranked Chemical Cost to Reduce Nickel Below 0.5 mg/l by Carbonate or Sulfide Precipitation [after 11]

Chemical Cost[a] *($/100,000 gal)*	*Carbonate (mg/l as $CaCO_3$)*	*Sulfide (mg/l)*	*Treatment pH*	*Achievable Nickel (mg/l)*
33.07	22	0	10.0	0.47
47.09	110	0	10.0	0.34
49.57	100	0	11.0	0.20
289.96	0	77	10.0	0.20
364.36	0	99	10.0	0.09
437.73	0	121	9.5	0.48

[a] Includes lime cost to increase initial pH 2.5 to indicated treatment pH.

mg/l. Ion exchange treatment requires waste stream segregation, particularly in the presence of cyanide, in order that the nickel regenerant can be recovered by precipitation. Nickel-cyanide permanently contaminates ion-exchange resins [48].

Cation-anion exchange has been used to treat nickel sulfate plating bath rinsewater containing 870 mg/l nickel [49]. Complete nickel removal was reported, and the reclaimed water was acceptable for reuse in the rinse tank. The unit processed 22 gpm. A second plant employed a scheme of cation exchange for nickel removal, followed by weak-base anion exchange [50]. Reclaimed concentrations of 60,000 mg/l nickel plus 24,000 mg/l sulfuric acid (the regenerant) were achieved upon regenerating the exchange resins. Chemical costs were given as approximately $0.52/lb nickel recovered [50]. A three-stage ion exchange plant has been described for one plating shop. The system comprises a carbon filter to remove suspended solids and protect the following anion exchange stage from organic fouling, a cation exchange column, and the anion-exchange column. The system processes 85 gpm of countercurrent rinse tank overflow [51].

With mixed metal wastes, recovery cannot be economically practiced. It has been demonstrated that for one waste flow (25 gpm), ion exchange would be more than twice as costly as chemical precipitation, in terms of operating plus capital costs [52]. Further, normal regeneration techniques would produce a dilute acidic effluent for which further recovery or treatment would be difficult. In one application of ion exchange to treat a dilute wastewater containing nickel (7–12 mg/l) and zinc (2–6 mg/l), the regeneration effluent at pH 1.5–2.0 contained only 180–250 mg/l nickel [52].

In a comparison of two nearly identical plating departments, where one employed ion-exchange treatment and the other employed continuous precipitation, capital costs were $1,075/1,000 gpd capacity for ion exchange vs $461/1,000 gpd for precipitation, and operating costs were 18.5¢/1,000 gpd vs 12.3¢/gal for precipitation [53]. The ion-exchange unit recycled 90% of the process water.

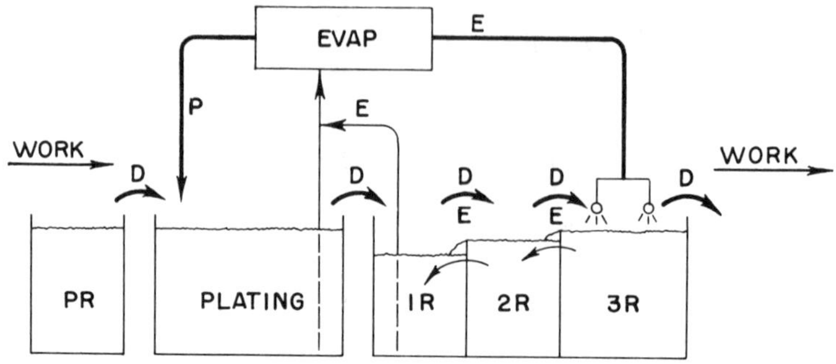

Figure 14.4 Simple closed-loop evaporative recovery system, incorporating countercurrent rinse. PR, prerinse tank; D, dragout; R, rinse tank; P, returned plating solution; E, rinse or distillate flow. (From Culotta and Swanton [54], courtesy of *Plating*.)

Evaporative Recovery

Evaporative recovery requires reduction in rinsewater volume, thereby increasing nickel concentrations in the rinsewaters to be processed by the recovery system. Countercurrent rinse serves this need. A typical evaporative system employing countercurrent rinse, plus nickel and distillate water recycle, is shown in Figure 14.4.

As early as 1955, the use of evaporative recovery for nickel-plating wastes was reported as economically feasible [55]. Capital costs reported were $10,920/1,000 gpd (1,000 gpd system) and operating-maintenance costs were $0.85/1,000 gal. However, nickel recovery value was estimated at $168/1,000 gal treated. Costs for nickel-plating solutions have been reported to range up to $3.68/gal [56], and formulas are available for comparison of precipitation treatment costs versus evaporative recovery costs [54, 57]. A commercially available evaporative recovery system, capable of treating 400 gal/hr, has been reported to cost $45,000, and could recover $100,000/yr in chemicals [58]. The principal operating cost for evaporative units is steam (Figure 14.5). This high energy cost makes consideration of multiple-effect or vapor-compression evaporators for larger capacity systems essential, from a net cost savings viewpoint [7].

Reverse Osmosis

There is a growing application of reverse osmosis for wastewater concentration and recovery of plating chemicals. Most of these existing commercial units are for treatment of nickel-plating solutions, since reverse osmosis is especially suited for handling the near-neutral nickel solutions, and the economics are favorable for recovery and reuse of expensive nickel chemicals. Generally, the permeate (purified water) is recycled to the rinse, and the concentrated salts to the plating bath. Table 14.11 summarizes the application of various reverse

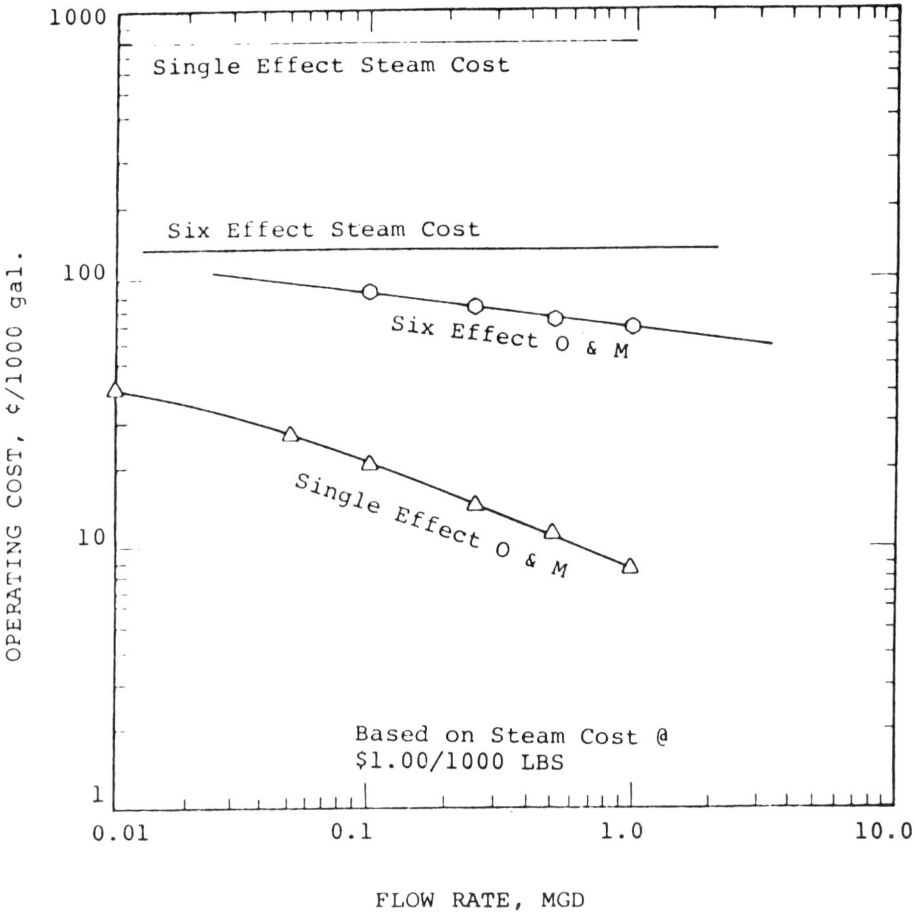

Figure 14.5 Steam cost and operation and maintenance cost (in 1973) for evaporation [47].

osmosis systems to nickel wastewaters. In all systems, boric acid permeates the membrane [59].

In pilot-plant scale studies, reverse osmosis has been found to give 99.6% rejection of plating rinsewater nickel ions [60, 61]. The system included three sequential countercurrent rinse tanks. Sufficient makeup water was added to the last tank to balance evaporative loss. The contents of the first, most concentrated rinse tank were treated by reverse osmosis, with concentrated nickel solution (12,000 mg/l) returned to the plating tank, and permeate returned to the first rinse tank. Rinse-tank nickel concentrations were: first tank 1,700 mg/l; second tank 50 mg/l; and third tank 12 mg/l [61].

In a comparison study of reverse osmosis systems application to nickel wastewater treatment, the relative efficiency of recovery (percent nickel rejection) was found to be dependent upon both type of membrane and module

Table 14.11 Application of Reverse Osmosis to Nickel Wastewaters [after 59]

Membrane Type	*Bath Type*	*Overall Performance*	*Comments*
Polyamide	Nickel sulfamate	Good	pH must be above 2.5
	Nickel fluoborate	Moderate	pH must be above 2.5
	Watts nickel	Good	
Cellulose acetate	Nickel sulfamate	Good	
	Nickel fluoborate	Good	

used [58, 62]. For nickel sulfamate bath rinsewater, rejection of nickel ranged from 84 to 98%, while nickel fluoborate bath rinse yielded rejections of 70–95%.

Other Processes

The application of electrodialysis has been reported at one nickel-plating facility [63]. Permeate is returned for rinsewater, and the relatively dilute concentrate is returned to the plating bath. Net evaporation in the bath compensates for the dilute nature of the recycled concentrate.

Nickel recovery during electrolytic destruction of cyanide has also been reported [64]. Nickel in the concentrated cyanide bath was reduced from 459 to 50–80 mg/l.

SUMMARY

The primary sources of nickel wastes are the plating and metal-processing industries. Thus nickel treatment is frequently combined with treatment for other contaminants, and specific costs of nickel removal are difficult to isolate. Treatment for nickel by lime precipitation at high pH, plus clarification, is capable of reducing residual nickel concentrations to less than 1.0 mg/l. Costs and efficiency of treatment can be substantially reduced by process control and modification. Filtration treatment for enhanced particulate nickel control will achieve an effluent nickel level below 0.5 mg/l. Carbonate or sulfide precipitation at high pH can yield even lower effluent nickel levels.

Implementation of process modifications also makes nickel recovery an attractive proposition when waste stream separation is practiced, as nickel is a relatively valuable metal. Several recovery systems have been demonstrated to be economically feasible, including precipitate recovery directly or after ion exchange, reverse osmosis, and evaporative recovery. For hot nickel-plating baths, process modifications have led to direct nickel recovery, by return of rinsewater to the plating bath as makeup for evaporative loss.

REFERENCES

1. Banerjee, N.G. and T. Banerjee. "Recovery of Nickel and Zinc from Refinery Waste Liquors: Part I—Recovery of Nickel by Electrodeposition," *J. Sci. Ind. Res.* 11B (1952): 77–78.
2. Dean, J.G., F.L. Bosqui, and K.H. Lanouette. "Removing Heavy Metals from Waste Water," *Environ. Sci. Technol.* 6 (1972): 518–522.
3. Wise, W.S. "The Industrial Waste Problem: IV. Brass and Copper Electroplating and Textile Wastes," *Sew. Ind. Wastes* 20 (1948): 96–102.
4. Battelle-Columbus Laboratories. *An Investigation of Techniques for Removal of Cyanide from Electroplating Wastes,* Water Pollution Control Research Series 12010 EIE 11/71, (Washington, DC: EPA, November 1971).
5. *State of the Art: Review on Product Recovery,* Water Pollution Control Research Services 17-7 ODJW 11/69 (Washington, DC: USDI, 1969).
6. Chalmers, R.K. "Pretreatment of Toxic Wastes," *Water Poll. Control* (London, 1970): 281–291.
7. Culotta, J.M. and W.F. Swanton. "Recovery of Plating Wastes: Selection of Lowest Cost Evaporator," *Plating* 57 (1970): 1221–1223.
8. Lowe, W. "The Origin and Characteristics of Toxic Wastes with Particular Reference to the Metals Industries," *Water Poll. Control* (London, 1970): 270–280.
9. Kennedy, I.F.T. and S.D. Gupta. "Electrochemical Removal of Trace Metals for Metal Wastes with Simultaneous Cyanide Destruction," *Fin. Manag.* (July 1978): 7–18.
10. Yeats, A.R. "Ion Exchange Selectively Removes Heavy Metals from Mixed Plating Wastes," paper presented at the 32nd Annual Purdue Industrial Waste Conference, West Lafayette, IN (May 1977).
11. McAnally, S., L. Benefield, and R. Reed. "Nickel Removal from a Synthetic and Actual Nickel Plating Wastewater Using Sulfide and Carbonate for Precipitation and Coprecipitation," paper presented at the 39th Annual Purdue Industrial Waste Conference, West Lafayette, IN (May 1984).
12. Robinson, A. and J. Sum. "Sulfide Precipitation of Heavy Metals," USEPA 600/2-80-139 (June 1980).
13. Yost, K.J. and D.R. Masarik. "A Study of Chemical-Destruct Waste Treatment Systems in the Electroplating Industry," *Plating Surface Fin.* (January 1977): 35–40.
14. Wing, R.E. and W.E. Rayford. "Heavy Metal Removal Processes for Plating Rinse Waters," *Proceedings of the 32nd Industrial Waste Conference, Purdue University* (1977), pp 838–852.
15. Nemerow, N.L. *Theories and Practices of Industrial Waste Treatment* (Reading, MA: Addison-Wesley, 1963).
16. Anderson, J.S. and E.H. Iobst, Jr. "Case History of Wastewater Treatment in a General Electric Appliance Plant," *J. Water Poll. Control. Fed.* 40 (1968): 1786–1795.
17. McElhaney, H.W. "Metal-Finishing Wastages Treatment at the Meadville, PA Plant of Talon, Inc," *Sew. Ind. Wastes* 25 (1953): 475–482.
18. Joster, T.L. and T.H. Taylor. "Industrial Waste Treatment at Scovill Manufacturing Company, Waterbury, CT," paper presented at the 28th Annual Purdue Industrial Waste Conference, West Lafayette, IN (May 1973).

19. Hockenbury, M.R. and A.W. Loven. "Treating Metal Forging and Processing Wastewater," *Ind. Wastes* (May/June, 1977): 45–49.
20. Cottrell, W.R.T. "Treatment of Effluent by Flotation Processes," *Effluent Water Treatment J.* 116 (1976): 563–567.
21. Winans, T.O. and D.E. Dreier. "Innovative Acid Pickling Wastewater Treatment at Universal-Cyclops Stainless Steel Plant," paper presented at the 34th Annual Purdue Industrial Waste Conference, West Lafayette, IN (May 1979).
22. *Treatability Studies for the Inorganic Chemicals Manufacturing Point Source Category,* USEPA 440/1-80/103 (July 1980).
23. Larsen, H.P., J.K.P. Shou, and L.W. Ross. "Chemical Treatment of Metal-Bearing Mine Drainage," *J. Water Poll. Control. Fed.* 45 (1973): 1682–1695.
24. Hill, R.D. "Control and Prevention of Mine Drainage," paper presented at the Environmental Resources Conference on Cycling and Control of Metals, Battelle-Columbus Laboratories, Columbus, OH (1972).
25. "Development Document for Interim Final Effluent Limitations Guidelines and New Source Performance Standards for the Coal Mining Point Source Category," USEPA 440/1-75/057, Group II (October 1975).
26. Chu, T-Y.J. and R.J. Ruane. "Wastewater Treatment for Coal-Fired Electric Generating Stations," paper presented at the 1978 WWEMA Industrial Pollution Conference, St. Louis, MO (April 1978).
27. Wangen, L. and J. Williams. "Control by Alkaline Neutralization of Trace Elements in Acidic Coal Cleaning Waste Leachates," *J. Water Poll. Control. Fed.* 54 (1982): 1302–1310.
28. Schmidt, J., L. Simovic, and E. Shannon. "Development Studies for Suitable Technologies for the Removal of Cyanide and Heavy Metals from Gold Milling Effluents," paper presented at the 36th Annual Purdue Industrial Waste Conference, West Lafayette, IN (May 1981).
29. Rosehart, R. and J. Lee. "Effective Methods of Arsenic Removal from Gold Mine Wastages," *Can. Min. J.* (June 1972): 53–57.
30. Kuppusamy, N. "Copper Removal from Power Plant Boiler Cleaning Waste," *Ind. Wastes* (March/April 1977): 43–45.
31. Rouse, J.V. "Removal of Heavy Metals from Industrial Effluents," *ASCE J. Environ. Eng. Div.* (October 1976): EE5 929–936.
32. Cawley, W., ed. *Treatability Manual,* vol. I. *Treatability Data,* USEPA 600-8-80-042-a (July 1980).
33. Jackson, D.V. "Metal Recovery from Effluents and Sludges," *Metal Fin. J.* 18 (211) (1972): 235–242.
34. "Development Document for Interim Final Effluent Limitations Guidelines and New Source Performance Standards for the Common and Precious Metals Segment of the Electroplating Point Source Category," USEPA 440/1-75/040, Group I, Phase II (April 1975).
35. "Development Document for Interim Final Effluent Limitations Guidelines and New Source Performance Standards for the Significant Inorganic Products Segment of the Inorganic Chemicals Manufacturing Point Source Category," USEPA 440/1-75/037, Group I, Phase II (May 1975).
36. Barnes, G.E. "Disposal and Recovery of Electroplating Wastes," *J. Water Poll. Control. Fed.* 40 (1968): 1459–1470.

37. Netzer, A., A. Bowes, and J. Norman. "Removal of Heavy Metals from Wastewater by Lime," *Proceedings of the 16th Conference on Great Lakes Research*, Huron, OH (1973).
38. Stone, E.H.F. "Treatment of Non-Ferrous Metal Process Waste," *Metal Fin. J.* 18 (1972): 280–290.
39. Tupps, C.C. "Treatment of Wastes for Automobile Bumper Finishing," Ind. Water Wastes 6 (1961): 111–114.
40. Hansen, N.H. and W. Zabban. "Design and Operation Problems of a Continuous Automatic Plating Waste Treatment Plant at Data Processing Division, IBM, Rochester, MN," *Proceedings of the 14th Industrial Waste Conference, Purdue University* (1959), pp 227–249.
41. "Effluent Treatment at B.S.R.," *Metal Fin. J.* 17 (1971): 201.
42. Kantawala, D. and H.D. Tomlinson. "Comparative Study of Recovery of Zinc and Nickel by Ion Exchange Media and Chemical Precipitation," *Water Sew. Works* 111 (1964): R281–R286.
43. Maruyama, T., S.A. Hannah, and J.M. Cohen. "Removal of Heavy Metals by Physical and Chemical Treatment Processes," *J. Water Poll. Control. Fed.* 47 (1975): 962–975.
44. "Development Document for Existing Sources Pretreatment Standards for the Electroplating Point Source Category," USEPA 440/1-79/003 (August 1979).
45. Patel, C. and D. Pearson. "Precipitate Densification by Recirculation," *Institute Chemical Engineers (Br.) Symposium* Series 52 (1977).
46. Lancy, L.E., W. Nohse, and D. Wystrach. "Practical and Economic Comparison of the Most Common Metal Finishing Waste Treatment Systems," *Plating* 59 (1972): 126–130.
47. Edwards, A.W., W.W. Eckenfelder, Jr., and J.A. Roth. *Industrial Water Cost Studies*, vol. I. Office of Research and Development, USEPA Contract 68-01-2485 (1975).
48. Silman, H. "Treatment of Rinse Water from Electrochemical Processes," *Metal Fin.* 69 (1971): 62–66.
49. Reents, A.C. and D.M. Stromquist. "Recovery of Chromate and Nickel Ions from Rinse Waters by Ion Exchange," *Proceedings of the 7th Industrial Waste Conference, Purdue University* (May 1952), pp 462–473.
50. Heidorn, R.F. and H.W. Keller. "Methods for Disposal and Treatment of Plating Room Solutions," *Proceedings of the 13th Industrial Waste Conference, Purdue University* (May 1958), pp 418–426.
51. "Treatment and Recovery of Plating Effluent," *Effluent Water Treatment J.* 13 (1973): 600.
52. Ross, R.D. *Industrial Waste Disposal* (New York: Reinhold, 1968).
53. Phasey, N.W. "Experience with Ion Exchange Resins for Effluent Treatment," *Prod. Fin.* 25 (1972): 28–34.
54. Culotta, J.M. and W.F. Swanton. "Case Histories of Plating Waste Recovery Systems," *Plating* 57 (1970): 251–255.
55. Hesler, J.C. "Practical Methods for Treatment of Metal Finishing Wastes," *Plating* 42 (1955): 1019–1029.
56. Smith, D. and C. Roy. "Disposal: What Does It Really Cost?", *Proceedings of the 4th Conference on Advanced Pollution Control for the Metal Finishing Industry*, USEPA 600/9-82-022 (December 1982).

57. Culotta, J.M. and W.F. Swanton. "The Role of Evaporation in the Economics of Waste Treatment for Plating Operations," *Plating* 55 (1968): 957–961.
58. Besselievre, E.B. *The Treatment of Industrial Wastes* (New York: McGraw-Hill, 1969).
59. Crampton, D. and P. Wilmoth. "Reverse Osmosis in the Metal Finishing Industry," *Metal Fin.* (March 1982).
60. Golomb, A. "Application of Reverse Osmosis to Electroplating Waste Water Treatment. Part II. The Potential of Reverse Osmosis in the Treatment of Some Plating Wastes," *Plating* 59 (1972): 316–319.
61. Golomb, A. "Application of Reverse Osmosis to Electroplating Waste Treatment, Part III. Pilot Plant Study and Economic Evaluation of Nickel Recovery," *Plating* 60 (1973): 482–486.
62. Donnelly, R.G., R.L. Goldsmith, K.J. McNulty, and M. Tan. "Reverse Osmosis Treatment of Electroplating Wastes," *Plating* 61 (1974): 432–442.
63. Millmam, W. and R. Heller. "Some Successful Applications of Electrodialysis," *Proceedings of the 4th Conference on Advanced Pollution Control for the Metal Finishing Industry,* USEPA 600/9-82-022 (December 1982).
64. Hillis, M. "Treatment of Cyanide Wastes by Electrolysis," *Trans. Inst. on Metal Fin.* (*Br.*) 53 (1975): 65–73.

15
Organic and Ammonia Nitrogen

Organic nitrogen occurs in the wastewaters of a number of industries. A principal form of organic nitrogen is as proteinaceous material. Organic nitrogen wastes include amino and nonamino chemicals such as urea, muranic acid, amino sugars, uric acid, aliphatic amines, and organic bases [1]. Some industrial sources discharge organic nitrogen compounds which are highly resistant to biological treatment. These include compounds such as pectin, chitin, and quaternary ammonium compounds.

Ammonia nitrogen is also a constituent of many industrial wastewaters. Its concentration varies widely not only among the wastes derived from various categories of industries, but also among different plants of the same industrial category. Industries discharging wastes that are characterized by significant concentrations of ammonia nitrogen include the iron and steel industry, petroleum refineries, fertilizer manufacturing, inorganic chemical manufacturing, ferroalloy manufacturing, glass manufacturing, meat processing, and feed lots. Other industries also discharge ammonia nitrogen in varying concentrations.

Ammonia nitrogen is indigenously generated as a pollutant in certain industries such as iron and steel making (by-product coke, blast furnace, ferromanganese operations), and meat processing [2]. In other industries, ammonia is used as a process chemical, such as in frosted glass production, where it is employed in the frosting solution [3]. Ammonia nitrogen can be initially present in low concentrations in some freshly generated wastes such as tannery wastes [4], egg-breaking wastes [5], and animal excreta [6], but its concentrations may increase rapidly upon storage of these wastes, due to deamination of the organic nitrogen present.

The concentration of organic or ammonia nitrogen in an industrial waste depends on several factors, among which are the nature of the raw materials, the production technology used, and the degree of water consumption and water reuse practices employed in a production facility. Tables 15.1 and 15.2 summarize the concentrations of organic and ammonia nitrogen reported for wastewaters of industrial origin.

Table 15.1 Industrial Sources and Levels of Organic Nitrogen

	Concentration as N (mg/l)		
Source	*Range*	*Avg*	*Reference*
Poultry manure	390–1,005		7
		3,170	8
Sugar factory	7–186		9
Textile waste		8	10
Boiler-cleaning solutions	10–2,600		11
Pharmaceutical and medicinal chemicals		500	12

TREATMENT TECHNOLOGY

There is limited information available on full-scale treatment systems installed specifically for control of industrial organic nitrogen wastewaters. Where such wastes also contain appreciable levels of organic carbon, biological treatment by lagoons, activated sludge, or trickling filtration is utilized to remove the organic carbon and simultaneously treat the organic nitrogen through bioassimilation plus biomineralization to ammonia nitrogen. Other processes reported as at least partially effective in organic nitrogen control include activated carbon adsorption; coagulation with lime, alum, or iron salts; and chlorine oxidation.

Ammonia removal from industrial wastewaters can be accomplished by various methods (Table 15.3). The choice of a treatment process for the removal of ammonia from a given wastewater depends primarily on (1) the nature of the waste, (2) the level of treatment desired, and (3) economics. An additional factor is the method of ultimate disposal of the effluent after treatment.

Although many processes have been identified as effective in the removal of ammonia, only a few methods are truly applicable for ammonia's direct control in wastewaters, because of their relative merits over the others in terms of ease of application, process stability and reliability, relevance to the nature of waste to be processed, and economics. Based on experience in the field to date, of the methods listed in Table 15.3, the major ammonia removal processes are: (1) biological nitrification, i.e., conversion of ammonia nitrogen to nitrite and nitrate, sometimes including nitrogen elimination by denitrification; (2) ammonia stripping; (3) breakpoint chlorination; and (4) ion exchange.

Methods for ammonia removal are best discussed in terms of the concentration of ammonia nitrogen. According to their ammonia nitrogen content, wastes may be characterized as: (a) concentrated (>500 mg NH_3-N/l), (b) intermediate (50–500 mg NH_3-N/l), and (c) dilute (<50 mg NH_3-N/l). In addition, the relative concentration of organic carbon is a primary consideration, when biological treatment of the waste is contemplated. The maintenance of optimal carbon:nitrogen ratios is essential for successful biological waste treatment

Table 15.2 Industrial Sources and Levels of Ammonia Nitrogen Wastewaters

Source	*Concentration as N (mg/l)*		*Reference*
Coke oven wastes			
Weak ammonia liquor	4,000–5,000		13
	8–150	(free NH_3)	14
	10–740	(fixed NH_3)	
	1,800–4,300		15
	1,300–2,000	(free NH_3)	16
	2,600–4,000	(fixed NH_3)	
	150–2,280		17
	6,000		18
	39–7,300		19
	7,000		20
	2,000–6,000		21
Ammonia still effluent	27–54		22
	0.04–2.5		20
	94		21
Iron Blast Furnace Wastes	1.41–12.3		13
	3.1		21
Iron-manganese blast furnace wastes	114		13
Gray iron foundry waste	3.8–7.5		23
Coal gasification wastes			9
Coke	5–1,000		
Anthracite/hard coal	5–1,000		
Brown coal	2,500		
Peat	1,000–10,000		
Wood	50–1,500		
Petroleum refinery wastes			
Refinery	100		24
	33		25
	590–1,400		26
	1,000–5,000		27
	0–120		28
Refinery sour water	135–6,550		29
Topping	0.05–20		30
Cracking	0.05–200		30
Lubrication	1–120		30
Petrochemical	4–300		30
	1,810		31
	100		24
Fertilizer wastes			
Ammonia and urea nitrogen manufacture	200–940		32
Ammonia-nitric acid waste	130		33
Fertilizer complex	600		34
Tantalum/Columbian plant waste	35,000		35

Table 15.2 (*continued*)

Source	*Concentration as N (mg/l)*	*Reference*
Electrolytic ferroalloys		
Electrolytic manganese	16–4,208	36
Manganese dioxide	75	36
Electrolytic chromium	1,076	36
Glass manufacturing	300–650	3
Agroindustries		
Poultry litter	196–4,197	6
Poultry waste	440	8
Cattle		
Litter	14–8,600	6
Urine	11–1,227 (fresh)	6
	486–7,163 (stored)	
Swine slurry	322–8,300	6
Swine waste	424	37
Food-processing wastes		
Poultry	3.1–11.6	38
	6.1	39
Soybean processing	0.2–12.7	40
Vinegar plant waste	0.7–28	41
Shrimp packing	1.8–13.8	42
Slaughter rendering	125	6
Meat packing	10–15.5	43
	7–50	2
Meat processing	52–80	9
Pea processing	35	6
Corn processing	45	6
Coffee wastes	11–78	6
Milk waste	33	6
	5–625	44
Potato starch plant wastes	5–40	9
Explosive industry wastes		45
Complex facilities	22.6–413.5	
On-site facilities	5–158	
Specialty and intermediate produced manufacturing wastes		
Nitrogen plants	16–532	
Nitroglycerin	0–33.7	
Dynamite mixhouse dust control	0–1,503	
Dynamite factory waste	1,270	
Distillery wastes	114–380	46
	5.4–18.5	46
Tannery wastes	83–159	9
Paint and ink formulating wastes	0–1.7	47
Secondary aluminum smelting	0.3–350	48
Synthetic rubber (butadiene-styrene) waste	14.7	9

Table 15.2 (*continued*)

Source	*Concentration as N (mg/l)*	*Reference*
Synthetic rubber wastes	31–76	49
Cuprammonium rayon wastes, blue water	800	9
Paper and pulp weak black liquor	264	25
Wood-preserving industry	32	50
Glue wastes	0.6	6
Plywood	397–450	51
Animal	78	52
Pharmaceutical wastes	91	9
Pharmaceutical and medicinal chemicals	475	12
Wool-processing wastes		
Scour water	160	53
Rinse water	5	53
Sodium bisulfide manufacture	260	54
Printed circuit board waste	300+	55
Power plant-cleaning waste	2830 (avg) 850–5,700	56

Table 15.3 Summary of Ammonia Removal Methods

Category	*Method*	*Reference*
Physical	Reverse osmosis	57–59
	Distillation	16
	Land application-irrigation[a]	
Chemical	Ion exchange	49
	Ammonia stripping	49
	Breakpoint chlorination	49
	By-products production of ammoniacal compounds	15
	Incineration	15
	Catalytic decomposition	15
	Electrodialysis	61
	Electrochemical treatment	61
Biological	Algal harvesting	62
	Biological nitrification	49

[a] Ammonia removal mechanisms may include plant uptake, ion exchange, and/or biological nitrification.

[63, 64]. If the ratio is low (low carbon with respect to nitrogen), volatilization of nitrogen in the form of ammonia may result. If the ratio is high (high carbon with respect to nitrogen) the biological stabilization of the organic carbon constituents of the waste will not be complete, unless nitrogen is supplemented. The use of physicochemical methods, alone or in combination with biological treatment methods, may be more economical in achieving a required quality effluent, when the wastewater under consideration does not have characteristics appropriate for direct biological treatment.

Organic Nitrogen

In biological treatment processes, organic nitrogen compounds which are susceptible to microbial degradation undergo biochemical transformations which result in either nitrogen assimilation into the cell mass, or mineralization to yield ammonia nitrogen. In some aerobic processes, the ammonia nitrogen product may be oxidized to nitrite and nitrate by autotrophic nitrifying bacteria. From full-scale experience, mineralization of organic nitrogen in most protein compounds is readily accomplished, with the proteinaceous material providing both the organic carbon and the nitrogen substrate.

Experience in soil application of nitrogeneous wastes has shown that neither a net increase nor a net decrease of inorganic nitrogen occurs when biodegradable wastes having an organic carbon:nitrogen (C:N) ratio of between 20 and 30:1 are applied to the soil [1]. Thus, compounds having a more narrow C:N ratio would favor mineralization, while those with a wider C:N ratio would favor immobilization of nitrogen.

Mineralization of organic nitrogen takes place under aerobic as well as anaerobic conditions. Proteins are initially hydrolyzed by the action of proteolytic bacteria to form amino acids, which are further converted to ammonia nitrogen. Various types of bacteria, fungi, and actinomycetes have the ability to bring about these transformations, although the rates differ among the various microbial types. Mineralization of organic nitrogen is known to occur within a temperature range of 2–65°C. The optimum range for ammonification seems to lie within 40–60°C [1]. Like many other microbially mediated reactions, it has an optimum pH range of 7–8 [65]. Simple retention of a wastewater for long periods of time can result in significant mineralization of organic to ammonia nitrogen. In one study, 18-hour retention of a waste in a clarifier-type basin resulted in a 38% mineralization of nonsettleable organic nitrogen [66].

The ammonification of urea, uric acid, amino compounds in casein, and organic nitrogen in chicken manure has been investigated [7]. Under aerobic conditions, the rate of mineralization of the convertible organic nitrogen portion of a urea-casein mixture (390–1,005 mg N/l) was found to be first order, and 0.111 per hour. In contrast, the rate of mineralization was found to be zero order under anaerobic conditions. The difference in reaction rate order between aerobic and anaerobic metabolism was speculated to be related to the energy requirements of the microorganisms, and differences in energy yields under the different conditions [7].

The mineralization of organic nitrogen has been studied with poultry wastewater slurries with the intention of controlling nitrogen and odor. In one study, it was reported that about 7% of the total organic nitrogen contained in fresh chicken manure suspension was readily converted to the inorganic form [67]. A similar observation was made in another study in which a chicken manure slurry having a C:N ratio of 6.3:1 was aerated for 40 days [8]. The initial organic nitrogen concentration of 3.17 g/l was reduced to 0.75 g/l, representing a removal of about 76%. Of the organic nitrogen mineralized, 71% volatilized as ammonia, while the other 5% remained in the slurry as ammonium ion.

Although it is known that bioadsorption quickly takes place and organics exerting BOD are removed in about 30–60 minutes when wastewater is brought into contact with returned activated sludge, little information is available as to the nature and quantity of the organic nitrogen bioadsorbed onto the activated sludge floc. During organic carbon oxidation, a portion of the organic nitrogen may be used for microbial uptake and synthesis, thus providing some reduction in the organic nitrogen pool. This reduction in the organic nitrogen of a primary clarified wastewater due to bacterial assimilation has been reported to be in the range of 30–40%. In algal-harvesting systems, the reduction in organic nitrogen of a secondary effluent was reported in the range of 50–90%. Whether the removal was due to mineralization and subsequent uptake by the algae, or to deposition of particulate matter containing organic nitrogen in algal ponds, is not indicated [49, 68].

The percent removals of organic nitrogen that have been observed in various biological wastewater treatment systems are given in Table 15.4. It should be noted that the removals of organic nitrogen presented in Table 15.4 are computed from organic nitrogen data often reported casually in various studies, where the principal objectives were different from organic nitrogen removal.

In contrast to the above results, one study was conducted specifically to investigate the fate of soluble organic nitrogen at three different municipal treatment plants. The concentration and percent removal of the total and soluble organic nitrogen are given in Table 15.5. It can be seen that the bulk of the organic nitrogen was removed through the treatment plants [83]. From the data of Table 15.5, the major reduction of both total and soluble organic nitrogen occurred in the activated sludge process. Effluent chlorination for disinfection yielded slight increases in organic nitrogen, perhaps due to cell lysis.

This study further explored the possibility of removing soluble organic nitrogen by various physical and chemical processes. The processes investigated were chemical coagulation, ion exchange and activated carbon adsorption, and chemical oxidation by compounds such as permanganate, ozone, chlorine, and hydrogen peroxide. These processes were evaluated individually,

Table 15.4 Observed Organic Nitrogen Removals in Waste Treatment Systems

Treatment System	*Raw Sewage*	*Settled Sewage*	*Effluent*	*Filtered Effluent*	*% Removal*	*Reference*
Septic Tank	33.5	15.7	4.7		86	69
Oxidation Ditch	14.2		0.2		99	70
	15.0		1.3		91	70
Orbal Plant	2.6		0.0		100	71
	1.6		0.4		75	71
	18.2		1.1		94	72
	18.7		0.75		96	72
	14.9		3.0		80	72
	15.8		2.9		82	72
Biophysical (Zimpro)	7.3	4.2	1.8	1.2	84	73
Two-Stage Nitrification-Attached Growth Denitrification	4.4	3.3	3	0.9	80	74
Nitrification-Denitrification	6.0		2.2	0.6	90	75
Three-Sludge Nitrification-Denitrification	8.6		1.0	0.5+	94	76
Pure Oxygen	4.3			0.81	81	77
Air-Activated Sludge	4.3			0.75	83	77
Submerged-Filter Denitrification						
60 min	3.6			1.5	58	78
30 min	5.7			2.5	56	78
Trickling Filter	16.6		7.3		56	79
	9.8		4.7		52	80
	6.3		2.5		60	80
	8.2		3.6		56	80
	9.9		2.2		78	80
	13.9		2.1		85	80
	11.3		8.9		21	81
	11.4		9.0		21	81
	13.3	9.9	6.1		54	82

All values in mg/l, except percent removals.

Table 15.5 Concentration of Total Organic and Soluble Organic Nitrogen in Wastewaters Processed Through Conventional Municipal Unit Operations [83]

	Palo Alto		*Santa Clara/ San Jose*		*Union City*	
Sample	*Total*	*Soluble*	*Total*	*Soluble*	*Total*	*Soluble*
Influent sewage	11.7	4.86	18.9	6.26	12.7	6.34
Primary effluent	7.44	4.37	13.4	5.46	10.7	4.92
Secondary effluent	2.77	1.23	3.3	1.78	4.6	1.86
Chlorinated effluent	2.51	1.42	3.3	1.87	4.9	1.88
Removal, %	78.5	70.8	82.5	70.1	61.4	70.3

All values in mg/l except percent removals.

as well as in combination. A summary of the removals obtained with the individual unit processes is given in Table 15.6. Among the processes tested, activated carbon adsorption was the most effective process in removing the soluble organic nitrogen [83].

A laboratory study was conducted in which a sequence of unit operations of an advanced waste treatment system was applied on Palo Alto secondary

Table 15.6 Organic Nitrogen Removals Achieved by Individual Unit Processes [83]

	Percent Soluble Organic Nitrogen Removed			*Percent Soluble COD Removed*		
Process	*Average*	*Maximum*	*Minimum*	*Average*	*Maximum*	*Minimum*
Lime coagulation (pH 11)	33	43	26	24	36	13
Alum coagulation (pH 6)	29	35	23	28	46	17
$FeCl_3$ coagulation (pH 6)	40	48	30	32	40	19
Anion exchange (Neut pH)	12	20	9	30	34	23
Cation exchange (Neut pH)	11	12	9	4	12	0
Cation exchange (pH 2)	42	50	32	22	26	18
Activated carbon adsorption	71	92	52	81	93	68
Chlorination (Neut pH)	42	48	39	18	28	10
Ozonation (Neut pH)	14	28	6	46	53	41

effluent to determine the percent removal of soluble organic nitrogen. The laboratory treatment sequence consisted of lime coagulation (300 mg CaO/l), followed by ammonia stripping (1,600 m^3 air/m^3 waste), recarbonation, mixed media filtration (240 m^3/m^2-day), breakpoint chlorination (8 mg Cl_2/mg NH_3-H), and activated carbon adsorption. The results of this study are summarized in Table 15.7. The unit processes which resulted in significant organic nitrogen reduction at these low influent nitrogen levels were lime coagulation and activated carbon adsorption.

Concentrated Ammonia Wastes

Wastewaters containing ammonia in excess of 500 mg/l as NH_3-N result primarily from coking, ferroalloy production, coal gasification, petroleum refining, agroindustries, fertilizer manufacturing, rayon making, and incandescent bulb making (see Table 15.2).

Methods for ammonia removal from coke oven effluents include quenching of hot coke by weak ammonia liquor, distillation and by-product recovery, incineration, deep-well disposal, catalytic decomposition, ion exchange, stripping, reverse osmosis, breakpoint chlorination, and biological treatment [15]. Of these, the established methods are (1) recovery of ammonia as ammonium sulfate after separation of oil and tar, (2) quenching of coke with ammonia liquor, and (3) air or steam stripping of ammonia in ammonia stills. In spite of impressive ammonia removals by each of the above methods, the resultant effluents may still contain excessive ammonia, for which additional treatment is required.

Generally, biological nitrification is most effectively applied to wastewaters containing intermediate to low levels of ammonia nitrogen. Its application has been attempted, however, for concentrated ammonia wastes. In the nitrification process, ammonia is biologically converted to nitrite and nitrate by aerobic autotrophic nitrifying bacteria belonging to the genera *Nitrosomonas* (ammonia to nitrite) and *Nitrobacter* (nitrite to nitrate). These autotrophic organisms have relatively precise environmental requirements, and the success of the biological nitrification process depends upon optimizing their performance in the wastewater treatment unit. Molecular dissolved oxygen is essential for the biochemical conversion to occur. Cold temperatures hamper the growth rate of the nitrifiers, and hence the rate of ammonia oxidation. Both wastewater alkalinity and pH are important. The complete nitrification of ammonia consumes approximately 7.1 mg carbonate alkalinity (as equivalent $CaCO_3$) per mg ammonia nitrogen oxidized, and sufficient alkalinity must be provided to avoid the induction of an acidic condition as the nitrification process continues. Wastes containing high concentrations of ammonia nitrogen and deficient in the required carbonate alkalinity cannot be nitrified unless supplemental inorganic carbon is added [84–86]. The effect of pH on nitrification has been

Table 15.7 Effect of Advanced Waste Treatment on the Removal of Organic Nitrogen from Secondary Effluent [83]

Sample	*Secondary Effluent*	*Lime Coagulation*	*Ammonia Stripping*	*Recarbonation*	*Filtration*	*Breakpoint Chlorination*	*Activated Carbon*	*Percent Removal*
1	2.32	1.21	1.1	—	1.18	1.06	0.49	79
2	3.12	0.87	0.83	—	0.79	0.63	0.03	99
3	1.95	0.90	0.85	0.93	0.82	0.79	0.27	86
Average	2.46	0.99	0.93	0.93	0.93	0.83	0.26	89
Incremental Removal, %		60	62	62	62	66	89	—

All values in mg/l except percent removals.

Table 15.8 Process Efficiency and Economics (1974) for Ammonia Control of Coke Wastes [19]

Treatment Level and Process Description	*Ammonia Nitrogen (mg/l)*		*Operating Cost ($/1,000 gal)*
	Influent	*Effluent*	
A. Ammonia still (stripping)	2,000	1,000	5.06
B. Level A with lime and steam addition	2,000	125	5.96
C. Level B plus effluent neutralization and breakpoint chlorination	2,000	10	9.18

reported to be somewhat variable, with the optimum value generally falling between pH 7.0 and 8.5 [87–89].

Results from studies on biological nitrification of weak ammonia liquor and ammonia-still blowdown have been variable [15]. When weak ammonia liquor (≈5,000 mg NH_3-N/l) was treated biologically, ammonia removal was reported to be negligible [90]. With another coke oven waste, laboratory scale single-stage and two-stage activated sludge systems were investigated and ammonia removals of 50–90% were obtained, depending on the loading rates and retention times [22]. In another pilot study, variable ammonia removals were reported in a system utilizing three separate stages, for carbon removal, nitrification, and denitrification [15]. Ammonia removal reached 90% when the waste was diluted to 12% of the original 3,800 mg NH_3-N/l and the carbon unit was functioning satisfactorily. Denitrification was able to remove 95–100% of the oxidized nitrogen.

In a bench study, the feasibility of a single-stage activated sludge process followed by ion exchange for the removal of residual ammonia from weak ammonia liquor was investigated [13]. This study indicated that ion exchange after biological treatment of the weak ammonia liquor yielded an average overall ammonia removal of 96%. The effluent, however, contained ammonia in the range of 0–200 mg/l. The capital and operational costs for a 400,000 gpd treatment plant were estimated (in 1972) at $1,750/1,000 gpd capacity and 27¢/1,000 gal of the waste treated, respectively.

A recent government study assessed the efficiencies and economics of alternative systems for treatment of coking wastes [19]. Three levels of physicochemical treatment were evaluated, as presented in Table 15.8. As these data demonstrate, the addition of lime for pH control, with steam stripping, was projected to yield above 90% reduction in ammonia nitrogen to a level near 100 mg/l. Further treatment, by breakpoint chlorination, is effective but at a high incremental operating cost.

Ammonia stripping is a process in which ammonium ions contained in a wastewater are converted by pH adjustment to un-ionized ammonia, which is subsequently desorbed by passing air or steam through the wastewater. The addition of steam raises the water temperature, and thereby increments the fraction of un-ionized ammonia available, at a given pH, to be stripped [13]. While ammonia is typically desorbed from dilute wastewater by air stripping at ambient temperatures, it is more often desorbed by steam stripping with concentrated wastes such as those originating from steel mills and fertilizer production.

Steam stripping is used to process wastes from steel, petrochemical, fertilizer, organic chemicals, and nonferrous metal-manufacturing operations, since these wastes contain very high concentrations of ammonia [16, 91]. The high costs of steam production are somewhat offset by the recovery and reuse of ammonia, the extent of which depends in large part on the concentration of ammonia in the waste. Anhydrous ammonia concentrations of 30% or higher can be recovered. Lime is generally used to raise the pH. Scale formation can be more easily controlled with steam stripping than with air stripping, and it is minimized by using NaOH. In the steam stripping process, the major cost element is steam production.

The factors that govern the efficiency of a steam stripping process are: (1) proper design of the steam stripper, (2) control of hydraulic rates of flow, (3) adequate steam, (4) pH levels of 11 or above, (5) temperatures of 200°F or above in the stripper, (6) sufficient droplet detrainment space, and (7) a compatible ammonia-condensing system [35].

Petroleum refinery wastes vary widely in ammonia concentration, and these wastes encompass both the high and intermediate strength categories. Depending on the nature of technology used, in-house water utilization, and recycle-reuse schemes employed, wastewater volumes vary, and this factor primarily dictates the concentration of ammonia in the wastewater.

For the treatment of high-strength ammonia petroleum refinery sour water (2,750 mg NH_3-N/l), the Chevron WWT process consisting of a two-stage ammonia and hydrogen sulfide-stripping process has been reported to produce an effluent containing less than 50 mg NH_3 and 5 mg H_2S/l, with recovery of high-quality anhydrous or aqueous ammonia and hydrogen sulfide [31].

Sour water strippers have been reported to produce effluents containing 1–100 mg NH_3-N/l, but generally the effluent concentration is held to about 50 mg/l, in order to provide sufficient residual nitrogen for a subsequent biological treatment step [29].

The cost of removing NH_3 and H_2S from sour water in a stripping unit has been reported for the Atlantic Richfield Refinery at Houston [91]. Capital costs (in 1973) of a unit treating a 0.58 MGD flow of sour water, containing about 1,800 mg NH_3/l, were \$1,450/1,000 gpd. The annual operating costs were \$155,000, or \$0.74/1,000 gal of sour water treated, based on a 364 days/year operation. The stripper bottoms contained 29 mg NH_3-N/l, and the process yielded a removal efficiency of 98%. Table 15.9 presents data on the performance of steam-stripping systems for concentrated wastes. Table 15.10 presents cost data.

Table 15.9 Treatment of Concentrated Ammonia Wastes by Steam Stripping

Wastewater Source	*Influent NH_3-N (mg/l)*	*% Removal*	*Reference*
Refinery sour water	1,800	98	92
Ion exchange column elutriant	600	98	19
Petrochemical waste	1,810	98	24
Nuclear fuel processing	1–3% NH_4F + 3–10% Free NH_3	99.4	35
Tantalum and columbium plant waste	35,000	99	35
Coke plant waste	2,000–6,000	95–98	21

The ion exchange process may also be used for the recovery of ammonia and hydrogen sulfide from sour waters [94]. In this process, the sour water passes through a bed of weak cation exchange resin. The ammonium ion is retained on the column and free H_2S is formed. Since the effluent H_2S is in undissociated form it can be easily stripped. The saturated cationic ion exchange resin is regenerated by using a mineral acid solution. A sour water containing about 390 mg NH_3-N/l was reported to yield an effluent containing 5 mg NH_3-N/l by the ion exchange process. The process cost (in 1973) $510/1,000 gpd treatment capacity, for a 0.2 MGD unit [94].

The fertilizer industry generates high-strength nitrogenous wastes, which are primarily inorganic in nature. However, urea nitrogen-manufacturing plant wastes may contain some carbonaceous matter, because of the organic nature of the urea [32]. The strength of these wastes may vary. Concentrations of ammonia as high as 1,800 mg/l for ammonium nitrate manufacturing has been reported (see Table 15.2).

Treatment for fertilizer wastewaters involves in-plant controls such as good housekeeping and incorporation of recycle-reuse water systems. Some plants include a closed-loop recycle system consisting of a retention pond for clarification and subsequent ammonia recovery [92, 95]. In other instances the fertilizer manufacturing wastes may be discharged into a local municipal treatment plant.

Steam stripping is used at a number of fertilizer plants. For several plants surveyed, ammonia in the wastewater fed to the stripping unit varied from 100

Table 15.10 Capital and Operating Cost (1982) of Steam-Operated Ammonia Stills [93]

Design Flow Rate (gpd)	*Capital $/1,000 gpd Capacity*		*Operating $/1,000 gal*
	Materials	*Total*	
200,000	$1,875	$13,750	$11.75
400,000	1,720	12,500	10.66

to 1,300 mg/l. Effluent concentrations ranged from 5 to 100 mg/l. When optimal conditions of temperature and pH are maintained, steam-stripping units will consistently yield effluents containing ammonia in the range of 20–30 mg/l [27, 95].

Proprietary processes are available for the treatment of urea-manufacturing wastes, which typically contain about 4,000 mg urea/l and 3,000 mg NH_3-N/l. In these processes, the urea is converted to NH_3 and CO_2 under high pressure and temperature [95]. The ammonia and CO_2 formed are stripped off and returned to the urea-processing plants. Effluents from these processes will typically contain 100 mg/l and 50 mg/l of urea and NH_3, respectively.

Agroindustrial wastes are similar to fertilizer industry wastes with respect to their nitrogen concentration. They are also highly carbonaceous, having a COD of 30,000–40,000 mg/l, with the actual concentration being dependent on the usage of water in the industry.

The performance of a waste treatment system utilizing algal uptake of nitrogen has been studied recently for the removal of nitrogen from poultry wastes. In this system, poultry excreta were flushed with water to a settling basin where the solids were retained and processed separately by anaerobic digestion. The supernatant was treated in the algae pond. By this process, about 45% of the total unoxidized nitrogen was removed in the system as a whole [62].

The "strong wastewater" from electrolytic manganese production units and the wastewater from electrolyte chromium plants contain high concentrations of NH_4-N [36]. Opportunities exist in some instances for the recovery of ferrous ammonium sulfate from these wastes. The treatment of these wastes is currently in lagoon systems by settling, aeration, or solar evaporation.

Intermediate Strength Wastewaters

There are many industrial wastewaters that fall into the intermediate strength category. A number of different treatment processes have been successfully applied to control intermediate ammonia nitrogen levels, depending upon overall wastewater characteristics including organic carbon content, and effluent discharge requirements. These processes include air and steam stripping, ion exchange, activated carbon adsorption, breakpoint chlorination, and biological nitrification.

Breakpoint chlorination is a process in which chlorine is added to a wastewater until a point is reached where the residual free chlorine is at a minimum and the ammonia concentration declines to zero. The chlorine added reacts with ammonia and eliminates it as N_2 gas. This point, which leaves a minimum of residual chlorine, is called breakpoint. Any chlorine added beyond this point will result in an increase in free residual chlorine. The actual requirement of chlorine in practicing breakpoint chlorination of wastewater depends on the temperature, pH, and concentration of reacting species, and the reaction may require as high as 9–10 mg chlorine for each mg NH_4-N oxidized.

Breakpoint-chlorinated effluents are frequently dechlorinated to remove the excess residual chlorine from the effluent before it is discharged. Activated carbon or SO_2 is frequently used for this purpose. About 0.9–1.0 mg SO_2 is required per mg residual chlorine. Hydrogen ions are generated in the dechlorination process, but the drop in pH is rarely of concern. About 2 mg alkalinity per liter (as $CaCO_3$) is consumed in removing 1 mg residual chlorine. Activated carbon columns may also be employed for dechlorination. In addition to achieving dechlorination, they have the added benefits of removing chlorinated as well as other organics.

Clinoptilolite is the ion exchange medium of choice for the selective removal of ammonia from moderate to dilute wastewaters. The selectivity of clinoptilolite for various cations is ranked as follows:

$$K^+ > NH_4^+ > Ba^{2+} > Na^+ > Ca^{2+} > Fe^{3+} > Al^{3+} > Mg^{2+} > Li^+.$$

Thus, clinoptilolite has a very high selectivity for NH_4^+, in comparison to other cations that are usually encountered in wastewaters. Ion exchange of NH_4^+ becomes progressively less effective as the pH is increased toward higher values. Within a pH range of 4–8, clinoptilolite ion exchange occurs optimally. Below a pH value of 4, hydrogen ion competes with the NH_4^+ ion, and above a pH value of 8, the NH_4^+ progressively shifts forward nonexchangeable unionized ammonia, i.e., NH_3 [96].

After pretreatment of a wastewater to remove most of the suspended solids and carbonaceous matter contained in it by either physicochemical or biological processes, the wastewater is passed through a bed of clinoptilolite (typically 4–6 ft bed depth containing the zeolite at a 20 × 50 mesh size) during a normal operating cycle. After the bed is exhausted or an unacceptable concentration of NH_4-N appears in the column effluent, the service cycle is terminated and the bed is regenerated. Ammonia from the spent regenerant is usually removed by either air or steam stripping at neutral or high pH, or by electrolysis. The regenerated brine is reused. Some regeneration techniques discharge the ammonia to atmosphere, while others recover the ammonia for use as fertilizer.

Table 15.11 presents removal efficiencies for physicochemical and biological processes that have been used for the treatment of some industrial wastes containing intermediate ammonia concentrations. Some of the activated sludge processes reported in Table 15.11 were not designed or operated specifically to remove NH_3-N from the wastes treated.

The bottoms from sour water strippers and waste streams from other petroleum-refining unit operations may produce wastewaters containing ammonia in the intermediate strength category. A summary of the performance of unit processes treating petroleum refinery wastewaters is given in Table 15.12. In the treatment of refinery wastewater, pretreatment for removal of oil

Table 15.11 Ammonia Nitrogen Removal for Intermediate Strength Wastes

Treatment Process	Industry	Concentration (mg/l) Influent	Effluent	Removal	Reference
Stream stripping	Fertilizer	185–211	25	86–98	95
Air stripping	Fertilizer	185–211	10	94–95	95
	Petrochemical waste	100		58–95	24
	Poultry waste	440		77	8
Ion exchange	Fertilizer	325–719	40	88–94	95
Activated carbon	Glue manufacture	78	6	92.3	52
Chlorination	Blast furnace	250	10	96	19
Soil filter	Synthetic rubber	28–112	0.05–2.5	90+	9
Land application	Tomato cannery			99+	97
	Potato processing			75	97
Irrigation	Dairy waste	75.5	0.4	99.4	37
	Swine waste	173.3	11.6	93	39
Trickling filter	Synthetic rubber	31–76	1–12	—	9
	Refinery			15–90	98
Aerated Lagoon	Refinery			10–45	98
	Pharmaceutical wastes	475	270	43	12
Oxidation pond	Refinery			0–15	98
Biological nitrification	Pulp and paper	186–270	20–62	76.5–89.6	25
	Phenolic waste	127	8	93.7	25
Nitrification-denitrification	Fertilizer	325–719	5	98–99	89
Activated sludge	Fertilizer	445	45	90	32
	Tannery	185	158	14.6	99
	Refinery	31.2–32.9	9.4–21.6	34–77.4	30

Table 15.12 Ammonia Concentrations in Treated Petroleum Refinery Effluents and Efficiency of Ammonia Removal by Various Unit Operations [30]

Treatment Method	*Effluent Concentration (mg/l)*	*% Removal*
Cooling towers (air stripping)	1–30	60–95
Activated carbon	10–140	7–33
	1–100	33–87
Oxidation pond	3–50	0–15
Aerated lagoon	4–25	10–45
Activated sludge	1–100	33–99
Trickling filter	25–100	15–90

and suspended solids reduces fouling of the activated carbon columns which are used to remove organics.

A number of different configurations of biological processes have been described for nitrification of ammonia (Table 15.13). These systems may be designed for combined removal of organic carbon plus ammonia nitrification, or for nitrification alone. It can be seen from Table 15.11 that variable ammonia removals occur in biological unit processes. This is often due to their inadequate design to achieve nitrification. The lack of a uniformly high percentage ammonia removal may also be due to improper environmental conditions, such as low temperature or high waste pH. High pH values facilitate the formation of free ammonia, which is detrimental to the nitrification process.

Dilute Ammonia Wastewaters

There are several waste streams that fall into this category. Most of the food-processing wastes such as cannery wastes, vegetable-processing wastes, some waste streams from paper and pulp mills, and starch and grain mill wastes, are low in ammonia concentration. Information on the removal of ammonia from specific industrial wastes containing low concentrations of NH_3-N is meager. The state of art for nitrogen control in domestic wastewaters, which also have

Table 15.13 Types of Biological Nitrification Processes

Category	*Type*
Suspended growth systems	Lagoons Activated sludge
Attached growth systems	Trickling filters Rotating biological disks Packed bed reactors

a low concentration of ammonia nitrogen (30 mg/l), is relatively more advanced and is based primarily upon biological processes [49]. Performance of these processes must be applied judiciously to industrial wastes which contain NH_3-N in the dilute concentration range. Important considerations are that the waste characteristics must be such that they do not inhibit the growth of the nitrifying organisms, and that they provide environmental conditions conducive for the initiation and maintenance of nitrification. In some situations, it may be necessary to supplement the desired alkalinity or subject the waste to a pretreatment step to make it more amenable for biological nitrification.

Biological nitrification has been demonstrated to be a feasible process for control of ammonia nitrogen, and effluent NH_3-N concentration of 1 mg/l has been achieved for wastewaters of initial low ammonia levels. Biological nitrification can be achieved in suspended growth systems such as single and two-stage activated sludge systems, and attached growth systems such as trickling filters and rotating biological discs. If total nitrogen reduction is required, the nitrified effluent can be subjected to denitrification.

Physicochemical methods can also be applied for the control of ammonia in dilute wastes. The major physicochemical processes that have been tested with domestic wastewater and that have potential application for the control of ammonia in dilute industrial wastes are air stripping, ion exchange, and breakpoint chlorination.

Breakpoint chlorination is used at one integrated steel mill to further reduce ammonia after biological (stabilization lagoon) treatment [100]. The treatment process, following the biological step, involves pH adjustment, breakpoint chlorination, deep-bed filtration, and dechlorination with sulfur dioxide. The process reduces an influent 3 mg/l ammonia to 0.7 mg/l. Chlorine dosage is 10 mg/mg ammonia, and sulfur dioxide demand is 1 mg/mg residual chlorine. Chlorinated organic by-products (including chloroform and 1,2-dichloropropane) have been detected at trace levels in the effluent [100].

Land disposal of cannery wastes and potato-processing wastes, containing low concentrations of nitrogen, has been reported [60]. In Woodland, California, tomato-canning wastes are combined with municipal wastes and subjected to lagoon treatment. The lagoon effluent, which contains about 11 mg/l total nitrogen, is used for irrigation. This nitrogen apparently is removed in the land disposal system, since no nitrogen contribution to groundwater is reported. Wastewater from a potato-processing plant in Idaho is used for spray irrigation. Of the 125 mg/l total nitrogen in the influent, 75% was removed and the rest apparently migrated to the groundwater. Soil disposal systems for the control of ammonia nitrogen can be successfully used if wastewater application rates are skillfully planned and appropriate resting periods are provided between applications of wastewater.

A summary of the various unit operations for removal of organic and ammonia nitrogen from dilute wastes is given in Table 15.14. These data are primarily for domestic wastewater.

Table 15.14 Selected Treatment Processes for Ammonia and Organic Nitrogen in Wastewater [49]

Treatment Process	*Effect On*		*Expected TN Removal (%)*
	Organic	*NH_3-N*	
Secondary treatment	15–25% removal	10% removed	10–20
Electrolysis	100% suspended organic N removed	85% removed	80–90
Land application			
Irrigation	Partly converted to NH_3/NH_4^+	Plant uptaken and nitrified	40–90
Infiltration/ percolation	Partly converted to NH_3/NH_4^+	Nitrified	5–10
Nitrification	Limited effect	Conversion to NO_3, 100% removed	5–10
Breakpoint chlorination	Uncertain removal	90–100% removed	80–95
Selective ion exchange	Uncertain removal	90–97% removed	80–95
Air stripping	No effect	90–95% removed	50–90
Algae stripping	Partial transformation to NH_3/NH_4^+	Algal uptake	50–80

SUMMARY

Ammonia and organic nitrogen are constituents in many industrial wastewaters. Their concentrations vary not only among effluents discharged from various industrial categories, but also within the effluents discharged from different plants belonging to the same category. The concentration in the industrial discharges depends primarily on (1) the nature of raw materials used and goods produced in a production facility, (2) the level of technology used in the manufacturing process, (3) wastewater management practices such as good housekeeping, incorporation of water recycle-reuse and by-product recovery concepts, and (4) the degree of in-house wastewater reduction and nature of end-of-pipe technology practices.

There is limited information on full-scale treatment of industrial organic nitrogen wastes. Where such wastes are amenable to biological assimulation and mineralization, conventional biological treatment processes appear effective. Activated carbon adsorption and chemical coagulation are also effective and would be applicable for refractory organic nitrogen constituents.

There is no single panacea that can be applied uniformly for the control of ammonia nitrogen contained in different industrial wastes. A judicious choice of ammonia nitrogen control technology, whether it is physicochemical, biological, or a combination of both, should be based on considerations such as opportunities for production technology modifications and substitution of factory inputs to the manufacturing process in order to achieve a reduction in wastewater quantity as well as in the concentration of pollutional parameters such as ammonia nitrogen; incorporation of optimal water utilization schemes, good housekeeping practices and by-product recovery schemes where possible; and evaluation of various process alternatives encompassing physicochemical and biological processes and the choice of a treatment process scheme that provides an economic solution for the control of ammoniacal nitrogen in the wastewater under consideration.

REFERENCES

1. Alexander, M. *Introduction to Soil Microbiology* (New York: Wiley, 1961).
2. "Development Document for Effluent Limitation Guidelines and New Source Performance Standards, Red Meat Process Segment of the Meat Product and Rendering Process Point Source Category," USEPA 440/1-74-012-a (1974).
3. "Development Document for Effluent Limitation Guidelines and New Source Performance Standards for the Pressed and Blown Glass Segment of the Glass Manufacturing Point Source Category," USEPA 440/1-75/034-a, Group I, Phase II (1975).
4. "Development Document for Effluent Limitation Guidelines and New Source Performance Standards for the Leather Tanning and Finishing Point Source Category," USEPA 440/1-74-016-a (March 1974).
5. Jewell, W.J., H.R. Davis, O.F. Johndrew, Jr., R.C. Loehr, W. Siderewicz, and R.R. Zall. "Egg Breaking and Processing Waste Control and Treatment," USEPA 660/2-75-019 (June 1975).
6. Loehr, R.C. *Agricultural Waste Management* (New York: Academic Press, 1974).
7. Wong-Chong, G.W. "The Kinetics of Microbial Nitrification as Applied to the Treatment of Animal Wastes," Ph.D. thesis, Cornell University (1974).
8. Loehr, R.C., T.B.S. Prakasam, E.G. Srinath, and Y.D. Joo. "Development and Demonstration of Nutrient Removal from Animal Wastes," Environmental Protection Technology Service, (Washington, DC: USEPA, 1973), p 340.
9. Koziorowski, B. and J. Kucharski. *Industrial Waste Disposal* (New York: Pergamon Press, 1972).
10. Osantowski, R. and K. Duchon. "Activated Carbon Treatment of Reverse Osmosis Brine for a Textile Waste," *Proceedings of the 3rd Annual Conference on Treatment and Disposal of Industrial Wastewaters and Residues*, Houston, TX (April 18–20, 1978).
11. Zieger, W. "Treatment Alternatives for Chelated Boiler Tube Cleaning Wastes," *Proceedings of the 3rd Annual Conference on Treatment and Disposal of Industrial Wastewaters and Residues*, Houston, TX (April 18–20, 1978).
12. Brown, G.E. "Land Application of High Nitrogenous Industrial Wastewater," *Proceedings of the National Conference on Management and Disposal of Res-*

idues from the Treatment of Industrial Wastewaters, Washington, DC (February 3–5, 1975).

13. Cousins, W.G. and A.B. Mindler. "Tertiary Treatment of Weak Ammonia Liquor," *J. Water Poll. Control. Fed.* 44 (1972): 607–18.
14. Pearce, A.S. and S.E. Punt. "Biological Treatment of Liquid Toxic Wastes, Part I," *Effluent Water Treatment J.* 15 (1975): 32–39.
15. Barker, J.E. and R.J. Thompson. "Biological Removal of Carbon and Nitrogen Compounds from Coke Plant Wastes," USEPA R2-73-167 (April 1973).
16. Jablin, R. and G.P. Chanko. "A New Process for Total Treatment of Coke Plant Waste Liquor," paper presented at the AIChE meeting, Advanced Water Pollution Control, New York (1972).
17. Biczysko, J. and J. Suchka. "Investigations of Phenolic Wastes Treatment in an Oxidation Ditch," *Proceedings of the Advanced Water Pollution Research Conference, Munich* (1966), 285–308.
18. Dunlap, R.W. and F.C. McMichael. *Environmental Impact of Coke Plant Wastewater Treatment and Disposal,* Report of American Institute of Iron and Steel (1975).
19. "Development Document for Effluent Limitation Guidelines and New Source Performance Standards for the Steel Making Segment of the Iron and Steel Manufacturing Point Source Category," USEPA 440/1-74-024-a (June 1974).
20. Fisher, C.W. "Coke and Gas," *Industrial Wastewater Control,* C.F. Gurnham, ed. (New York: Academy Press, 1965), pp 221–233.
21. Wong-Chong, G. and A. Dequittner. "Physical-Chemical Treatment of Coke Plant Wastewater at Crucible Steel, Midland, PA," paper presented at the Conference of the Water Pollution Control Federation, Las Vegas, NV (September 1980).
22. Adams, C.E., R.M. Stein, and W.W. Eckenfelder, Jr. "Treatment of Two Coke Plant Wastewaters to Meet Guideline Criteria," *Proceedings of the 29th Industrial Waste Conference, Purdue University* (May 1974), pp 864–880.
23. Osantowski, R. and J. Ruppersberger. "Upgrading Foundry Wastewater Treatment," paper presented at the 39th Annual Purdue Industrial Waste Conference, West Lafayette, IN (May 1984).
24. Prather, B.V. "Wastewater Aeration May Be Key to More Efficient Removal of Impurities," *Oil Gas J.* 57 (1959): 78–79.
25. Adams, C.E. and W.W. Eckenfelder, Jr. "Nitrification Design Approach for High Strength Wastewater," paper presented at the Water Pollution Control Federation Annual Meeting, Cleveland, OH (1973).
26. Garrison, W.E., J.G. Kremmer, and J. Murk. "Improved Hypochlorination Techniques and Problems in Disinfection of Municipal Wastewater Containing Refinery Thiosulfate," *Proceedings of the 28th Annual Purdue Industrial Waste Conference,* West Lafayette, IN (May 1973), pp 309–322.
27. Melin, G.A., J.L. Niedzwiecki, and A.M. Goldstein. "Computer Improves Sour Water Stripping," *Oil Gas J.* 73 (1975): 96–98.
28. Reynolds, J.H., E.J. Middlebrooks, and D.B. Procella. "A Continuous Flow Kinetic Model to Predict the Effects of Temperature on the Toxicity of Oil Refinery Waste to Algae," *Proceedings of the 28th Industrial Waste Conference, Purdue University* (May 1973), pp 12–37.
29. Beychok, M.R. *Aqueous Wastes from Petroleum Plants* (New York: Wiley, 1967).

30. "Development Document for Effluent Limitation Guidelines and New Source Performance Standards of the Petroleum Refining Point Source Category," USEPA 440/1-74-014-a (April 1974).
31. Klett, R.J. "Treat Sour Waters for Profit," *Hydrocarb. Proc.* (October 1972).
32. Hutton, W.C. and S.A. LaRocca. "Biological Treatment of Concentrated Ammonia Wastewaters," *J. Water Poll. Control. Fed.* 5 (1975): 989–997.
33. Adams, C.E., P.A. Krenkel, and E.C. Bingham. "Investigations into the Reduction of High Nitrogen Concentration," paper presented at the 5th International Water Pollution Conference, San Francisco, CA (1970).
34. Arnold, D.W. and W.E. Wolfram. "Ammonia Removal and Recovery from Fertilizer Complex Wastewaters," paper presented at the 30th Annual Purdue Industrial Waste Conference, West Lafayette, IN (May 1975).
35. Struzeski, E.J. "Ammonia Steam Stripping at Miscellaneous Non-ferrous Metals Plants, Case Histories," paper presented at the 33rd Annual Purdue Industrial Waste Conference, West Lafayette, IN (May 1978).
36. "Development Document for Effluent Limitations for Interim Final Guidelines and Proposed New Source Performance Standards for the Electrolytic Ferroalloys Segment of the Ferroalloy Manufacturing Point Source Category," USEPA 440/1-75/038-a, Group I, Phase II (February 1975).
37. Erickson, A.E., B.G. Ellis, J.M. Tiedje, A.R. Wolcott, C.M. Hansen, F.R. Peabody, E.E. Miller, and J.W. Thomas. "Soil Modification for Denitrification and Phosphate Reduction of Feedlot Waste," USEPA 660-2-74-057 (1974).
38. Nemerow, N.L. "Baffled Biological Basins for Treating Poultry Plant Wastes," *J. Water Poll. Control. Fed.* 41 (1969): 1602–1612.
39. Thibodeaux, L.J., D. Mitchell, and B. Krisch, Jr. "Methodology of a Regional Wastewater Audit," *Proceedings of the 28th Industrial Waste Conference, Purdue University* (May 1973), pp 358–386.
40. Anderson, D.R., R.E. Benson, J.A. Lofteon, and T.S. Satchell. "Soybean Processing Oil Refining Wastewaters, Characteristics and Treatment," *Proceedings of the 28th Industrial Waste Conference, Purdue University* (May 1973), pp 38–51.
41 Malhotra, W.K. and M.J. Miller. "Biological Treatment of Vinegar Plant Wastes," *Proceedings of the 28th Industrial Waste Conference, Purdue University* (May 1973), pp 676–687.
42. Horn, C.R. and F.G. Pohland. "Characterization and Treatability of Selected Shellfish Processing Wastes," *Proceedings of the 28th Industrial Waste Conference, Purdue University* (May 1973), pp 819–831.
43. Witherow, J.L. "Small Meat Packers Waste Treatment Systems," *Proceedings of the 28th Industrial Waste Conference, Purdue University* (May 1973), pp 994–1009.
44. Zall, R.R. "Characteristics of Milking Centre Wastage Effluents from New York State Dairy Farms," *J. Milk Food Technol.* 35 (1972): 53–55.
45. Patterson, J.W. and R.A. Minear. "State-of-the-Art for Inorganic Chemicals Industry: Commercial Explosives," USEPA 600/2-74-009-b (1975).
46. Basu, A.K. "Characterization of Distillery Wastewater," *J. Water Poll. Control. Fed.* 47 (1975): 2184–2190.
47. "Development Document for Effluent Limitation Guidelines and New Source Performance Standards for the Oil Base Solvent Wash Subcategories of the Paint

Formulating and the Ink Formulating Point Source Category," USEPA 440/1-75/050-a, Group II (July 1975).

48. "Development Document for Effluent Limitation Guidelines and New Source Performance Standards for the Secondary Aluminum Smelting Subcategory of the Nonferrous Metals Manufacturing Point Source Category," USEPA 440/1-74-019-e (1974).
49. *Process Design Manual for Nitrogen Control* (Washington, DC: USEPA, 1975).
50. Middlebrooks, E.J. "Wastes from the Preservation of Wood," *ASCE J. Sanitary Eng. Div.* SA1 5785, 41 (1968).
51. Graham, J.L. "Aerobic Secondary Treatment of Plywood Glue Wastes," USEPA R2-73-195 (1973).
52. Wang, L.K., R.P. Leonard, J.G. Michalovic, and D.W. Goupil. "Glue Treatment—Pick a Way," *Water Wastes Eng.* 10 (1973): 20–24.
53. Hatch, L.T., T.N. Sargent, R.E. Sharpin, and W.T. Wirtanen. "Chemical/Physical and Biological Treatment of Wool Processing Wastes," *Proceedings of the 28th Industrial Waste Conference, Purdue University* (May 1973) pp 558–603.
54. *Treatability Studies for the Inorganic Chemicals Manufacturing Point Source Category,* USEPA 400/1-80/103 (July 1980).
55. Jashnani, I. and R. Green. "Pretreatment System for a Circuit Board Plant," paper presented at the 8th National Conference on Municipal Sludge Management, Municipal and Industrial Sludge Management, Disposal and Utilization, Miami Beach, FL (March 19–21, 1979).
56. Chu, T-Y, G. Steiner, and C. McEntyre. "Removal of Complex Copper-Ammonio Ions from Aqueous Wastes with Fly Ash," paper presented at the 32nd Annual Purdue Industrial Waste Conference, West Lafayette, IN (May 1977).
57. Newkirk, R.W. and P.J. Schroeder. "RO Can Help with Wastes," *Hydrocarb. Proc.* (October 1972): 103–106.
58. Douglas, A.S., M. Tagami, and C.E. Milstead. "Membrane Materials for Wastewater Reclamation by Reverse Osmosis," USEPA ORD-17040-EF006/70 (1970).
59. Smith, J.M., A.N. Masse, and R.P. Miele. "Renovation of Municipal Wastewater by Reverse Osmosis," USEPA 17040-05/70 (1970).
60. Pound, C.E. and R.W. Crites. *Wastewater Treatment and Reuse by Land Application,* vols. I and II, *Tech. Serv.* EPA 660/1-73-006-a, b (1973).
61. Elaissen, R. and G. Tchobanoglous. "Chemical Processing of Wastewater for Nutrient Control," *J. Water Poll. Control. Fed.* 40 (1968): 171–180.
62. Dugan, G.L., C.G. Gouleke, and W.J. Oswald. "Recycling System for Poultry Wastes," *J. Water Poll. Control. Fed.* 44 (1972): 432–440.
63. Helmers, E.N., J.D. Frame, A.E. Greenburg, and C.N. Sawyer. "Nutritional Requirements in the Biological Stabilization of Industrial Wastes, Part III, Treatment Water Supplementing Wastes," *Sew. Ind. Wastes* 24 (1952): 496.
64. Haltrich, W. "Influence of Nitrogen Supply on Biological Treatment," *Proceedings of the 24th Industrial Waste Conference, Purdue University* (May 1969), pp 1185–1194.
65. Delwiche, C.C. "Biological Transformation of Nitrogen Compounds," *Ind. Eng. Chem.* 48 (1956): 1421–1427.
66. Stones, T. "Nitrogenous Changes in Sewage During Settlement," *Effluent Water Treatment J.* (April 1978).
67. Hashimoto, A.G. "Aeration of Poultry Wastes for Odor and Nitrogen Control," *Trans. Am. Soc. Agricul. Eng.* 17 (1974): 978–982, 986.

68. Metcalf and Eddy, Inc. *Wastewater Engineering, Treatment, Disposal, Reuse,* 2nd ed. (New York: McGraw-Hill, 1979).
69. Sack, W.A. and S.A. Phillips. "Evaluation of the Biodisc Treatment Process for Summer Camp Application," USEPA 670/2-73-022 (1973).
70. Matsche, N.F. "Removal of Nitrogen by Simultaneous Nitrification-Denitrification in an Activated Sludge Plant with Mammoth Rotor Aeration," *Prog. Water Technol.* 8 (1977): 625–637.
71. Applegate, C.S., B. Wilder, and J.R. Deshaw. "Total Nitrogen Removal in a Multichannel Orbal Plant, Huntsville, Texas" paper presented at the 51st Annual Conference of the Water Pollution Control Federation, Anaheim, CA (1978).
72. Drews, R.J.L.C. and A.M. Greeff. "Nitrogen Elimination by Alternation of Aerobic/'Anoxic' Conditions in Orbal Activated Sludge Plants," *Water Res.* 7 (1973): 1183–1194.
73. Shuckrow, A.J. and J.L. Culp. "Appraisal of Powdered Activated Carbon Process for Municipal Wastewater Treatment," USEPA 600/2-77-156 (1977).
74. Wilson, T.E. "Process Design for Nitrogen Control NSSD, Peoria, Tampa," paper presented at the USEPA Technology Transfer National Conference on Nitrogen Control (1976).
75. Barth, E.F. and B.W. Ryan. "Nutrient Control by Plant Modification at El Lago, Texas," *J. Water Poll. Control. Fed.* 50 (1978): 1768–1785.
76. Schwinn, D.E. and G.K. Tozer. "Largest Advanced Wastewater Treatment Plant in U.S. and the World," *Environ. Sci. and Technol.* 8 (1974): 10.
77. Heidemann, J.A. "An Experimental Evaluation of Oxygen and Air Activated Sludge Nitrification Systems With or Without pH Control," EPA Report for Contract 68. 03-0349 (1975) cited in reference 87.
78. McHarness, D.C. and P.L. McCarty. "Field Study of Nitrification with the Submerged Filter," USEPA R2-73-158 (1973).
79. Grantham, G.R., E.B. Phelps, W.T. Calaway, and D.L. Emerson. "Progress Report on Trickling Filter Studies," *Sew. Ind. Wastes* 22 (1950): 867–874.
80. Osborn, D.E. "Operating Experiences with Double Filtration in Johannesburg, Part 3," *J. Inst. Sew. Purif.* (1965): 272–281.
81. Stenquist, R.J., D.S. Parker, and T.J. Dosh. "Carbon Oxidation-Nitrification in Synthetic Media Trickling Filter," *J. Water Poll. Control. Fed.* 46 (1974): 2327–2339.
82. Beckman, W.J., R.J. Avendt, T.J. Mulligan, and G.J. Kehrberger. "Combined Carbon Oxidation Nitrification," *J. Water Poll. Control. Fed.* 44 (1972): 1916–1931.
83. Randtke, S.J., G.E. Parkin, J.V. Keller, and P.L. McCarty. "Soluble Organic Nitrogen Characteristics and Removal," USEPA 600/2-78-030 (1978).
84. Mulbarger, M.C. "Nitrification and Denitrification in Activated Sludge Systems," *J. Water Poll. Control. Fed.* 43 (1971): 2059–2070.
85. Lue-Hing, C., A.W. Obayashi, D.R. Zenz, B. Washington, and B.M. Sawyer. "Biological Nitrification of Sludge Supernatant by Rotating Discs," *J. Water Poll. Control. Fed.* 48 (1976): 25–46.
86. Anthonisen, A.C., R.C. Loehr, T.B.S. Prakasam, and E.G. Srinath. "Inhibition of Nitrification by Ammonia and Nitrous Acid," *J. Water Poll. Control. Fed.* 48 (1976): 835–852.
87. Wild, H.E., C.N. Sawyer, and T.C. McMahon. "Factors Affecting Nitrification Kinetics," *J. Water Poll. Control. Fed.* 43 (1971): 1845–1854.

88. Rimer, A.E. and R.L. Wookward. "Two Stage Activated Sludge Pilot Plant Operations at Fitchburg, Massachusetts," *J. Water Poll. Control. Fed.* 44 (1972): 101–116.

89. Bringmann, G. "Vollustandige Biologische Stickstoffe Eliminierung aus Klarwasseren im Auschluss an ein Hockeleistuns-Nitrifikations-Verfahren," *Gesundheitsinzenieur* 82 (1961): 233–235.

90. Kostenbader, P.D. and J.W. Flecksteiner. "Biological Oxidation of Coke Plant Weak Ammonia Liquor," *J. Water Poll. Control. Fed.* 41 (1967): 199–207.

91. Rodriguez, D.G. "Sour Water Stripper: Its Design and Application," *Water, 1973, AIChE Symp. Ser.* 70 (1973): 667–670.

92. "Development Document for Effluent Limitation Guidelines and New Source Performance Standards for the Formulated Fertilizer Segment of the Fertilizer Manufacturing Point Source Category," USEPA 440/1-75/042-a (1975).

93. Ganczanczyk, J. and M. Kelleher. "Costs of Pretreatment of Coke-Plant Effluents," paper presented at the USEPA/AISI Symposium on Pollution Abatement Technology, Pittsburgh, PA (November 1982).

94. Kunin, R. "Denitrification by Ion Exchange," *Proceedings of Correspondence (at home) Conference on Denitrification of Municipal Waters, University of Massachusetts* (1973).

95. "Development Document for Effluent Limitation Guidelines and New Source Performance Standards for the Basic Fertilizer Chemicals Segment of the Fertilizer Manufacturing Point Source Category," USEPA 440/1-74-001-a (1974).

96. Koon, J.H. and W.J. Kaufmann. "Ammonia Removal from Wastewater by Ion Exchange," *J. Water Poll. Control. Fed.* 47 (1975): 448–465.

97. Kepple, L.G. "New Ammonia Removal and Recovery Processes," *Water Sew. Works* 121 (1974): 4, 42.

98. Prettyman, R.W. and L.G. Kepple. "Ammonia Removal and Recovery of Industrial Wastes," *ASHRAE J.* (September 1976): 41–46.

99. Thackston, E.L. "Secondary Treatment of Wastes from an Alum Tannery," *Proceedings of the 28th Industrial Waste Conference, Purdue University* (May 1973).

100. Radigan, P. and J. Manda. "Wastewater Treatment at an Integrated Steel Mill," *Ind. Wastes* (July/August 1981).

16
Nitrite and Nitrate Nitrogen

Industries discharging wastewaters containing high concentrations of nitrite or nitrate include fertilizer manufacturing, iron and steel production, munition manufacturing, cattle feed lots, swine and poultry processing, electronic component manufacturing, and nuclear fuel processing and uranium oxide fuel fabrication. In addition, there are many other categories of industries which discharge wastewaters containing nitrite or nitrate.

Table 16.1 summarizes the ranges of nitrate concentration observed in various industrial categories. Data on wastewater nitrite levels are less frequently reported than are data on nitrate. However, where wastewaters contain significant levels of nitrate, nitrite is also commonly present but typically at a lower concentration than the nitrate. For example, the wastewater from one potassium nitrate-manufacturing facility was reported to contain nitrate at 2,640 mg/l and nitrite at 640 mg/l [18]. Industrial wastes may not contain nitrite or nitrate initially, but they may be potential sources of these nitrogen compounds if they contain ammonia and are treated biologically.

The production practices, raw materials used, and water utilization practices in a given industry dictate the concentration of nitrite or nitrate in the wastewater generated. For example in munition manufacturing, the nitrate contained in the wastewater is contributed by the ammonium nitrate, nitric acid, and sodium nitrate used in the production of the explosives products [9]. Wastewaters from the uranium purification process may contain as high as 33% nitrate, as a result of solvent extraction procedures in which nitric acid is used as a principal reagent [4, 22].

When very high concentrations of nitrate are found in wastewaters, advantage may be taken of this fact and schemes incorporated into the manufacturing of the primary products for recovery of by-products from the waste streams. In addition to recovering the chemicals, relatively clean water can be obtained which can be used either for make up or cooling purposes. In the production of ammonium nitrate, wastes containing as high as 15–33 g/l of nitrate nitrogen may result, and this high concentration permits complete recovery of the ammonium nitrate by recycling the waste to the associated nitric acid plant. Any resultant wastewater discharge from the nitric acid plant is due to leaks and spills and can be controlled by good housekeeping and waste reduction management practices.

Table 16.1 Levels of Nitrate Nitrogen Reported in Industrial Wastewaters

Industrial Source	*Nitrate-Nitrogen Concentration (mg/l)*	*Reference*
Fertilizer manufacturing		
Ammonia-nitric acid waste	500	1
Ammonium nitrate	6,000	1
Ammonium nitrate process condensate	15–542	2
Ammonia and urea manufacturing	10–135	3
Nitrate fertilizer waste	1000	4
Ammonium sulfate manufacturing	44	5
Calcium nitrate manufacturing	15–180	6
Iron and steel manufacturing		
Open hearth plant wastes	20–33	7
Vacuum degassing	3–25	7
Munitions manufacturing		
Nitroglycerin	167	8
Nitroglycerin	344–12,500	9
TNT wastes	200	10
TNT wastes	53–213	11
PETN wastes	49–172	9
Explosives manufacture	92–1150	9
Nitrogen plants	73–564	8
Dynamite factory	540	12
Dynamite mix house-dust control wastes	7–2500	9
Blasting caps	13.4–30.4	9
Nitrocellulose waste	320	13
Nitrocellulose waste	565	14
Agroindustries		
Cattle waste	0.2–1,400	15
Dairy waste	9.6	16
Chicken waste	0.02–3,000	15
Swine waste	2–300	15
Horse waste	600	15
Food processing		
Milking parlor	0.65–1.6	15
Cheese processing	0.2–24.4	15
Coffee wastes	6–29	
Agricultural and industrial		
Chemical manufacturing	300	17
Potassium nitrate manufacture	2,640	18
Plywood glue	23–75	19
Penicillin fermentation	28	6
Electronic components	80–1,000	20
Electronic components	2,000	21
Nuclear fuel processing	1,000	4
Spent nitric acid extractant liquor	15,400–33,000	22
Zinc bonderizing	84	23
Galvanizing waste	10–120	6
Spent metal-descaling bath	100,000	24

Table 16.2 Summary of Treatment Processes for Nitrate Removal

Treatment Method	*% Removal*
Biological denitrification	<100
Land application	<95
Chemical reduction	<90
Selective ion exchange	75–95
Reverse osmosis	85
Electrodialysis	40
Evaporative recovery	99+

TREATMENT TECHNOLOGY

A review of the technologies available for the treatment of wastewaters containing nitrite and nitrate indicates that a variety of treatment methods have been investigated. Few methods yield high percentage removals, however. Expected nitrate removal efficiencies of various treatment processes are given in Table 16.2. The choice of a method for the removal of nitrate from industrial wastewaters depends primarily on the nature of the waste, the degree of treatment required, and economics. If wastes containing ammonia and nitrate are subjected to biological nitrification, the effluent nitrate concentration will be higher than the influent nitrate concentration. Hence it is important to consider the influent ammonia concentration, in cases where the wastes are treated by aerobic biological systems to achieve carbon and ammonia oxidation, prior to a denitrification step.

Nitrate Recovery

A limited number of studies have been conducted to determine the feasibility of recovery processes for nitrate in fertilizer wastes [1, 3, 25]. It is possible to recover nitrate as a by-product, if the wastewater has a sufficiently high concentration. For example, ammonium nitrate can be economically recovered from ammonia nitrate-process condensate wastewater due to the high concentration contained in the wastewater. This wastewater can be used as a feed to the nitric acid plant, thereby allowing an internal recycling of the condensate and at the same time increasing the product yield [2]. Recovery processes may incorporate preconcentration steps, such as ion exchange or evaporation.

Staebler [24] has proposed the treatment of a concentrated hydrofluoric-nitric acid waste by calcium precipitation for fluoride removal followed by thin-film evaporative recovery of calcium nitrate for its fertilizer value. Pilot-plant evaporation results are presented in Table 16.3, for a feed containing 38,000 mg/l nitrate-nitrogen.

Table 16.3 Pilot Studies in Evaporative Concentration of a Nitrate Waste [24]

Distillation Temperature (°F)	*Distillate Nitrate (mg/l)*	*Percent Evaporation*
201–208	26	68
214–225	22	60
288–296	25	81

Biological Denitrification

Successful nitrification of high-strength ammonia wastewaters and denitrification of high-strength nitrate wastewaters has been demonstrated, for wastes having concentrations of nitrate and ammonia similar to those of fertilizer wastes [7, 26]. Anaerobic lagoons and rock filters also have potential application in the removal of nitrate by biological denitrification, provided that adequate hydrogen donors are available. There are meager data on full-scale application of these latter processes for treating nitrate wastes, however.

Denitrification involves the biochemical reduction of nitrite and nitrate to nitrogen gas. A number of heterotrophic microorganisms are able to accomplish denitrification, under anoxic conditions. The control of pH is important, since nitrifiers are sensitive to low pH. Nitrate reduction is optimum between pH 6.5 and 8.5 [22]. Temperature is also an important parameter, with maximum denitrification reported to occur at 60°C [27]. The denitrifying organisms require a source of organic carbon. Methanol is often used, due to availability and economics.

Munition-manufacturing wastes are rich in nitrate nitrogen, because of the use of potassium nitrate and nitric acid as key ingredients, and discharge of nitrogen-containing compounds such as nitroaromatic derivatives. Nitrate levels in the munition wastes vary widely (see Table 16.1), but in general the wastes may be characterized as concentrated.

Nitrogen control for munition wastes can be achieved either by biological or by physiochemical processes. Ammonia nitrogen is present in some wastes and can be nitrified biologically. The resulting oxidized nitrogen will increment the concentration of nitrate already present, and thus influence the size of a biological denitrification unit. Studies conducted by the U.S. Army suggest that denitrification, ion exchange, and reverse osmosis offer the greatest potential for nitrate removal from these wastes [9].

Denitrification of munition wastes can be accomplished by both suspended and attached growth denitrification systems. In a pilot-scale study (5,000 gpd), an average influent nitrate concentration of 558 mg/l was reduced to 47 mg/l at a detention time of 18 hr, when the mixed-liquor suspended solids concentration was within the range of 1,000–3,000 mg/l [28]. In another study,

Table 16.4 Performance of Biodenitrification Units [21]

Unit Process	*Influent Nitrate Concentration (mg/l)*	*Detention Time (hr)*	*Temperature (°C)*	*MLVSS (mg/l)*	*Nitrate Removal (%)*
Nuclear fuel-processing waste					
Modified activated sludge	1,100	27	30	3,000 (MLSS)	99
	1,100	96	22–25	—	85
	3,600	28	29	4,350	44
Packed-bed reactor	1,000	1.1	28	—	95
Anaerobic columns	1,000	1.26	28	—	99
	3,500	0.71	28	—	66
Electronic component waste					
Completely stirred reactor and plug-flow filter series system	1,000 (CSR)	0.36–5.9	27 ± 2	4,000 (CSR)	
		0.26–6.7 (PFF)		15,000 (PFF)	up to 95

where an anaerobic column was used for denitrification of munition wastes, almost complete removal (99.9%) of the influent nitrate nitrogen concentration of 100 mg/l was achieved at a detention of about 2.5 hr. This high efficiency may be hampered if proper choices of column media and of operation of the units are not made. Problems arising out of faulty design and operation include short circuitry of influent flow because of accumulation of solids in the column, gas binding, and odor generation [8].

Results have been published from biological denitrification studies for the removal of nitrate from the highly nitrogenous wastes discharged from nuclear-processing and electronic component-manufacturing plants [4, 20, 29]. These studies have revealed that completely mixed anaerobic suspended-growth and plug flow-type anaerobic filters were able to remove more than 90% of the nitrate nitrogen contained in these wastes, under appropriate operating conditions. An optimal design configuration for treatment of the electronic component wastes was found to be a series operation consisting of a completely stirred reactor (CSR) followed by a plug-flow submerged filter (PFF) [21]. A summary of the results of these investigations is given in Table 16.4.

Clark et al [22] have reported on the denitrification treatment of a concentrated wastewater from the solvent extraction purification of uranium. The wastewater, containing 33% (wt/wt) nitrate, plus aluminum, calcium, and other constituents, is partially diluted, mixed with calcium acetate (the organic carbon source) plus inorganic nutrients, and treated in a 25,000 gal stirred-bed anaerobic reactor on a semi-batch basis. The typical reactor feed (wt basis) is 15.4% nitrate, 12.3% calcium acetate, 70.9% water, and 1.4% miscellaneous

Table 16.5 Capital and Operating Costs (in 1975) for Denitrification of Domestic Waste Water [34]

Plant Capacity (MGD)	*Capital Cost ($/1,000 gpd)*	*Operating and Maintenance Costs ($/1,000 gal)*
1	180	5.6
5	88	3.2
10	69	2.7

constituents including inorganic nutrients. Effluent nitrate levels below 100 mg/l are achieved, at feed levels up to 7.26 g NO_3/1-day. Large amounts of heat are evolved in the biochemical process.

Many industrial wastes contain low concentrations of nitrate, comparable to that of raw and nitrified domestic wastewater. The presence of nitrate at low concentrations in wastes does not normally represent a significant problem. Nitrate removal by biological denitrification is well documented for nitrified domestic waste waters. Suspended-growth systems for achieving carbon and ammonia oxidation, combined with denitrification of the nitrified wastewater have been reported [30, 31]. In addition to these suspended-growth systems, attached-growth systems such as anaerobic filters have also been shown to achieve very high removals of nitrate [32, 33]. Supplementation with a hydrogen donor such as methanol is generally required to effect complete removal of nitrate. Based on the information available from domestic wastewater denitrification studies, methanol-based systems can achieve 90–95% nitrogen removal, as opposed to 75–90% for systems utilizing carbon oxidation—ammonia oxidation—denitrification unit operations [30].

Information on cost estimates for denitrification of industrial wastes containing low concentrations of nitrate is sparse. However, some cost estimates are available for denitrification of domestic wastewater, which was assumed to contain 29 mg oxidized nitrogen/liter. These estimates are given in Table 16.5.

Ion Exchange

A process that can be used when a high degree of nitrate removal or recovery is desired is ion exchange. This process has been successfully used for recovery of nitrogen in fertilizer wastes containing ammonium nitrate [25]. The ammonium nitrate wastewater is first passed through a strongly acidic cationic resin for the removal of ammonium ion. The effluent from this ion exchange column contains nitric acid, because of the reaction of the nitrate ions with the hydrogen ions of the ion exchange resin. The ammonia-free effluent from the cationic exchange column is then treated in an anion exchange column, to remove nitrate. The effluent from this second column will contain very low concentrations of ammonia and nitrate, and can be used as makeup water.

The spent cationic resin column is regenerated using nitric acid, and the anionic resin column is regenerated with a solution of ammonium hydroxide. In this regeneration process, ammonium nitrate is formed. The regenerant liquor is much more concentrated than the incoming wastewater and can be used directly or concentrated further. Using this type of ion exchange and regenerating process, the Farmers' Chemical Company in Tennessee has been able to remove both the ammonia and the nitrate in ion exchange columns. The spent regenerants, which are rich in ammonium nitrate (19% w/v), are recovered as products. An influent containing nitrate at 1,240 mg/l, when processed through this ion exchange system, yielded an effluent nitrate of 7–11 mg/l.

Weak-base anion exchange resins can be operated on the salt cycle, using sodium chloride as a regenerant instead of ammonium hydroxide. It is reported that such an anion exchange column, when operated on a salt cycle by regenerating with 5 lb salt/ft^3, will reduce the nitrate level from 70 to 10 mg/l, if the total salinity in the waste is less than 1,000 mg/l [29]. A similar ion exchange process was demonstrated for concentrating pollutants from zinc-bonderizing waste rinsewaters. These waters contained about 85 mg/l nitrate, and 75-fold concentration was achieved.

Resin selectivity for nitrate is a potentially serious problem when sulfate is present in the wastewater. Sulfate is reported to be preferred with a selectivity ratio of more than 2:1 over nitrate [35]. The reported sequence of selectivity, for both weak- and strong-base resins, is sulfate > nitrate > chloride > bicarbonate. Because of this, resin regeneration costs, and therefore operating costs, are more closely associated with the wastewater sulfate than with nitrate content.

Preliminary studies have indicated that the ion exchange process can be used for the removal of nitrate from munition wastes, in spite of the high concentrations of sulfate often contained in these wastes. A process using Dowex resin MWA-1 reduced nitrate concentration from 1,650 to 20 mg/l. An estimated investment cost (in 1964) of $1.65 million was given for treating a flow of 3 MGD [28].

A cost comparison for the removal of nitrate from the waste generated from a 1,000 ton/day ammonium nitrate fertilizer-manufacturing plant has been reported [2]. Included were cost and performance data for ion exchange, ammonium nitrate recovery by absorption into nitric acid, and biological nitrification-denitrification. Results are given in Table 16.6. Although the biological nitrification-denitrification process appears to be most economical, and produces an effluent lower in nitrate concentration than the ion-exchange unit effluent, it has been reported in another study that biological processes did not perform well with similar fertilizer waste [25].

Sorg [36] has compared the economics of ion exchange and reverse osmosis for nitrate removal from groundwater. Capital cost comparisons are pre-

Table 16.6 Performance and Cost Estimate (in 1974) of Nitrate Removal [2]

Process	*Capital Cost ($)*	*Annual Operating Cost ($)*	*Effluent NO_3^- Concentration (mg/l)*
Nitrate removal by ion exchange	580,000	416,700	40
Ammonium nitrate effluent utilization	182,020	57,580	—
Nitrification-denitrification	110,000	55,950	5

sented in Table 16.7; and operating costs, exclusive of regeneration and brine disposal, in Table 16.8. Regeneration costs, as a function of influent nitrate and sulfate levels, are described in Table 16.9. At all treatment plant capacities evaluated, ion exchange represented a lower capital investment than did reverse osmosis. Operating costs were also lower for ion exchange (Table 16.8), although the economic advantage narrows with increasing influent sulfate and nitrate levels (Table 16.9).

Table 16.7 Capital Cost Comparison (in 1980) for Nitrate Removal from Groundwater [36]

Capacity (gpd)	*Capital, $/1,000 gpd Capacity*	
	Ion Exchange	*Reverse Osmosis*
50,000	$1,300	$1,560
100,000	740	1,310
250,000	392	980
500,000	258	860

Table 16.8 Operating Cost Comparison (in 1980) for Nitrate Removal from Groundwater [36]

Capacity (gpd)	*Operating, 1980 $/1,000 gal Treated*	
	Ion Exchange[a]	*Reverse Osmosis*
50,000	$0.68	$1.17
100,000	0.41	0.92
250,000	0.22	0.69
500,000	0.15	0.57

[a] Excluding regeneration and brine disposal.

Table 16.9 Regeneration Costs (in 1980) for Ion Exchange Treatment of Nitrate, ¢/1,000 gal [36]

Influent Sulfate (mg/l)	*Influent Nitrate*		
	100 mg/l	*150 mg/l*	*200 mg/l*
50	5.2	7.3	9.2
150	7.8	9.8	11.8
250	9.9	11.8	14.4

Other Processes

Other processes for nitrite and nitrate control include chemical reduction, land application, and reverse osmosis. Several chemicals have been investigated for reducing nitrate to nitrogen gas. Only ferrous iron has been reported to be economically feasible, and no full-scale plants use this technology. The process requires a catalyst (copper) for denitrification, and must be operated at alkaline pH. Disadvantages include limited efficiency (70% nitrate reduction), and large ferrous iron demand [36].

In South Africa a full-scale land disposal system has been used for the disposal and utilization of a dynamite factory waste. The fertilizer value of the waste provided the opportunity for growing grass, by irrigation, on 2,000 acres of land. The waste contained 1,270 mg/l ammonia nitrogen and 500 mg/l nitrate nitrogen. The application rate represented 500 lb N/acre/year. No untoward effects were reported with respect to nitrate contamination of ground or surface waters [12]. The success of full-scale application of land disposal system for treatment of wastes, however, depends on the availability of land, climate, and economics.

In a reverse osmosis pilot study, no removal of nitrate could be achieved when the pH of the munition waste was 1.5; however, as the pH was increased to 7, about 90% of the influent nitrate was removed [9]. One of the disadvantages of the reverse osmosis process is that it cannot produce effluents containing less than 20 mg nitrate N/l. Hence it has to be followed by a polishing unit, such as biological denitrification, to achieve complete nitrogen removal. Thus, the application of reverse osmosis for nitrate removal on a large scale may not prove to be an economic process.

SUMMARY

Nitrate and nitrite are present in many industrial wastewaters, at varying concentrations. Nitrate nitrogen in industrial wastes can be controlled by in-house modifications of the production process, by water utilization schemes, and by end-of-pipe treatment which may be either physicochemical or biological. Pro-

Table 16.10 Nitrate Treatment Methods [9]

Method	*Removal Efficiency (%)*	*Approximate Cost ($/MG)*
Biodenitrification	70–95	3.45–30
Algae harvesting	50–90	20–35
Ion exchange	80–99	170–300
Electrodialysis	30–50	100–250
Chemical reduction	33–90	—
Reverse osmosis	50–96	100–600
Distillation	90–98	400–1,000
Land application	5–15	—

cesses such as electrodialysis and reverse osmosis have nitrate removal efficiencies in the range of 40–85%, but their use for nitrate removal from industrial wastes is limited due to high operating costs and the required pretreatment of the wastewaters. Ion exchange is more widely used, although principally as either a recovery process, or for nitrate removal from drinking water supplies.

Nitrate removal by denitrification is by far the most widely studied process. It is adaptable to many wastes which are amenable to biological waste treatment. Extremely high removals (>98%) can be obtained in properly designed attached- or suspended-growth systems. Attached-growth systems such as upflow anaerobic filters and upflow fluidized anaerobic sand filters yield very high rates of nitrate removal. In both types of unit, supplementation with an organic carbon source such as methanol is required. A summary of the removal efficiencies and estimated costs of various processes for the removal of nitrogen from munition wastes is presented in Table 16.10.

REFERENCES

1. Adams, C.E., P.A. Krenkel, and E.C. Bingham. "Investigations into the Reduction of High Nitrogen Concentration," paper presented at the 5th International Water Pollution Conference, San Francisco, CA (1970).
2. "Development Document for Effluent Limitation Guidelines and New Source Performance Standards for the Basic Fertilizer Chemicals Segment of the Fertilizer Manufacturing Point Source Category," USEPA 440/1-74-011-a (1974).
3. Hutton, W.C. and S.A. LaRocca. "Biological Treatment of Concentrated Ammonia Wastewaters," *J. Water Poll. Control. Fed.* 5 (1975): 989–997.
4. Francis, C.W. and M.W. Callahan. "Biological Denitrification and Its Application in Treatment of High Nitrate Wastewater," *J. Environ. Quality* 4 (1975): 153–162.
5. "Development Document for Effluent Limitation Guidelines and New Source Performance Standards for the Formulated Fertilizer Segment of the Fertilizer Manufacturing Point Source Category," USEPA 440/1-75-042-a (1975).

6. Koziorowski, B. and J. Kucharski. *Industrial Waste Disposal* (New York: Pergamon Press, 1972).
7. "Development Document for Effluent Limitation Guidelines and New Source Performance Standards for the Steel Making Segment of the Iron and Steel Manufacturing Point Source Category, USEPA 440/1-74-024-a (June 1974).
8. Tucker, D.O, C.W. Randall, and P.H. King. "Column Denitrification of a Munition Manufacturing Wastewater," *Proceedings of the 29th Industrial Waste Conference, Purdue University* (May 1974), pp 165–175.
9. Patterson, J.W. and R.A. Minear. "State-of-the-Art for the Inorganic Chemicals Industry, Commercial Explosives," USEPA 600/2-74-009-b (1975).
10. Nay, M.W., Jr., C.W. Randall, and P.H. King. "Factors Affecting Color Development During the Treatment of TNT Waste," *Proceedings of the 27th Industrial Waste Conference, Purdue University* (May 1972), pp 983–993.
11. Schulte, G.R., R.C. Hoehn, and C.W. Randall. "The Treatability of a Munitions Manufacturing Waste with Activated Carbon," *Proceedings of the 28th Industrial Waste Conference, Purdue University* (May 1973), pp 150–162.
12. Lever, N.A. "Disposal of Nitrogenous Liquid Effluent from Modderfontein Dynamite Factory," *Proceedings of the 21st Industrial Waste Conference, Purdue University* (May 1966), pp 902–925.
13. Mudrack, K. "Nitro-cellulose Industrial Waste," *Proceedings of the 21st Industrial Waste Conference, Purdue University* (May 1966), pp 656–664.
14. Wang, L., M. Pressman, W. Shuster, R. Shade, F. Bilgen, and T. Lynch. "Treatment of Nitrocellulose Manufacturing Wastewater," *Proceedings of the 3rd Annual Conference on Treatment and Disposal of Industrial Wastewaters and Residues,* Houston, TX (April 18–20, 1978).
15. Loehr, R.C. *Agricultural Waste Management* (New York: Academic Press, 1974).
16. Erickson, A.E., B.G. Ellis, J.M. Tiedje, A.R. Wolcott, C.M. Hansen, F.R. Peabody, E.C. Miller, and J.W. Thomas. "Soil Modification and Phosphate Reduction of Feedlot Waste," USEPA 660/2-74-057 (June 1974).
17. Woodey, R.D. and T.F. Brown. "Continuous Flow Organic Wastes Pilot Plant Design," *Proceedings of the 28th Industrial Waste Conference, Purdue University* (May 1973), pp 185–191.
18. Shindale, A., M. Bond, and D. Truax. "Treatment of Wastes Generated from the Production of Methyl Parathion and Potassium Nitrate," *Proceedings of the 3rd Annual Conference on Treatment and Disposal of Industrial Wastewaters and Residues,* Houston, TX (April 18–20, 1978).
19. Graham, J.L. "Aerobic Secondary Treatment of Plywood Glue Wastes," USEPA R2-73-195 (April 1973).
20. Jewell, W.J. and R.J. Cummings. "Denitrification of Concentrated Nitrate Wastewater," paper presented at the Water Pollution Control Federation Annual Meeting, Cleveland, OH (1973).
21. Michael, R.P. and W.J. Jewell. "Optimization of the Denitrification Process, unpublished manuscript."
22. Clark, F.E., J.M. Napier, and R.B. Bustamante. "Biological Treatment of Concentrated Nitrate Waste," *Proceedings of the 1977 National Conference on Treatment and Disposal of Industrial Wastewaters and Residues,* Houston, TX (April 26–28, 1977).
23. McGarvey, F.X. and S.A. Fisher. "Effluent Control in Zinc Bonderizing Pro-

cesses: The Recycle of Rinse Waters via an Ion Exchange Process," *Plating* 62 (1975): 231–234.

24. Staebler, C.J. Jr. "Treatment and Recovery of Fluoride and Nitrate Industrial Wastes," *Proceedings of the National Conference on Management and Disposal of Residues from the Treatment of Industrial Wastewaters,* Washington, DC (February 3–5, 1975).
25. Bingham, E.C. "Fertilizer Maker Stops Nitrogen," *Water Wastes Eng.* 9 (1972): F4–F8.
26. Anthonisen, A.C. "The Effects of Free Ammonia and Free Nitrous Acid on the Nitrification Process," Ph.D. thesis, Cornell University (1974).
27. Bremmer, J. and K. Shaw. "Denitrification in Soil," *J. Agricul. Sci.* 51 (1958): 40–52.
28. Eskelund, G. "Elimination of Nitrate Wastes," Technical Memorandum 2130, Picatinny Arsenal, Dover, NJ (1964).
29. Clark, F.E., H.C. Francis, and J.W. Strohecker. "Biological Denitrification of High Nitrate Waste Solution," paper presented at the 30th Annual Purdue Industrial Waste Conference, West Lafayette, IN (May 1975).
30. *Process Design Manual for Nitrogen Control* (Washington, DC: USEPA, 1975).
31. Barth, E.G., Brenner, R.C., and R.F. Lewis. "Chemical-Biological Control of Nitrogen in Wastewater Effluent," *J. Water Poll. Control Fed.* 40 (1968): 2040–2054.
32. Tamblyn, T.A. and B.R. Sword. "The Anaerobic Filter for the Denitrification of Agricultural Subsurface Drainage," *Proceedings of the 24th Industrial Waste Conference, Purdue University* (May 1960), pp 1135–1150.
33. Jeris, J.S. and R.W. Owens. "Pilot Scale High Rate Biological Denitrification at Nassau County, NY," paper presented at the New York Water Pollution Controllers Association, New York, NY (1974).
34. Van Note, R.H., P.V. Herbert, R.M. Patel, C. Chupek, and L. Feldman. "A Guide to the Selection of Cost-Effective Wastewater Treatment Systems," USEPA 430/9-75-002 (July 1975).
35. Clifford, D. and W.J. Weber, Jr. "A Nitrate-Removal Ion-Exchange Process with a Land-Disposable Regenerant," *Proceedings of the 1977 National Conference on Treatment and Disposal of Industrial Wastewaters and Residues,* Houston, TX (April 26–28, 1977).
36. Sorg, T. "Compare Nitrate Removal Methods," *Water Wastes Eng.* (December 1980): 26–31.

17
Oil and Grease

Oily waste materials are often measured in terms of their hexane or freon extractability, for purposes of pollution evaluation. Hexane and freon are the primary solvents employed to extract oily compounds from wastewaters, although ethyl ether, carbon tetrachloride, and petroleum ether continue to be used despite attempts to standardize analytical methodology. Analytical results for oil and grease in wastewater may vary, depending upon the specific extraction technique, particular extraction solvent used, and final analytical technique employed [1–5]. Oily wastes include greases, as well as many types of oils. Grease is not a specific chemical compound, but a rather general group of semiliquid materials which may include fatty acids, soaps, fats, waxes, and other similar extractable materials. Unlike some industrial oils, which represent precise chemical composition, greases are, in effect, defined by the analytical method employed to separate them from the aqueous phase of a waste [6].

Oil and grease can be characterized in three ways: by polarity, biodegradability, and physical characteristics [7]. Polar greases and oils are normally derived from animal and vegetable materials, and are the characteristic form found in food-processing wastewaters. Nonpolar oils and greases are derived from petroleum or mineral sources. Generally, polar oils and greases are biodegradable, while nonpolar forms are considered bioresistant.

Tabakin et al [8] have proposed five categories to describe the physical forms of oil in wastewater:

1. *Free oil.* That which rises rapidly to the surface under quiescent conditions.
2. *Mechanical dispersions.* Fine droplets ranging in size from microns to a few millimeters in diameter, which are stabilized by electrical charges or other forces but not through the influence of surface active agents.
3. *Chemically stabilized emulsions.* Oil droplets similar to mechanical dispersions but with enhanced stability resulting from surface active agents at the oil/water interface.
4. *"Dissolved" oil.* Truly soluble species in the chemical sense plus very finely divided oil droplets (typically less than 5 μ diameter). This form generally defies removal by normal physical means.
5. *Oil-wet solids.* Oil adhered to the surface of particulate material in the wastewater.

There are many industrial sources of oil wastes. Table 17.1 summarizes reported industrial oily waste concentrations. Total plant effluent-oil concentrations are not always indicative of the contaminant problem, because of dilution by other plant waste streams such as cooling water [21], or intermittent discharges such as batch dumps leading to high instantaneous oil values [48].

Major industrial sources of oily waste include petroleum refineries, metals manufacture and machining, and food processors. Petroleum refineries produce large quantities of oil and oily emulsion wastes, and the American Petroleum Institute (API) has exerted a good deal of effort in developing and publishing methods of oily waste control [e.g., 16, 51]. In the metals industry two major sources of oily wastes are steel manufacture and metal working. Oily wastes include both emulsified and free oils. In steel manufacture, steel ingots are rolled into desired shapes in either hot or cold rolling mills. Oily wastes from hot rolling mills contain primarily lubricating and hydraulic pressure fluids. In cold strip rolling, however, the steel ingot is usually oiled prior to rolling, to lubricate and reduce rusting. Additional oil-water emulsions are sprayed during rolling to act as coolants. After shaping, the steel is rinsed to remove the adhering oil. Rinse and coolant waters from the cold rolling mills may contain several thousand mg/l oil, of which 25% or more may be emulsified and thus difficult to separate from the wastewater. One source reports that emulsified oil in hot rolling mill effluents rarely exceeds 20 mg/l [16].

Metal working produces shaped metal pieces such as pistons and other machine parts. Oily wastewaters from metal-working processes contain grinding oils, cutting oils, and lubrication fluids. Coolant oil-water emulsions are also employed in many metal-working processes. Soluble and emulsified oil content of wastewaters may vary from 100 to 5,000 mg/l [33, 52]. One aluminum-rolling facility reported effluent oil levels ranging from 30 to 500 mg/l [25].

A third major source of oily wastes, and particularly greases, is the food-processing industry. In the processing of meat, fish, and poultry, oily and fatty materials are produced primarily during slaughtering, cleaning, and by-product processing. The major grease source is the rendering area, in particular from the wet (or steam) rendering process [16]. Grease content in meat packinghouse waste streams may run several thousand mg/l [29, 30, 53], and waste from one fish-processing plant has been reported to contain up to 13,700 mg/l fish oil [28].

TREATMENT TECHNOLOGY

Treatment of oily waste is similar in concept to treatment of domestic sewage. In domestic sewage treatment a primary level of treatment is employed to separate the easily settleable solids from the liquid—and in treatment of oily waste, primary treatment separates the floatable (free and nonemulsified) oils from the water and emulsified oil. A secondary treatment phase is then required to break the oil-water emulsion and separate the remaining oil and water [54].

Table 17.1 Concentrations of Oily Materials in Industrial Wastewaters

Industrial Source	*Oily Waste Concentrations (mg/l)*	*Reference*
Petroleum refinery[a]	40–154	9
	35–178	10
	20	11
	10–500	12
	200–1,800 (570 avg)	4
	16–3,200 (410 avg)	13
	10–146	14
Steel-rolling mill	7,200	15
Steel mill		
Hot rolling	20	16
Cold rolling	700	17
Cold rolling	100–2,000	18
Free oil	500	
Cold rolling	60–500	19
Cold rolling coolant	2,088–48,742	20
Free oil	2,035–36,664	
Cold rolling rinse	113–3,034	20
Free oil	83–2,284	
By-product coking ovens	2–240	21
Coke plant	21.4	22
Continuous iron-casting plants	20–22	21
Blast furnace	102	22
Steel wire production	716	23
Metal forging and processing		
Forge shop wastewater	35	24
Hydraulic power system wastewater	300–9,300 (5,000 avg)	
Aluminum rolling	5,000–50,000	25
Can production		
Forming operations	200,000	26
Washing operation	819	
Food processing	3,830	27
Food processing (fish)	520–13,700	28
Seafood processing	30–546	1,2
Rendering plants	14–3,551	29,30
Edible oil refining	4,000–6,000	31,32
Metal finishing	100–5,000	33
	665	34
Electrolytic copper industry	0–436	35
Secondary copper industry	1.8–9.5	21

Table 17.1 (*continued*)

Industrial Source	*Oily Waste Concentrations (mg/l)*	*Reference*
Automotive plating		
Total wasteflow	326	36
Roll building	1,123	
Compressor cooling water	33	
Cooling-tower blowdown	159	
Oil field brine	25–50	37
Oil drilling and production wastes	7–1,300	3
Crude oil field tanker ballast	32–1,050	38
Crude oil tank ballast	3–72	39
Road tanker washing	32–91	40
Paint manufacture	1,900	27
	9–999	41
Aircraft maintenance	500–1,200	27
Free oil	250–500	
Transport vehicle maintenance	490	15
Textile mill	21	42
Wool scouring	1,605–12,260	43
Yarn scouring	2,300–8,160	43
Soap manufacture	250	44
Tannery wastes, hide curing	40,200	45
Tannery wastes, general waste	up to 850	45
Sheepskin tannery	102–350 (280 avg)	46
Fabricated and reclaimed rubber plants	>1–130	47
Tire and innertube manufacturing	38–928	48
Emulsion crumb rubber plant	7–91	48
Solution crumb rubber plant	0.8–194	48
Latex rubber plant	28	48
Glass container manufacturing	10–25	49
	50–100	15
Wood preserving	55–1,360	50

[a] API oil separator effluent.

The performance of any oil-water separation process will depend on the distribution of the oil among the five physical categories listed previously. Furthermore, since many oily wastes contain other constituents whose presence may have an impact on the treatment process performance, the successful treatment approach may be unique to the particular waste involved [5, 8, 13].

Where a significant fraction of the oil is adhered to the surface of settleable solids (oil-wet solids), simple gravity sedimentation as in a primary clarifier

Table 17.2 Performance of Gravity Sedimentation in Removal of Suspended Solids plus Oil and Grease [after 55]

	Suspended Solids (mg/l)		*Oil and Grease (mg/l)*	
Industry	*Influent*	*Effluent*	*Influent*	*Effluent*
Adhesives and sealants	10,600	2,260	2,200	522
Copper foundry	52	20	30	6.2
Ferrous foundry	1,500	64	14	2.7
Ink manufacture	1,600	110	2,400	260
Steel cold rolling	260	30	619	7
Steel hot forming	185	39	120	14
Leather tanning and finishing	3,170	945	490	57
Paint manufacture	15,600	1,400	2,400	160

can result in appreciable reductions in the wastewater oil concentration. Table 17.2 presents performance data for gravity sedimentation of suspended solids and oil and grease, for several types of industry wastewater. The percent efficiency of removal of the suspended solids and oil are comparable.

Dissolved air flotation (DAF) is normally considered a secondary level treatment technology, since in order to assure maximum effectiveness it is necessary to incorporate chemical coagulation and flocculation. The chemical treatment aspects of DAF are extremely important with respect to the removal of colloidal or emulsified oil [5].

Primary Treatment

Primary treatment takes advantage of the difference in specific gravity of oils and greases versus water. Gravity type separators are the most common devices employed in oily waste treatment. The treatment process normally involves retaining the oily waste in a holding tank and allowing gravity separation of the oily material, which is then skimmed from the wastewater surface. In theory, the separation process in gravity separators can be predicted by Stoke's law, but turbulence and shortcircuiting are common. Thus, in practice, the effectiveness of a gravity separator depends upon proper hydraulic design, and design period of wastewater retention. Longer retention times allow better separation of the floatable oils from the water. The influence of retention time on separator efficiency for a refinery wastewater is indicated in Figure 17.1. Short detention times of less than 20 minutes resulted in less than 50% oil-water separation, while more extended holding periods improved oil separation from the waste stream.

Gravity separators are equally effective in removing both greases and nonemulsified oils. The standard unit in refinery waste treatment is the API

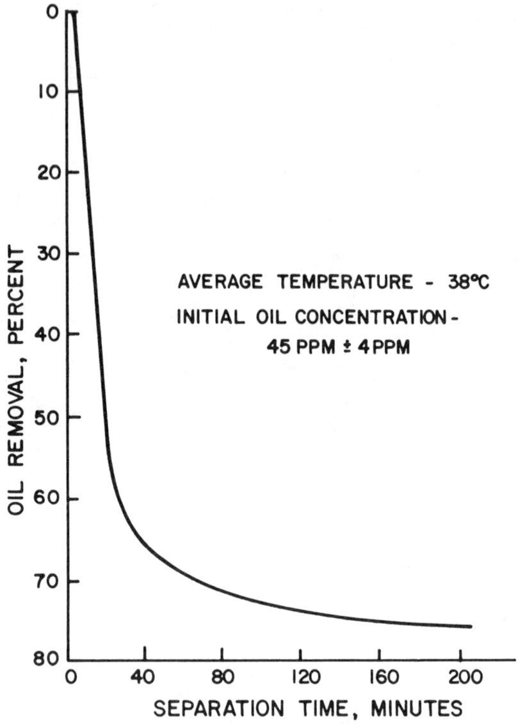

Figure 17.1 Effect of detention time on oil removal by gravity separation. (From Wallace et al [56], courtesy of Purdue University Industrial Waste Conference.)

separator, based upon design standards published by the American Petroleum Institute [51].

Two other gravity separators, both referred to as plate separators, are the corrugated plate interceptor (CPI) and the parallel plate interceptor (PPI). The latter unit is not in common use. These units differ from the API separator in that placement of inclined plates in the separation chamber effectively decreases the vertical rise of the oil globule. Coalescence of the globules into larger drops can occur at the plate surface [5, 8]. The CPI and PPI units offer one distinct advantage over API units: they require only 15–20% of the API installation area [5] and may even have minor cost advantages [13]. One major petroleum refiner having experience with both types of unit (API and plate separators) indicated that smaller size can be a disadvantage in being unable to accommodate slugs of oil, lacking sufficient retention time for breaking of emulsions, and experiencing severe fouling of the plates in some situations [13].

Table 17.3 presents the efficiencies of several oil separation processes, including the API separator. The values given in this table are based upon

Table 17.3 Efficiencies of Oil Separation Processes [57]

Treatment	Source of Influent	% Removal Free Oil	% Removal Emulsified Oil
API separator	Raw waste	60–99	na
Air flotation, without chemicals	API effluent	70–95	10–40
Air flotation, with chemicals	API effluent	75–95	50–90
Chemical coagulation and sedimentation	API effluent	60–95	50–90

treatment of petroleum refinery waste. However, efficiencies presented are also generally applicable to oily wastes from other sources undergoing the indicated treatment. For example, 90% separation of free oil has been reported in a skimming tank receiving an influent oil concentration of 700 mg/l (470 mg/l free and 230 mg/l emulsified) [17]. The waste stream originated from a cold steel rolling mill. The flow stream volume was 3,000 gpm. Additional treatment was required for the emulsified oil, in order to meet an effluent requirement of 15 mg/l oil. Peoples et al [58] have reported API separator oil effluent levels ranging from 35 to 178 mg/l, for a refinery waste, with an average influent oil concentration of 60 mg/l. For a similar refinery waste, API separator effluent oil levels averaging about 20 mg/l have been cited [11].

Grease content of the waste stream from one meat-packing plant was 2,850 mg/l, for a flow volume ranging from 1 to 7 MGD [53]. A 90-min detention period in the gravity separator, and skimming of the floating greases, reduced the waste stream grease content by about 75%, to approximately 750 mg/l. Grease was recovered for reprocessing and sale. Additional data on gravity separator performance are contained in Table 17.4.

Recovery of skimmed oil or grease from all major types of oily waste is increasingly common, due to the value of the recoverable oil. Frequently a substantial savings is possible through recovery of recycle of oil. Recovery of edible oil by combined skimming and air flotation yielded (in 1975) a \$300/day credit (60–80% of treatment cost) for one refinery. Net treatment costs were 20–40¢/1,000 gal [32]. Addition of a slop oil recovery system to recover oil from API and air flotation treatment yielded a net value of \$375,000/yr for a small (100 bbl/day) refinery. Net value was based on in-house fuel value plus decreased waste hauling costs, minus oil recovery system costs [61]. One metals-manufacturing complex using ultra filtration recovery has reported a fuel credit of \$20,850/yr [52]. An aluminum-rolling mill has projected cost recovery for reclaimed oil of \$5.30/1,000 gal of water treated (20¢/gal of oil at 80% recovery) for both chemical and ultrafiltration treatment [25]. Additional credit for water recovery of \$0.05/1,000 gal and \$1.37/1,000 gal was reported for ultrafiltration alone and ultrafiltration plus reverse osmosis, respectively. If recovered oils cannot be reused, they are typically disposed of by burial or

Table 17.4 Performance of Gravity-Type Oil Separators

Industry	*Treatment*	*Oil Concentration (mg/l)*		*% Removal*	*Reference*
		Influent	*Effluent*		
Petroleum refinery	API Separator	30	12	60	38
	API Separator	50	8	84	38
	API Separator	100	20	80	38
	API Separator[a]	—	9–915 (range)	—	55
			109 (avg)	—	
	CPI Separator	150–500	50–86	—	59
	CPI Separator	500–700	178–330	—	
Petrochemical complex	API Separator	190	132	31	38
Gasoline refinery	API Separator	253	17	93	38
Edible oil refinery	API Separator	3,416–6,100	1,050–1,125	—	31
Animal and vegetable oil refinery	Skimmer	3,984	3,195	20	32
Tannery	Clarifier	490	57	88	45
Windshield manufacture	Clarifier	1,700	10[b]	99	60
Cold steel rolling	Skimmer	1,860	1,350	27	55
Forge shop wastewater	Oil Retention Lagoon	35	10.6	70	24
Timber products processing	Skimmer	—	70	—	55

[a] Summary for 20 plants.
[b] Design performance.

Table 17.5 Capital Costs (in 1967) for Refinery Oily Waste Treatment [57]

Process	*Flow (MGD)*	*Capital Cost ($)*	*Cost ($/1,000 gpd)*
API Separator	3.0	46,500	15.50
	7.5	105,000	14.00
	15.0	210,000	14.00
Air Flotation	3.0	111,500	37.16
	7.5	222,750	29.70
	15.0	371,250	24.75

incineration. Odor- and nuisance-free oil sludge incineration has been reported [62, 63].

Refinery oily waste treatment costs have been reported for both capital and maintenance and operation [57]. These data, for a typical level of refinery technology, are summarized in Tables 17.5 and 17.6. Equivalent data, on capital costs for the meat-processing industry, and maintenance and operation costs for grease and oil treatment of food industries wastewaters, are summarized in Tables 17.7 and 17.8. Costs for treatment of oily wastes in the fabricated and reclaimed rubber industry (Table 17.9) are estimated to be extremely high, due in part to the small wastewater volumes involved. One survey of the wood-preserving industry indicates that 95% of existing plants currently employ gravity oil separation, at an estimated industry average capital cost (in 1974) of $2,126/1,000 gpd and operating cost of $1.17/1,000 gal wastewater treated [50].

Secondary Treatment

Unlike primary treatment, which consists only of gravity separation plus skimming, any of several different processes are directed toward breaking the oil-

Table 17.6 Annual Maintenance and Operating Costs (in 1967) for Refinery Oily Waste Treatment [57]

Process	*Flow (MGD)*	*Maintenance and Operating Cost ($/yr)*	*Cost ($/1,000 gpd)*[a]
API Separator	3.0	23,000	0.02
	7.5	36,000	0.01
	15.0	55,000	0.01
Air Flotation	3.0	37,000	0.03
	7.5	74,000	0.03
	15.0	117,750	0.02

[a] Based on 360 day/year operation.

Table 17.7 Capital Costs (in 1967) for Treatment of Meat Processing Oily Wastes [64]

Process	*Flow (MGD)*	*Capital Cost ($)*	*Cost ($/1,000 gpd)*
Gravity Separator	0.125	12,000	96.00
	0.54	35,000	64.81
	1.68	250,000	14.88
Air Flotation	0.125	[a]	[a]
	0.54	60,000	111.11
	1.68	150,000	89.28

[a] Not reported.

Table 17.8 Annual Maintenance and Operating Costs (in 1967) for Treatment of Meat Processing Oily Wastes [64]

Process	*Flow (MGD)*	*Maintenance and Operating Cost ($/yr)*	*Cost ($/1,000 gpd)*[a]
Gravity separator	0.125	1,000	0.03
	0.54	10,000	0.07
	1.68	18,000	0.04
Air flotation	0.125	[b]	[b]
	0.54	13,000	0.10
	1.68	30,000	0.07

[a] Based on 250 days/year operation.
[b] Not reported.

Table 17.9 Estimated Costs (in 1974) for Oil Waste Treatment in the Fabricated and Reclaimed Rubber Industry [47]

Process Waste Flow (gpm)	*Capital Costs ($/1,000 gpd)*		*Operating Costs ($/1,000 gal)*	
	(I)	*(II)*	*(I)*	*(II)*
2.7	20,256	21,538	30.51	33.08
14.0	8,150	8,550	11.32	12.55
17.5	8,379	8,814	11.28	12.64

I. Segregation of process water, with API separator treatment (influent oil 61 mg/l, effluent oil 30 mg/l).
II. I + Filter coalescence (effluent oil, 10 mg/l).

water emulsion which has passed through the primary separator, and separating the demulsified oil from the water phase.

Emulsions may be broken by chemical, electrical, or physical methods. Chemical methods are in widest use for treatment of oily wastewaters. The electrical process is generally directed toward emulsions containing mainly oil, with small quantities of water. Most oily wastes in contrast are primarily water with lesser amounts of oil. Physical emulsion-breaking methods include heating, centrifugation, and precoat filtration, with the latter two most common. Centrifugation breaks oil emulsions by separating the oil and water phases under the influence of centrifugal force. Centrifugation is best applied to oily sludges and is generally not used in treatment of the typical dilute oily waste stream, unless the volume is small [16].

In one treatment process, a mobile heating plus centrifugation process has been applied to treat an oily waste at 10,000 gal/week [65]. The oil, from a metal-working plant, is primarily emulsified. In the process, magnesium chloride is added, the waste is heated to 95°C for 3–15 minutes, and centrifuged. Oil is reduced from initial levels of 2,000–4,000 mg/l to effluent values of 22–89 mg/l. The process is claimed to be rapid, and to produce an effluent of near-neutral pH [65].

Filtration/Coalescence

Oil removal by filtration involves a combination of two mechanisms: direct removal based on droplet size, and induced coalescence [5]. Factors affecting performance are many and include the character of the oily waste, concentration of oil and droplet size, waste solids content, and hydraulic variation [5, 8]. Operational problems include "poisoning" of the media, which may necessitate high unit replacement and maintenance costs. Good effluent quality can be achieved under proper circumstances, with a range of 1–50 mg/l effluent oil reported [8]. However, as demonstrated by the data of Table 17.10, some oily wastewaters are not susceptible to treatment by simple filtration.

Filtration has been investigated on a 12 gpm pilot scale, for treating a cold rolling steel mill waste containing 230 mg/l emulsified oil. The mill waste volume was 3,000 gpm. The waste stream first passed through an oil-water separator to remove the floating oil by skimming [17]. The overall goal of treatment was to produce an effluent containing less than 15 mg/l oil. Three alternative levels of secondary treatment were evaluated with respect to achieving the required level of oil removal. These were:

- a high rate sand and gravel filter (HRF)
- addition of a polyelectrolyte coagulant prior to high-rate filtration
- Passage of the HRF effluent through a diatomaceous earth (DE) filter

Table 17.11 lists the pilot plant effluent quality (for an average influent oil content of 230 mg/l) versus projected full-scale (3,000 gpm) treatment system

Table 17.10 Performance of Multimedia Filtration in Oil and Grease Removal [after 55]

	Oil and Grease (mg/l)		
Industry	*Influent*	*Effluent*	*% Removal*
Commercial laundries	76	46	39
	8	1	87
	16	33	(−106)
Aluminum foundry	8,500	9,940	(−17)
Steel continuous casting	22	<0.5	98+
Steel hot forming	8.8	6.7	24
Paint manufacture	1,000	1,140	(−14)
Petroleum refining	8	8	0
	35	6	83
	10	8	20
	12	12	0
	18	11	39
	27	17	37
Textile mill	303	476	(−57)

capital and operation costs for the various treatment alternatives. Other applications of HRF to oily wastes are summarized in Table 17.12. Coalescence filtration has been described as being particularly suited to mechanical emulsions [38]. Table 17.13 contains a summary of performance for monograded sand coalescers.

Ultrafiltration/Reverse Osmosis

Ultrafiltration is a filtration process which is effective down to molecular dimensions (10–100 angstroms). The membranes used are truly porous, and separation is a physical process requiring elevated pressure to achieve passage of fluid through the filter. Essentially, the water molecules (and other low molecular-weight solutes) pass through the membrane pores while larger molecules or aggregates such as oil micels are rejected. Actual performance is a function of variables such as membrane composition, structure and thickness, applied pressure, temperature, and waste concentration [67]. Table 17.14 pre-

Table 17.11 Filtration Treatment of an Emulsified Oily Waste (1971) [17]

Treatment	*Effluent Oil (mg/l)*	*Capital Cost ($)*	*Operating Costs ($/day)*
HRF	20	750,000	300
Coagulant and HRF	10	775,000	372
HRF and DE	5	1,000,000	440

Table 17.12 High Rate Filtration Treatment of Oily Wastewaters

Wastewater Source	*Flow (gpm)*	*Initial Oil (mg/l)*		*Effluent Oil (mg/l)*		*Reference*
		Range	*Avg*	*Range*	*Avg*	
Petroleum refinery[a]	12–18	35–178	60	7–17	11	58
Oil field brine	77–112	4–203	29.8	3.3–17	7.2	37
	4,000	—	9.85	—	0.78	37
	—	15–30	—	2–3	—	66

[a] Results from pilot plant.

sents pilot-scale ultrafiltration performance data on several industrial wastewaters. Effluent oil levels below 10 mg/l appear difficult to achieve, and certain categories of industry waste (e.g., adhesives and sealants) yield high effluent-oil levels. Direct ultrafiltration treatment of oily waste (after pH adjustment) from an aluminum-rolling facility achieves reduction of influent oil at 80,000 mg/l to 244 mg/l, at a 100,000 gpd facility. The permeate is treated by subsequent reverse osmosis, which yields further oil reduction to 1.2 mg/l. Although in operation for one year and achieving performance predicted during a 15-month pilot study, a number of problems not experienced at pilot scale were encountered during startup [25]. A full-scale ultrafiltration unit following oil recovery by centrifugation has been in operation for more than three years at a multiple metal products manufacturing plant [52].

Another membrane process, reverse osmosis, can selectively reject solutes (molecules and ions) of the same size order as water molecules [67]. Both processes have been applied to the separation of oil-water emulsions, providing both oil and water recovery. Bench scale ultrafiltration/reverse osmosis tests

Table 17.13 Performance Data for Coalescence Filtration Through Monograded Sand [38]

Influent Oil (mg/l)	*Effluent Oil (mg/l)*	*% Removal*
35	16	54
42	13	69
36	9	75
55	23	58
32	9	72
255	29	89
201	15	93
99	14	86
870	23	97
1,050	21	98
489	23	95

Table 17.14 Pilot Scale Performance of Ultrafiltration in Oil and Grease Removal [after 55]

Industry	*Oil and Grease (mg/l)*		*% Removal*
	Influent	*Effluent*	
Adhesives and sealants	113	131	(−16)
	522	162	69
	478	184	62
Commercial laundries	600	<9	98+
	749	28	96
	795	10	99
	7,890	38	99+
Synthetic rubber manufacture	12	5	58
	28	11	61
Timber products processing	2,160	55	97

on a synthetic rubber waste reduced influent oil levels of 5–11 mg/l to an average effluent level below 5 mg/l [55, 68]. Pretreatment requirements for application of reverse osmosis to oil removal are much more critical than for ultrafiltration.

Table 17.15 presents a cost comparison for a 100,000 gpd facility treating a highly concentrated oily wastewater. A comparison of chemical treatment with ultrafiltration and ultrafiltration/reverse osmosis indicates a small capital cost advantage but significant operating cost advantages for the latter two technologies.

Chemical Treatment

Chemical treatment of an emulsion is usually directed toward destabilizing dispersed oil droplets or destroying any emulsifying agents present in a first stage, followed by removal of the separated oil. The process usually consists of rapidly mixing coagulant chemicals with the wastewater, followed by flocculation, and flotation or settling. Acidification may also be effective in breaking an oil-water emulsion. Alternative chemical demulsifying processes include [69]:

Table 17.15 Cost Comparison (in 1978) for Oily Wastewater Treatment [25]

Technology	*Capital Cost ($)*	*Operating Cost ($/1,000 gpd)*
Chemical	1,730,000	11.33
Ultrafiltration	1,450,000	7.73
Ultrafiltration plus reverse osmosis	1,550,000	8.81

- adding of coagulating salts
- adding of acids
- adding of salts and heating the emulsion
- adding of salts and treatment by electricity
- adding of acids plus organic cleaving agents

The use of coagulant salts is most common in industrial wastewater treatment, with oil removal by sedimentation or flotation. Polymers may be added, to enhance demulsification or flocculation.

It has been emphasized that coagulation with aluminum or iron salts is generally effective for demulsifying oily wastes [70]. However, the aluminum or iron precipitate hydroxide sludges, which may prove difficult to dewater. Acids generally cleave emulsions more effectively than do coagulant salts, but are more expensive, and the resultant acidic wastewater must be neutralized after oil-water separation. The pH required for acid demulsification depends upon the nature of the waste, with pH as low as 2–3 reported in one application [16] and a pH of 5.0–5.5 being effective in another [71]. Oily wastes from an aluminum can-forming process were reduced from 200,000 mg/l to 10,000 mg/l by acidification to pH 2 with sulfuric acid [26]. Elevation of temperature to 140°F further reduced the residual oil content to 1,000 mg/l. Acid pickling wastewater may be employed if available, to break the oily emulsion. The use of waste hydrochloric and sulfuric acids from plating operations has also been reported [72].

A machining plant wastewater is treated by sulfuric acid additon to pH 2–3, to break the oil water emulsion. After 6–12 hours of acid digestion, calcium chloride and caustic soda are added to coagulate the oil. The process reduces oil from an average of 21,225 to 6.0 mg/l [73]. An advantage claimed for the process, in addition to efficiency exceeding 99%, is that the floated precipitate can be burned with waste oil in the powerhouse. Capital investment for the treatment system (in 1973) was $1,800,000, for a wastewater volume of 200,000 gpd [73].

The "salting out" of emulsified oil by adding large quantities of an inorganic salt may create additional pollution problems by significantly increasing the dissolved solids in the effluent. Organic demulgaters are extremely effective demulsifying agents, but due to their high cost they are considered impractical at high rinsewater flow rates and low oil concentrations [70]. Thus acid or coagulant salts find the widest acceptance as demulsifying agents in industry. Frequently, coagulant aides are added to assist in flocculating the coagulant particles of oil sludge.

Chemical coagulation followed by gravity sedimentation is employed in a number of industries. One company treats oily waste from a ball- and roller-bearing plant by coagulation with sodium carbonate, lime, and a polyelectrolyte flocculating agent to achieve grease and oil reduction from 302 mg/l to 28 mg/l. This exceeds 90% removal, in a process treating 30,000–60,000 gpd. Coagulation with alum (35 mg/l) followed by sedimentation reduces API separator

Table 17.16 Removal of Oil and Grease by Chemical Coagulation plus Gravity Sedimentation [after 55]

		Oil and Grease (mg/l)		
Industry	*Treatment Chemical*	*Influent*	*Effluent*	*% Removal*
Paint manufacture	sodium aluminate	1,260	22	98
	alum	1,810	11	99
	alum + lime	830	16	98
	alum + lime + ferric chloride	393	91	77
	alum + lime + polymer	980	22	98
	alum + polymer	1,700	880	48
	alum + polymer	642	8	99
	alum + polymer	1,200	153	87
Commercial laundry	alum + polymer	15	4	73
Steel pickling	lime	3	1	66
Steel pickling	lime + polymer	650	6	99
Steel pipe fabrication	lime + polymer	5	4	20
Paint manufacture	polymer	1,100	22	98

refinery effluent oil from initial levels of 50–100 mg/l to a final effluent oil content consistently less than 15 mg/l. Treatment is on a waste flow of 300 gpm [74]. Other results reported for chemical coagulation plus sedimentation are presented in Table 17.16.

Following chemical destabilization of an oily emulsion, air flotation is commonly employed as an alternative to sedimentation, to separate the oil and water. Air flotation consists either of saturating a portion of the wastewater with dissolved air under high pressure (dissolved air flotation, DAF) or of dispersing air bubbles by a rotor-dispenser mechanism (induced air flotation, IAF). In DAF the pressure is then abruptly released, resulting in the formation of thousands of microscopic air bubbles which attach to oil droplets and float them to the surface. In IAF a rotor forces the liquid through the dispenser orifice, creating a negative pressure that draws air down into the liquid, generating the desired small air bubbles. IAF normally represents a lower capital investment and smaller space requirements than does DAF.

In air flotation the oil-air bubble mixture forms a froth layer at the surface, which is skimmed. Chemical flocculating agents such as salts of iron and aluminum, with or without organic polyelectrolytes, are particularly helpful in improving the effectiveness of the air flotation process. Dissolved air flotation has been cited as the most effective step for waste load reductions in the rendering industry, with expected grease removals of 80–90% without the addition of chemicals to the process [29]. With chemical addition, greater than 90%

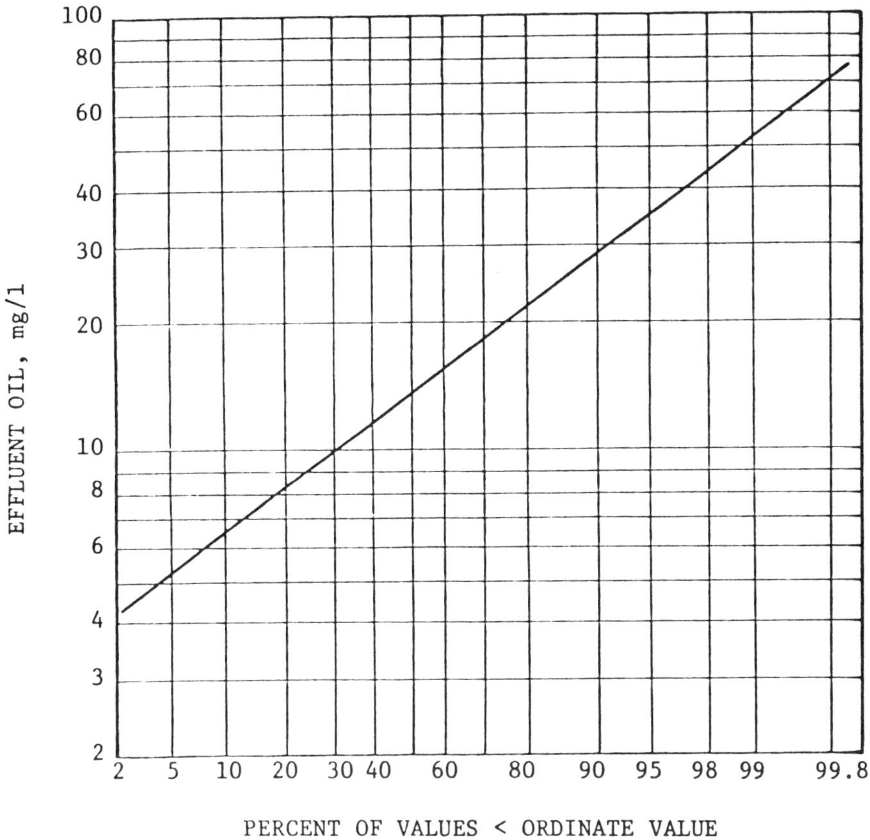

Figure 17.2 Cumulative plot of ten flotation units effluent oil concentrations [3].

removal is expected, and one plant, which also employed pH and chemical feed rate control, consistently achieves 90–95% grease removal [29].

Expected efficiencies of air flotation, with and without the use of chemical aids, were presented in Table 17.3. Coagulant aids can double the effectiveness of air flotation in removing emulsified oils. Figure 17.2 provides a cumulative frequency of observed effluent oil concentrations from DAF units. Capital and operating costs of air flotation as a function of waste stream were presented in Tables 17.5–17.8. Additional capital and operating costs for DAF are presented in Table 17.17. Generalized capital and operation-maintenance costs are presented in Figures 17.3 and 17.4 for air flotation systems, as a function of hydraulic capacity and daily solids removal.

In one application of DAF, 62% oil removal was achieved by direct air flotation, and 94% removal upon addition of 25 mg/l aluminum sulfate coagulant [54]. For an oil refinery wastewater, 70% oil removal resulted by air flotation. Addition of polyelectrolyte and bentonite clay increased oil removal to 95%

Table 17.17 Capital and Operating Costs for Dissolved Air Flotation Units [75]

Flow Capacity (gpm)	*Size Unit (ft^2)*	*Capital Cost ($)*	*Operating Cost (1982$/year)*
60–180	50	30,000	920–1,370
360–720	300	70,000	3,170–5,870

[54]. For another refinery, 79% oil removal was obtained by flotation, increased to 87% at an aluminum sulfate coagulant dosage of 25 mg/l [54]. In a third refinery operation, an air flotation efficiency of 70–80% resulted, with up to 90% oil removal upon addition of aluminum sulfate at concentrations of 30–70 mg/l. Lime, added at 75–100 mg/l provided even better removal, to 95% [54]. One cold rolling steel mill scrapped an existing treatment system to install a flotation system which enabled recovery of 15,000–20,000 gal oil per month. The system cost (1972 basis) $579/1,000 gpd capacity [18].

Air flotation treatment has been applied to many industrial wastes, both with and some without chemical demulsifying aides. Treatment data are summarized in Table 17.18. The best effluents resulted from pretreatment with chemical coagulants, prior to air flotation.

One aluminum can producer treated oily rinsewaters containing 819 mg/l oil and grease by alum, caustic, and polymer addition followed by DAF. The system was fully automated and pH controlled. Effluent oil and grease of 14 mg/l was reported [26]. An industrial pilot process achieved 93% grease reduction of a meat-packing waste by air flotation [53]. Grease content of the waste was reduced from 1,944 mg/l to 142 mg/l, with 300 mg/l aluminum sulfate added. The process resulted in decreased volumes and qualities of recoverable grease, however. Cost of chemicals was 13.4¢/1,000 gal waste treated, and capital costs were estimated at $30,000 for a 1.0 MGD unit. One rendering plant reduces grease loadings of up to 29,000 mg/l to 1,000–2,000 mg/l with air flotation and follows the flotation unit with precoat vacuum filtration to achieve effluent grease concentrations of 50 mg/l [30].

Electrical Processes

Two processes utilize electricity for the destabilization of oil in wastewater: electroflotation and electrocoagulation. The former technology is an alternative to air flotation to induce small gas bubble formation through the electrolysis of water to oxygen and hydrogen gas. The recent development of lead dioxide electrodes has improved the economics of the electroflotation process. Application of this technology to treatment of meat and poultry preparation and packing wastes for fat and grease reduction has been reported. Effluent oil content of 30–35 mg/l required less than 1.5 kwh/1,000 gal treated effluent [40]. In another application of electroflotation, to a tannery wastewater, the power requirement was 0.3 kwh/1,000 gal treated [46]. This is in contrast to a Russian

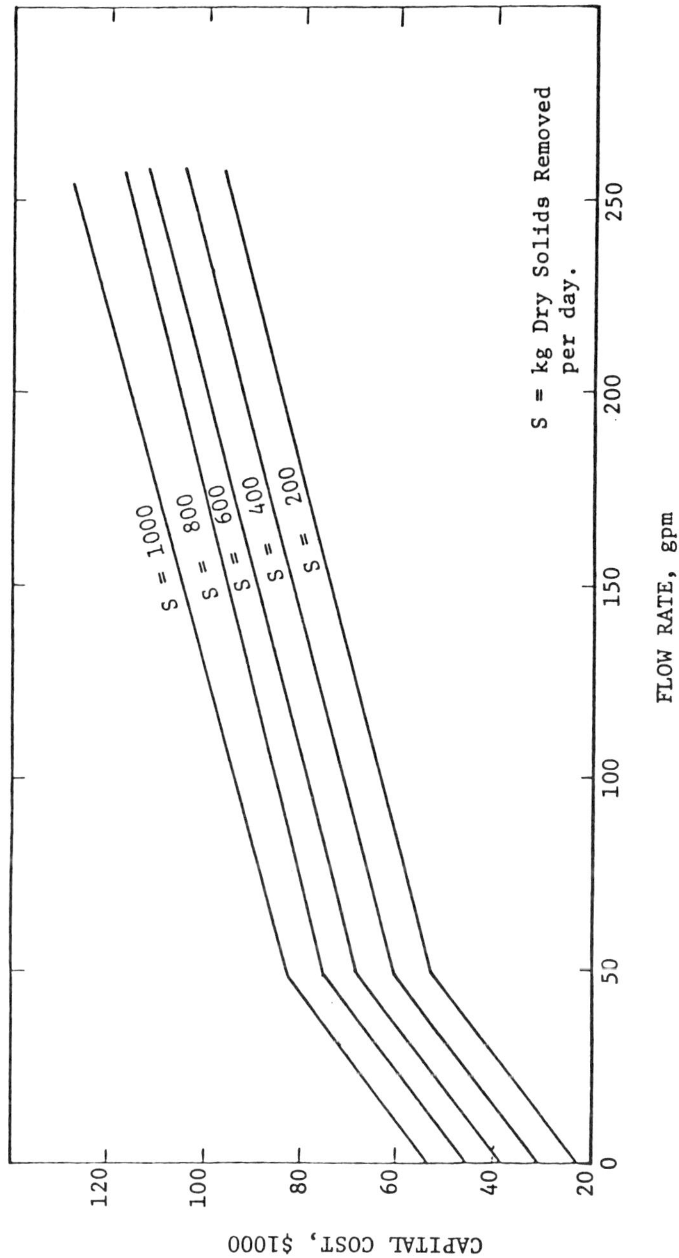

Figure 17.3 Capital costs (in 1975) for wastewater air flotation systems [2].

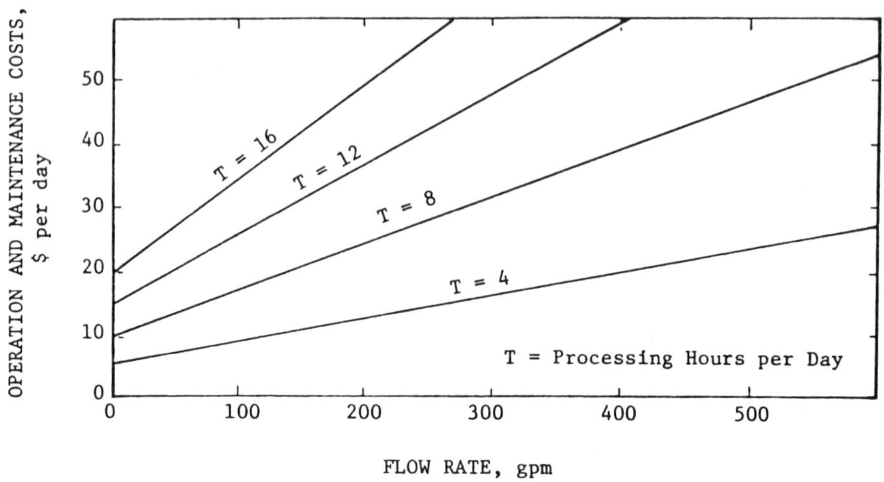

Figure 17.4 Operation and maintenance costs (in 1975), including chemicals, for air flotation systems [2].

application to oil removal from coal-tar industry wastes, which required 3.0 kwh/1,000 gal [76].

The electrocoagulation process utilizes consumable electrodes such as scrap iron. Voltage applied to the system oxidizes the electrode to release a metallic coagulant such as the ferrous ion. One patented system of this type operates at an average 20 volts, with current ranging from 15–35 amps [77]. The waste stream requires sufficient conductivity for cell operation and to prevent passivation of the electrode material. In one application of electrocoagulation followed by electroflotation, a tannery wastewater initial oil level of 280 mg/l was reduced to 14 mg/l [46]. Electrocoagulation unit power requirement was 0.7 kwh/1,000 gal. A cost comparison with other oil treatment technologies is presented in Table 17.19.

Biological Treatment

Oil- and grease-laden waters are, on occasion, treated in lagoons and other biological treatment processes. Within biological treatment systems, the polar biodegradable oils and greases are decomposed, and effluent levels of this class of material should remain between 2 and 8 mg/l [78]. Nonpolar oils and greases are either removed through primary clarification or incorporated into the biological floc. These adsorbed materials are then removed with the waste sludge [7]. Table 17.20 presents data on activated sludge and aerated lagoon treatment of oily wastes. In one application, for packinghouse wastes, an anaerobic lagoon achieved 88% grease removal [79]. Influent grease content ranged from 425 to 1,270 mg/l, while effluent grease concentrations of 69–147 mg/l were recorded.

Table 17.18 Dissolved Air Flotation Treatment of Oily Wastewaters

		Oil Concentration (mg/l)			
Industrial Source	*Coagulant*	*Influent*	*Effluent*	*% Removal*	*Reference*
Refinery	None	125	13	72	27
	100 mg/l alum	100	10	90	
	None	154	40	74	9
	$FeCl_3$	34	3	95	38
	—	72	7	90	38
	Polyelectrolyte (5–15 mg/l)	16–3200 (410 avg)	1–206 (27 avg)	93	13
Oil tanker ballast water	100 mg/l alum (+1 mg/l polymer)	133	15	89	27
Petrochemical	$FeCl_3$	100	14	86	38
	$FeCl_3$	180	14	92	38
	Alum	165	22	87	38
Steel mill	Lime + $FeSO_4$	100–2,000	8–275 (76 avg)	—	18
	Lime + $FeSO_4$	100–2,000	3–188 (20 avg)	—	18
Glass container manufacture	"Single chemical addition"	100	1	99	15
Paint manufacture	150 mg/l alum (+1 mg/l polymer)	1,900	0	100	27
Aircraft maintenance	30 mg/l alum	250–700	20–50	90+	27
Transport vehicle maintanance	"Single chemical addition"	490	5	99	15
Commercial laundry	Alum + polymer	205	76	63	55

Commercial laundries	Calcium chloride + polymer	703	143	80	55
		440	190	57	
		760	170	78	
		1,600	230	86	
		230	84	63	
		513	53	90	
	Ferrous sulfate + lime + polymer	915	28	97	55
	Ferric sulfate + polymer	341	101	70	55
	Polymer	39	16	59	55
Meat packing	None	3,830	270	93	27
	None	4,360	170	96	27
	300 mg/l alum	1,944	142	93	53
Edible oil without API	Alum + polymer	4,250	1,570	63	31
Edible oil	—	5,700	630	89	31
Edible oil with API	Alum + polymer	950	250	74	31
Edible oil with API + equalization	Alum + polymer	1,107	95	91	31
Edible oil with skimmer	Alum + polymer	558	245	56	32
Seafood processing	Lime + polymer	3,195	357	89	1
Seafood processing—tuna (pilot scale)	Lime + $FeCl_3$	558	106	81	1
Seafood processing—gulf shrimp (pilot scale)	Alum + polymer	852	128	85	1
Yarn scouring	Aluminum chlorohydrate	2,300–8,160	40–248	88–99	51
Wool scouring	Aluminum chlorohydrate	1,650–12,260	35–287	82–99+	51

Table 17.19 Cost Comparison (in 1982) for Electrocoagulation and Other Oil Removal Technologies [77]

Process	*Capital Cost*[a] *($)*	*Operating Cost ($/1,000 gal)*
Chemical coagulation + DAF	75,000	>0.85
Ultrafiltration	175,000	—
Electrocoagulation + DAF	60,000	<0.68

[a] Based on treatment plant capacity of 45,000 gpd and influent oil of 1,000 mg/l.

The lagoon treated 528,000 gal/day, at a capital cost (in 1969, excluding land) of $44,000.

Treatment of a petroleum refinery waste by a process involving API separator-air flotation-biological (oxidation pond) treatment has been reported [9]. Effluent from the air flotation unit contained 40 mg/l oil. Effluent from the oxidation pond, which had a minimum 20-day detention time, contained 18 mg/l or less oil. Another refinery built a 5 MGD extended-aeration activated-sludge plant to treat refinery waste [52, 80]. The unit receives API separator effluent containing about 20 mg/l oil. At design loadings, the activated sludge unit effluent oil is 2–10 mg/l. Operating costs for the unit total 4.8¢/1,000 gal. Included is a somewhat high maintenance cost of 1.7¢/1,000 gal. Annual maintenance (in 1974–1975) at $28,765, represented 1.4% of installed cost. Capital costs were $400/1,000 gpd.

Activated sludge treatment of by-product coking wastes reduced influent oil concentrations from 240 to 2.5 mg/l. Capital costs for the overall treatment system (0.79 MGD) were (in 1974) $885/1,000 gpd and operating costs were $0.84/1,000 gal [21]. An edible oil refinery using API separation, air flotation, and aerated lagoon followed by activated sludge treatment, achieves a final effluent oil level of 11 mg/l. Raw waste concentrations of 3,000–6,000 mg/l are

Table 17.20 Biological System Performance for Control of Oil and Grease [after 55]

Type of System	*Industry*	*Oil and Grease (mg/l)*		*% Removal*
		Influent	*Effluent*	
AS[a]	By-product coke manufacture	240	5	98+
AS	Leather tanning and finishing	171	91	47
		247	35	86
		553	17	97
		413	25	94
AL[b]	Leather tanning and finishing	720	17	97
AS	Textile mill	324	303	6

[a] Activated sludge.
[b] Aerated lagoon.

Table 17.21 Performance of Combined Municipal-Industrial Systems in Oil and Grease Removal [after 7]

Type of System[a]	*Number of Plants*	*Average Oil and Grease (mg/l)*		*% Removal*
		Influent	*Effluent*	
AS-C	18	40.8	4.0	90
AS-C	5	57.3 (36.7)[b]	15.4	73
AS-S	1	162.5	9.4	94
AS-O	1	47.3 (13.4)[b]	4.3	91
AS-CS	3	38.5	7.1	82
AS-HR	4	26.6	3.0	89
AS-EA	4	21.9	2.9	87
TF-HR	15	34.1	7.8	77

[a] AS = activated sludge; TF = trickling filter; C = conventional or complete mixed; S = step aeration; O = pure oxygen; CS = contact stabilization; HR = high rate; EA = extended aeration.

[b] Polar oil and grease concentration.

reduced to 95–250 mg/l through the flotation unit, and to 64–100 mg/l through the aerated lagoon. Maximum flow for the system is 25.9 MGD [31].

Twedell et al [81] have reported the use of mutant bacterial inoculums into an industrial activated sludge plant, to enhance oil and grease removal during periods of high influent oil and grease. The plant influent fluctuates between 100 and 1,500 mg/l oil. Batch reactor studies indicated that at an initial 400 mg/l oil, the mutant population could achieve 61% reduction within 12 hours and 95% reduction within 72 hours. Upon inoculation of the 0.5 MGD full-scale system, effluent oil levels of 600–1,100 mg/l were reduced to 100–400 mg/l [81].

Young [7] has reported on the removal of oil and grease in combined municipal-industrial biological treatment plants. A summary of performance data is presented in Table 17.21. Correlation analysis indicated a strong inter-relationship between effluent oil and grease, biochemical oxygen demand (BOD), and suspended solids [7].

Other Processes

Activated carbon has been reported effective in the removal of oil and grease, although its use specifically for this function is limited. Its application in refinery wastewater treatment has been reported capable of achieving an effluent oil and grease level below 8 mg/l [5]. It has also been proposed for treatment of reverse osmosis brine from a textile mill water recovery process [42]. Per-

Table 17.22 Removal of Oil and Grease by Granular Activated Carbon Treatment [after 55]

Industry	*Scale*	*Oil and Grease (mg/l)*		*Removal*
		Influent	*Effluent*	
Commercial laundry	Pilot	20.4	<9	56+
	Full	4	8	(−100)
Gum and wood chemicals	Full	28	2.2	92
Petroleum refining	Pilot	8.5	7.5	12
		8.3	7.1	14
		6.3	6.0	5
		12	8.7	28
		11	14	(−27)
		17	13	24
		12	1.8	85

formance data were not supplied. Pilot studies on the treatment of a synthetic rubber wastewater by dual media filtration plus granular activated carbon achieved an effluent oil and grease level below 5 mg/l [68]. Table 17.22 presents additional data on activated carbon treatment of industrial wastewaters.

SUMMARY

Treatment of floating oils is achieved economically and efficiently by gravity separation and skimming. Oil removals as high as 99% have been reported. Treatment of emulsified oil-water mixtures is more complex and costly, and represents a secondary phase of treatment after the primary gravity separation

Secondary treatment technologies include filtration and coalescence, ultrafiltration and reverse osmosis, chemical coagulation followed by air flotation or sedimentation, and electrical technologies. Biological and activated carbon treatment is also utilized for secondary oil and grease removal. Due to the unique characteristics of individual industrial oily wastewaters, the relative efficiencies and economics of the treatment technologies vary widely, and no single system has overriding advantages over the other alternatives available.

REFERENCES

1. "Development Document for Effluent Limitations Guidelines and New Source Performance Standards for the Catfish, Shrimp and Tuna Segment of the Canned and Preserved Seafood Processing Point Source Category," USEPA 440/1-74-020-a (June 1974).

2. "Development Document for Interim Final Effluent Limitations Guidelines and Proposed New Source Performance Standards for the Fish Meal, Salmon, Bottom Fish, Sardine, Herring, Clam, Oyster, Scallop and Abalone Segment of the Canned and Preserved Seafood Processing Point Source Category," USEPA 440/1-74/041, Group I, Phase II, (January 1975).
3. "Development Document for Interim Final Effluent Limitations Guidelines and New Source Performance Standards for the Offshore Segment of the Oil and Gas Extraction Point Source Category," USEPA 440/1-75/055, Group II (September 1955).
4. Luthy, R.G., R.E. Selleck, and T.G. Galloway. "Removal of Emulsified Oil with Organic Coagulants and Dissolved Air Flotation," *J. Water Poll. Control. Fed.* 50 (1978): 331–346.
5. Ford, D.L. and R.L.Elton. "Removal of Oil and Grease from Industrial Wastewaters," *Chem. Eng.* (Deskbook Issue, October 17, 1977): 49–56.
6. *Standards Methods for the Examination of Water and Wastewater,* 14th ed. (New York: American Public Health Association, 1976).
7. Young, J. "Removal of Grease and Oil by Biological Treatment Processes," *J. Water Poll. Control. Fed.* 51 (1979): 2071–2087.
8. Tabakin, R.B., R. Trattner, and P.N. Cheremisinoff. "Oil/Water Separations: The Options Available. Part 1 and Part 2," *Water Sew. Works* (July 1978): 74–77; (August 1978): 72–75.
9. Wigren, A.A. and F.L. Burton. "Refinery Wastewater Control," *J. Water Poll. Control. Fed.* 44 (1972): 117–128.
10. "Humble Oil Treats Wastes at Baytown," *Water Sew. Works* 118 (1971): IW4–5.
11. Rose, W.L. and G.E. Gorringe. "Activated Sludge Plant Handles Loading Variations," *Oil Gas J.* 70 (1972): 62–65.
12. "Development Document for Effluent Limitations Guidelines and New Source Performance Standards for the Petroleum Refining Point Source Category, USEPA 440/1-74-014-a (April 1974).
13. Davies, B.T. and R.W. Vose. "Custom Designs Cut Effluent Treating Costs. Case Histories at Chevron USA Inc.," *Proceedings of the 32nd Industrial Waste Conference, Purdue University* (May 1977), pp. 1035–1060.
14. Mahmud, Z. and W.C. Thanh. "Biological Treatment of Refinery Wastes," paper presented at the 33rd Annual Purdue Industrial Waste Conference, West Lafayette, IN (May 1978).
15. Katnik, K.E. and A.M. Pavilcius. "A Novel Approach for the Treatment of Oily Wastewaters," paper presented at the 33rd Annual Purdue Industrial Waste Conference, West Lafayette, IN (May 1978).
16. *Industrial Oily Waste Control* (New York: American Petroleum Institute, n.d.).
17. Symons, C.R. "Treatment of Cold Mill Wastewaters by Ultrahigh-Rate Filtration," *J. Water Poll. Control. Fed.* 43 (1971): 2280–2286.
18. Schuldt, A.A. and V. A. Suffoletta. "Recovery and Treatment of Spent Steel Mill Solutions," *Ind. Wastes* 1 (1975): 20–23.
19. Donovan, E.J. Jr. "Treatment of Wastewater for Steel Cold Finishing Mills," *Water Waste Eng.* (November 1970): F22–F25.
20. "Treatment of Wastewater—Waste Oil Mixtures," USEPA 12010 EZV 02/70 (1970).

21. "Development Document for Effluent Limitations Guidelines and New Source Performance Standards for the Iron and Steel Manufacturing Point Source Category," USEPA 440/1074-024-a (June 1974).
22. Wong-Chong, G. and A. Dequittner. "Physical-Chemical Treatment of Coke Plant Wastewater at Crucible Steel, Midland, PA," paper presented at Conference, Water Pollution Control Federation, Las Vegas, NV (September 28–October 3, 1980).
23. Reece, C. "Steel Welding Wire Manufacturer Meets Stringent Effluent and Sludge Disposal Criteria," paper presented at the 37th Annual Purdue Industrial Waste Conference, West Lafayette, IN (May 1982).
24. Hockenburg, M.R. and A.W. Loven. "Treating Metal Forging and Processing Wastewater," *Ind. Wastes* 3 (1977): 45–59.
25. Sonksen, M.K., F.M. Sittig, and E.F. Maziarz. "Treatment of Oily Wastes by Ultrafiltration/Reverse Osmosis. A Case History," paper presented at the 33rd Annual Purdue Industrial Waste Conference, West Lafayette, IN (May 1978).
26. Ettelt, G.A. "Can Forming Plant Removes Oils and Phosphates; Controls pH, Part 1 and Part 2," *Ind. Wastes* 2 (January/February 1976): 20–22; (March/April 1976): 26–31.
27. Boyd, J.L. and G.L. Shell. "Dissolved Air Flotation Application to Industrial Wastewater Treatment," paper presented at the 45th Annual Conference of Water Pollution Control Federation, Atlanta, GA (1972).
28. Chun, M.J., R.H.F. Young, and N.C. Burbank, Jr. "A Characterization of Tuna Packing Waste," *Proceedings of the 23rd Industrial Waste Conference, Purdue University* (May 1968), pp 768–805.
29. "Development Document for Effluent Limitations Guidelines and New Source Performance Standards for the Renderer Segment of the Meat Products and Rendering Processing Point Source Category," USEPA 440/1-74/031-d, Group I, Phase II (January 1975).
30. Smedley, J.B. and R. Blank. "Grease Removal with a Precoat Filter: A Case History," unpublished manuscript, ENVIROCON LIMITED, 1111 West Georgia St., Vancouver, BC, Canada V6E 367.
31. Walsh, T.J., J.R. Keigher, and J.J. Joseph. "Oil Removal and Biological Treatment of the Effluent from an Edible Oil Refinery," paper presented at the 48th Annual Meeting, Water Pollution Control Federation, Miami Beach, FL (October 8, 1975).
32. Seng, W.C. and G.M. Kreutzer. "Resume of Total Operation of Waste Treatment Facility for Animal and Vegetable Oil Refinery," *J. Am. Oil. Chem. Soc.* 52 (1975): 9A–13A.
33. Brink, R.J. "Operating Costs of Waste Treatment at General Motors," *Proceedings of the 19th Industrial Waste Conference, Purdue University* (May 1964), pp 12–16.
34. Germain, J.E., C.A. Vath, and C.F. Griffin. "Solving Complex Waste Disposal Problems in the Metal Finishing Industry," paper presented at Georgia Water Pollution Controllers Association Conference, Atlanta, GA (September 1968).
35. "Development Document for Interim Final Effluent Limitations Guidelines and Proposed New Source Performance Standards for the Primary Copper Smelting Subcategory and the Primary Copper Refining Subcategory of the Copper Segment of the Nonferrous Metals Manufacturing Point Source Category," USEPA 440/1-75/032-b, Group I, Phase II (February 1975).

36. Cullinane, M.J. Jr. "Treatment of Plating Wastes from the Automotive Industry," *Ind. Wastes* (March/April 1978): 29–31.
37. Wallace, J.T. Jr. and J.W. Brown. "Deep Bed Filtration of Oilfield Produced Water," paper presented at the 27th Annual Purdue Industrial Waste Conference, West Lafayette, IN (May 1972).
38. Berne, F. and L.P. Aggarwal. "Effluent Treatment in the Petroleum Industry," *Effluent Water Treatment J.* 14 (1974): 26–29.
39. Lindsay, J.T. and B.U. Prather. "Solving Petroleum Refinery Wastewater Problems," *J. Water Poll. Control. Fed.* 49 (1977): 1779–1785.
40. Cottrell, W.R.T. "Treatment of Effluent by Flotation Processes," *Effluent Water Treatment J.* 16 (1976): 563–567.
41. "Development Document for Effluent Limitations Guidelines and New Source Performance Standards for the Oil Base and Solvent Wash Subcategories of the Paint Formulating and the Ink Formulating Point Source Category," USEPA 440/1-75/050-a, Group II (July 1975).
42. Osantowski, R. and K. Duchon. "Activated Carbon Treatment of Reverse Osmosis Brine for a Textile Waste," *Proceedings of the 3rd Annual Conference on Treatment and Disposal of Industrial Wastewaters and Residues,* Houston, TX (April 18–20, 1978).
43. Evans, B.R. "Treatment of Scouring Liquor by Electroflotation," *Effluent Water Treatment J.* 14 (1974): 85–88.
44. "Development Document for Effluent Limitations Guidelines and New Source Performance Standards for the Soap and Detergent Manufacturing Point Source Category," USEPA 440/1-74-018-a (April 1974).
45. "Development Document for Effluent Limitations Guidelines and New Source Performance Standards for the Leather Tanning and Finishing Point Source Category," USEPA 440/1-74-016-a (March 1974).
46. Ramirex, E.R., L.K. Barker and O.A. Clemens. "Physio-Chemical Treatment of Tannery Wastewater by Electrocoagulation," *Proceedings of the 32nd Industrial Waste Conference, Purdue University* (May 1977), pp 183–188.
47. "Development Document for Effluent Limitations Guidelines and New Source Performance Standards for the Fabricated and Reclaimed Rubber Segment of the Rubber Processing Point Source Category," USEPA 440/1-74/030-a, Group I, Phase II (December 1974).
48. "Development Document for Effluent Limitations Guidelines and New Source Performance Standards for the Tire and Synthetic Segment of the Rubber Processing Point Source Category," USEPA 440/1-74-013-a (February 1974).
49. "Development Document for Effluent Limitations Guidelines and New Source Performance Standards for the Pressed and Blown Glass Segment of the Glass Manufacturing Point Source Category," USEPA 440/1-75/034-a, Group I, Phase II (January 1975).
50. "Development Document for Effluent Limitations Guidelines and New Source Performance Standards for the Plywood, Hardboard and Wood Preserving Segment of the Timber Products Processing Point Source Category," USEPA 440/1-74-023-a (April 1974).
51. *Manual on Disposal of Refinery Wastes,* 7th ed. (New York: American Petroleium Institute, 1963).

52. Roberts, D.A. and D.S. Yeaple. "Oil Treatment and Reclamation," paper presented at the 48th Annual Meeting, Water Pollution Control Federation, Miami Beach, FL (October 6, 1975).
53. Garrison, V.M. and R.J. Geppert. "Packinghouse Waste Processing Applied Improvement of Conventional Methods," *Proceedings of the 15th Industrial Waste Conference, Purdue University* (May 1960), pp 207–217.
54. Quigley, R.E. and E.L. Hoffman. "Flotation of Oily Wastes," *Proceedings of the 21st Industrial Waste Conference, Purdue University* (May 1966), pp 527–533.
55. Cawley, W., ed. *Treatability Manual,* vol. III. *Technologies for Control/Removal of Pollutants,* USEPA 600-8-80-042-c (July 1980).
56. Wallace, A.T., G.A. Rohlich, and J.R. Villemonte. "The Effect of Inlet Conditions of Oil-Water Separators at SOHIO's Toledo Refinery," *Proceedings of the 20th Industrial Waste Conference, Purdue University* (May 1965), pp 618–625.
57. *The Cost of Clean Water,* vol. III. *Industrial Waste Profiles, No. 5—Petroleum Refining*. (Washington, DC: USDI, 1967).
58. Peoples, R.F., P. Krishnan and R.N. Simonsen. "Nonbiological Treatment of Refinery Wastewater," *J. Water Poll. Control. Fed.* 44 (1972): 2120–2128.
59. Morrison, J. "Tilted-Plate Separators for Refinery Wastewater," *Oil Gas J.* 68 (1970): 86–88.
60. "Development Document for Effluent Limitations Guidelines and New Source Performance Standards for the Flat Glass Segment of the Glass Manufacturing Point Source Category," USEPA 440/1-74-001-c (January 1974).
61. "Emulsion Breaking Solves Oily-Waste Disposal," *Oil Gas J.* (November 22, 1971) (reprint).
62. Dreier, D.E. and J.D. Walker. "Grease Incineration," *Proceedings of the 19th Industrial Waste Conference, Purdue University* (May 1964), pp 161–166.
63. "Fluid Bed Incineration of Petroleum Refinery Wastes," USEPA 12050 EKT 03/171 (1971).
64. *The Cost of Clean Water,* vol. III. *Industrial Waste Profiles No. 8—Meat Products,* (Washington, DC: USDI, 1967).
65. Chalmers, R.K. "Treatment of Wastes from Metal Finishing and Engineering Industries," *Process in Water Technology,* vol I (New York: Pergamon Press, 1972).
66. Bleakley, W.B. "Wilmington Attacks Water Problems," *Oil Gas J.* 70 (1972), 71–72.
67. Klinkowski, R.R. "Ultrafiltration: An Emerging Unit Operation," *Chem. Eng.* (May 8, 1978), 164–173.
68. Kleper, M. and R. Turner. "Assessment of Wastewater Treatment Alternatives for the Synthetic Rubber Manufacturing Industry," *Proceedings of the 3rd Annual Conference on Treatment and Disposal of Industrial Wastewaters and Residues,* Houston, TX (April 18–20, 1978).
69. Mentens, A. "Treatment of Wastes Originating from Metal Industries," *Proceedings of the 22nd Industrial Waste Conference, Purdue University* (May 1967), pp 908–925.
70. Barker, J.E., V.W. Foltz, and R.J. Thompson. "Treatment of Waste Oil-Waste Water Mixtures," paper presented at the Annual Conference of AIChE, Chicago, IL (November 1970).
71. Stoner, L.B. "Waste Treatment Facilities for Jones and Laughlin Steel Corporation Hennepin Works, *Proceedings of the 26th Industrial Waste Conference, Purdue University* (May 1971).

72. Werner, H.W. "Treatment of Metal Finishing Wastes by Robertshaw Controls Company," paper presented at the 27th Annual Purdue Industrial Waste Conference, West Lafayette, IN (May 1972).
73. Vaughn, S.H. and R.S. McCurday. "Wastewater Treatment at Ford's Windsor Complex," *Ind. Waste* 19 (1973): 34–43.
74. McPhee, W.T. and A.R. Smith. *From Refinery Wastes to Pure Water,* (New York: Hydrotechnic Corp., n.d.).
75. Granczarczyk, J. and M. Kelleher. "Costs of Pretreatment for Coke Plant Effluents," paper presented at the Symposium on Pollution Abatement Technology for 1982, sponsored by USEPA and AISI, Pittsburgh, PA (November 1982).
76. Mikheeva, E.P., V.M. Kagasov, F.S. Yugai, A.A. Mamakov, P.V. Akulov, and Yu A. Zhilyaev. "Electroflotation of Oils from Waste Waters from Coal-Tar Industry, III. Selection of Apparatus Layout," *Elektron. Obrab. Mater.* 5 (1976); 55–59. *Chem. Abstr.* 86 (1977): 160585y.
77. Yarema, D. "Oily Wastewater Treatment by Electrochemical Techniques," *Proceedings of the 4th Conference on Advanced Pollution Control for the Metal Finishing Industry*, USEPA 600/9-82-022 (December 1982).
78. McCarty, P.L. "Treatability of Oily Wastewaters from Food Processing and Soap Manufacture," *Proceedings of the 27th Industrial Waste Conference, Purdue University,* (May 1972), pp 141, 867.
79. Saucier, J.W. "Anaerobic Lagoons Versus Aerated Lagoons in the Treatment of Packinghouse Wastes," *Proceedings of the 24th Industrial Waste Conference, Purdue University* (May 1969), pp 534–541.
80. Rose, W.L. and R.E. Gorringe. "Deep-Tank Extended Aeration of Refinery Wastes," *J. Water Poll. Control. Fed.* 46 (1974): 393–403.
81. Twedell, A., J. Piskura, and T. Zitrides. "Mutant Bacteria for the Control of an Oil Spill in an Advanced Wastewater Treatment Facility," *Proceedings of the 3rd Annual Conference on Treatment and Disposal of Industrial Wastewaters and Residues,* Houston, TX (April 18–20, 1978).

18
Toxic Organics

Historically, regulated pollutants have been traditional parameters, such as are listed in Table 18.1. With the exception of generalized categories of organic materials (e.g., BOD, oil and grease, phenolic compounds) the predominance of the regulated parameters, particularly for industry, were inorganic pollutants. However, in settlement of several cases filed against the U.S. Environmental Protection Agency (EPA) in the early 1970s by environmental groups, the EPA agreed to promulgate effluent standards for a group of 65 specific and general classes of organic and inorganic compounds. This responsibility was specified in a consent decree accepted by the EPA (*NRDC* vs *Train*), and ultimately legislated by the 1977 amendment to the Clean Water Act. EPA subsequently expanded that list of 65 compounds to encompass 114 organic compounds (Table 18.2), plus cyanide, asbestos, and 13 metals. The resulting 129 specific elements and compounds are termed *priority pollutants*.

This chapter provides an overview of industrial sources and treatment technologies for the organic priority pollutants (except for phenolics, Chapter 20). Due to the recent regulation of these pollutants, there are limited data available for many of them.

The occurrence of an organic priority pollutant in an industrial wastewater may be due to one or more of the following causes:

- The pollutant is a raw material used in the manufacturing process.
- The pollutant is a contaminant of the raw material.
- The pollutant is a product of the manufacturing process.
- The pollutant is a reaction by-product of the manufacturing process.
- The pollutant is inadvertently generated during wastewater pretreatment or during treatment prior to effluent discharge.

Control options available, depending upon the cause or source of the pollutant, include raw material prepurification, manufacturing or wastewater treatment process modifications, raw material or product recovery and reuse, or application of specific treatment technology for the target pollutant.

The organic priority pollutants can be grouped by type of compound, as in Table 18.3, which also lists typical industrial sources and wastewater concentrations of these pollutants. Certain pollutants (e.g., acenaphthene) are characteristically present in the effluents of many diverse industries, although often

Table 18.1 Typical Traditional Regulated Pollutants

General	*Industry-Unique*
Suspended solids	Extreme pH
BOD	Cyanide
Coliform bacteria	Heavy metals
Oil and grease	Phenolics
Phosphorus	Temperature
Nitrogen	COD

at low concentrations. Others (e.g., naphthalene) are equally diverse in source, with some sources generating relatively concentrated pollutant levels. Table 18.4 contains additional information on sources and wastewater levels of toxic organics.

TREATMENT TECHNOLOGY

Although it has been suggested that many of the toxic organics are subject to treatment by conventional treatment technologies [8], there is in actuality scant reliable treatment data from full-scale systems. Most of the full-scale data involve performance of systems not specifically designed or operated for removal of the toxic organics, and thus represent incidental removal performance. The principal technologies available are physical processes such as air or steam stripping or adsorption, and biological treatment. Other technologies include chemical oxidation by chlorine or ozone, ion exchange, and incineration for concentrated wastes.

Stripping

Both air and steam stripping have been described for treatment of organic priority pollutants [9–12]. Although reported as effective treatment mechanisms, few full-scale treatment results have been presented.

The extent of removal of a compound by air stripping is governed by many factors, including air and liquid pollutant concentrations, air and liquid temperatures, air-to-liquid ratio, and liquid matrix effects such as the presence of surfactants. Both diffusion and mass transfer rates are important, and are a function of power input, packing medium, and depth (for packed tower strippers), hydraulic regime (e.g., Reynolds number), air bubble or water droplet size (for surface or diffused aeration) and other factors [13, 14]. Busch [15] has emphasized that the mass transfer coefficient is not a constant, but a variable which reflects the design and operating regime of a particular system. Roberts et al [13] have related the mass transfer coefficient for surface aeration stripping

Table 18.2 EPA List of Organic Priority Pollutants

Compound Name
1. Acenaphthene[a]
2. Acrolein[a]
3. Acrylonitrile[a]
4. Benzene[a]
5. Benzidine[a]
6. Carbon Tetrachloride[a] (tetrachloromethane)
Chlorinated benzenes[a] (other than dichlorobenzenes)
7. Chlorobenzene
8. 1,2,4-Trichlorobenzene
9. Hexachlorobenzene
Chlorinated ethanes[a] (including 1,2-dichloroethane 1,1,1-trichloroethane and hexachloroethane)
10. 1,2-Dichloroethane
11. 1,1,1-Trichloroethane
12. Hexachloroethane
13. 1,1-Dichloroethane
14. 1,1,2-Trichloroethane
15. 1,1,2,2-Tetrachloroethane
16. Chloroethane (ethyl chloride)
Chloroalkyl ethers[a] (chloromethyl, chloroethyl and mixed ethers)
17. bis(chloromethyl) ether
18. bis(2-chloroethyl) ether
19. 2 Chloroethyl vinyl ether (mixed)
Chlorinated napthalene[a]
20. 2-Chloronapthalene
Chlorinated phenols[a] (other than those listed elsewhere; includes trichlorophenols and chlorinated cresols)
21. 2,4,6-Trichlorophenol
22. Para-chloro-meta-cresol
23. Chloroform (trichloromethane)[a]
24. 2-Chlorophenol[a]
Dichlorobenzenes[a]
25. 1,2-Dichlorobenzene
26. 1,3-Dichlorobenzene
27. 1,4-Dichlorobenzene
Dichlorobenzidine[a]
28. 3,3′-Dichlorobenzidine
Dichloroethylenes[a] (1,1-Dichloroethylene and 1,2-dichloroethylene)
29. 1,1-Dichloroethylene
30. 1,2-Trans-dichloroethylene
31. 2,4-Dichlorophenol[a]
Dichloropropane and dichloropropene[a]
32. 1,2-Dichloropropane
33. 1,2-Dichloropropylene (1,2-dichloropropene)
34. 2,4-Dimethylphenol[a]
Dinitrotoluene[a]
35. 2,4-Dinitrotoluene
36. 2,6-Dinitrotoluene
37. 1,2-Diphenylhydrazine[a]
38. Ethylbenzene[a]
39. Fluoranthene[a]
Haloethers[a] (other than those listed elsewhere)
40. 4-Chlorophenyl phenyl ether
41. 4-Bromophenyl phenyl ether
42. Bis(2-chloroisopropyl) ether
43. Bis(2-chloroethoxy) methane
Halomethanes[a] (other than those listed elsewhere)
44. Methylene chloride (dichloromethane)
45. Methyl chloride (chloromethane)
46. Methyl bromide (bromomethane)
47. Bromoform (tribromomethane)
48. Dichlorobromomethane
49. Trichlorofluoromethane
50. Dichlorodifluoromethane
51. Chlorodibromomethane
52. Hexachlorobutadiene[a]
53. Hexachlorocyclopentadiene[a]
54. Isophorone[a]
55. Naphthalene[a]
56. Nitrobenzene[a]

Table 18.2 (*continued*)

Compound Name	*Compound Name*
Nitrophenols[a] (including 2,4-dinitrophenol and dinitrocresol)	87. Trichloroethylene[a]
57. 2-Nitrophenol	88. Vinyl chloride[a] (chloroethylene)
58. 4-Nitrophenol	Pesticides and Metabolites
59. 2,4-Dinitrophenol[a]	89. Aldrin[a]
60. 4,6-Dinitro-o-cresol	90. Dieldrin[a]
	91. Chlordane[a] (technical mixture and metabolites)
Nitrosamines[a]	
61. N-Nitrosodimethylamine	
62. N-Nitrosodiphenylamine	DDT and metabolites[a]
63. N-Nitrosodi-n-propylamine	92. 4-4′-DDT
64. Pentachlorophenol[a]	93. 4,4′-DDE (p,p′-DDX)
65. Phenol[a]	94. 4,4′-DDD (p,p′-TDE)
Phthalate esters[a]	Endosulfan and metabolites[a]
66. Bis(2-ethylhexyl) phthalate	95. α-Endosulfan-Alpha
67. Butyl benzyl phthalate	96. β-Endosulfan-Beta
68. Di-n-butyl phthalate	97. Endosulfan sulfate
69. Di-n-octyl phthalate	
70. Diethyl phthalate	Endrin and metabolites[a]
71. Dimethyl phthalate	98. Endrin
	99. Endrin aldehyde
Polynuclear aromatic hydrocarbons (PAH)[a]	
	Heptachlor and metabolites[a]
72. Benzo(a)anthracene (1,2-benzanthracene)	100. Heptachlor
	101. Heptachlor epoxide
73. Benzo(a)pyrene (3,4-benzopyrene)	
74. 3,4-Benzofluoranthene	Hexachlorocyclohexane (all isomers)[a]
75. Benzo(k)fluoranthene (11,12-benzofluoranthene)	102. α-BHC-Alpha
	103. β-BHC-Beta
76. Chrysene	104. γ-BHC (lindane)-Gamma
77. Acenaphthylene	105. δ-BHC-Delta
78. Anthracene	
79. Benzo(ghi)perylene (1,12-Benzoperylene)	Polychlorinated biphenyls (PCB)[a]
	106. PCB-1242 (Arochlor 1242)
80. Fluorene	107. PCB-1254 (Arochlor 1254)
81. Phenanthrene	108. PCB-1221 (Arochlor 1221)
82. Dibenzo(a,h)anthracene (1,2,5,6-dibenzanthracene)	109. PCB-1232 (Arochlor 1232)
	110. PCB-1248 (Arochlor 1248)
83. Indeno (1,2,3-cd) pyrene (2,3,-o-phenylenepyrene)	111. PCB-1260 (Arochlor 1260)
	112. PCB-1016 (Arochlor 1016)
84. Pyrene	113. Toxaphene[a]
85. Tetrachloroethylene[a]	114. 2,3,7,8-Tetrachlorodibenzo-p-dioxin (TCDD)[a]
86. Toluene[a]	

[a] Specific compounds and chemical classes as listed in the consent degree.

Table 18.3 Industrial Sources of Organic Priority Pollutants [after 1]

Compound[a]	*Industrial Source*	*Concentration (μg/l)*	
		Avg	*Range*
Polynuclear Aromatic Hydrocarbons (PAH)			
1. Acenaphthene	Textile mills	8.7	0–12
	Timber products processing	170	0–55,000
	Petroleum refining	280	37–522
	Pharmaceuticals	12	0–100
	Foundries	21	0–200
	Nonferrous metals	7.4	0–100
	Iron and steel	9	0–53
20. 2-Chloronaphthalene	Auto and other laundries	1	0–17
	Steam electric power	10	—
	Iron and steel	22	—
39. Fluoranthene	Coal mining	4	0–16
	Timber products processing	1,600	0–35,000
	Petroleum refining	270	—
	Auto and other laundries	7	0–14
	Coil coating	20	0–130
	Foundries	40	0–390
	Iron and steel	110	0–3,100
	Nonferrous metals	55	0–3,000
55. Naphthalene	Coal mining	57	0–402
	Textile mills	44	0–110
	Petroleum refining	400	0–1,000
	Timber products processing	3,500	0–45,000
	Paint and ink formulation	1,500	0–18,000
	Auto and other laundries	240	0–4,800
	Pharmaceuticals	2.8	0–14
	Foundries	27	0–160
	Leather tanning and finishing	26	0–67
	Nonferrous metals	110	0 5,000
	Iron and steel	2,300	0–29,000
72. Benzo(a)anthracene	Coal mining	9.2	0–35
	Timber products processing	157	0–7,700
	Auto and other laundries	120	—
	Coil coating	33	0–160
	Foundries	3,300	—
	Iron and steel	90	0–2,000
	Nonferrous metals	13	0–180
73. Benzo(a)pyrene	Coal mining	4.6	0–12
	Timber products processing	420	0–2,700
	Foundries	16	0–57
	Nonferrous metals	17	0–570
	Iron and steel	21	0–510
74. Benzo(b)fluoranthene	Coal mining	3.6	0–12
	Timber products processing	87	0–1,680
	Foundries	14	0–36
	Coil coating	18	0–35
	Nonferrous metals	19	0–260
	Iron and steel	28	0–200

Table 18.3 *(continued)*

Compound[a]	Industrial Source	Concentration (μg/l) Avg	Range
75. Benzo(k)fluoranthene	Coal mining	3.6	0–12
	Timber products processing	27	0–3,900
	Coil coating	18	0–35
	Nonferrous metals	20	0–210
	Iron and steel	40	0–360
76. Chrysene	Coal mining	9.2	0–35
	Timber products processing	98	0–4,700
	Coil coating	17	0–160
	Foundries	1,100	0–13,000
	Nonferrous metals	160	0–10,000
	Iron and steel	94	0–2,200
77. Acenaphthylene	Timber products processing	930	0–2,060
	Petroleum refining	46	4–87
	Foundries	15	0–62
	Nonferrous metals	8.2	0–120
	Iron and steel	290	0–6,400
78. Anthracene	Coal mining	23	0–132
	Petroleum refining	170	0–1,100
	Auto and other laundries	23	0–470
	Pharmaceuticals	1.8	0–7
	Coil coating	28	0–1,400
	Foundries	64	0–470
	Leather tanning and finishing	62	0–140
	Nonferrous metals	43	0–3,000
	Iron and steel	91	0–2,800
79. Benzo(ghi)perylene	Coal mining	3.6	0–12
	Timber products processing	6	0–315
	Nonferrous metals	12	0–150
	Iron and steel	16	—
80. Fluorene	Coal mining	8	0–47
	Timber products processing	2,300	0–48,000
	Petroleum refining	270	—
	Pharmaceuticals	3.5	0–14
	Foundries	66	0–800
	Coil coating	17	0–300
	Nonferrous metals	5	0–94
	Iron and steel	85	0–2,500
81. Phenanthrene	Timber products processing	3,800	0–39,000
	Auto and other laundries	28	0–470
	Pharmaceuticals	1.8	0–7
	Coil coating	28	0–1,400
	Foundries	63	0–470
	Leather tanning and finishing	62	0–140
	Nonferrous metals	46	0–3,000
	Iron and steel	99	0–2,800
82. Dibenzo(ah)anthracene	Iron and steel	8	0–16
	Nonferrous metals	8.2	0–110

Table 18.3 *(continued)*

Compound[a]	Industrial Source	Concentration (μg/l) Avg	Range
83. Indeno(1,2,3-cd)pyrene	Timber products processing	130	0–5,500
	Nonferrous metals	18	0–350
	Iron and steel	7	—
84. Pyrene	Coal mining	4	0–19
	Timber products processing	1,000	0–22,000
	Petroleum refining	35	30–40
	Auto and other laundries	6	0–11
	Foundries	44	0–1,100
	Coil coating	13	0–50
Nitrogen Compounds			
3. Acrylonitrile	Textile mills	1,600	—
	Rubber processing	23,000	—
	Iron and steel	6,900	0–21,000
5. Benzidine	Leather tanning and finishing	4	—
	Nonferrous metals	1.2	0–6
28. 3,3′-Dichlorobenzidine	—	—	—
37. 1,2-Diphenylhydrazine	Textile mills	22	—
	Pharmaceuticals	2	0–10
61. N-Nitrosodimethylamine	—	—	—
62. N-Nitrosodiphenylamine	Coal mining	3.8	0–30
	Textile mills	15	0–72
	Rubber processing	35	0–53
	Auto and other laundries	95	0–1,800
	Pharmaceuticals	12	0–30
	Foundries	250	0–1,400
	Leather tanning and finishing	36	—
63. N-Nitrosodi-n-propylamine	Foundries	60	0–210
Aromatic Compounds			
4. Benzene	Coal mining	2.6	0–15
	Textile mills	<5	0–200
	Timber products processing	350	0–2,800
	Petroleum refining	>100	—
	Paint and ink formulation	1,200	0–9,900
	Gum and wood chemicals	180	0–710
	Rubber processing	610	0–3,400
	Auto and other laundries	840	0–23,000
	Pharmaceuticals	220	0–2,100
	Ore mining and dressing	2.1	0–4.2
	Steam electric power	45	—
	Foundries	200	—
	Leather tanning and finishing	19	0–150
	Nonferrous metals	11	0–160
	Iron and steel	2,000	0–43,000

Table 18.3 (*continued*)

Compound[a]	Industrial Source	Concentration (μg/l)	
		Avg	Range
7. Chlorobenzene	Textile mills	25	0–300
	Auto and other laundries	6	0–12
	Pharmaceuticals	67	0–600
	Leather tanning and finishing	1.4	—
	Nonferrous metals	3.9	0–40
	Iron and steel	<10	—
8. 1,2,4-Trichlorobenzene	Textile mills	410	0–2,700
	Steam electric power	10	—
	Foundries	7	—
	Nonferrous metals	22	0–260
9. Hexachlorobenzene	Textile mills	1.3	0–2
	Nonferrous metals	220	0–5,000
25–27. Dichlorobenzenes	Coal mining	<1.2	0–10
	Textile mills	2	0–280
	Auto and other laundries	30	0–1,100
	Steam electric power	20	10–26
	Leather tanning and finishing	69	0–200
35. 2,4-Dinitrotoluene	Pharmaceuticals	12	0–49
	Foundries	15	0–50
	Nonferrous metals	1.7	0–16
	Iron and steel	710	—
36. 2,6-Dinitrotoluene	Textile mills	54	—
	Foundries	17	0–50
	Iron and steel	25	0–47
38. Ethylbenzene	Textile mills	54	0–2,800
	Timber products processing	37	0–2,100
	Petroleum refining	>100	—
	Paint and ink formulation	3,400	0–15,000
	Gum and wood chemicals	4,700	20–28,000
	Rubber processing	500	0–1,500
	Auto and other laundries	550	0–17,000
	Pharmaceuticals	16	0–86
	Ore mining and dressing	4.4	0–8.8
	Foundries	39	—
	Leather tanning and finishing	65	—
	Nonferrous metals	2.5	0–21
	Iron and steel	61	0–39,000
56. Nitrobenzene	Paint and ink formulation	100	0–180
	Foundries	94	0–280
	Leather tanning and finishing	73	—
	Nonferrous metals	11	0–160
	Iron and steel	1,400	—
86. Toluene	Coal mining	1.1	0–8
	Textile mills	26	0–620
	Timber products processing	180	0–3,200
	Petroleum refining	>100	—

Table 18.3 *(continued)*

Compound[a]	Industrial Source	Concentration (μg/l) Avg	Range
	Paint and ink formulation	10,000	0–260,000
	Gum and wood chemicals	2,000	20–17,000
	Rubber processing	77	0–350
	Pulp, paper, and paperboard mills	<4	0–44
	Auto and other laundries	1,300	0–51,000
	Pharmaceuticals	2,400	0–17,000
	Foundries	120	—
	Leather tanning and finishing	67	0–400
	Nonferrous metals	6.9	0–55
	Iron and steel	590	0–17,000
Halogenated Hydrocarbons			
6. Carbon tetrachloride	Paint and ink formulation	1,900	0–30,000
	Rubber processing	180	0–350
	Auto and other laundries	95	0–1,700
	Pharmaceuticals	460	0–6,000
	Inorganic chemicals manufacture	—	0–200
	Foundries	45	0–480
	Iron and steel	40	—
	Nonferrous metals	81	0–2,300
10. 1,2-Dichloroethane	Paint and ink formulation	100	0–420
	Rubber processing	93	—
	Auto and other laundries	250	0–500
	Pharmaceuticals	8.7	0–74
	Steam electric power	44	—
	Inorganic chemicals manufacture	—	0–620
	Foundries	83	0–170
	Nonferrous metals	20	0–180
11. 1,1,1-Trichloroethane	Coal mining	2.5	0–10
	Textile mills	7.8	0–17
	Paint and ink formulation	350	0–1,000
	Timber products processing	30	0–90
	Gum and wood chemicals	280	0–830
	Pulp, paper, and paperboard mills	106	1–414
	Auto and other laundries	520	0–6,600
	Pharmaceuticals	10	0–130
	Steam electric power	15	13–26
	Coil coating	12	0–1,300
	Foundries	1,400	0–16,000
	Leather tanning and finishing	1.4	—
	Nonferrous metals	3.6	0–40
	Iron and steel	44	0–420
12. Hexachloroethane	Inorganic chemicals manufacture	—	0–90
	Nonferrous metals	1.5	0–23
13. 1,1-Dichloroethane	Textile mills	14	—
	Paint and ink formulation	10	0–33
	Leather tanning and finishing	<3	—
	Foundries	55	—
	Coil coating	18	—
	Nonferrous metals	20	0–180
	Iron and steel	8	—

Table 18.3 (*continued*)

Compound[a]	*Industrial Source*	*Concentration (μg/l)* Avg	Range
14. 1,1,2-Trichloroethane	Paint and ink formulation	290	0–2,800
	Auto and other laundries	1,500	0–3,000
	Pharmaceuticals	95	0–1,300
	Leather tanning and finishing	1.4	—
	Foundries	10	—
	Nonferrous metals	1.2	0–29
15. 1,1,2,2-Tetrachloroethane	Rubber processing	1.5	—
	Pharmaceuticals	2	0–10
	Foundries	18	—
	Leather tanning and finishing	5.4	—
	Nonferrous metals	2.8	0–35
16. Chloroethane	Gum and wood chemicals	170	0–520
	Rubber processing	4,900	—
23. Chloroform	Textile mills	48	0–640
	Timber products processing	—	0–20
	Paint and ink formulation	120	0–900
	Gum and wood chemicals	<680	<10–1,400
	Rubber processing	27	0–270
	Petroleum refining	13	0–15
	Pulp, paper, and paperboard mills	<6	1–17
	Auto and other laundries	780	0–35,000
	Pharmaceuticals	300	0–1,600
	Ore mining and dressing	2.9	1.1–4.6
	Steam electric power	14	—
	Inorganic chemicals manufacture	—	0–690
	Leather tanning and finishing	15	0–41
	Nonferrous metals	61	0–1,800
	Iron and steel	64	0–1,400
29. 1,1-Dichloroethylene	Coal mining	<1.4	0–10
	Textile mills	<5	—
	Paint and ink formulation	76	0–620
	Auto and other laundries	12	0–23
	Pharmaceuticals	8.9	0–95
	Steam electric power	16	—
	Coil coating	15	0–240
	Nonferrous metals	200	0–6,100
	Iron and steel	<10	—
30. 1,2-trans-Dichloroethylene	Paint and ink formulation	72	0–260
	Auto and other laundries	230	0–460
	Steam electric power	11	—
	Coil coating	<10	0–82
	Foundries	<6	—
	Leather tanning and finishing	6	—
	Nonferrous metals	17	0–480
	Iron and steel	8	0–10
32. 1,2-Dichloropropane	Textile mills	100	—
	Paint and ink formulation	180	0–970

Table 18.3 *(continued)*

Compound[a]	*Industrial Source*	*Concentration (μg/l)*	
		Avg	*Range*
33. 1,3-Dichloropropene	Iron and steel	<10	—
44. Methylene chloride	Coal mining	14	0–82
	Textile mills	47	0–110
	Timber products processing	—	0–20
	Petroleum refining	>45	30–90
	Paint and ink formulation	17,000	0–210,000
	Gum and wood chemicals	4,000	190–16,000
	Rubber processing	19	0–70
	Pulp, paper, and paperboard mills	<1.9	<1–4
	Auto and other laundries	64	0–640
	Pharmaceuticals	2,600	0–20,000
	Ore mining and dressing	5.3	2.5–8
	Steam electric power	2,300	65–9,400
	Foundries	100	0–2,400
	Nonferrous metals	680	0–88,000
	Iron and steel	50	0–140
45. Methyl chloride	Textile mills	<5	—
	Timber products processing	77	0–2,600
	Pharmaceuticals	300	0–1,500
46. Methyl bromide	Pharmaceuticals	3	0–15
47. Bromoform	Pulp, paper, and paperboard mills	14	—
	Steam electric power	210	13–580
	Nonferrous metals	6.8	0–65
48. Dichlorobromomethane	Textile mills	6.6	—
	Petroleum refining	110	—
	Paint and ink formulation	27	—
	Auto and other laundries	10	0–33
	Inorganic chemicals manufacture	—	0–310
	Leather tanning and finishing	<3	—
49. Trichlorofluoromethane	Textile mills	<5	0–2,100
	Auto and other laundries	120	—
	Nonferrous metals	12	0–100
50. Dichlorodifluoromethane	—	—	—
51. Chlorodibromomethane	Paint and ink formulation	43	—
	Petroleum refining	110	—
	Auto and other laundries	6	0–12
	Steam electric power	59	—
	Nonferrous metals	5.4	0–81
52. Hexachlorobutadiene	—	—	—
53. Hexachlorocyclopentadiene	—	—	—

Table 18.3 (*continued*)

Compound[a]	*Industrial Source*	*Concentration (μg/l)*	
		Avg	*Range*
85. Tetrachloroethylene	Textile mills	<5	0–2,100
	Petroleum refining	<50	—
	Paint and ink formulation	920	0–4,900
	Auto and other laundries	9,600	0–93,000
	Pharmaceuticals	3.5	0–36
	Steam electric power	78	—
	Inorganic chemicals manufacture	—	0–196
	Foundries	70	0–370
	Leather tanning and finishing	12	—
	Nonferrous metals	14	0–310
	Iron and steel	42	0–1,100
87. Trichloroethylene	Coal mining	<1.4	0–10
	Textile mills	47	0–840
	Paint and ink formulation	960	0–5,000
	Auto and other laundries	90	0–800
	Pharmaceuticals	8	0–62
	Foundries	109	0–280
	Leather tanning and finishing	2	—
	Coil coating	<10	0–310
	Nonferrous metals	59	0–900
	Iron and steel	13	—
88. Vinyl chloride	Textile mills	11	—
Ethers			
17. Bis(chloromethyl)ether	—	—	—
18. Bis(2-chloroethyl)ether	Pharmaceuticals	19	0–170
	Foundries	4.5	—
19. 2-Chloroethyl vinyl ether	—	—	—
40. 4-Chlorophenyl ether	Coal mining	0.4	0–33
	Iron and steel	210	—
41. 4-Bromophenyl phenyl ether	—	—	—
42. Bis(2-chloroisopropyl) ether	Foundries	3.0	—
43. Bis(2-chloroethoxy) ether	Coal mining	11	0–50
	Foundries	20	—
Phenolic Compounds			
21. 2,4,6-Trichlorophenol	Textile mills	20	0–29
	Timber products processing	250	0–530
	Steam electric power	30	—
	Leather tanning and finishing	3,000	0–25,000
	Foundries	67	0–350
	Iron and steel	39	—
22. p-Chloro-m-cresol	Textile mills	168	—
	Foundries	42	0–280
	Iron and steel	200	0–4,300

Table 18.3 *(continued)*

Compound[a]	Industrial Source	Concentration (μg/l) Avg	Range
24. 2-Chlorphenol	Coal mining	12	0–86
	Textile mills	78	—
	Timber products processing	15	0–42
	Pharmaceuticals	2.4	0–22
	Steam electric power	27	—
	Foundries	53	0–210
	Iron and steel	1,400	0–36,000
31. 2,4-Dichlorophenol	Textile mills	26	0–41
	Pulp, paper, and paperboard mills	2	—
	Auto and other laundries	2	0–10
	Pharmaceuticals	1	0–5
	Steam electric power	240	—
	Leather tanning and finishing	16	—
	Iron and steel	35	—
34. 2,4-Dimethylphenol	Coal mining	4.8	0–22
	Timber products processing	1,300	0–6,600
	Petroleum refining	100	—
	Auto and other laundries	20	0–460
	Ore mining and dressing	270	—
	Coil coating	21	—
	Foundries	81	0–1,100
	Nonferrous metals	7	0–14
	Iron and steel	4,000	0–84,000
57. 2-Nitrophenol	Coal mining	2.7	0–19
	Rubber processing	9.4	—
58. 4-Nitrophenol	Auto and other laundries	13	0–15
	Pharmaceuticals	400	0–3,500
	Foundries	400	—
59. 2,4-Dinitrophenol	Auto and other laundries	19	—
	Iron and steel	33	0–440
60. 4,6-Dinitro-o-cresol	Coal mining	28	0–194
	Foundries	16	0–70
	Iron and steel	130	0–970
64. Pentachlorophenol	Textile mills	52	0–940
	Timber products processing	13,000	0–310,000
	Paint and ink formulation	3,300	0–27,000
	Rubber processing	17	0–31
	Pulp, paper, and paperboard mills	<4	<1–9
	Auto and other laundries	3	0–9
	Pharmaceuticals	4.4	0–62
	Foundries	120	0–1,600
	Leather tanning and finishing	1,700	—
	Nonferrous metals	1.5	0–17
	Iron and steel	76	—
65. Phenol	(*See* Chapter 20)	—	—

Table 18.3 (*continued*)

Compound[a]	*Industrial Source*	*Concentration (μg/l)*	
		Avg	*Range*
Phthalates			
66. Bis(2-ethylhexyl) phthalate	Coal mining	11	0–50
	Textile mills	26	0–860
	Explosives manufacture	72	—
	Timber products processing	280	0–1,500
	Petroleum refining	300	180–700
	Paint and ink formulation	7,300	0–87,000
	Rubber processing	170	0–530
	Pulp, paper, and paperboard mills	<16	<1–66
	Auto and other laundries	1,000	0–18,000
	Pharmaceuticals	37	0–170
	Ore mining and dressing	12	0–50
	Steam electric power	27	<10–36
	Inorganic chemicals manufacture	—	0–270
	Coil coating	75	0–1,100
	Foundries	21,000	0–817,000
	Leather tanning and finishing	25	—
	Nonferrous metals	237	0–7,000
	Iron and steel	451	0–10,000
67. Butyl benzyl phthalate	Textile mills	42	0–73
	Pulp, paper, and paperboard mills	51	2–100
	Auto and other laundries	34	0–1,500
	Pharmaceuticals	33	0–360
	Ore mining and dressing	22	0–66
	Steam electric power	10	—
	Coil coating	57	0–300
	Foundries	29	0–180
	Nonferrous metals	11	0–98
	Iron and steel	34	0–340
68. Di-n-butyl phthalate	Coal mining	<9	3.3–10
	Textile mills	16	0–67
	Petroleum refining	1.3	—
	Paint and ink formulation	4,100	0–69,000
	Auto and other laundries	29	0–820
	Pharmaceuticals	10	0–90
	Ore mining and dressing	45	0–140
	Steam electric power	18	<10–48
	Coil coating	10	0–160
	Foundries	160	0–5,400
	Nonferrous metals	25	0–390
	Iron and steel	33	0–520
69. Di-n-octyl phthalate	Auto and other laundries	24	0–410
	Coil coating	16	0–180
	Foundries	710	—
	Nonferrous metals	8.4	0–95
	Iron and steel	13	0–120
70. Diethyl phthalate	Coal mining	<3.7	0–10
	Textile mills	6	0–86
	Petroleum refining	12	—

Table 18.3 *(continued)*

Compound[a]	Industrial Source	Concentration (μg/l) Avg	Range
	Pulp, paper, and paperboard mills	<15	<1–29
	Auto and other laundries	4	0–25
	Pharmaceuticals	8	0–31
	Ore mining and dressing	2.9	0–9.6
	Steam electric power	11	10–11
	Coil coating	53	0–600
	Foundries	58	0–730
	Nonferrous metals	4.7	0–83
	Iron and steel	7	0–10
71. Dimethyl phthalate	Textile mills	12	0–14
	Rubber processing	5	0–14
	Auto and other laundries	1	0–5
	Coil coating	10	0–14
	Foundries	72	0–2,200
	Leather tanning and finishing	17	—
	Nonferrous metals	4.7	0–56
	Iron and steel	16	0–20
Pesticides			
89. Aldrin	Ore mining and dressing	<10	—
	Foundries	<2.5	—
	Nonferrous metals	0.6	0–7
	Iron and steel	<10	—
90. Dieldrin	Foundries	4	—
	Iron and steel	<10	—
91. Chlordane	Foundries	12	—
	Iron and steel	<10	—
92. 4,4′-DDT	Foundries	6	—
	Iron and steel	<10	—
93. 4,4′-DDE	Foundries	10	—
	Iron and steel	<10	—
94. 4,4′-DDD	Foundries	9	—
	Iron and steel	<10	—
95. α-Endosulfan	Foundries	2.5	—
	Nonferrous metals	11	0–15
96. β-Endosulfan	Coal mining	<1.1	0–<6.7
	Foundries	<1.7	—
	Nonferrous metals	1.6	0–15
	Iron and steel	<10	—
97. Endosulfan sulfate	Foundries	<1.3	—
	Iron and steel	<10	—
98. Endrin	—	—	—

Table 18.3 (*continued*)

Compound[a]	*Industrial Source*	*Concentration (μg/l)* Avg	*Concentration (μg/l)* Range
99. Endrin aldehyde	Foundries	5.2	0–20
	Iron and steel	<10	—
100. Heptachlor	Foundries	<7.5	—
	Iron and steel	<10	—
101. Heptachlor epoxide	Iron and steel	<10	—
102. α-BHC	Foundries	7.2	—
	Iron and steel	<10	—
103. β-BHC	Foundries	25	—
	Iron and steel	<10	—
104. γ-BHC	Coal mining	<1.7	0–<10
	Foundries	7	—
	Iron and steel	<10	—
105. δ-BHC	Foundries	9.3	—
	Iron and steel	<10	—
113. Toxaphene	Iron and steel	<10	—
Polychlorinated Biphenyls (PCB)			
106–112. Aroclor compounds	Petroleum refining	—	0.1–5
	Pulp, paper, and paperboard mills	—	1–18
	Foundries	—	0–1,400
	Nonferrous metals	—	0–52
	Iron and steel	<10	—
Miscellaneous Compounds			
2. Acrolein	—	—	—
114. TCDD	—	—	—

[a] Numbers refer to compound numbers in Table 18.2.

of volatile chlorinated hydrocarbons to the oxygen mass transfer coefficient. Commercial aeration equipment transfer rates range from about 2.5 to 6 lb oxygen/hp-hr [15].

One principal factor in air stripping is the tendency of the compound to establish an equilibrium between the gas and the liquid phases. This is the principle of Henry's law, which states that the partial pressure of the compound (a measure of the concentration in the gas phase) at equilibrium is equal to a constant (Henry's law constant) times the concentration of the compound in the liquid phase. Henry's law constant can be considered a partition coefficient which describes the relative tendency for the compound to partition between the gas and liquid at equilibrium, in accordance with the following equation:

$$H = \frac{C_g}{C_l} \tag{18.1}$$

Table 18.4 Other Industrial Sources of Organic Priority Pollutants

Compound[a]	*Industrial Source*	*Concentration (μg/l)*	*Reference*
4. Benzene	Kraft bleach plant	4	2
	Petroleum refinery	9,058	3
6. Carbon tetrachloride	Pesticide manufacture	24,500	4
11. 1,1,1-Trichloroethane	Research center	46	5
12. Hexachloroethane	Pesticide manufacture	350	4
20. 2-Chloronaphthalene	Pesticide manufacture	575	4
21. 2,4,6-Trichlorophenol	Kraft bleach plant	160	2
22. p-Chloro-m-cresol	Petroleum refinery	157	3
23. Chloroform	Pesticide manufacture	3,880	4
	Kraft bleach plant	5,500	2
	Research center	>19	5
38. Ethylbenzene	Petroleum refinery	18,430	3
52. Hexachlorobutadiene	Pesticide manufacture	1,320	4
53. Hexachlorocyclopentadiene	Pesticide manufacture	6,810	4
55. Naphthalene	Pesticide manufacture	585	4
	Petroleum refinery	12,328	3
73. Benzo(a)pyrene	Shale oil	2–312	6
	Coke by-products	12–16	6
	Petroleum refinery	0.1–5	6
77. Acenaphthalene	Petroleum refinery	101	3
80. Fluorene	Petroleum refinery	1,447	3
81. Phenanthrene	Petroleum refinery	2,013	3
85. Tetrachloroethylene	Pesticide manufacture	77	4
	Research center	34	5
86. Toluene	Pesticide manufacture	3,060	4
	Research center	32	5
	Petroleum refinery	15,220	3
	Munitions manufacture	4,000	7
87. Trichloroethylene	Research center	450	5
	Kraft bleach plant	3	2
89. Aldrin	Pesticide manufacture	410	4
90. Dieldrin	Pesticide manufacture	700	4
91. Chlordane	Pesticide manufacture	2,610	4
98. Endrin	Pesticide manufacture	2,210	4
100. Heptachlor	Pesticide manufacture	530	4
101. Heptachlor epoxide	Pesticide manufacture	68	4

[a] Numbers refer to numbers in Table 18.2.

where H = dimensionless Henry's law constant
C_g = concentration in gas phase (e.g., g/m^3)
C_l = concentration in liquid phase (e.g., g/m^3)

In the literature, Henry's Constants are often present for Equation 18.2:

$$H' = \frac{P_g}{C} \tag{18.2}$$

where H' = Henry's constant, atm/mol/m^3
P_g = gas partial pressure, atm
C = liquid concentration, mol/m^3

The relationship between the two forms of Henry's constants is [13]:

$$H' = H(RT) \tag{18.3}$$

where R = gas constant = 8.2×10^{-5} atm – m^3/mol – °K
T = temperature, °K

Henry's law constant is a property of the compound which can be calculated by dividing the compound's vapor pressure (P) by its solubility (S). H can also be determined experimentally. Values for H and solubility, for most of the 114 organic priority pollutants, are shown in Table 18.5. Recently published data on several organic compounds have shown fairly close agreement between calculated and experimentally determined values [16]. Temperature obviously affects the value of H, since both solubility and especially vapor pressure are temperature dependent. The presence of electrolytes, surface ac-

Table 18.5 Henry's Law Constant and Compound Solubility for Selected Priority Pollutants [9, 16]

Compound[a]	*Henry's Law Constant (dimensionless)*	*Solubility (mg/l)*
1. Acenaphthene	0.010*(c)	3.88
2. Acrolein	0.0046(b)	200,000
3. Acrylonitrile	0.0030(b)	93,000
4. Benzene	0.23*(c)	1780
6. Carbon tetrachloride (tetrachloromethane)	1.26*(c)	800
7. Chlorobenzene	0.16*(c)	448
8. 1,2,4-Trichlorobenzene	0.059*(c)	30
9. Hexachlorobenzene	0.071*(c)	0.006
10. 1,2-Dichloroethane	0.046*(c)	8700

Table 18.5 *(continued)*

Compound[a]	*Henry's Law Constant (dimensionless)*	*Solubility (mg/l)*
11. 1,1,1-Trichloroethane	0.21*(c)	4400
12. Hexachloroethane	0.41*(c)	8
13. 1,1-Dichloroethane	0.23*(c)	5100
14. 1,1,2-Trichloroethane	0.037(c)	4420
15. 1,1,2,2-Tetrachloroethane	0.020(c)	3000
16. Chloroethane	0.73(c)	5700
23. Chloroform (trichloromethane)	0.14*(c)	7840
24. 2-Chlorophenol	0.001(b)	28,000
25. 1,2-Dichlorobenzene	0.081*(c)	100
26. 1,3-Dichlorobenzene	0.11*(c)	123
27. 1,4-Dichlorobenzene	0.11*(c)	79
29. 1,1-Dichloroethylene	0.63*(b)	5000
30. 1,2-Trans-dichloroethylene	0.22*(b)	6300
31. 2,4-Dichlorophenol	0.002	4500
32. 1,2-Dichloropropane	0.123*(b)	2700
33. 1,2-Dichloropropylene (1,3-dichloropropene)	0.15*(c)	2700
38. Ethylbenzene	0.27*(c)	152
42. Bis(2-chloroisopropyl) ether	0.005	1700
44. Methylene chloride (dichloromethane)	0.13*(c)	16,700
45. Methyl chloride (chloromethane)	0.38(c)	5380
46. Methyl bromide (bromomethane)	4.39(b)	900
47. Bromoform (tribromomethane)	0.022*(c)	3190
48. Dichlorobromomethane	0.088*(c)	—
49. Trichlorofluoromethane	2.43*(c)	1100
50. Dichlorodifluoromethane	98.8(c)	280
51. Chlorodibromomethane	0.033*(c)	—
52. Hexachlorobutadiene	0.43*(b)	2
53. Hexachlorocyclopentadiene	0.68*(c)	0.810
54. Isophorone	0.0002(b)	12,000
55. Naphthalene	0.017(c)	30
56. Nitrobenzene	0.0001*(c)	1900
57. 2-Nitrophenol	0.0036(b)	2100
60. 4,6-Dinitro-o-cresol	6×10^{-5}*(c)	130
64. Pentachlorophenol	0.0001(b)	14

Table 18.5 (*continued*)

Compound[a]		*Henry's Law Constant (dimensionless)*	*Solubility (mg/l)*
65.	Phenol	6×10^{-5}*(c)	82,000
68.	Di-n-butyl phthalate	0.0030(c)	500
71.	Dimethyl phthalate	0.00002(b)	5000
77.	Acenaphthylene	5×10^{-3}*(c)	3.9
78.	Anthracene	0.067(c)	0.075
80.	Fluorene	5×10^{-3}*(c)	1.90
81.	Phenanthrene	0.006(c)	1.18
85.	Tetrachloroethylene	1.2*(c)	150
86.	Toluene	0.25*(c)	515
87.	Trichloroethylene	0.49*(c)	1000
88.	Vinyl chloride (chloroethylene)	301.(c)	60
89.	Aldrin	0.021*(c)	0.027
90.	Dieldrin	2.4×10^{-3}*(c)	0.19
91.	Chlordane	2.0×10^{-3}*(c)	1.85
92.	4,4′-DDT	0.0016(c)	3.1×10^{-3}
100.	Heptachlor	0.062*(c)	0.056
101.	Heptachlor epoxide	1.3×10^{-3}*(c)	0.275
102.	α-BHC-Alpha	0.094(c)	10
103.	β-BHC-Beta	0.53(c)	5
104.	γ-BHC (lindane)-Gamma	1.5×10^{-5}(c)	10
106.	PCB-1242 (Arochlor 1242)	0.023(c)	0.24
107.	PCB-1254 (Arochlor 1254)	0.35*(c)	0.212
110.	PCB-1248 (Arochlor 1248)	0.14(c)	5.4×10^{-2}
111.	PCB-1260 (Arochlor 1260)	0.29(c)	2.7×10^{-3}
113.	Toxaphene	0.20*(c)	3.0

[a] Numbers refer to compound numbers in Table 18.2.

(a) = 288°K (15°C) (c) = 298°K (25°C)

(b) = 293°K (20°C) (d) = 303°K (30°C)

* Experimental determined values [16]. All other Henry's constants are calculated [9].

tive agents, lipids, and particulate matter in the wastewater may all affect the equilibrium to some extent. A volatile compound will typically have a value of Henry's constant greater than 0.1 [17].

The Henry's law constant H can be determined by fairly simple laboratory equilibration tests, providing that analytical methods are available for the compound of interest in wastewater [16]. Thus, the amenability of a compound to removal by stripping from any wastewater can be evaluated.

The removal of an air-strippable compound depends not only on its equilibrium partition ratio but also on the contact opportunity, or intensity and duration of aeration. This is demonstrated by the pilot results presented in Table 18.6. In general, the efficiency of stripping corresponds to the Henry's constant value. However, increased air : liquid ratio or increased aeration detention time (e.g., 47, bromoform) can only increase stripping efficiency up to the equilibrium status defined by Henry's law.

In a diffused air system, the maximum possible removal of a compound by stripping can be estimated, assuming that the air leaving the process is saturated with the compound in accordance with Henry's law, with the process at steady state. The resulting expression is:

$$\frac{C_e}{C_i} = \frac{1}{1 + (Qa/Qi)\,H} \tag{18.4}$$

where C_e = the effluent concentration
C_i = the influent concentration
Qa/Qi = the air-to-water ratio

In agitated aqueous systems, even without air stripping, the rate of loss of some of the organic priority pollutants may be quite rapid due to pollutant volatilization. Bench-scale studies on the volatilization (evaporation) rate of several of the more important chlorinated solvents demonstrate this phenomenon [11]. In stirred reactors, at initial organic compound levels of 1 mg/l, 50% of the concentrations were stripped by volatilization within less than 30 minutes and 90% within 85 or fewer minutes (Table 18.7). The observed time values, for these stirred reactors, were between 20- and 60-fold longer than those predicted for a diffused air system based upon Henry's law calculations.

The principal index used to estimate steam-stripping capability is the boiling point of the organic compound. A compound predicted to have good steam-stripping characteristics would exhibit a relatively low boiling point (<150°C) plus an acceptable Henry's law constant value. In one study on steam stripping of carbon tetrachloride, initial compound levels of 100–200 mg/l were reduced to 0.16 mg/l [12]. Other results are presented in Table 18.8. Steam stripping is an established chemical-engineering technique, with many proved applications in industry. Steam stripping of organic priority pollutants produces a contaminated condensate, which may be suitable for recovery or which may require further treatment.

Table 18.6 Pilot Results for Air Stripping of Toxic Organics

Compound[a]	*Type System*	*Influent Concentration (mg/l)*	*Air:Liquid Ratio (Qa/Qi)*	*Detention Time (min)*	*Effluent Concentration (mg/l)*	*Percent Removal*	*Reference*
4. Benzene	Packed tower	15.1	25	2.6	4.1	73	18
		10.5	50	2.4	2.9	72	
		105	50	2.4	24.2	77	
		11.8	100	4.0	2.9	75	
		92.5	100	4.0	9.3	90	
6. Carbon tetrachloride	Packed tower	5.4	25	2.6	1.7	69	18
		84.2	50	2.4	15.5	82	
		15.3	100	4.0	2.0	87	
		67.4	100	4.0	7.6	89	
7. Chlorobenzene	Packed tower	13.3	25	2.6	4.6	65	18
		9.2	50	2.4	2.6	72	
		22.8	50	2.4	6.4	72	
		7.8	100	4.0	1.8	77	
		29.6	100	4.0	3.3	89	
		3.6	200	6.9	0.1	97	
11. 1,1,1-Trichloroethane	Packed tower	0.090	27	—	0.031	66	19
		0.040	51	—	0.025	38	
		0.090	101	—	0.031	66	
		0.225	146	—	0.041	82	
		0.273	154	—	0.057	79	
		1.332	156	—	0.143	89	

	Diffused air	0.750	62	26	0.020	97	
		0.750	62	46	0.018	98	
		0.580	125	46	0.008	99	
		0.680	22	60	0.110	84	
		0.180	23	60	0.006	97	
		0.220	50	60	0.007	97	
23. Chloroform	Packed tower	0.258	15	6.6	0.005	98	14
		0.248	22	6.6	0.004	98	
		0.258	30	6.6	0.004	98	
		0.660	30	6.6	0.053	92	
		1.612	30	6.6	0.169	90	
		2.371	30	6.6	0.224	90	
	Packed tower	0.12	38	1.6	0.026	78	20
		0.12	100	3.2	0.019	84	
		0.11	100	6.4	0.010	91	
		0.12	100	12.8	0.004	97	
	Packed tower	15.1	25	2.6	4.1	73	18
		9.6	50	2.4	2.4	75	
		248	50	2.4	54.6	78	
		25.4	100	4.0	4.4	83	
		225	100	4.0	15.6	93	
	Packed tower	0.050	27	—	<0.002	96+	19
	Diffused air	0.009	62	26	0.000	100	19
		6.00	88	60	0.60	90	21

Table 18.6 (*continued*)

Compound[a]	*Type System*	*Influent Concentration (mg/l)*	*Air:Liquid Ratio (Qa/Qi)*	*Detention Time (min)*	*Effluent Concentration (mg/l)*	*Percent Removal*	*Reference*
25. 1,2-Dichlorobenzene	Packed tower	13.3	25	2.6	4.6	65	
		9.2	50	2.4	2.6	72	
		22.8	50	2.4	6.4	72	
		29.6	100	4.0	3.3	89	
		10.2	200	6.9	0.7	93	
26. 1,3-Dichlorobenzene	Packed tower	7.4	25	2.6	2.7	64	18
		10.3	50	2.4	3.1	70	
		26.6	50	2.4	7.5	70	
		35.0	100	4.0	3.6	90	
		5.7	200	6.9	0.3	95	
27. 1,4-Dichlorobenzene	Packed tower	14.2	25	2.6	5.7	60	18
		5.3	50	2.4	2.9	45	
		27.8	50	2.4	8.3	70	
		24.0	100	4.0	6.2	74	
		11.0	200	6.9	0.7	97	
30. 1,2-Trans-dichloroethylene	Packed tower	5.4	25	2.6	1.7	69	18
		3.2	50	2.4	0.9	72	
		114	50	2.4	27.9	76	
		7.3	100	4.0	1.5	80	
		105	100	4.0	16.6	84	
32. 1,2-Dichloropropane	Packed tower	0.050	27	—	<0.005	90+	19
	Diffused air	0.270	62	26	<0.005	98+	19

38.	Ethyl benzene	Diffused air	0.008	62	26	0.000	100	19
44.	Methylene chloride	Packed tower	0.490	27	30	0.018	96	19
			1.770	154	30	<0.005	99+	
		Diffused air	1.800	62	26	0.120	93	19
					46	0.077	96	
			1.390	125	46	0.003	99+	
47.	Bromoform	Packed tower	0.12	32	1.6	0.07	45	20
			0.287	50	2.4	0.159	45	18
			0.261	50	4.0	0.105	60	
			0.275	100	4.0	0.079	71	
			0.224	100	6.8	0.045	80	
			0.254	200	6.9	0.044	83	
			0.253	200	11.8	0.020	92	
48.	Dichlorobromo-methane	Packed tower	0.12	32	1.6	0.03	74	20
			0.144	50	2.4	0.056	61	18
			0.204	50	4.0	0.038	82	
			0.204	100	4.0	0.037	82	
			0.170	100	6.8	0.018	90	
			0.116	200	6.9	0.012	90	
			0.225	200	11.8	0.005	98	
51.	Chlorodibromo-methane	Packed tower	0.090	32	1.6	0.04	59	20
			0.208	50	2.4	0.092	56	18
			0.237	50	4.0	0.062	74	
			0.229	100	4.0	0.050	78	
			0.197	100	6.8	0.025	87	
			0.201	200	6.9	0.021	90	
			0.230	200	11.8	0.008	97	

Table 18.6 *(continued)*

Compound[a]	*Type System*	*Influent Concentration (mg/l)*	*Air:Liquid Ratio (Qa/Qi)*	*Detention Time (min)*	*Effluent Concentration (mg/l)*	*Percent Removal*	*Reference*
55. Napthalene	Packed tower	4.7	25	2.6	3.0	36	18
		8.9	50	2.4	2.9	67	
		14.9	50	2.4	7.8	48	
		24.0	100	4.0	6.2	74	
		3.9	200	6.9	0.4	91	
56. Nitrobenzene	Packed tower	4.2	25	2.6	4.0	5	18
		111	50	2.4	99	11	
		115	100	4.0	97	16	
		3.9	200	6.9	2.8	28	
86. Toluene	Packed tower	3.4	25	2.6	0.9	74	18
		5.3	50	2.4	1.2	77	
		39.7	50	7.4	9.1	77	
		37.1	100	4.0	2.8	93	
		2.7	200	6.9	0.1	96	
87. Trichloroethylene	Packed tower	0.080	51	—	0.014	83	19
		0.273	154	—	0.057	79	
		1.332	156	—	0.143	89	
	Diffused air	0.400	62	26	0.010	98	19
		0.400	62	46	0.010	98	
		1.080	1.1–7.5	30	0.250	77	22

[a] Numbers refer to compound numbers in Table 18.2.

Table 18.7 Evaporation Rates of Chlorinated Compounds from Dilute Aqueous Solutions [11]

	Evaporation Time (min)	
Compound[a]	*50%*	*90%*
11. 1,1,1-Trichloroethane	20	69
23. Chloroform	21	71
44. Methylene chloride	21	69
85. Tetrachloroethylene	26	83
87. Trichloroethylene	21	69

[a] Numbers refer to compound numbers in Table 18.2.

Adsorption

Adsorption is a well established technology for the control of organic pollutants. Activated carbon has had the widest application. Porous polymeric resins also have potential application, although their principal use to date has been for systems designed for organics recovery for reuse [23]. Most activated carbon systems involve continuous flow columns containing granular carbon, with thermal regeneration upon carbon exhaustion. Batch slurry powered carbon systems are less widely used.

The use of activated carbon has been shown to be a feasible unit process for removal of some toxic organic compounds from water and wastewater. An additional benefit of the use of activated carbon is the safe ultimate disposal of the toxic compound when the exhausted carbon is thermally regenerated for

Table 18.8 Efficiency of Steam Stripping of Toxic Organic Pollutants [after 1]

Compound[a]	*Effluent Concentration (μg/l)*	*% Removal*
10. 1,2-Dichloroethane	33.0	97–99+
11. 1,1,1-Trichloroethane	42.0	9
14. 1,1,2-Trichloroethane	<0.05	99+
15. 1,1,2,2-Tetrachloroethane	32.0	40–99+
23. Chloroform	13.0	89–99+
28. 1,2-Dichloroethylene	340.0	76–99+
44. Methylene chloride	160.0	75–87
85. Tetrachloroethylene	2.3	78–99+
87. Trichloroethylene	16.0	61–99+

[a] Numbers refer to compound in Table 18.2.

recycle, for those compounds exhibiting thermal decomposition within the temperature range of carbon regeneration.

Activated carbon is a highly porous material having a surface area of approximately 1,000 square meters per gram. Adsorption is a phenomenon in which the molecules are adsorbed through association with the surface of the carbon. A number of factors influence the adsorption process. These include [9, 24]:

Adsorbent characteristics
- Carbon surface area
- Carbon pore structure and size distribution
- Carbon particle size
- Surface polarity

Adsorbate characteristics
- Compound solubility
- Molecule size
- Compound ionization
- Compound polarity

Aqueous system characteristics
- Temperature
- System pH
- Competing solutes
- Dissolved solids

The quantitative effect of the factors listed above can be expressed by the Freundlich adsorption equation:

$$X/M = KC_f^{1/n} \tag{18.5}$$

where $X = C_o - C_f$, the amount of compound adsorbed from a given volume of solution
M is the weight of carbon
C_o is the initial amount of compound in the untreated solution
C_f is the amount of compound remaining after carbon treatment
K and $1/n$ are empirical constants

Graphically, K is the X/M intercept of the isotherm plot at $C_o = 1$, and $1/n$ is the slope of the line, when the equation is plotted on logarithmic paper. The intercept is an approximate indicator of adsorption capacity and the slope of adsorption intensity. The concentration of compound on the carbon in equilibrium with a residual soluble concentration, C_f, is expressed by the X/M value. Since X/M values are dependent on the initial concentration of compound, comparisons among compounds must be made at similar concentrations.

Table 18.9 presents Freundlich isotherm data on carbon adsorption of most of the 114 organic priority pollutants. These data were generated using pulverized Calgon Filtrasorb 300 activated carbon, at initial chemical concentrations of 1.0 mg/l, and at the pH indicated in Table 18.9. Table 18.10 presents adsorption data reported from three other studies.

Efficiency of adsorption of organic compounds depends on the relative adsorbability of the individual components. Based on the factors listed earlier, some generalizations can be made regarding the adsorbability of certain compounds. Solubility of the organic contaminant in water is a very important factor [21, 29]. As solubility decreases, adsorption capacity increases. Factors such as pH concentration, temperature, and ionic strength, which affect solubility, will also affect adsorption. Verschueren [29] has reported a strong correlation between compound solubility and ratio of sorbed to residual soluble concentration, for a wide range of organic compounds. The empirical relationship takes the form

$$\log P = 8 - 0.66 \log S \tag{18.6}$$

where $P = (X/M)/C_o$
S = solubility in μg/l

Table 18.9 Summary of Freundlich Parameters [25]

Compound[a]	*pH*	*K*[b] *(mg/g)*	*1/n*
1. Acenaphthene	5.3	190	0.36
2. Acrolein	5.2	1.2	0.65
3. Acrylonitrile	5.3	1.4	0.51
4. Benzene	5.3	1.0	1.6
	7.0	0.7	2.9
5. Benzidine	3.0	1.0	0.35
	7.0–9.0	220	0.37
6. Carbon tetrachloride	5.3	11.1	0.83
	7.0	40	0.84
7. Chlorobenzene	7.4	91	0.99
8. 1,2,4-Trichlorobenzene	5.3	157	0.31
9. Hexachlorobenzene	5.3	450	0.60
10. 1,2-Dichloroethane	5.3	3.6	0.83
11. 1,1,1-Trichloroethane	5.3	2.5	0.34
12. Hexachloroethane	5.3	97	0.38
13. 1,1,-Dichloroethane	5.3	1.8	0.53
14. 1,1,2-Trichloroethane	5.3	5.8	0.60
15. 1,1,2,2-Tetrachloroethane	5.3	10.6	0.37
16. Chloroethane	5.3	0.6	0.95
18. Bis(2-chloroethyl)ether	5.3	0.09	1.84
19. 2-Chloroethyl vinyl ether	5.3	3.9	0.80
20. 2-Chloronaphthalene	5.5	280	0.46
21. 2,4,6-Trichlorophenol	3.0	219	0.29
	6.0	155	0.40
	9.0	130	0.39

Table 18.9 *(continued)*

Compound[a]	*pH*	*K*[b] *(mg/g)*	*1/n*
22. Para-chloro-meta-cresol	3.0	122	0.29
	5.5	124	0.16
	9.0	99	0.42
23. Chloroform	5.3	2.6	0.73
	7.0	11	0.84
24. 2-Chlorophenol	3.0–9.0	51	0.41
25. 1,2-Dichlorobenzene	5.5	129	0.43
26. 1,3-Dichlorobenzene	5.1	118	0.45
27. 1,4-Dichlorobenzene	5.1	121	0.47
28. 3,3-Dichlorobenzidine	7.2–9.1	300	0.20
29. 1,1-Dichloroethylene	5.3	4.9	0.54
30. 1,2-Dichloroethylene	6.7	3.1	0.51
31. 2,4-Dichlorophenol	3.0	147	0.35
	5.3	157	0.15
	9.0	141	0.29
32. 1,2-Dichloropropane	5.3	5.9	0.60
33. 1,2-Dichloropropylene	5.3	8.2	0.46
34. 2,4-Dimethylphenol	3.0	78	0.44
	5.8	70	0.44
	9.0	108	0.33
35. 2,4-Dinitrotoluene	5.4	146	0.31
36. 2,6-Dinitrotoluene	5.4	145	0.32
37. 1,2-Diphenylhydrazine	5.3	16,000	2.0
38. Ethylbenzene	7.3	53	0.79
39. Fluoranthene	5.3	664	0.61
40. 4-Chlorophenyl phenyl ether	5.3	111	0.26
41. 4-Bromophenyl phenyl ether	5.3	144	0.68
42. Bis(2-chloroisopropyl) ether	5.4	24	0.57
43. Bis(2-chloroethoxy) methane	5.3	11	0.65
44. Methylene chloride	5.8	1.3	1.16
47. Bromoform	5.3	19.6	0.52
	7.0	200	0.83
48. Dichlorobromomethane	5.3	7.9	0.61
	7.0	19	0.76
49. Trichlorofluoromethane	5.3	5.6	0.24
51. Chlorodibromomethane	5.3	4.8	0.34
	7.0	63	0.93
52. Hexachlorobutadiene	5.3	258	0.45
	7.0	360	0.63
53. Hexachlorocyclopentadiene	5.3	370	0.17
54. Isophorone	5.5	32	0.39
55. Naphthalene	5.6	132	0.42
56. Nitrobenzene	7.5	68	0.43
	3.0	101	0.26
	5.5	99	0.34
57. z-Nitrophenol	9.0	85	0.39
58. 4-Nitrophenol	3.0	80	0.17
	5.4	76	0.25
	9.0	71	0.28

Table 18.9 *(continued)*

Compound[a]	*pH*	K^b *(mg/g)*	*1/n*
59. 2,4-Dinitrophenol	3.0	160	0.37
	7.0	33	0.61
	9.0	41	0.25
60. 4,6-Dinitro-o-cresol	3.0	237	0.32
	5.2	169	0.35
	9.0	43	0.90
61. N-Nitrosodimethylamine	7.5	6.8×10^{-5}	6.6
62. N-Nitrosodiphenylamine	3.0–9.0	220	0.37
63. N-Nitrosodi-n-propylamine	3.0–9.0	24	0.26
64. Pentachlorophenol	3.0	260	0.39
	7.0	150	0.42
	9.0	100	0.41
65. Phenol	3.0–9.0	21	0.54
66. Bis(2-ethylhexyl) phthalate	5.3	11,300	1.50
67. Butyl benzyl phthalate	5.3	1,520	1.26
68. Di-n-butyl phthalate	3.0	220	0.45
70. Diethyl phthalate	5.4	110	0.27
71. Dimethyl phthalate	3.0–9.0	97	0.41
73. Benzo(a) pyrene	7.1	34	0.44
75. Benzo(k) fluoranthene	7.1	181	0.57
77. Acenaphthylene	5.3	115	0.37
78. Anthracene	5.3	376	0.70
79. Benzo(ghi)perylene	7.0	11	0.37
80. Fluorene	5.3	330	0.28
81. Phenanthrene	5.3	215	0.44
82. Dibenzo(ah)anthracene	7.1	69	0.75
85. Tetrachloroethylene	5.3	51	0.56
86. Toluene	5.6	26	0.44
87. Trichloroethylene	5.3	28	0.62
	7.0	21	0.50
89. Aldrin	5.3	651	0.92
90. Dieldrin	5.3	606	0.51
91. Chlordane	5.3	245	0.38
92. DDT	5.3	322	0.50
93. DDE	5.3	232	0.37
95. α-Endosulfan	5.3	194	0.50
96. β-Endosulfan	5.3	615	0.83
97. Endosulfan sulfate	5.3	686	0.81
98. Endrin	5.3	666	0.80
100. Heptachlor	5.3	1,220	0.95
101. Heptachlor epoxide	5.3	1,038	0.70
102. α-BHC	5.4	303	0.43
103. β-BHC	5.4	220	0.49
104. γ-BHC	5.3	256	0.49
108. PCB-1221	5.3	242	0.70
109. PCB-1232	5.3	630	0.73

[a] Numbers refer to compound numbers in Table 18.2.
[b] K at C_o = 1 mg/l.

Table 18.10. Freundlich Parameters for Selected Organic Priority Pollutants

Pollutant[a]		*Type Carbon*	*Initial Concentration (mg/l)*	*Freundlich Values* K	*Freundlich Values* 1/n	*Reference*
1.	Acenaphthene	Filtrasorb	3.93[b]	624	0.46	26
5.	Benzidine	Darco	1.0	85	0.25	27
		Filtrasorb	1.0	173	0.29	
37.	1,2-Diphenylhydrazine	Darco	1.0	92	0.26	27
		Filtrasorb	1.0	135	0.16	
55.	Napthalene	Darco	1.0	58	0.28	27
		Filtrasorb	1.0	176	0.52	
		Filtrasorb	31.7[b]	277	0.43	26
61.	Nitrosodimethylamine	Darco	1.0	1.4×10^{-6}	8.15	27
		Filtrasorb	1.0	6.8×10^{-5}	6.57	
72.	Benz(a)anthracene	Filtrasorb	0.014	216	0.50	26
76.	Chrysene	Filtrasorb	0.002[b]	716	0.46	26
77.	Acenaphthylene	Filtrasorb	16.1[b]	266	0.30	26
78.	Anthracene	Filtrasorb	0.073[b]	330	0.62	26
80.	Fluorene	Filtrasorb	1.98[b]	674	0.60	26
81.	Phenanthrene	Filtrasorb	1.29[b]	273	0.41	26
84.	Pyrene	Filtrasorb	0.135[b]	389	0.39	26
89.	Aldrin	—	0.01	3.5	—	28
90.	Dieldrin	—	0.01	9.3	—	28
92.	DDT	—	0.01	7.3	—	28
94.	DDD	—	0.01	27	—	28
98.	Endrin	—	0.01	24	—	28
106.	PCB-1242	—	0.01	19	—	28
107.	PCB-1254	—	0.01	6	—	28

[a] Numbers in column refer to compound numbers in Table 18.2.
[b] Aqueous solubility at 25°C. Tests used saturated solutions.

Molecular weight and polarity have a pronounced effect on adsorption. Usually an increase in molecular weight improves adsorption. Nonpolar molecules are more strongly adsorbed than are polar molecules. Molecular structure is another important factor. The influence of substitute groups on adsorbability can be described in general terms. Hydroxyl usually reduces adsorbability because of increased polarity. Amino groups have a similar but greater effect than hydroxyl. Many amino acids are, in fact, not adsorbed to any appreciable extent. Carbonyl groups have a variable effect depending on the host molecule. Sulfonic groups are polar and decrease adsorbability. Nitro groups often increase adsorbability. Generalizations based on molecular structure can also be made. Aromatic and substituted aromatic compounds are, in general, more adsorbable than are aliphatic compounds. Amines, ethers, and halogenated aliphatic compounds adsorb more efficiently than do low molecular-weight alcohols, glycols, or low molecular-weight straight-chain unsubstituted aliphatic compounds.

Full-scale activated carbon treatment systems do not achieve the maximum carbon loadings predicted by isothermal experiments, due to solute competition, mass transfer limitations, and wastewater matrix effects. In one study on the removal of volatile organic compounds in pilot carbon columns, carbon loadings at column breakthrough were less than 20% of the calculated capacities [30]. Performance data are summarized in Table 18.11. Table 18.12 presents additional data on activated carbon treatment of toxic organics.

Activated carbon treatment costs vary widely, depending upon carbon exhaustion and attrition rates, and plant size. Table 18.13 summarizes costs for a granular activated carbon column system designed to treat relatively dilute concentrations of volatile organic compounds. Table 18.14 presents the details of a system including preaeration and on-site carbon regeneration.

Polymeric resin systems have also been employed for adsorption treatment of toxic organics. Adsorption is via the relatively weak van der Waals binding force. Pyrolyzed polymeric resins are also available, some designed specifically to remove nonpolar low molecular-weight organics such as trichloroethylene [22]. Shelby et al [4] compared XAD-4 polymeric resin (Rohm and Haas) with activated carbon for the treatment of a complex pesticide manufacturing wastewater. The results, for effluent values over 3,000 bedvolumes of throughput, are presented in Table 18.15. The resin and carbon systems were operated in parallel, and the effluent values reflect the variable influent concentrations as well as the comparative efficiency of the resin and carbon.

Table 18.11 Performance of Pilot Granular Activated Carbon Columns in Removing Volatile Organic Pollutants [30]

Compound[a]	*Influent Concentrations (μg/l)*	*Breakthrough (bed volumes)*	*Observed Capacity (mg/g)*
4. Benzene	47.5	2,500	0.74
11. 1,1,1-Trichloroethane	9.5	1,800	0.09
15. Tetrachloroethane	22.6	3,000	0.35
23. Chloroform	69.3	0	0.43
30. 1,2-Trans-dichloroethylene	64.9	500	0.52
44. Methylene chloride	21.7	0	0.05
88. Vinyl chloride	7.7	400	0.03

[a] Numbers refer to compound numbers in Table 18.2.

Table 18.12 Activated Carbon Column Treatment of Toxic Organic Compounds

Compound[a]	*Concentration (μg/l)* *Influent*	*Effluent*	*% Removal*	*Reference*
1. Acenapthene	—	<0.04	93+	1
4. Benzene	28,000	<10	99+	1
		73	48–80+	31
6. Carbon tetrachloride	61,000	<10	99+	31
7. Chlorobenzene	—	<0.2	96+	1
	50,000	12	99+	31
8. 1,2,4-Trichlorobenzene	—	<0.04	93+	1
	870	nd[b]	—	31
9. Hexachlorobenzene	110	nd	—	31
10. 1,2-Dichloroethane	—	230,000	86–99+	1
	52	nd	—	31
11. 1,1,1-Trichloroethane	—	<10	99+	1
	22	15	32	5
	23	nd	—	31
13. 1,1-Dichloroethane	—	8,100	89–99+	1
	55	nd	—	31
14. 1,1,2-Trichloroethane	—	<10	99+	1
	780	<10	98+	31
15. 1,1,2,2-Tetrachloroethane	70,000	<10	99+	31
16. Chloroethane	—	71,000	58–99+	1
20. 2-Chloronaphthalene	150	nd	—	31
21. 2,4,6-Trichlorophenol	85	<10	88+	31
23. Chloroform	—	<11	67–99+	1
	19	6	68	5
	44,000	<10	99+	31
	19	9.3	51	32
25. 1,2-Dichlorobenzene	—	<0.05	99+	1
	1,300	nd	—	31
29. 1,1-Dichloroethylene	16	nd	—	31
30. 1,2-Trans-dichloro-ethylene	3,200	<10	—	31
31. 2,4-Dichlorophenol	5,100	nd	—	31
32. 1,2-Dichloropropane	—	5.4	64–99+	1
	130	nd	—	31
34. 2,4-Dimethylphenol	—	<0.1	89+	1
38. Ethylbenzene	590	<10	98+	31
39. Fluoranthene	—	<0.02	82–90+	1
44. Methylene chloride	—	140	31–92	1
	140	46	67	31
45. Methyl chloride	370	nd	—	31
47. Bromoform	1.8	0.2	89	32
48. Dichlorobromomethane	8.8	2.1	76	32

Table 18.12 *(continued)*

Compound[a]	*Concentration (μg/l)* *Influent*	*Effluent*	*% Removal*	*Reference*
51. Chlorodibromomethane	29	nd	—	31
	4	0.6	85	32
52. Hexachlorobutadiene	1,500	nd	—	31
62. N-Nitrosodiphenylamine	—	<0.07	82+	1
64. Pentachlorophenol	—	13	63–97+	1
66. Bis(2-ethylhexyl) phthalate	—	65	18–66	1
67. Butyl benzyl phthalate	—	5.7	83–99+	1
68. Di-n-butyl phthalate	—	1.3	62–99+	1
69. Di-n-octyl phthalate	—	110	76–96+	1
84. Pyrene	—	<0.01	83–97+	1
85. Tetrachloroethylene	100	32	68	1
	44,000	12	99+	31
	270	<10	99+	31
86. Toluene	—	80	38–99+	1
	25,000	<10	99+	31
87. Trichloroethylene	—	2.8	29–58	1
	32	6	81	5
	5,000	nd	—	31
	240	<10	99+	31
100. Heptachlor	573	<0.01	99+	31
102. α-BHC	—	<1	47+	1
	184	<0.01	99+	31
103. β-BHC	548	<0.01	99+	31
104. γ BHC	392	0.12	99+	31

[a] Numbers refer to compound numbers in Table 18.2.
[b] Nondetectable.

Table 18.13 Treatment Costs for Granular Activated Carbon Column Systems [33]

Cost	*Plant Size (MGD)* *1*	*10*	*50*
Capital, $1,000	239	1,151	1,900
Operating and maintenance, $1,000/yr	58	176	130
Carbon, $1,000/yr	150	1,058	4,493
Total Annual, ¢/1,000 gal	90.2	52.6	37.4

Table 18.14 Itemized Cost for Preaeration and Granular Activated Carbon System with On-Site Carbon Regeneration [22]

Item	*Cost (1981)*
Capital[a]	
Aeration system	$ 25,000
Carbon columns	300,000
Carbon	20,900
Regeneration facility	504,000
Miscellaneous	38,600
Engineering and contingency	223,400
Total capital	$1,111,900
Operating and maintenance	
Power	$ 16,000
Labor	35,000
Replacement carbon	2,100
Miscellaneous	7,500
Total O&M	$ 60,600

[a] Treatment capacity is 450 gpm.

Until breakthrough, the resin and carbon performance was comparable, with carbon normally achieving somewhat lower effluent concentrations. For most compounds the carbon capacity was greater than that of the resin, with breakthrough occurring at a higher bedvolumes throughput. Activated carbon was thermally regenerated, while the resin was regenerated with methanol in order to assess the feasibility of methanol recovery by distillation, and reuse. Methanol elution was efficient, and 95% elution was completed within 10 bedvolumes of regenerant. Distillation recovery of the methanol resulted in carryover of organics with boiling points similar to methanol, plus light organics forming azeotropes with methanol. Resin steam stripping pretreatment for light organics, prior to methanol regeneration, resolved this difficulty.

Singhal [22] has compared the cost of pyrolyzed resin (Ambersorb XE-340) treatment with granular carbon including on-site regeneration, for trichloroethylene treatment. Resin regeneration was by steam stripping, with once-per-year ethanol washing. Capital costs (1980 basis) were $0.86 and $1.72/gpd treatment capacity for resin and carbon, respectively. The regeneration facilities represented 11% and 45% of the capital costs, respectively. Operating costs were $0.29 and $0.26/1,000 gal treated.

Resin systems have had wide application in recovery processes, including for chlorinated pesticides, phenols, aliphatic chlorinated hydrocarbons, and benzene, toluene, and xylene [23]. Successful organic solvents employed as regenerants include acetone, formaldehyde, methanol, and isopropanol. Inorganic regenerants include steam, caustic for adsorbed weak organic acids, and acids for adsorbed weak bases.

Table 18.15 Effluent Values for XAD-4 Resin and Carbon Column Treatment [4]

Compound[a]	Influent[b]		1,000 bv[c]		2,000 bv		3,000 bv	
	Avg	*Range*	*Resin*	*Carbon*	*Resin*	*Carbon*	*Resin*	*Carbon*
6. Carbon tetrachloride	20,450	1,700–150,000	900	1,200	490	1,500	55,000	560
12. Hexachloroethane	104	12–730	0.7	0.2	0.1	0.3	33	1.7
20. 2-Chloronaphthalene	18	3–210	<3	<3	<3	<3	<3	<3
23. Chloroform	1,430	46–7,100	35	27	670	1,200	1,100	790
52. Hexachlorobutadiene	265	23–3,500	0.3	0.7	0.6	0.1	<0.1	0.5
53. Hexachlorocyclopentadiene	1,127	70–7,500	2.2	3.7	1.5	0.8	2.3	4.2
55. Napthalene	529	3–1,600	<3	760	17	<3	<3	<3
85. Tetrachloroethylene	34	0.9–290	0.3	<0.1	2.2	<0.1	5.5	<0.4
86. Toluene	2,360	50–8,900	10	<3	45	22	1,200	170
89. Aldrin	84	21–320	1.0	2.4	1.7	0.1	0.3	0.3
90. Dieldrin	28	1–130	0.6	1.6	0.2	0.2	4.1	2.2
91. Chlordane	217	1–1,300	<0.1	13	0.2	<0.1	2.3	2.7
98. Endrin	123	7–810	1.9	7.6	1.2	0.9	2.5	5.1
100. Heptachlor	40	55–190	1.7	3.5	0.8	0.8	1.4	3.8
101. Heptachlor epoxide	11	0.6–63	<0.1	<0.1	1.3	<0.1	<0.1	<0.1

[a] Numbers refer to compound numbers in Table 18.2.
[b] All concentrations in μg/l.
[c] Bed volumes.

Many of the current applications of resins in treatment recovery systems are based upon economic advantages over other alternatives such as carbon adsorption, solvent extraction, and steam stripping. In comparison with activated carbon using pure chemical solutions, resin adsorptive capacity is often less than carbon. However, in complex, mixed-organics waste streams, the resin may demonstrate greater selectivity for the target pollutants than does carbon. Carbon capacity is then lost due to adsorption of nontargeted organics, and in such systems, the operating capacity of the resin may be greater due to the more rapid loss of carbon capacity associated with the poorer selectivity of the carbon.

In one XAD-4 resin application, endrin plus heptachlor at 0.1–2.0 mg/l is reduced to below 3 μg/l in a single resin bed treating about 100 gpm. Isopropanol is used as the regenerant, and recovered by distillation for reuse. The concentrated still bottoms are disposed off-site. In another application, toxaphene is reduced from 70–2,600 μg/l to below 2 μg/l by resin adsorption [23]. Regeneration is required twice per year [23]. Other resin systems are described in Table 18.16.

Biological Treatment

Although the organic priority pollutants are generally considered to be toxic, the extent of toxicity is a function of both the relative toxicity of the individual compounds and of the absolute concentration (exposure level). The introduc-

Table 18.16 Summary of Applications of Resin Adsorption Systems [after 23]

Process[a]	*Wastewater Characteristics*	*Flow (gpm)*	*Influent Concentration (mg/l)*	*Regenerant*	*Capital Cost*
T	Endrin and heptachlor	100	0.1–2.0	Isopropanol	—
T	Toxaphene	250	0.07–2.6	—	—
T	Dichlorophenol	—	1,500	Caustic	—
T	Pesticide wastes	50	10	Methanol	$220,000
R	Phenoxy herbicides	70–180	20–1,500	Methanol	—
R[b]	Dichloroethylene, chloroform and carbon tetrachloride	—	0.1–7.0	Steam	—
R	Chlorinated hydrocarbons	250	300	Steam	490,000
R	Chlorinated hydrocarbons	60	500	Steam	285,000
R	Chlorinated hydrocarbons	60	7,500	Steam	285,000
R[b]	Benzene, toluene and xylene	250	300	Steam	400,000

[a] T = treatment; R = recovery of pollutants for recycle.
[b] Pilot scale system.

tion of a waste stream containing one or more organic priority pollutants into either an aerobic or an anaerobic biological treatment process can yield any one or combination of the following types of responses [34].

- Inhibition—the organic compound interferes with the proper functioning of the biological process, and overall treatment efficiency deteriorates.
- Nonbiodegradability—the organic compound does not effect the treatment efficiency, but is recalcitrant and passes substantially unchanged through the treatment system.
- Chemical conversion—the biological process transforms the organic compound into a different chemical which no longer responds to the specific analytical test procedure.
- Biodegradation—the organic compound is mineralized into oxidized forms such as carbon dioxide and water, and cellular mass.
- Acclimation-degradation—initial introduction of the organic compound does not show degradation but continued exposure causes a population shift or adaptation which eventually results in degradation of the toxicant.
- Sorption—the organic compound is removed from the waste stream by sorption onto soil particles, primary sludge, or mixed liquor particles, without biodegradation occurring.

In real systems it is difficult to distinguish the relative influence of the last four listed types of responses, since each results in the disappearance of the subject compound from the soluble phase of the wastewater. An additional mechanism for the disappearance of a priority pollutant during biological treatment is volatilization. This mechanism can predominate over others for many of the volatile organic priority pollutants [35]. Petrasek [36] has reported a strong correlation between the Henry's law constant and the fraction of priority pollutant found in activated sludge aeration-basin off-gas.

The results of static flask acclimation-degradation studies on most of the organic priority pollutants are summarized in Table 18.17. In these studies, bacterial inocula from settled domestic sewage were dosed individually with each priority pollutant at 5 mg/l and incubated for 7 days. Then, and each 7 days thereafter, a subculture was redosed, through three successive subcultures. Biodegradation data on all pollutants which demonstrated less than 90% biodegradation in the original culture are presented in Table 18.17. Pollutants not presented in Table 18.17 demonstrated 90% or greater biodegradation in the original culture, unless otherwise specified in the footnotes to the table.

Three distinct modes of bioperformance are apparent from the data of Table 18.17. Some compounds, such as benzene (no. 4) exhibited partial degradation in the original inoculum, but through acclimation were more completely degraded by subsequent subcultures. A second pattern is demonstrated by hexachlorobenzene (no. 9), where inhibition of degradation suggests toxic effects. A few compounds, such as 1,2-diphenylhydrazine (no. 37) exhibited

Table 18.17 Results of Acclimation/Degradation Studies on Priority Pollutants [after 37]

	% Biodegradation	
Pollutant[a,b,c]	*Original Culture*	*Third Subculture*
4. Benzene	49	100
6. Carbon tetrachloride	87	100
7. Chlorobenzene	89	100
8. 1,2,4-Trichlorobenzene	54	24
9. Hexachlorobenzene	56	5
10. 1,2-Dichloroethane	26	63
11. 1,1,1-Trichlorobenzene	29	83
13. 1,1-Dichloroethane	50	91
14. 1,1,2-Trichloroethane	6	44
15. 1,1,2,2-Tetrachloroethane	0	29
19. 2-Chloroethyl vinyl ether	76	100
22. Para-chloro-meta-cresol	78	100
23. Chloroform	49	100
24. 2-Chlorophenol	86	100
25. 1,2-Dichlorobenzene	45	29
26. 2,3-Dichlorobenzene	59	35
27. 1,4-Dichlorobenzene	55	16
29. 1,1-Dichloroethylene	78	100
30. 1,2-Trans-dichloroethylene	67	95
32. 1,2-Dichloropropane	42	89
33. 1,3-Dichloropropylene	55	85
35. 2,4-Dinitrotoluene	77	27
36. 2,6-Dinitrotoluene	82	29
37. 1,2-Diphenylhydrazine	80	77
39. Fluoranthene	0	100
40. 4-Chlorphenyl phenyl ether	0	1
41. 4-Bromophenyl phenyl ether	0	0
42. Bis-(2-chloroisopropyl)ether	85	100
43. Bis-(2-chloroethoxy)methane	0	0
47. Bromoform	11	48
48. Dichlorobromomethane	35	59
49. Trichlorofluoromethane	59	73
51. Chlorodibromomethane	25	39
59. 2,4-Dinitrophenol	60	100
60. 4,6-Dinitro-o-cresol	52	51

Table 18.17 *(continued)*

Pollutant[a,b,c]	*% Biodegradation*	
	Original Culture	*Third Subculture*
61. Nitrosodimethylamine	71	57
62. N-Nitrosodiphenylamine	87	100
63. N-Nitroso-di-n-propylamine	27	50
64. Pentachlorophenol	19	100
66. Bis-(2-ethyl hexyl)phthalate	0	95
69. Di-n-octyl phthalate	0	92
72. Benzo(a)anthracene	16	0
76. Chrysene	0	38
78. Anthracene	43	92
80. Fluorine	82	77
84. Pyrene	71	100
85. Tetrachloroethylene	45	87
87. Trichloroethylene	64	87
106. PCB-1242	37	66
107. PCB-1254	11	0
110. PCB-1248	0	0
111. PCB-1260	0	0
112. PCB-1016	44	48

[a] All priority pollutants not listed in this table, except for those identified in note b, exhibited 90% or greater degradation by the original inoculum.

[b] Priority pollutants not tested were

benzo(a)pyrene	benzidine
3-4-benzofluoranthene	3,3-dichlorobenzidine
benzo(k)fluoranthene	bis (2-chloromethyl)ether
benzo(g,h)perylene	vinyl chloride
Dibenzo(a,h)anthracene	chloroethane
indeno(1,2,3-cd)pyrene	methyl chloride
toxaphene	methyl bromide
dioxin	dichlorodifluoromethane
endrin aldehyde	

All pesticides tested (compounds 89–105, Table 18.2) exhibited zero degradation through the third subculture.

[c] Numbers in this column refer to compound numbers in Table 18.2.

consistent but incomplete removal, with neither improvement nor deterioration in extent of degradation through the subculturing acclimation process.

Data on the occurrence and removal of organic priority pollutants in full-scale biological treatment systems are limited. Table 18.18 summarizes the data available on removal of organic priority pollutants in biological systems. In general, the static biodegradation results of Table 18.17 are consistent with the full-scale results of Table 18.18. For several pollutants, such as some of the

Table 18.18 Removal of Organic Priority Pollutants in Biological Treatment Systems

Pollutant[a]	*Type of Treatment*[b]	*Influent Level (μg/l)*	*Effluent Level (μg/l)*	*Percent Removal*	*Reference*
1. Acenaphthene	AS-M	0.9	—	95	38
	AS-I(10)	15	—	84+	1
	AS-I(5)	84	5	94	39
	AL-I(1)	4.0	—	0	1
2. Acrolein	AS-I[c]	62,000	50	99+	40
	AS-I[c]	31,000	30	99+	41
3. Acrylonitrile	AS-I(5)	10,300	65	99+	39
	AS-I[c]	152,000	50	99+	40
	AS-I[c]	31,000	30	99+	41
4. Benzene	AS-M	8.8		66	38
	AS-M	7.7		92	38
	AS	—		90–100	10
	AL	6–53		(−17)–96	42
	AS-M	6.8	3.5	49	43
	PACT	160	—	99+	42
	PACT	105	0.9	99	44
	AS-I(9)	10,250	—	60+	1
	AS-I(5)	581	6	99+	39
	AS-I[c]	153,000	40	99+	40
	AS-I[c]	39,000	<50	99+	41
	AL-I(2)	46.9	—	84+	1
5. Benzidine	AS-I	4	—	0	1
	AL-I	11.9	—	41	1
6. Carbon tetrachloride	AS-M	1.0	nd[d]	100	43
	PACT	94	1.4	95	44
	PACT	95	—	94	42
	AS-I(2)	250	—	98+	1
	AS-I(5)	51	nd	100	39
	AS-I[c]	7,000	7	99+	41
7. Chlorobenzene	PACT	1,720	30	98	44
	PACT	1,900	—	99+	38
	AS-M	177	—	99+	38
	AS-I(6)	15.2	—	67+	1
	AS-I(5)	20	16	20	39
8. 1,2,4-Trichlorobenzene	PACT	523	169	66	44
	AS-M	0.6	—	83	38
	AS-I(11)	285	—	67+	1
	AS-I(5)	234	39	83	39
9. Hexachlorobenzene	AS-I(4)	0.75	—	47+	1
	AL-I	10	—	0	1
10. 1,2-Dichloroethane	AS-M	1.5	—	65	38
	AS-M	8.4	—	85	42
	AS-I(5)	524	9	98	39
	AS-I[c]	63,000	1,100	98	41
	AS-I[c]	258,000	3,700	99	40

Table 18.18 (*continued*)

Pollutant[a]	*Type of Treatment*[b]	*Influent Level* (μg/l)	*Effluent Level* (μg/l)	*Percent Removal*	*Reference*
11. 1,1,1-Trichloroethane	PACT	13	0.6	89	44
	PACT	18	—	99+	38
	AS-M	88.4	—	70	38
	AS-M	1,791	—	98	38
	AS-M	14	5.2	83	43
	AS-I(6)	9.2	—	74+	1
	AL-I	550	—	96	1
	AS-I(5)	7	5	29	39
	AS-I[c]	60,000	94	99+	41
	RBC-I	42	22	48	5
13. 1,1-Dichloroethane	AS-M	1.7	—	6	38
	AS-M	6.7	—	92	28
	AS-I(2)	7.1	—	9+	1
	AS-I(5)	9	5	44	39
14. 1,1,2-Trichloroethane	AS-M	5.2	nd	100	43
	AS-I	11	—	9+	1
	AS-I(5)	12	5	58	39
15. 1,1,2,2-Tetrachloroethane	AS-I(2)	12.8	—	22+	1
	AS-I(5)	17	5	71	39
	AS-I[c]	52,000	2,000	96	41
	AS-I[c]	201,000	13,000	94	40
16. Chloroethane	PACT	280	12	94	44
	AS-I(5)	12	5	58	39
17. Bis(chloromethyl)ether	AS-I	59	—	83	1
18. Bis(2-chloroethyl)ether	AS-M	0.12	—	02	27
	AS-I	19	—	47+	1
20. Naphthalene	AS[c]	2,000	—	100	10
	AS-I	2	—	50	1
	TF-I	2		0	1
	AL-I	19	—	>47	1
21. 2,4,6-Trichlorophenol	AS-I(10)	703	—	36	1
	AS-I(5)	100	53	47	39
	AL-I	1,000	—	99	1
23. Chloroform	PACT	201	21	81	44
	AS-M	37.5	16	57	43
	AS-M	4.2	—	5	38
	AS-M	17.1	—	98	38
	AS-M	37.5	—	49	38
	AS-I(16)	33	—	61+	1
	AS-I(5)	348	13	96	39
	AS-I[c]	14,000	105	99+	41
	TF-I	19	—	0	1
	AL-I(3)	531	—	36+	1
	AL-I	425–2,645	—	(−1–86)	38
	RBC-I	8	4	50	5

Table 18.18 *(continued)*

Pollutant[a]	*Type of Treatment*[b]	*Influent Level (μg/l)*	*Effluent Level (μg/l)*	*Percent Removal*	*Reference*
24. 2-Chlorophenol	PACT	11	1.6	95	44
	AS-I(2)	10	—	46	1
	AS-I(5)	53	35	34	39
25–27. Dichlorobenzene	PACT	720	—	35	42
25. 1,2-Dichlorobenzene	PACT	259	120	44	44
	AS-M	1.8	—	28	38
	AS-M	4.8	—	28	38
	AS-M	43.9	—	97	38
	AS-I(12)	28.8	—	74+	1
	AS-I(5)	331	28	92	39
	AS-I[c]	10,000	50	99+	41
26. 1,3-Dichlorobenzene	AS-M	3.1		86	38
	AS-I[c]	30,000	2.5	99+	41
	AS-I[c]	83,000	50	99+	40
27. 1,4-Dichlorobenzene	AS-I(8)	30	—	82+	1
	AL-I	53	—	81+	1
29. 1,1-Dichloroethylene	AS-M	43.2	—	97	38
	AS-M	3.2	5.4	(−69)	43
30. 1,2-Trans-dichloroethylene	AS-M	3.0	1.3	57	43
	AS-I(5)	42	5	88	39
31. 2,4-Dichlorophenol	AS-I(2)	13.3	—	25+	1
	AS-I(5)	347	24	93	39
	AS-I[c]	24,100	5,500	77	41
	AS-I[c]	75,000	3,600	95	40
32. 1,2-Dichloropropane	AS-I(2)	16	—	67+	25
	AS-I(5)	138	7	95	38
	AS-I[c]	48,000	170	99+	41
	AS-I[c]	182,000	1,810	99	40
33. 1,3-Dichloropropene	AS-I(5)	148	6	96	39
34. 2,4-Dimethylphenol	AS-I(3)	39.7	—	32+	1
	AS-I(5)	270	5	98	39
35. 2,4-Dinitrotoluene	PACT	2,000	—	75	42
	PACT	1,900	243	65	44
	AS-I[c]	32,000	250	99+	41
	AL-I	3	—	0	1
36. 2,6-Dinitrotoluene	PACT	1,640	575	64	39
	PACT	1,900	—	76	42
	AS-I	390	—	0	1
	AL-I	12	—	83+	1
37. 1,1-Diphenylhydrazine	AS[c]	2,400	—	100	27
	AS-I	340	—	0	1
	AL-I	14	—	0	1

Table 18.18 *(continued)*

Pollutant[a]	*Type of Treatment*[b]	*Influent Level (μg/l)*	*Effluent Level (μg/l)*	*Percent Removal*	*Reference*
38. Ethylbenzene	PACT	41	1.7	94	39
	PACT	29	—	78	42
	AS-M	148	—	99	38
	AS-M	25.5	5.1	80	43
	AS-I(24)	882	—	83	1
	AS-I(5)	283	5	98	39
	AS-I[c]	19,300	<10	99+	41
	AL-I(3)	45	—	78+	1
39. Fluoranthene	AS-M	0.6		92	38
	AS-I	2	—	0	1
	AL-I	2	—	0	1
	AS-I(5)	17	6	65	39
41. 4-Bromophenyl phenyl ether	AS-I	360	—	95	1
42. Bis(2-chloroisopropyl)ether	AS-M	0.16	—	62	38
	AL-I	2	—	0	1
43. Bis(2-chloroethoxy)methane	AL-I	25	—	60+	1
44. Methylene chloride	AS-M	7.1	7.8	(−10)	43
	AS-M	9.0	—	49	38
	AS-M	293	—	0	38
	AS-M	647	—	72	38
	AS-I(5)	144	—	34	1
	AS-I(5)	17	9	47	39
	AS-I[c]	56,000	50	99+	41
	AS-I[c]	180,000	51	99+	40
	TF-I	1	—	0	1
	AL-I(3)	1,114	—	65	1
45. Methyl chloride	PACT	1,770	nd	100	39
	AL-I	56	—	91+	1
46. Bromomethane	AS-I(5)	1,250	nd	100	39
47. Bromoform	PACT	910	—	89	42
	AS-I	3	—	0	1
48. Dichlorobromomethane	PACT	54	—	99+	42
	AS-I	5.8	—	0	1
	AS-I(5)	20	5	75	39
49. Trichlorofluoromethane	PACT	155	3.0	95	44
	PACT	920	—	99	42
	AS-I(5)	556	—	19	1
	TF-I	48	—	79+	1

Table 18.18 (*continued*)

Pollutant[a]	*Type of Treatment*[b]	*Influent Level (μg/l)*	*Effluent Level (μg/l)*	*Percent Removal*	*Reference*
51. Chlorodibromo-methane	PACT	81	—	99+	42
	AS-I(5)	6	nd	100	39
54. Isophorone	AS-I(2)	10	—	>0	1
	AS-I(5)	660	nd	100	39
	AL-I	3.0	—	33	1
55. Naphthalene	AS-M	3.2	3.2	0	43
	AS-M	1.3	—	35	38
	AS-M	10.9	—	62	38
	AS-I(26)	50	—	64+	1
	AS-I(5)	802	6	99	39
	AS-I[c]	2,300	<50	97+	41
	TF-I	55	—	0	1
	AL-I(2)	14.1	—	29+	1
56. Nitrobenzene	PACT	454	2	99+	44
	AS-I(5)	3,000	91	97	39
	AS-I[c]	37,500	24,800	34	41
	AS-I[c]	100,000	2,200	98	40
	AL-I	10	—	0	1
57. 2-Nitrophenol	AS-I	40	—	99+	1
	AS-I(5)	40	9	78	39
58. 4-Nitrophenol	PACT	56	—	99+	42
	PACT	1,020	10	97	44
	AS-I	90	—	99+	1
	AL-I	13	—	23	1
59. 2,4-Dinitrophenol	PACT	161	5	98	44
	AS-I(5)	673	59	91	39
60. 2,6-Dinitro-o-cresol	PACT	11	—	99	42
61. Nitrosodimethylamine	AS[c]	2,000	—	95+	27
62. N-Nitroso-diphenylamine	AS-I	5.3	—	84+	1
	AL-I	3	—	67	1
63. N-Nitrosodi-n-propylamine	AS-M	0.5	—	88	38
	AS-M1	6.7	—	99	38
	AS-I(2)	11	—	0	1
64. Pentachlorophenol	AS-I(15)	5,333	—	70+	1
	AS-I(5)	216	115	47	39
	AS-I[c]	3,700	83	98	41
65. Phenol	TF	105		91	45
	AS	105		>99	12
	PACT	440		96	42
	AS-I(30)	335		>77	1
	TF-I	37		0	1
	AL-I(3)	25.4		>41	1

Table 18.18 *(continued)*

Pollutant[a]	*Type of Treatment*[b]	*Influent Level (μg/l)*	*Effluent Level (μg/l)*	*Percent Removal*	*Reference*
66. Bis(2-ethylhexyl) phthalate	PACT	2.4	—	67	42
	AS-M	25.5	12.0	53	43
	AS-M	1.3	—	31	38
	AS-M	50	—	78	38
	AS-I(38)	102	—	37	1
	TF-I	35	—	83	1
	AL-I(5)	37	70+	1	
	TL-I(2)	26	—	58+	1
	AS-I(5)	24	28	(−16)	39
	AS-I[c]	350	50	86	41
67. Butyl benzyl phthalate	AS-M	1.7	3.2	(−88)	43
	AS-M	18	—	91	38
	AS-I	11	—	0	1
	AS-I(5)	86	6	93	39
	AL-I	6	—	0	1
68. Di-n-butyl phthalate	AS-M	4.3	4.3	0	43
	AS-M	5.6	—	68	38
	AS-M	13	—	62	38
	AS-I(9)	23	—	>60	1
	AS-I(5)	86	6	93	39
	TF-I	15	—	25	1
	AL-I	1	—	0	1
69. Di-n-octyl phthalate	AS-I	5,000	—	0	1
	AS-I(5)	28	5	82	39
70. Diethyl phthalate	AS-M	6.6	—	98	38
	AS-M	1.4	—	79	38
	AS-I(17)	15	—	56	1
	AS-I(5)	134	9	93	39
	TF-I	140	—	0	1
	AL-I	4	—	0	1
71. Dimethyl phthalate	AS-I(9)	60	—	60	1
	AS-I(5)	46	5	89	39
	AL-I	8	—	25	1
73. Benzo(a)pyrene	Al-I	3	—	33	1
	AS-I(5)	13	18	(−38)	39
77. Acenaphthylene	AS-M	0.2	—	80	38
	AS-I(5)	65	5	92	39
	AL-I	5	—	0	1
78. Anthracene	AS-I(7)	8.5	—	>60	1
80. Fluorene	AS-M	1.7	—	94	38
	AS-I(2)	2	—	99+	1
	AS-I(5)	56	5	91	39
81. Phenanthrene	AS-M	3.2	—	81	38
	AL-I	3	—	0	1
	AS-I[c]	560	<10	98+	41

Table 18.18 (*continued*)

Pollutant[a]	*Type of Treatment*[b]	*Influent Level (μg/l)*	*Effluent Level (μg/l)*	*Percent Removal*	*Reference*
82. Dibenzo(a,h) anthracene	AS-I(5)	15	75	(−400)	39
84. Pyrene	AS-I(5)	17	6	65	39
	AS-I(5)	2.3	0	16	1
	AL-I	3	—	67	1
85. Tetrachloroethylene	PACT	24	1.7	93	44
	PACT	62	—	88	42
	AS-M	6.4	—	0	38
	AS-M	49	—	80	38
	AS-M	560	—	99+	38
	AS-M	41.5	0.7	84	43
	AS-I(11)	25.6	—	75+	1
	RBC-I	23	16	30	5
	AL-I	25	—	60+	1
86. Toluene	PACT	519	1.7	99+	44
	PACT	680	—	99+	42
	AS-M	70	—	99+	38
	AS-M	112	—	99	38
	AS-M	15.5	6.5	58	43
	AS-I(31)	119	—	52	1
	AS-I(5)	4,500	7	99+	39
	AS-I[c]	16,800	<10	99+	41
87. Trichloroethylene	AS-M	20	0.3	69	43
	AL	10–120	—	(−23)–73	10
	PACT	60	—	90	42
	AS-I(12)	31.3	—	68	1
	TF-I	1	—	0	1
	RBC-I	210	31	85	5
88. Vinyl chloride	AS-I(5)	6	nd	100	39
89. Aldrin	AS + MMF[c]	—	—	100	46
90. Dieldrin	AS + MMF[c]	—	—	72	46
92. DDT	AS + MMF[c]	—	—	100	46
93. DDE	AS + MMF[c]	—	—	39	46
94. DDD	AS + MMF[c]	—	—	100	46
	AS-M	0.3	—	37	38
	AS-M	0.17	—	17	38
100. Heptachlor	AS-I	6.3	—	76	1
104. Landane	AS + MMF[c]	—	—	0	46
106–112. PCB	TF-I	524	6.5	99	47
	TF-M	30–470	10–50	33–89	48

Table 18.18 (*continued*)

[a] Numbers in this column refer to compound numbers in Table 18.2.

[b] AS = activated sludge; M = municipal facility; I = industrial facility; AL = aerated lagoon; TF = trickling filter; MMF = mixed media filtration; PACT = activated sludge with powdered activated carbon added, patented duPont process; TL = tertiary polishing lagoon; RBC = rotating biological contactor. Numbers in parentheses represent number of treatment plants for which data were averaged.

[c] Pilot-scale data.

[d] Nondetectable.

benzoic compounds, full-scale performance exhibits much greater removal than would be inferred from Table 18.17. Without question some compounds, such as phenol, are readily degraded in full-scale systems, although in the case of this particular compound, the extent of degradation is strongly influenced by both opportunity for acclimation and by treatment temperature [45].

Table 18.19 is a synopsis of removal data in activated sludge systems, by treatment stage. Most of the data are from pilot systems, and multiple sets of data for individual pollutants represent different systems. Many of the compounds demonstrate significant removal across the primary clarifier. This can be due to volatilization (e.g., at the weir) or due to removal via sorption onto primary sludge solids.

Table 18.19 Synopsis of Toxic Organics Removal in Activated Sludge System Treatment Stages [after 6, 36, 49–51]

		Percent Removal	
Compound[a]	*Influent* (μg/l)	*Primary*	*Total*
1. Acenaphthene	39.8	(−35)	>97
2. Benzene	33.7	19	99
	18	7	80
6. Carbon tetrachloride	38.2	11	99
	160	13	>99
7. Chlorobenzene	171	13	>99
8. 1,2,4-Trichlorobenzene	17	12	82
11. 1,1,1-Trichloroethane	43.3	15	97
	20	13	82
13. 1,1-Dichloroethane	60.5	37	>99
15. 1,1,2-Trichloroethane	65.7	27	69
23. Chloroform	19	21	89
	73	1	56
	47.6	20	97
26. 1,3-Dichlorobenzene	3	14	40
27. 1,4-Dichlorobenzene	6	0	88
30. 1,2-Trans-dichloroethylene	1	0	100

Table 18.19 (*continued*)

Compound[a]	*Influent* (μg/l)	*Percent Removal*	
		Primary	*Total*
32. 1,2-Dichloropropane	54.5	19	97
34. 2,4-Dimethylphenol	95.7	36	99
38. Ethylbenzene	47.8	29	99
	23	0	89
39. Fluoranthene	30.6	(−30)	>94
44. Methylene chloride	106	39	99
	89	16	55
48. Dichlorobromomethane	28.6	24	98
51. Chlorodibromomethane	37.2	37	73
55. Naphthalene	12	0	92
59. 2,4-Dichlorophenol	6	2	46
64. Pentachlorophenol	7.6	(−71)	>47
66. Bis(2-ethylhexyl)phthalate	51.7	(−1)	78
	12	0	77
68. Di-n-butylphthalate	43.8	(−24)	>94
	6	40	44
69. Di-n-octylphthalate	28.2	(−22)	>83
70. Diethylphthalate	46.4	(−24)	>97
	4	0	36
71. Dimethylphthalate	47.3	>21	>98
72. Benzo(a)anthracene	23.8	(−5)	>97
	31.4	81	>99
73. Benzo(a)pyrene	34.5	96	>99
75. Benzo(k)fluoranthene	8.1	73	>99
76. Chrysene	38.9	6	>97
78. Anthracene	34.8	3	>97
79. Benz(ghi)perylene	8.7	86	>99
80. Fluorene	37.9	(−36)	>98
81. Phenanthrene	40.4	(−10)	>98
83. Indeno(1,2,3-cd)pyrene	15.0	80	>99
84. Pyrene	11.8	81	>99
	30.4	(−29)	>93
85. Tetrachloroethylene	52	25	91
	7	(−29)	57
86. Toluene	321	10	87
	195	24	95
100. Heptachlor	31.7	>10	>93
104. Lindane	45.5	8	>43
107. PCB-1254	33.5	(−240)	>91
113. Toxaphene	47.4	(−85)	>94

[a] Numbers refer to compound numbers in Table 18.2.

The role of sorption removal may be quite significant. Sorption appears to be associated with the degree of lyophobia of individual compounds since a strong correlation has been reported between the extent of sorption and the compound octanol-water partition coefficient [17, 36, 51]. Convery [46], in a review of the available literature on removal of pesticides, reported that DDT, dieldrin, and lindane tend to concentrate in the primary sludge. DDT is also bioconverted to DDE and DDD. Other organic pollutants reported by Convery to concentrate in either the primary or secondary sludge are chlordane and PCB. In another study, on removal of organic priority pollutants in combined industrial-municipal treatment facilities, biosorption was observed for a number of pollutants [38]. Table 18.20 lists those compounds for which a combined sludge priority pollutant concentration (wet-weight basis) of five or greater times the influent pollutant concentration was observed.

Where significant sorption does occur, highest concentration factors are usually associated with the primary sludge, and much lower concentration factors are observed for the waste-activated sludge (Table 18.21). The reason for

Table 18.20 Priority Pollutants Bioconcentrating on Sludge by a Factor of Five or Greater [after 38]

Compound[a]	*Concentration Factor*[b]
1. Acenaphthene	24
25. 1,2-Dichlorobenzene	7–8
26. 1,3-Dichlorobenzene	5–28
39. Fluoranthene	14–45
55. Naphthalene	8–29
63. N-nitrosodi-n-propylamine	100
66. Bis(2-ethylhexyl)phthalate	30 76
67. Butyl benzyl phthalate	172
68. Di-n-butyl phthalate	15–21
69. Di-n-octyl phthalate	22
71. Dimethyl phthalate	6
73. Benzo(a)pyrene	48
75. Benzo(k)fluoranthene	20
76. Chrysene	20
77. Acenaphthylene	100
80. Fluorene	9–30
81. Phenanthrene	23–31
82. Pyrene	31
85. Tetrachloroethylene	6
86. Toluene	48

[a] Numbers refer to compound numbers in Table 18.2.

[b] Concentration factor = sludge (wet-weight) pollutant concentration divided by influent wastewater pollutant concentration.

Table 18.21 Priority Pollutant Concentration Factors for Primary and Waste Activated Sludge [after 36, 51]

Compound[a]	*Sludge Concentration Factor*[b]	
	Primary	*Secondary*
1. Acenapthene	111	1
4. Benzene	5	<1
7. Chlorobenzene	5	<1
34. 2,4-Dimethylphenol	85	<1
38. Ethylbenzene	13	<1
39. Fluoranthene	138	5
55. Naphthalene	60	<1
64. Pentachlorophenol	111	2
66. Bis(2-ethylhexyl)phthalate	74	18
67. Butylbenzylphthalate	248	3
68. Di-N-butylphthalate	89	4
69. Di-N-octylphthalate	73	17
70. Diethylphthalate	12	8
76. Chrysene	76	7
78. Anthracene	128	2
80. Fluorene	62	1
81. Phenanthrene	119	<1
84. Pyrene	160	3
87. Trichloromethylene	10	<1
100. Heptachlor	85	18
104. Lindane	21	4
107. PCB-1254	37	47

[a] Numbers refer to compound numbers in Table 18.2.

[b] Concentration factor = sludge (wet-weight) pollutant concentration divided by influent wastewater pollutant concentration. For primary sludge, factor is calculated for primary sludge and primary influent. For secondary sludge, factor is calculated for return activated sludge and primary effluent.

this disproportionate distribution is unclear, since the method of calculating the concentration factor eliminates the influence of sludge solids concentration.

In summary, biological treatment is an effective technology for removal of many of the toxic organics from wastewaters. Any one or combination of the removal mechanisms of volatilization, biodegradation, or sorption may contribute to the degree of removal achieved.

Other Processes

Several other processes have been evaluated for control of toxic organics, although most results available are from laboratory or pilot-scale studies. Processes described in the literature include oil extraction, chemical oxidation, and ion exchange. Fox and Merrick [47] have reported on the efficiency of a full-scale oil-water separation system in removing PCBs from an industrial wastewater. PCBs have a high octanol-water partition coefficient, and would therefore tend to extract from the wastewater aqueous phase into the oil phase. The process reduced PCBs from 1,464 to 31 μg/l [47].

Chemical oxidation, by both ozone and chlorine, has been reported effective for some toxic organics. Harrison et al [6] have summarized studies on chlorination efficiency for several polynuclear aromatic hydrocarbons. Six-hour reaction at a chlorine dosage of 6 mg/l yielded the reductions presented in Table 18.22. Longer reaction times, up to 20 hours, appear to increase the efficiency of chlorine oxidation of these compounds [6].

Ozone has been shown to be at least partially effective in destroying some organic priority pollutants. Several factors influence the effectiveness of the oxonation process, including the reactivity of the ozone itself with the target compound, the rate of reactivity (slow for some compounds), the ozone demand to achieve a desired degree of treatment, the extent of incidental stripping associated with ozone dispersion, and other treatment variables such as pH and temperature. For example, in ozone treatment of phenol, treatment proceeds at pH 11 at a rate approximately twice as fast as at pH 7 [52]. However, at the higher pH, ozone self-decomposition also accelerates, resulting in a higher net ozone consumption. Data on ozone treatment, most at bench scale, are summarized in Table 18.23.

Limited data have been reported on ion exchange removal of volatile chlorinated organic toxicants [2, 5]. Trace removals are observed, but the efficiency of the ion exchange process is quite poor for this class of compounds.

SUMMARY

A broad array of treatment technologies is available to control the organic priority pollutants, and each technology has advantages and disadvantages. No

Table 18.22 Efficiency of Chlorine Oxidation of Selected Polynuclear Aromatic Hydrocarbons [6]

Polynuclear Aromatic Hydrocarbon[a]	*Initial Level (μg/l)*	*% Reduction*
72. Benzo(a)anthracene	16.1	53
73. Benzo(a)pyrene	53.1	98
74. Benzo(b)fluoranthene	5.2	34
75. Benzo(k)fluoranthene	68.7	56
84. Pyrene	27.1	24

Numbers refer to compound numbers in Table 18.2.

Table 18.23 Ozone Treatment of Selected Organic Priority Pollutants

Pollutant[a]	*Conditions*	*Initial Concentration (μg/l)*	*Treatment Time (min)*	*% Removal*	*Reference*
20. Napthalene	pH 8	7,360	70	84	27
30. 1,2-Trans-dichloroethylene	—	2.1	—	0	1
37. 1,1-Diphenylhydrazine	—	9,900	2	99+	27
39. Fluoranthene	—	0.2	—	6	1
44. Methylene Chloride	—	38	—	0	1
61. Nitrosodimethylamine	pH 7	35,000	130	<1	27
65. Phenol	pH 7	1.0×10^6	52	48	52
	pH 11	0.96×10^6	28	47	52
66. Bis(2-ethylhexyl) phthalate	—	100	—	0	1
67. Butyl benzyl phthalate	—	1.0	—	97	1
68. Di-n-butyl phthalate	—	12	—	77	1
72. Benzo(a)anthracene	—	—	—	95	6
73. Benzo(a)pyrene	—	—	—	61	6
		0.2	—	90	1
74. Benzo(b)fluoranthene	—	0.1	—	80	1
82. Dibenzo(ah)anthracene	—	—	—	94	6
84. Pyrene	—	—	—	85	6
	—	0.3	—	67	1
86. Toluene	—	1.1	—	15	1
87. Trichloroethylene	—	0.9	—	6[b]	1
90. Dieldrin	O_3 = 11 mg/l	10	—	15	46
	O_3 = 32 mg/l	10	—	50	
92. DDT	O_3 = 11 mg/l	10	—	<10	46
	O_3 = 36 mg/l	10	—	30	
104. Lindane	O_3 = 11 mg/l	10	—	<10	46
	O_3 = 36 mg/l	10	—	55	

[a] Numbers in this column refer to compound numbers in Table 18.2.
[b] Compound concentration increased after ozonation.

technology is uniformly applicable to all the organic toxicants. Among the physicochemical processes, air stripping is most economical (Table 18.24), although it results in atmospheric release of volatile organics. Steam stripping can be effective but expensive; in addition, unless recovery is the objective, it results in a contaminated condensate requiring further treatment or disposal.

Table 18.24 Total Treatment Cost for Control of Volatile Organic Hydrocarbons [20]

Treatment Process	*Cost ($/1,000 gal)*
Powdered activated carbon	40
Granular activated carbon	16
Packed tower aeration	3

Activated carbon treatment is effective for many of the organic toxicants, but is also expensive. Polymeric resin adsorbents have potential, particularly where recovery and reuse can be implemented. Biological treatment is also effective for many of the toxicants and is often the least costly option. Risks associated with biological treatment include volatilization and consequent atmospheric pollution, bioconversion to unacceptable end products, and concentration by sorption onto the sludges to levels which prevent cost-effective sludge management.

REFERENCES

1. Cawley, W., ed. *Treatability Manual*, vol. I. *Treatability Data*, USEPA 600-8-80-042-a (July 1980).
2. Fritch, J. "Evaluation of Ion Exchange Technology for Toxic and Nonconventional Pollutant Reduction in Bleach Plant Effluents," Summary Report on USEPA Contract 68-03-2605, *Indust. Environ. Res. Lab.* NTIS PB81-208175 (1981).
3. Robertson, J. and D. Tierney. "Fate of Selected Trace Organics in Canadian Petroleum Refinery Wastewater Treatment Processes," paper presented at the 56th Water Pollution Control Federation Conference, Atlanta, GA (October 1983).
4. Shelby, S., J. Koon, D. Marks, and H. Scott. "Adsorption of Chlorinated and Nonchlorinated Organics from a Pesticide Manufacturing Plant Wastestream," paper presented at the 53rd Water Pollution Control Federation Annual Conference, Las Vegas, NV (October 1980).
5. Giunta, J., J. Mueller, and W. Riznychok. "Total Reuse Industrial Wastewater Treatment System at IBM Research Center," paper presented at the 56th Water Pollution Control Federation Annual Conference, Atlanta, GA (October 1983).
6. Harrison, R., R. Perry, and R. Wellings. "Polynuclear Aromatic Hydrocarbons in Raw, Potable and Waste Waters," *Water Research* 9 (1975): 331–346.
7. Bell, B. and G. Hardcastle. "Treatment of a High Strength Industrial Waste in a Continuously Fed, Intermittently Operated Activated Sludge System," paper presented at the 56th Water Pollution Control Federation Annual Conference, Atlanta, GA (October 1983).
8. Schaffer, R.B. and J. Riley. "Occurrence of Priority Pollutants in Industrial Effluents," paper presented at the Annual Conference, Water Pollution Control Federation, Anaheim, CA (October 1978).
9. "Innovative and Alternative Technology Assessment Manual—Draft," EPA 430/9-78-009 (1979).
10. Patterson, J. "Evaluation of Treatability of Industrial Organic Pollutants Associated with the Muskegon County Wastewater System, Michigan," Report to the Enforcement Division, USEPA Region V (January 1980).
11. Dilling, W.L., N.B. Tefertiller, and G.J. Kallos. "Evaporation Rates and Reactivities of Methylene Chloride, Chloroform, 1,1,1-Trichloroethane Trichloroethylene, Tetrachloroethylene and Other Chlorinated Compounds in Dilute Aqueous Solutions," *Environ. Sci. Technol.* 9 (1975): 833–838.
12. Strier, M.P. "Pollutant Treatability: A Molecular Engineering Approach," *Environ. Sci. Technol.* 14 (1980): 28–31.

13. Roberts, P., P. Dandliker, C. Matter, and C. Munz. "Volatilization of Trace Organic Contaminants during Surface Aeration: Model Studies," Technical Report 257, Department of Civil Engineering, Stanford University, CA (n.d.)
14. Singly, J. and L. Bilello. "Advances in the Development of Design Criteria for Packed Column Aeration," Technical Paper, *Environmental Science and Engineering, Inc*. Gainesville, FL (May 1982).
15. Busch, A. "A Practical Approach to Designing Aeration Systems," *Chem. Eng.* 90 (1983): 76–84.
16. Warner, H., J. Cohen, and J. Ireland. "Determination of Henry's Law Constants of Selected Priority Pollutants," Internal EPA Report, Municipal Environmental Research Laboratory, Cincinnati, OH (April 1982).
17. Matter-Muller, C., W. Gujer, W. Giger, and W. Stumm. "Non-Biological Elimination Mechanisms in a Biological Sewage Treatment Plant," *Prog. Water Technol.* 12 (1980): 299–314.
18. Mumford, R. and J. Schnoor. "Air Stripping of Volatile Organics in Water," unpublished undated report, Department of Civil and Environmental Engineering, University of Iowa.
19. Pekin, T. and A. Moore. "Air Stripping of Trace Volatile Organics from Wastewater," unpublished undated report, Camp, Drosser and McKee, Inc., Boston, MA.
20. Umphres, M., R. Trussell, and C. Tate. "A Study of Tribalomethane Removal by Packed Tower Aeration," unpublished undated report, James M. Montgomery Consulting Engineers, Pasadena, CA.
21. McGuire, M., I. Suffet, and J. Radziul. "Assessment of Unit Processes for the Removal of Trace Organic Compounds from Drinking Water," *J. Am. Water Works Assoc*. 70 (1978): 565–572.
22. Singhal, A. "Tying up TCE," *Water/Eng. Manag*. (September 1981): 56–63.
23. Fox, C. "Removing Toxic Organics from Waste Water," *Chem. Eng. Prog*. (August 1979): 70–77.
24. McGuire, M. and I. Suffett. "Adsorption of Organics from Domestic Water Supplies," *J. Am. Water Works Assoc*. 70 (1978): 621–636.
25. Dobbs, R.A., R.J. Middendorf, and J.M. Cohen. "Carbon Adsorption Isotherms for Toxic Organics," USEPA 600/8-80-023 (April 1980).
26. Walters, R. and R. Luthy. "Equilibrium Adsorption of Polycyclic Aromatic Hydrocarbons from Water onto Activated Carbon," *Environ. Sci. Technol*. 18 (1984): 395–403.
27. Fochtman, E.G. and W. Eisenberg. "Treatability of Carcinogenic and Other Hazardous Organic Compounds," EPA-600/2-79-097 (1979).
28. Bernandin, F., Jr. and E.M. Froelich. "Practical Removal of Toxicity by Adsorption," paper presented at the 30th Annual Purdue Industrial Waste Conference, West Lafayette, IN (May 1978).
29. Verschueren, K. *Handbook of Environmental Data on Organic Chemicals* (New York: Van Nostrand Reinhold, 1983).
30. McManus, A., P. Werthman, and J. Westendorf. "Granular Activated Carbon Removal of Priority Pollutants in a Combined Municipal/Industrial Wastewater," undated, unpublished report, Malcolm Pirnie, Inc., White Plains, NY.
31. McDougall, W. and R. Fusco. "Containment and Treatment of the Love Canal Landfill Leachate," *J. Water Poll. Control. Fed*. 52 (1980): 2914–2924.

32. *Water Reuse and Recycling*, vol. 2. *Evaluation of Treatment Technology*, USDI OWRT/RU-79/2 (1979).
33. Thompson, J. "Treatment Methods and Costs for Organic Contaminant Removal," *Water Sew. Works*, (1979): R94–98.
34. Patterson, J. and P. Kodukula. "Biodegradation of Hazardous Organic Pollutants," *Chem. Eng. Prog.* 1 (1981): 48–55.
35. Pellizzari, E. "Volatile Organics in Aeration Gases at Municipal Treatment Plants," USEPA 600/S2-82-056 (August 1982).
36. Petrasek, A. "Removal and Partitioning of the Volatile Priority Pollutants in Conventional Wastewater Treatment Plants—A Capsule Report," unpublished USEPA report, Municipal Environmental Research Laboratory, Cincinnati, OH (October 1981).
37. Tabak, H.H., S.A. Quave, C.I. Mashni, and E.F. Barth. "Biodegradability Studies with Organic Priority Pollutant Compounds," *J. Water Poll. Control. Fed.* 53 (1981): 1503–1518.
38. Convery, J.J., J.M. Cohen, and D.F. Bishop. "Occurrence and Removal of Toxics in Municipal Wastewater Treatment Facilities," paper presented at the Seventh Joint US/Japan Conference, Tokyo (May 1980).
39. "CMA/EPA Five Plant Study" (Austin, TX: Engineering-Science, Inc., 1982).
40. Stover, E. and D. Kincannon. "Biological Treatability of Specific Organic Compounds Found in Chemical Industry Wastewaters," *J. Water Poll. Control. Fed.* 55 (1983): 97–109.
41. Kincannon, D., A. Weinert, R. Padorr, and E. Stover. "Predicting Treatability of Multiple Organic Priority Pollution Wastewaters from Single Pollutant Treatability Studies," paper presented at the 37th Annual Purdue Industrial Waste Conference, West Lafayette, IN (May 1982).
42. Hutton, D.G. "Removal of Priority Pollutants," *Ind. Wastes* (March/April 1980): 22–26.
43. Robertson, J., W. Cowen, and J. Longfield. "Water Pollution Control," *Chem. Eng.* 87 (1980): 102–119.
44. Heath, H., Jr. "Full-Scale Demonstration of Industrial Wastewater Treatment Utilizing du Pont's PACT Process," USEPA 600/S2-81-159 (December 1981).
45. Vela, G.R. and J.R. Ralson. "The Effect of Temperature on Phenol Degradation in Wastewater," *Can. J. Microbiol.* (1978): 1366.
46. Convery, J.J. "Occurrence and Treatability of Toxic Pollutants in Municipal Wastewater," Internal EPA/MERL-Cincinnati Memorandum (December 21, 1973).
47. Fox, L. and N. Merrick. "Controlling Residual Polychlorinated Biophenyls in Wastewater Treatment through Conventional Means," paper presented at the 37th Annual Purdue Industrial Waste Conference, West Lafayette, IN (May 1982).
48. Bergh, A. and R. Peoples. "Distribution of Polychlorinated Biphenyls in a Municipal Wastewater Treatment Plant and Environs," *Sci. Total Environ.* 8 (1977): 197–204.
49. Lurker, P., C. Clark, and V. Elia. "Atmospheric Release of Chlorinated Organic Compounds from the Activated Sludge Process," *J. Water Poll. Control. Fed.* 54 (1982): 1566–1573.
50. Crawford, M. and R. Southworth. "Fate of Priority Pollutants in a POTW," USEPA Effluent Guidelines Division Report, Washington, DC (n.d.)

51. Petrasek, A.C., I. Kugelman, B. Austern, T. Pressley, L. Winslow, and R. Wise. "Fate of Toxic Organic Compounds in Wastewater Treatment Plants," submitted to *J. Water Poll. Control. Fed.* (November 1982).
52. Anderson, G.C. "Ozonation of High Levels of Phenol in Water," AIChE Symposium Series 166(73) (1977).

19
pH Control

Industrial wastes of many types, and from a variety of processes, exhibit extreme pH values. Table 19.1 lists representative industries, and gives wastewater pH characteristics associated with those industries. Extreme pH wastes may be released on a batch basis as acid or alkaline processing baths are dumped, or released as a continuous waste stream resulting from a specific industrial process (e.g., rinsing). The treatment of concentrated spent acidic or alkaline solutions, such as cleaning or pickling baths, can present unique problems. These solutions can contain several percent by weight of acid or alkali, plus other constituents such as oil, detergents, or soluble metals [15]. The volume and composition of continuous-waste streams containing acidic or basic compounds can be quite variable. Dickerson and Brooks [16] have described the waste stream from a nitrocellulose explosive production line as varying in volume from 3,000 to 9,000 gpm with acidity, measured as free sulfuric acid, ranging from 30 to 1,100 mg/l. Wastewater from one chrome pigments plant has been reported to vary from pH 1 to 13 [10].

In addition to the free acids and bases typically encountered in industrial waste streams, some industries produce "combined" acidic salts, which form weak acids by hydrolysis upon dilution of the waste by receiving water. Typical of such industrial effluents are those originating from steel mills and other metal-processing industries, where acid or alkaline cleaning solutions are employed. Control of effluent and stream pH requires treatment not only for the free acids and bases in the waste, but also for the acidic and/or basic or alkaline cleaning solutions employed. Waste treatment processes employed for the free acids and bases are normally effective in controlling acid and base salts.

TREATMENT TECHNOLOGY

There are many acceptable methods for treating acidic and alkaline wastes. These have been discussed in detail in the technical literature [16–20]. Treatment of both types of waste is based upon chemical neutralization, usually to pH 6–9. Methods include: mixing of compatible acid and alkaline wastes so that the net effect is a near-neutral pH; passing acid wastewaters through beds of limestone; mixing acid wastes with lime or dolomite lime slurries; adding the proper amounts of concentrated caustic soda (NaOH) or soda ash (Na_2CO_3)

Table 19.1 Example Sources and Characteristics of Extreme pH Wastewaters

Source	*pH*	*Reference*
Spent pickle liquor (H_2SO_4)	5% acid	1
Spent pickle liquor (HCl)	<0.1	2
Steel pickle rinsewater (HCl)	3.2	2
Spent copper-plating bath	0.2	2
Copper-plating rinsewater	2.5	2
Plating rinsewater	1.0–1.3	3
Chrome-plating rinsewater	2.0–2.5	4
Nickel-plating rinsewater	7.4	4
Printed circuit board rinsewater	5.2	5
Alkaline cleaner bath	12.5	6
Aluminum etching bath	13.3–13.8	7
Aluminum anodizing prep bath	0.55	7
Aluminum cleaning, anodizing, and plating rinsewater	1.7	5
Aluminum anodizing and plating rinsewater	7.1–10.6	5
Chlor-alkali plant	11.5–11.8	8, 9
Chrome pigments manufacture	1–3	10
Chrome pigments manufacture	7.9–8.2	9
Copper sulfate manufacture	3.0–3.3	9
Hydrofluoric acid production	3.1	9
Nickel sulfate manufacture	2.1–7.3	9
Secondary lead and battery manufacture	1.4	11
Gold ore extraction	11.7	12
Coke plant effluent	8.6–12.0	13
Acidic coal-cleaning leachate	2.3	14

to acid wastewaters; bubbling waste boiler-flue gas through alkaline wastes; adding compressed CO_2 to alkaline wastes; adding strong acid to alkaline wastes.

All the above techniques are well established, and have been employed in treating acid or alkaline industrial wastes. The chemical for and method of neutralization is selected on a basis of overall cost, as chemical costs vary widely, and the equipment for utilizing various neutralizing agents will differ with the treatment method employed. Other factors which influence reagent selection are speed of reaction and reaction by-products formed.

Process wastes may be treated either continuously or on a batch basis. Usually the flow rate will determine this except where treatment also involves removal of some other contaminant, as when pH neutralization is associated with the precipitation of some toxic metal. Due to its complexity, continuous-

Table 19.2 Comparison of Costs (1974–1975) of Neutralizing Chemicals for Alkaline Wastes [24, 25]

Chemical	*Relative Cost per ton*	*lb Required/100 lb Waste Alkali*		
		$CaCO_3$	*$Ca(OH)_2$*	*NaOH*
Sulfuric acid (50° Be)	1.0	126	170	158
Hydrochloric acid (20° Be)	2.6	232	314	250
Nitric acid (39.5° Be)	3.5	210	284	263

flow neutralization is generally used only when flow rates exceed 70 gpm [21].

Control of pH in continuous-flow treatment systems is one of the greatest challenges in treatment facility design and operation, due to the nonlinearity of the pH scale and titration curve. Variable influent wastewater flow rate and pH exacerbate the problem. Numerous automated control systems have been utilized, including feedback and feedforward controllers, feedback adaptive feedforward control, and multistage neutralization [22, 23]. The more simple but economical of these systems cannot achieve accurate pH control for fluctuating influent conditons. The more sophisticated systems give more reliable performance, but at much greater cost.

Alkaline Wastes

The choice of reagent for neutralization of an alkaline wastewater is generally between sulfuric acid and hydrochloric acid [18]. Sulfuric acid is usually selected because of its lower cost (Table 19.2). Hydrochloric acid often has the advantage of soluble reaction end products, which may, however, cause the waste to contain high effluent dissolved solids.

Carbon dioxide, primarily due to its greater cost, has been employed much less frequently to neutralize alkaline wastes than have sulfuric or hydrochloric acids. Treatment by flue gas, containing up to 14% by volume CO_2, is more common. Capital and operating costs of a flue gas-neutralization system employed on a 2.5 MGD flow have been reported [26]. The system, which involved two-stage pH adjustment of effluent from an ammonia stripping tower, consisted of first-stage adjustment from pH 11.0 to 9.6 by compressed and scrubbed stack gas, settling of calcium carbonate precipitate, and second-stage pH adjustment by stack gas to the range 6.8–7.5. Total operating costs were (in 1972) \$4.40 per million gallons.

One facility has reported switching from sulfuric acid to carbon dioxide, for pH control of a chlor-alkali wastewater [8]. Advantages claimed for the carbon dioxide system include protection against overshooting the pH adjustment and lower capital cost than for other alternatives evaluated. The gas is

Table 19.3 Comparison of Costs (1973–1974) of Neutralizing Chemicals for Acid Wastes [24, 25]

Chemical	*Relative Cost per ton*	*lb Required/100 lb Waste Acid*		
		H_2SO_4	*HCl*	*HNO_3*
Lump limestone, high Ca	0.6	110	148	86
Lump limestone, dolomitic	0.5	94	127	73
Pulverized limestone, high Ca	0.8	110	148	86
Pulverized limestone, dolomitic	0.7	94	127	73
Hydrated lime, high Ca	1.5	79	107	62
Hydrated lime, dolomitic	1.2	65	87	51
Pebble lime, high Ca	1.0	60	80	46
Pebble lime, dolomitic	0.9	54	73	42
Pulverized quicklime, high Ca	1.1	60	80	46
Pulverized quicklime, dolomitic	0.9	54	73	42
Sodium bicarbonate	10.0	173	233	135
Soda ash	6.3	119	160	93
Caustic soda (50%)	4.8	164	220	127

diffused into a 10-ft deep flowthrough treatment tank. Adequate depth is necessary to ensure maximum solution of the gas. The pH of the 17 MGD wastestream is reduced from 11.8 to 8.3 [8].

Acidic Wastes

The selection of a caustic agent to neutralize an acidic waste is usually between sodium hydroxide (caustic soda, NaOH), sodium carbonate (soda ash, Na_2CO_3), and various limes (calcium oxide compounds). The important factors in selection of a caustic reagent include purchase price, neutralization rate and capacity, storage and equipment costs, ease of handling, and neutralization end products. The relative costs of various caustic reagents are summarized in Tables 19.3 and 19.4. Although it is more expensive than other commonly used alkalis, liquid sodium hydroxide is frequently selected due to composition uniformity, ease of storage and feedings, rapid reaction rate, and solubility of end products. Sodium carbonate is not as reactive as sodium hydroxide and can produce foaming problems due to release of carbon dioxide [18].

Dolomite limestones are inexpensive. However, they are relatively insoluble in water. In spite of this they are widely used, sometimes as a pretreatment before alkalis which give closer pH control. At proper hydraulic loading rates, limestone-packed bed systems appear capable of achieving an effluent pH of about 5.0–6.5 [27]. Head loss through packed-bed reactors is a major design consideration. Dolomite limestone reacts much more slowly than

Table 19.4 Cost Comparison of Commercial Caustic Agents

Agent	*Chemical Formula*	*Basicity Factor*[a]	*Relative Treatment Cost*[b]
Sodium hydroxide	NaOH (50%)	0.687	6.24
Sodium carbonate	Na_2CO_3	0.507	10.98
High-calcium hydrate	$Ca(OH)_2$	0.710	1.86
Dolomite hydrate	$Ca(OH)_2 \cdot MgO$	0.912	1.18
High-calcium quicklime	CaO	0.941	1.00
Dolomite quicklime	$CaO \cdot MgO$	1.110	0.76
High-calcium limestone	$CaCO_3$	0.489	1.40
Dolomite limestone	$CaCO_3 \cdot MgCO_3$	0.564	1.05

[a] A measure of the alkali available in the commercial product for neutralization based upon grams equivalent of CaO per gram reagent [18].

[b] Based upon one unit of cost for CaO and the treatment effectiveness (i.e., "Basicity Factor") of each reagent.

does pure limestone, since the presence of magnesium carbonate at levels in excess of 1% retards the rate of limestone dissolution [28].

Limestone or lime slurries used to treat sulfuric acid wastes have a tendency to form sludges of insoluble sulfates, and reaction rates are slow [20]. Because of the tendency for formation of surface cakes of gypsum ($CaSO_4$) in treating sulfuric acid wastes with crushed limestone, alternate but more expensive neutralizing chemicals are sometimes selected. However, if that difficulty can be overcome, limestone neutralization has another advantage in addition to economy. Reaction of the limestone with acid yields a bicarbonate buffering effect and near-neutral effluent pH, without the need for automatic pH monitoring and chemical feed controls [1].

In one application of limestone neutralization, to a sulfuric acid rinsewater from steel pickling, a continuous-flow limestone treatment process has been successfully employed [1]. The process unit is a shallow, slowly stirred, upflow reactor. All calcium-based alkalis are characterized by their low solubility in water, and much of their reactivity results from maintaining a large, clean surface area. It is therefore necessary to ensure that the limestone particles do not remain coated with precipitated calcium sulfate. Stirring of the reactor maintains clean particle surfaces. The upflow mode of treatment flushes evolved carbon dioxide gas from the bed, as well as fine suspended materials which might blind the reactor. This unit has successfully treated a waste sulfuric acid of up to 5% w/v, to achieve a final effluent pH of 6.6–7.0 [11]. Others have recommended aeration of the limestone bed, to provide agitation and to flush evolved CO_2 from the waste liquid [29, 30]. Advantages claimed are increased reactor rate, and higher effluent pH due to CO_2 removal from the effluent.

There are several types of lime available (see Tables 19.3, 19.4). Quicklime is more economical than hydrated lime per ton of neutralizing chemical,

Table 19.5 Lime Treatment of an Acidic Wastewater [after 3]

Month	*Volume Treated (gal)*	*Wastewater pH*		*Lime Demand (lb/gal)*
		Influent	*Effluent*	
1	63,000	1.0	9.9	0.37
2	46,000	1.0	10.1	0.50
3	43,500	1.0	10.0	0.51
4	55,000	1.0	9.9	0.37
5	40,000	1.3	10.1	0.20
6	63,000	1.0	10.2	0.23

but requires the installation of on-site slaking equipment. Dry lime tends to bulk and bridge in storage, therefore requiring special agitation systems in the storage silo. Hydrated lime is more costly, but does not require slaking equipment and can be fed from storage directly to the neutralization reactor [31].

Caustic is available in either liquid or dry form. Liquid is usually chosen, for ease of handling, and the normal liquid form purchased is as 50% NaOH. However, at 50%, caustic freezing (crystallization) occurs at 50°F, requiring either indoor storage or heated outdoor storage tanks (frequently buried). In some instances, caustic shipped at 50% is diluted at the plant site, since the freezing temperature of 20% caustic is −20°F [31].

In one neutralization application, for a steel wire-pickling and copper-plating rinsewater, caustic was abandoned in favor of dry lime [2]. The treatment system involves neutralization and precipitation of metals. Neutralization by sodium hydroxide yielded large quantities of sludge with poor settling and thickening properties, resulting in short-circuiting in the clarifier. The modified system uses hydrated lime, fed via a screw auger from a 30-ton storage silo. Improved sludge properties have resulted from the switch to lime.

The use of sulfuric acid plus lime has been reported in treating a metal-finishing waste [32]. Both acid and alkaline wastes are produced in the industrial process, and an effluent pH of approximately 9.0 is routinely achieved. When lime is used, it is usually fed as a hydrated lime slurry, due to ease of addition and more rapid reactivity.

It has been pointed out that acid wastes which contain iron salts, as many metal-plating and ferrous industry wastewaters do, may require 3–10 times the amount of neutralizing caustic agent required by acidic solutions of equivalent pH which do not contain the iron salts [33, 34]. These salts of weak acids or bases influence the pH of effluents and receiving waters and exert a chemical demand for additional neutralizing agent.

Table 19.5 summarized the treatment efficiency and neutralizing agent demand for an acid waste originating from cleaning, plating, and other metal-finishing operations. The acid waste was overneutralized to an alkaline pH, in order to promote subsequent precipitation of waste heavy metals. Thus the large quantities of lime used resulted from four requirements:

- Neutralization of the free acids
- Neutralization of the combined acids (i.e., acid salts)
- Overneutralization to an alkaline pH, in order to achieve conditions suitable for heavy metal-hydroxide precipitation
- The hydroxyl ion required to react with the heavy metals in forming insoluble hydroxide precipitates.

Because of the lime demand, operating costs for this acid-waste treatment process were extremely high (in 1950) at $26.20/1,000 gal. Capital cost of the acid-waste treatment facility was $106,387, exclusive of the waste collection system [3].

Spent Pickle Liquor

Spent acids, used in pickling for removal of oxides from steel surfaces, typically contain 2–7% free acid, and 5–15% ferrous iron. Historically, spent pickling-bath liquors have been disposed by batch neutralization, or deep-well injection. More recently a market has developed for use of these acidic ferrous salt solutions for phosphorus control at municipal wastewater treatment plants [35].

In discussing the problem of spent pickle liquor from steel processes, Heynike and von Reiche [36] state that the classical method of using the concentrated acid until it is almost spent, and then neutralizing it with lime, is outmoded. Dewatering the sludges produced is impractical, and large sludge-holding lagoons require valuable space. Today, steel producers appear to be turning more to acid recovery and acid regeneration methods, effectively negating the need to treat the concentrated acidic waste [36, 37].

One additional method of utilization of spent concentrated hydrochloric acid is as feedstock for electrolysis plants, to produce chlorine. Several such plants are operating in Europe and Japan, and one U.S. plant became operational in 1972 [38].

Acid Mine Drainage

Increasing attention is being directed toward neutralization treatment of acid mine drainage, which may have a pH of 2.0 or less [39]. A typical neutralization system utilizes sodium carbonate, sodium hydroxide, limestone, or lime, with lime being most common. Except for limestone, it is the cheapest reagent (see Table 19.3), and also reacts rapidly. Its disadvantage is that voluminous sludge is produced. Limestone produces a denser sludge, but is not effective above pH 6.5 because of its slow reactivity at higher pH. Two-stage treatment with limestone followed by lime appears to offer the advantages of both reagents [39].

SUMMARY

Neutralization techniques for acid or alkaline wastes are well established and consist generally of use of sulfuric acid for alkaline solutions and one of several alkali chemicals for acid waste streams. Costs for treatment of wastes containing only free acids or bases are low, and normally represent only a fraction of the total waste treatment costs of a typical industrial plant. However, if the waste, in addition to free acids, contains other contaminants such as acid or base salts (e.g., ferrous sulfate) or heavy metals which will precipitate simultaneously with pH adjustment, the treatment chemicals demand is sharply increased, as well as the associated cost of waste treatment.

REFERENCES

1. Pilot, J. "Treatment of Pickling Plant Effluent at BSC Whitehead Works," *Effluent Water Treatment J.* 14 (1974): 260–263.
2. Reece, C. "Steel Wire Welding Manufacturer Meets Stringent Effluent and Sludge Disposal Criteria," paper presented at the 37th Annual Purdue Industrial Waste Conference, West Lafayette, IN (May 1982).
3. Cooper, J.E. "How to Dispose of Acid Wastes," *Chem. Ind.* (1950): 684–685.
4. Yeats, A. "Ion Exchange Selectively Removes Heavy Metals from Mixed Plating Wastes," paper presented at the 32nd Annual Purdue Industrial Waste Conference, West Lafayette, IN (May 1977).
5. Robinson, A. and J. Sum. "Sulfide Precipitation of Heavy Metals," USEPA 600/2-80-139 (June 1980).
6. Pinto, S. and L. Gould. "Ultrafiltration System Pays for Itself," *Ind. Wastes* (September/October 1980): 16–18.
7. Saunders, F., M. Sezgin and J. Medero. "Segregated Neutralization for Treatment of Concentrated Aluminum Finishing Wastes," in *Proceedings of the 4th Conference on Advanced Pollution Control for the Metal Finishing Industry,* USEPA 600/9-82-022 (December 1982).
8. Griffith, M. and D. Burgener. "Carbon Dioxide Neutralization of an Alkaline Effluent," *Ind. Wastes* (March/April 1982): 18–25.
9. *Treatability Studies for the Inorganic Chemicals Manufacturing Point Source Category,* USEPA 440/1-80-103 (July 1980).
10. Campbell, H.J., N.C. Scrivner, K. Batzar, and R.F. White. "Evaluation of Chromium Removal from a Highly Variable Wastewater Stream," *Proceedings of the 32nd Industrial Waste Conference, Purdue University* (May 1977), pp 102–115.
11. Mezey, E. "Characterization of Pollutants from a Secondary Lead and Battery Manufacturing Facility," USEPA 600/2-79-039 (January 1979).
12. Schmidt, J., L. Simovic, and E. Shannon. "Development Studies for Suitable Technologies for the Removal of Cyanide and Heavy Metals from Gold Milling Effluents," paper presented at the 36th Annual Purdue Industrial Waste Conference, West Lafayette, IN (May 1981).
13. Wong-Chong, G. and A. Dequittner. "Physical-Chemical Treatment of Coke Plant Wastewater at Crucible Steel, Midland, PA," paper presented at the Water Pollution Control Federation Conference, Las Vegas, NV (September 1980).

14. Wangen, L. and J. Williams. "Control by Alkaline Neutralization of Trace Elements in Acidic Coal Cleaning Waste Leachates," *J. Water Poll. Control. Fed.* 54 (1982): 1302–1310.
15. Day, R. and E. Lee. "Characteristics of Spent Concentrated Acid and Alkaline Cleaning Baths," *Proceedings of the 3rd Annual Conference on Treatment and Disposal of Industrial Wastewaters and Residues,* Houston, TX (April 18–20, 1978).
16. Dickerson, B.W. and R.M. Brooks. "Neutralization of Acid Wastes," *Ind. Eng. Chem.* 42 (1950): 599–605.
17. Nemerow, N.L. *Theories and Practices of Industrial Waste Treatment* (Reading, MA: Addison-Wesley, 1963).
18. Ross, R.D. *Industrial Waste Disposal* (New York: Reinhold, 1968).
19. Cooper, J.E. "How to Dispose of Acid Wastes," *Chem. Ind.* (1950): 684–685.
20. "Water Pollution Control," *Chem. Eng.* (Deskbook Issue) 78 (1971): 65–75.
21. Williams, E.T. "pH Neutralization Process Considerations," *Plating* 58 (1971): 967–970.
22. Gray, D. "New Solutions to pH Control Problems," *Poll. Eng.* (April 1984): 54–56.
23. Hoffman, F. "Control for Neutralizing Acid Wastes—An Overview," *Ind. Wastes* (July/August 1979): 25–31.
24. "Development Document for Effluent Limitations Guidelines and New Source Performance Standards for the Major Inorganic Products Segment of the Inorganic Chemicals Manufacturing Point Source Category," USEPA 440/1-74-007-a (March 1974).
25. Edwards, A.W., W.W. Eckenfelder, Jr., and J.A. Roth. *Industrial Water Cost Studies,* vol. I, Office of Research and Development, USEPA Contract 68-01-2485 (1975).
26. Evans, D.R. and J.C. Wilson. "Capital and Operating Costs—AWT," *J. Water Poll. Control. Fed.* 44 (1972): 1–13.
27. Bisogni, J. and G. Vaillancourt. "Neutralization of Dilute Acidified Streams with Limestone Packed Bed Reactors," paper presented at the Water Pollution Control Federation Conference, Las Vegas, NV (September 1980).
28. Pearsen, F.J. and A.J. McDonnell. "Use of Crushed Limestone to Neutralize Acid Wastes," *ASCE J. Environ. Eng. Div.* 1975: EE1 139–158.
29. Ohsawa, K. "Adjustment of pH of Waste Water," *Chem. Abstr.* 84 (1976): 249.
30. Krylenko, V.I. "Increasing the Neutralization Rate of Acidic Water with Ground Limestone," *Chem. Abstr.* 85 (1976): 264.
31. Lanouette, K.H. "Heavy Metals Removal," *Chem. Eng.* (Deskbook Issue) (1977): 73–80.
32. Hanson, N.H. "Design and Operation Problems of a Continuous Automatic Plating Waste Treatment Plant at the Data Processing Division, IBM Rochester, MN," *Proceedings of the 14th Industrial Waste Conference, Purdue University* (May 1959), pp 227–249.
33. Omya, S.A. "Treatment of Residual Acid Effluent," *Water Waste Treatment* 12 (1968): 27–28.
34. MacDougall, H. "Waste Disposal at a Steel Plant: Treatment of Sheet and Tin Metal Wastes," *ASCE Separate no. 493* (September 1954).
35. Howell, J. "Beneficial Disposal of Pickle Liquor," *Ind. Wastes* (May/June 1982): 21–22.

36. Heynike, J.J.C. and F.V.K. von Reiche. "Water and Pollution Control in the Iron and Steel Industry with Special Reference to the South African Iron and Steel Industrial Corporation," *Water Poll. Control* (London, 1969): 569–573.
37. Dempster, J.H. and H.M. Gomaa. "Recycling of Waste Acid from Steel Pickling Operations," *Ind. Waste,* (January 1973): 26–27.
38. "Recovering Chlorine from Waste HCl," *Environ. Sci. Technol.* 9 (1975): 16–17.
39. Mill, E.S. "Control and Prevention of Mine Drainage," paper presented at the Conference on Control and Recycling of Metals, Battelle-Columbus Laboratories, Columbus, OH (1972).

20
Phenol

Although described in the technical literature simply as *phenols*, this waste category may include a variety of similar chemical compounds among which are various substituted and polyphenols, chlorophenols, and phenoxyacids. In terms of pollution control, reported concentrations of phenol are thus often the result of a standard analytical methodology which measures a general group of similar compounds rather than being based upon explicit identification of the single compound, phenol (hydroxybenzene).

Many industries are characteristic sources of phenolic pollutants, including oil refineries, chemical plants, explosives manufacture, resin manufacture, and coke ovens [1]. Detailed studies of coke plant waste streams have demonstrated that while weak ammonia liquor is the major phenol source, other waste streams also contain significant phenol levels [2–4]. Furthermore, a variety of recovery configurations exist in the by-product recovery segment of coking plants. The concentration and nature of phenols produced vary, depending upon the temperature of carbonization and the type of coal used in the process [4]. The older, Beehive, process generates little wastewater, and phenol levels, when observed, are low [2].

Other sources of phenols include aircraft maintenance [5, 6], foundry operations [7], Orlon manufacture [8], caustic air scrubbers in paper-processing plants [9], rubber reclamation plants [10, 11], nitrogen works [12], stocking factories [13], textile mills [14], and fiberglass manufacturing [15–17]. Table 20.1 summarizes the levels of phenol in wastewaters of various industries.

TREATMENT TECHNOLOGY

Treatment technologies are available for reduction of all levels of initial phenol concentration, and frequently there is a region of overlap of application among methods. Both physicochemical and biological treatment are in successful full-scale industrial use, and high efficiencies of treatment are reported. However, any particular industrial system may encounter difficulties in employing a specific treatment method, depending upon the overall composition of the waste stream. Phenolic wastes often contain large concentrations of other waste constituents which require special treatment procedures. This is particularly true for refinery and coke oven wastes. Oil and cyanides are commonly present in

Table 20.1 Levels of Phenol Reported in Industrial Wastewaters

Industrial Source	*Phenol Concentration (mg/l)*	*Reference*
Coke ovens		
Weak ammonia liquor, without dephenolization	3,350–3,900	18
	1,400–2,500	19
	2,500–3,600	20
	3,000–10,000	21
	580–2,100	22
	700–12,000	13
	600–800	23
Undistilled ammonia liquor	505–1,670	24
Weak ammonia liquor, dephenolated	28–332	25
	10	21
	104	26
	10–30	19
	4.5–100	12
Distilled ammonia liquor	250–632	24
Wash oil still wastes	30–150	27
Coal conversion		
Synthane by-product water	1,700–6,600	28
	5,300–7,000	29
CO_2-acceptor process	9–27	29
Hygas process	3,400–6,800	29
Petroleum refineries		
Sour water	80–185 (140 avg)	27
General waste stream	40–80	30
Poststripping	80	31
General (catalytic cracker)	40–80	32
Mineral oil wastewater	100	12
API separator effluent	0.3–11	33
	0.35–6.8 (2.7 avg)	34
	8–36	35
	6–100	36
API separator/DAF effluent	28 avg (43 max)	37

Table 20.1 (*continued*)

Industrial Source	*Phenol Concentration (mg/l)*	*Reference*
Petroleum refineries		
General wastewater	30	38
	10–70	39
	10–100	40
	40–80	30
	1.4–55	41
Petrochemical		
General petrochemical	50–600	42
	31–800	43
Benzene refineries	210	12
Nitrogen works	250	12
Benzoic acid, benzaldehyde, phenol (toluene oxidation)	700–1,220	44
Tar distilling	300	12
Aircraft maintenance	200–400	5, 6
Herbicide manufacturing (includes chloro derivatives and phenoxy acids)	210	45
	239–524	46
Pesticide chemicals	2.8–129	41
Other		
Blast furnace raw waste loads	0.09–0.59	2
	0.47	26
Ferroalloy plants	5.6–9.1	47
Rubber reclamation	3–10	10, 11
Orlon manufacturing	100–150	8
Plywood manufacturing		48
log conditioning	0.20–4.7	
glue washwaters	0.14–4.2	
Wood preserving		48
general (13 plants)	50–953	
creosote	134	
	47	49
creosote + pentachlorophenol (5 plants)	82–258	
Pulp and paper	22	41
Wet-process hardboard	0.7–1.0	48
Gum and wood chemicals	4.7	41
Adhesives and sealants	47–244	41
Leather tanning and finishing	4.4–5.5	41
Phenolic resin production	1,600	50
Stocking factory	6,000	13

Table 20.1 *(continued)*

Industrial Source	*Phenol Concentration (mg/l)*	*Reference*
Other		
Synthetic phenol, plastics, resins	369	51
Phenolic resin	1,270–1,345	52
Resin intermediate production (*p*-nitrophenol)	2,400–25,000	52
Fiberglass manufacturing	40–400	16
Insulation fiberglass	4.1–363	53
Insulation fiberglass (recirculation system)	240–2,564	53
Paint manufacture	1.1	41
Paint stripping	1–4	54
Tank truck washwater	29 avg	55

large quantities in such phenolic wastewaters, and their removal, with attendant economical disadvantage, prior to phenol removal may be required. In general, however, phenol removal is successfully practiced by industry.

Treatment methods for phenol can be best discussed in terms of the phenol concentration in the wastewater to be treated. It is apparent from Table 20.1 that industrial wastes may contain an extremely wide range of phenolic material. Therefore treatment technology is discussed below under three headings—concentrated, intermediate, and dilute phenolic wastewaters. Recycle of process wastewater, with or without phenol removal, is also used with apparent success and is discussed separately.

Concentrated Wastewaters

High phenol concentrations (above 500 mg/l) in a waste stream made recovery an attractive consideration in the past, particularly in the by-product coking process. Recent studies, however, have recommended abandoning dephenolization in favor of the implementation of more complete treatment systems. Changing economics, through competition from other industries, have made recovery of phenols and other constituents unprofitable for many coke plants [2].

Phenol recovery value from coke plant ammonia liquor is estimated in 1969 at 20¢/ton of coal processed [21]. The more common and successful methods employed for phenol recovery from concentrated wastes are based upon extractive recovery into an imiscible organic solvent. Efficiencies are extremely high, with recoveries of 98–99% reported. However, even such high

Table 20.2 Recovery Processes for Concentrated Phenolic Wastewaters [12]

Process	*Influent Phenol (mg/l)*	*Effluent Phenol (mg/l)*
Benzene-caustic dephenolization process	3,000	210–240
Countercurrent Podbielniak extractors	2,000	100
Pulsed column extractors	2,200	30
Phenosolvan dephenolization (Luigi)	1,570	4.5
	2,465	9.6
IFAWOL dephenolization process (Carl Still)	4,000	40

percent efficiencies can leave significant residual phenol in the treated effluent of a concentrated waste stream. The problem of solvent loss due to slight solubility in the wastewater, versus efficiency of phenol removal, must be balanced economically, for solvent costs can be high. In addition, any solvent dissolved in the extracted water represents increased organic matter in the waste. Commercial recovery processes are summarized in Table 20.2, accompanied by reported operating data. Pilot scale results from other solvent extraction studies are presented in Table 20.3.

The Koppers light-oil extraction process has been reported to reduce 1,500–2,000 mg/l phenol to 10–30 mg/l [19]. Another plant reduced influent phenol concentrations of 450 mg/l to 0.5 mg/l by dephenolization [2]. Vapor recirculation, involving steam stripping and countercurrent caustic extraction, is also used for dephenolization in the by-product coking process, with reported recoveries of 95% [2]. Other processes are also available [56]. Cost data for three by-product coke plants, using ammonia stripping plus dephenolization, and attendant treatment results, are present in Table 20.4.

Successful treatment of concentrated phenolic waste by activated carbon has been reported [57]. The waste, initially at 2,235 mg/l phenol, was reduced to 1,950 mg/l by chemical coagulation and filtration. Activated carbon treatment yielded an effluent phenol level below 0.1 mg/l. However, rapid phenol breakthrough in the three series carbon columns was of concern, relative to process

Table 20.3 Pilot-Scale Results on Solvent Extraction for Phenol Recovery [after 41]

		Phenol (mg/l)		
Industry	*Solvent*	*Influent*	*Effluent*	*% Recovery*
Organic chemicals	Isobutane	68	66	3
Organic chemicals	Isobutylene	67	63	6
Organic chemicals	Isobutylene	580	160	72
Petroleum refining	Isobutylene	23,000	9,600	58

Table 20.4 Cost and Performance Data for Ammonia Stripping/Dephenolization Treatment, By-product Coke Ovens [23]

Parameter	*Plant A*	*Plant B*	*Plant C*
Waste flow, MGD	0.59	0.24	0.086
Influent phenol (mg/l)	450	310	540
Effluent phenol (mg/l)	1.4	219	4.6
Capital costs ($\$ \times 10^6$)	2.35	4.00	2.00
Operating cost ($/1,000 gal)	5.59	19.40	19.60

economics. Rapid phenol breakthrough has been confirmed in a three-carbon column pilot study on a phenolic resin production wastewater [52].

It has been reported that phenol solutions of 7,000 mg/l can be incinerated, for fuel costs in 1971 of 1.5¢/1,000 gal. Capital costs for the incinerator were estimated at $15,000 [58].

Intermediate Strength Wastewaters

Wastewaters containing phenol in the range of 5–500 mg/l may be considered of intermediate strength. Processes applicable to such wastewaters include biological treatment, activated carbon, and chemical oxidation, among others.

In the absence of high concentrations of toxic substances, or in instances of their successful prior removal, biological treatment is widely employed for treatment of wastewaters containing intermediate phenol levels. The biological processes include lagoons, oxidation ditches, trickling filters, and activated sludge. Table 20.5 summarizes reported performance of various biological processes employed for treatment of intermediate and dilute strength phenol in wastewaters. The activated sludge system is reported preferable to other biological processes because of its high treatment efficiency and ease of control [31]. Vela and Ralston [49] have reported that the efficiency of activated sludge degradation of phenol is not temperature dependent above 10°C, but declines with temperature below 10°C.

Phenol concentrations up to 500 mg/l are generally considered suitable for treatment by biological processes, but as illustrated in Table 20.5, higher levels have been successfully subjected to biological treatment. The exact concentration of phenol that can be directly treated biologically will depend to some extent upon other contaminants in the waste, and whether or not these contaminants are toxic to the microorganisms employed in the biological process. The cumulative effect of toxic constituents in phenolic wastes has been discussed by Reid and Libby [6], wherein chromium accumulation gradually decreased the biological activity in a phenol treatment system.

Biological systems are also subject to fluctuations in organic and hydraulic

Table 20.5 Performance of Biological Systems in the Treatment of Phenolic Wastes

Treatment Process	*Source of Industrial Waste*	*Phenol (mg/l)*		*Reference*
		Influent	*Effluent*	
Biooxidation pond	Refinery wastewater	30	1	38
	Paint manufacture	1.1	0.003	41
	Leather tanning and finishing	4.4	0.0	51
Aerated lagoon	Municipal plus herbicide manufacturing waste	0.8–1.65	0.2–0.45	46
	(Same–phenoxyacids)	2.15–4.14	1.51–1.73	46
Stabilization pond	Effluent from aerated lagoon above			
	Phenols	0.2–0.45	0.06–0.13	46
	Phenoxyacids	1.51–1.73	1.10–1.48	46
Oxidation ditch	Coke oven	990	8.5	22
Trickling filter	Oil refinery	30–40	0.5–0.7	25
		—	0.04	59
		9–25	0.6–3.8	27
	Petrochemical plant	83	55	43
		800	336	43
		48–52	11–19	43
		500	250	43
		31	2	43
		49	1	43
	Wood preserving[a]	31	<1	41
	Aircraft maintenance wastes	8	0.02	5
Activated sludge	Coke oven	400–700	10	23
		6.5	0.008	60
		350	0.06	41
		9–10	0.9	12
		890	0.65	25
		110–200	0–11	13
	Coke oven[a]	3,350–3,390	0.2–0.8	18
	Coke oven with dephenolization	135	7	13
	Coke oven without dephenolization	1,400	15	13
		350	0.064	2

Table 20.5 *(continued)*

Treatment Process	*Source of Industrial Waste*	*Phenol (mg/l)*		*Reference*
		Influent	*Effluent*	
Activated sludge	Oil refinery	80	<0.5	31
		8–10.5	<0.01–0.055	33
		0.9–9	<0.01–0.25	33
		0.3–1.1	<0.01–0.6	33
		3–8	.007–0.05	53
		40–80	≤1.0	30
		15–20	0.01–0.03	61
	Gas liquor	1,200	12	62
		200	0.2	62
	Stocking manufacture	183–455	0.1–1.3	13
	Leather tanning and finishing	5.5	1.4	41
	Creosote waste	47	4	49
	Fiberglass manufacturing	40–400	0.1–2.6	16
		199	0.8	53
	Refinery and petrochemical	10–100	<1	40
	Benzoic acid, benzaldehyde, phenol manufacturing	700–1220	0.23–0.84	44
Activated sludge plus trickling filter	Oil refinery	8–11	<0.01–0.5	33
Activated sludge plus Dow-Pac filter	Oil refinery	40–80	0.35	30
Activated sludge plus powdered activated carbon	Organic chemicals	1.0	0.01	63
	Oil refinery[a]	6.0	0.01	41
	Oil refinery[a]	6.0	0.01	41
		2.4	<0.01	41
		1.4	0.01	41
		55	0.06	41
Aero accelerator (activated sludge)	Coke oven	150	≤1.0	52
	Foundry wastes	3,000	30–150	53

[a] Pilot plant data.

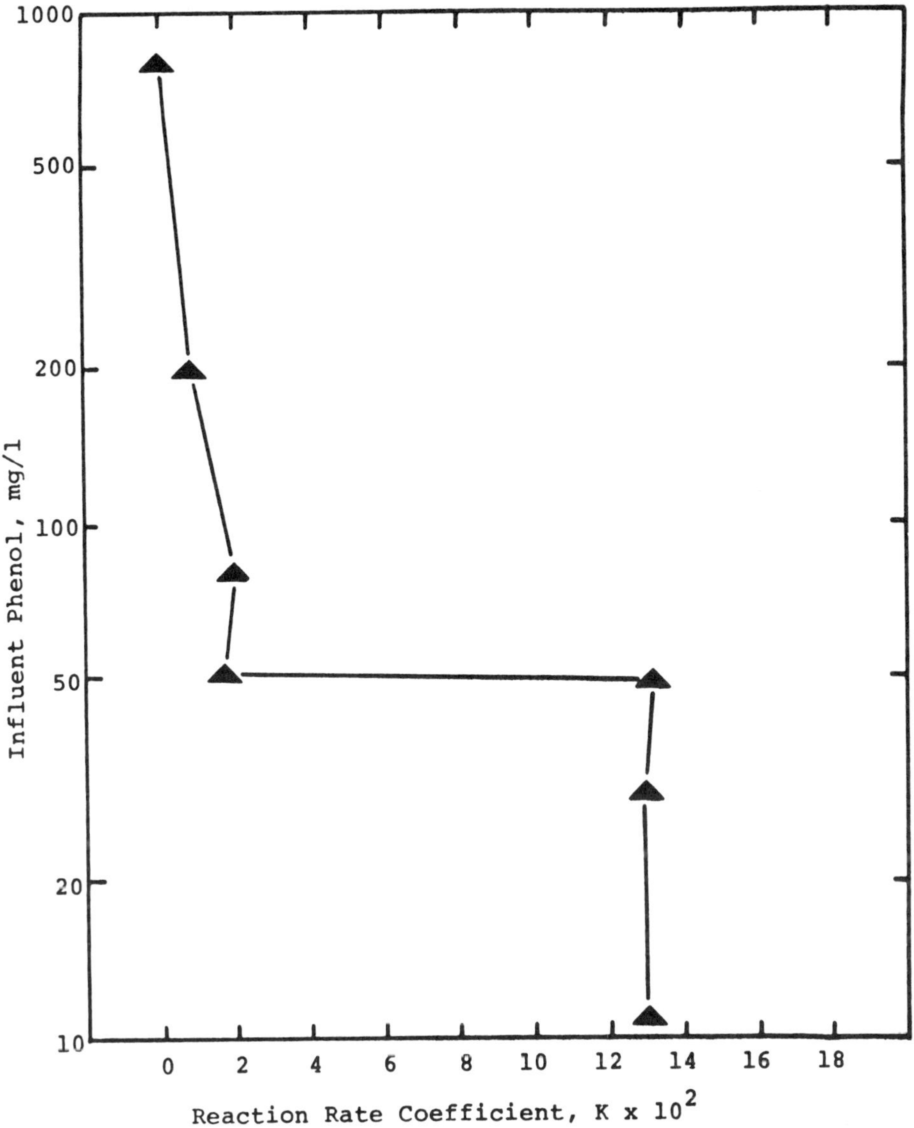

Figure 20.1 Influence of phenol concentrations on trickling filter reaction rate coefficient. (From Joyce [43], courtesy of American Institute of Chemical Engineers.)

loadings, and effluent values will vary accordingly. Joyce [43] has reported a correlation between phenol loadings for trickling filtration treatment of phenolic petrochemical wastes, and efficiency of phenol removal. His data also suggest a possible toxicity effect on the trickling filter population, at influent phenol values above 50 mg/l (Figure 20.1).

Table 20.6 Treatment Costs Reported for Biological Systems Treating Phenolic Wastewaters

	Design Flow (MGD)	*Capital Cost ($/1,000 gpd)*	*Treatment Cost (¢/1,000 gal)*	*Reference*
Activated sludge	0.072	4,513 (1974)[a]	—	44
	2.16	463 (cumulative 1965–1970)	26	64
	0.48	1,456 (1974)	84	2
	0.15	7,733 (1971)	—	53
	0.11	2,818 (1962)	—	7
	1.92	417 (1965)	22	18
Activated sludge, sand filtration	2.88	868 (1970)	22	34
Air flotation—holding pond, trickling filter	6.5	154 (1972)	—	40

[a] Year of cost data.

Problems with highly variable phenol loadings in a refinery wastewater being treated by activated sludge, plus a change in overall waste characteristics between design and construction of the plant, have been reported [9]. A strong process stability dependency upon sulfate concentration has been reported for one full-scale activated-sludge plant [44]. The relative concentrations of phenol and sulfides in a refinery waste have also reportedly influenced treatment efficiency [64]. Limited, bench-scale data on phenol effects on anaerobic digestion indicate little toxicity over a phenol range of 1–100 mg/l [65]. Some phenolic wastes, such as wet scrubber effluents in the ferroalloy industry, may not be amenable to biological treatment due to overall low BOD in the wastewater, despite containing appreciable (5–10 mg/l) phenol [47]. Cost data reported for biological systems treating phenolic wastewaters are presented in Table 20.6.

Other treatment methods have been employed for phenol removal. One of these is activated carbon. Several synthetic resin manufacturers are reported using activated carbon for phenol removal [66]. Table 20.7 summarizes treatment results achieved with activated carbon for phenol removal. For most applications, extremely efficient treatment is achieved. However, there have been reports of rapid phenol breakthrough in carbon columns [52, 57].

Treatment costs for activated carbon are given in Table 20.8. Actual costs for phenol removal by activated carbon depend upon the total quantity of other adsorbable materials in the waste stream. Generalized capital and operating costs for activated carbon are given in Figure 20.2. In many cases, activated

Table 20.7 Performance of Activated Carbon in Phenol Removal [after 41, 67]

Industry	*Phenol (mg/l)* Influent	*Phenol (mg/l)* Effluent	*% Removal*
Organic chemicals	36	0.001	>99
	0.52	0.05	>99
	0.12	0.003	>99
	0.32	0.001	>99
	12.8	0.001	>99
Dye formulation	6.0	0.01	>99
Plastics manufacture	9.8	0.013	>99
	8.5	0.006	>99
	26.5	0.005	>99
Gum and wood chemicals	4.7	0.6	88
Pesticide chemicals	77.9	2.3	97
	2.8	<0.7	>75
	129	4.3	97
Explosives	16.6	0.023	>99
Petroleum refinery	44.0	0.001	>99
Inorganic chemicals	5,325	0.25	>99

carbon is preceded by granular media filtration. Figures 20.3 and 20.4 provide generalized capital and operating costs for granular media filtration.

Incremental cost beyond biological treatment for activated carbon treatment of an insulation fiberglass waste (effluent phenol reduced from 0.8 mg/l to 0.05 mg/l) was estimated in 1974 to be $1,066/1,000 gpd capacity, with $0.48/1,000 gal operating costs. Plant capacity was 0.15 MGD [53]. Cost data esti-

Table 20.8 Costs (1971–1972) for Activated Carbon Treatment of Phenolic Wastewaters

Phenol (mg/l)	*Flow (MGD)*	*Capital Costs ($/1,000 gpd)*	*Treatment Costs ($/1,000 gal)*	*Reference*
6	2.88	694	0.18	34
210	0.15	2,000	0.36	45
370	0.50	2,600	1.57	51

All systems include on-site carbon regeneration.

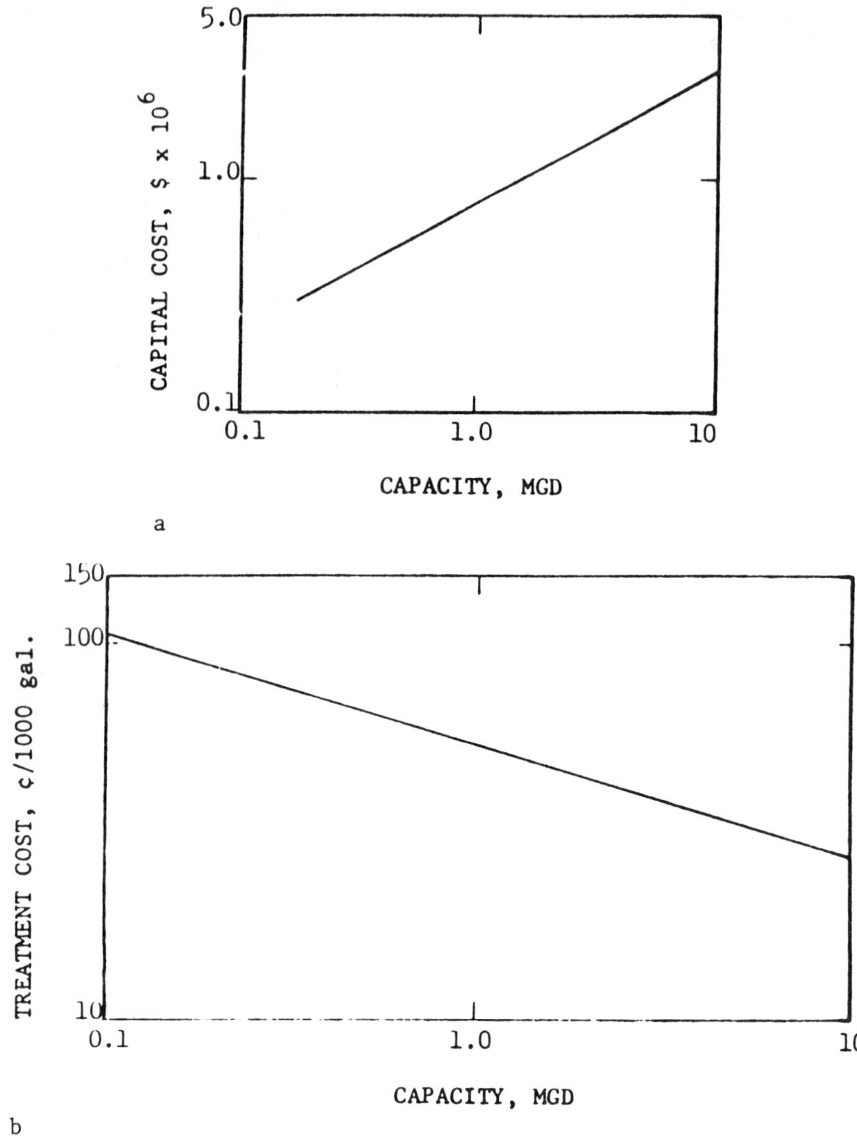

Figure 20.2 Capital and operating costs (in 1974) for activated carbon treatment [68].

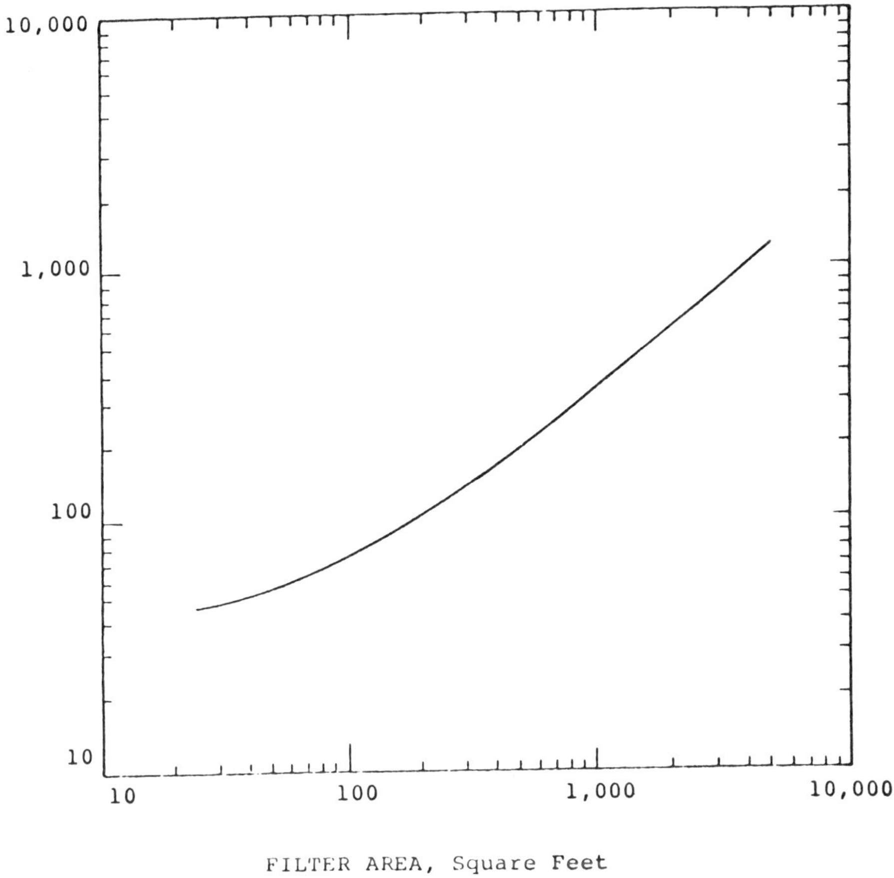

Figure 20.3 Capital cost (in 1975) of granular media filtration [69].

mated for activated carbon treatment of petroleum refinery wastes, following biological treatment and granular media filtration, are given in Table 20.9.

Several applications of chemical oxidation of phenols have been reported [11, 26, 41, 70–72]. Oxidants employed include permanganate, chlorine and chlorine dioxide, ozone, and hydrogen peroxide. In one application of permanganate to a wastewater at initial phenol concentration of 125 mg/l, phenol removal of only 62% was achieved [71].

The reactions of phenol with chlorine have been extensively investigated, due to the strong taste and odor problems which can result from incomplete reaction and resultant yield of chlorophenolic compounds [73]. With chlorination oxidation, extremely high concentrations of chlorine must be dosed to effect complete phenol oxidation [48, 70, 71]. A primary interference is the presence of other oxidizable substances, which also exert a chlorine demand. Figure 20.5 presents treatment data for chlorination of phenolic wastes. Chlo-

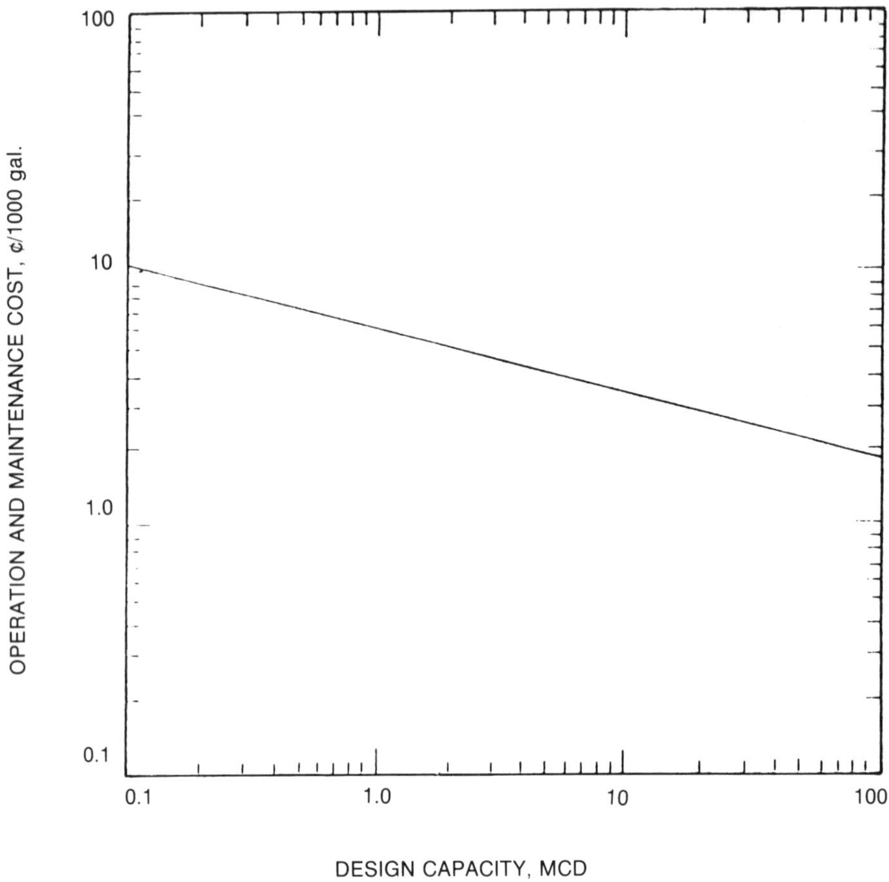

Figure 20.4 Operation and maintenance cost (in 1975) for granular media filtration [69].

rination must be carried to completion and at pH less than 7, or toxic chlorophenols are produced [67]. Using chlorination, zero residual phenol, reduced from approximately 5 mg/l, resulted for rubber reclamation wastes [11]. Chlorination for phenol and cyanide removal have been proposed for wet scrubber wastes in the ferroalloy industry, where wastes contain insufficient organics to sustain biological treatment [47]. Simple lagooning plus chlorination reduced 7.27 mg/l phenol to 0.5 mg/l.

Ozone and chlorine dioxide have also been reported as successful oxidants for phenol removal [11, 15, 23, 25, 30, 31, 48, 70]. The stoichiometric relationships between oxidant dose and phenol removals are shown in Figures 20.6 and 20.7, respectively. Chemical oxidation of both concentrated and intermediate strength phenol wastes by ozone has been proposed [74, 75]. No full-scale applications of the technology have been reported, however. Ozonation

Table 20.9 Estimated Incremental Treatment Cost (in 1974) Beyond Biological Treatment for Activated Carbon Treatment of Petroleum Refining Wastes [36]

Type of Plant	*Waste Flow (MGD)*	*Capital Cost ($/1,000 gpd)*	*Operating Cost ($/1,000 gal)*
Crackling	0.38	3,888	4.06
	1.88	1,920	1.84
	3.75	1,432	1.37
Petrochemical	0.75	2,733	2.77
	3.0	1,567	1.50
	6.0	1,108	1.06
Lubrication	1.13	2,400	2.37
	4.95	1,232	1.18
	11.25	844	0.81
Integrated	3.12	1,540	1.51
	7.3	1,077	1.04
	15.6	673	0.65

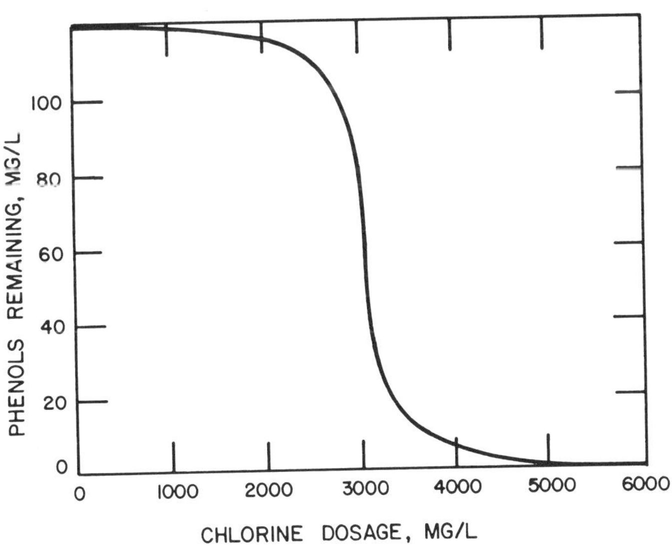

Figure 20.5 Phenol reduction by chlorine treatment. (From Cleary and Kinney [70], courtesy of Purdue University Industrial Waste Conference.)

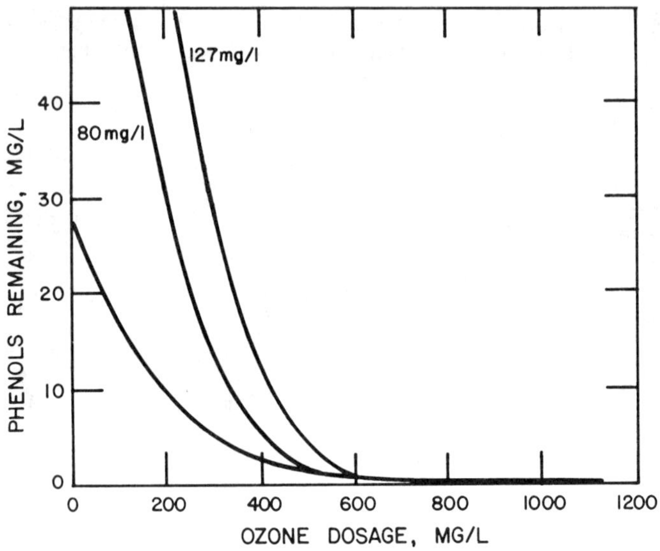

Figure 20.6 Phenol reduction by ozonation. Initial phenol concentration as indicated. (From Cleary and Kinney [70], courtesy of Purdue University Industrial Waste Conference.)

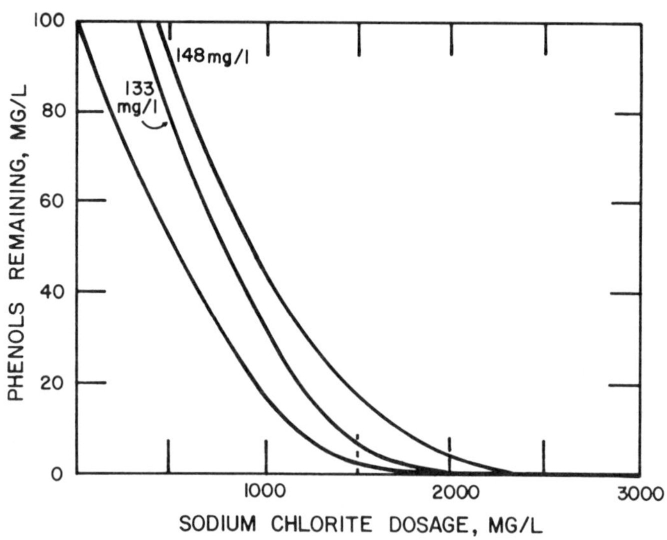

Figure 20.7 Phenol reduction by application of chlorine dioxide (as sodium chlorite). Initial phenol concentrations as indicated. (From Cleary and Kinney [70], courtesy of Purdue University Industrial Waste Conference.)

of phenol produces carbon dioxide as an end product, with catechol and orthoquinine as intermediates [74]. The oxidation reaction rate is first order with respect to both phenol and ozone concentrations, and the reaction rate increases with pH [74, 75]. In one bench-scale application of ozonation to a wastewater of the adhesives and sealants industry, phenol was reduced from 47 to 0.13 mg/l [41]. Although ozone is capable of efficient phenol oxidation, operating and capital costs may be extremely high, and ozone would likely be considered only as a low level or polishing treatment. Besselievre [76] reports that 1.5–2.5 parts ozone to 1 part phenol is effective for phenol removals to 3 μg/l.

Wong-Chong and Dequittner [26] have suggested that hydrogen peroxide has several advantages over other chemical oxidants for phenol removal. Disadvantages of chlorine include risk of formation of chlorinated phenolic compounds, and excess chlorine demand associated with other wastewater constituents. Ozone requires a high capital investment for the ozone generator. Potassium permanganate requires solid-feed equipment, has a high chemical cost per unit of oxidizing capacity, and produces a sludge not associated with alternative oxidant chemicals.

Hydrogen peroxide is reported to readily oxidize phenol in the presence of ferrous ion catalyst [72]. At sufficient peroxide dosage, oxidation to carbon dioxide end product is complete. The process is reported in use by a sanitizer chemical formulator and in the petroleum-refining and chemicals industries [72], as well as in coke-plant waste treatment [26]. Iron catalyst concentrations up to 300 mg/l have been reported, with reaction completed within five minutes [77]. Optimum treatment pH is reported to be pH 3–4 [72, 77], with sharp reductions in efficiency at higher and lower pH [26]. Pure phenol solutions require 3.0–4.0 mg peroxide/mg phenol [72]. However, other constituents in a wastewater can also exert a peroxide demand. Complete treatment of a coke plant waste requires 15–20 mg peroxide/mg phenol, due to peroxide reaction with wastewater ammonia [26].

Other methods of control of intermediate concentration phenol wastewaters have been reported. Ultrafiltration has been tested at pilot scale on wastewater from the adhesives and sealants industry. Influent phenol levels of 84–244 mg/l were only partially removed, to yield effluent phenol of 13–102 mg/l [41]. Klemetson and Scharbow [78] have reported that ultrafiltration membrane rejection, and therefore increased phenol removal efficiency, increases with increasing pH. Rejections near zero were observed at pH 6.2, and in excess of 70% at pH 10.

A wastewater hydrocarbon stripper reduced initial phenol levels of 5–6 mg/l to less than 1 mg/l [79]. Stripped gases were combusted. Capital costs in 1966 were $139/1,000 gpd for a 0.18 MGD wastewater flow, and $74/1,000 gpd for a 1.1 MGD wastewater flow. Operating costs for the fully automated units were 9.3¢ and 7.0¢/1,000 gallons, respectively.

Dilute Wastes Treatment

Biological treatment generally is capable of reduction of phenol down to the 0.5–1 mg/l level, except in instances of very high influent phenol concentrations (see Table 20.5). Removal below the 1.0 mg/l level can be achieved by other processes. Normally, chemical or physicochemical methods replace the biological processes for treatment to these low phenolic values.

Biologically treated refinery effluent with a phenol residual of 0.16–0.35 mg/l was reduced to 0.003 mg/l with ozone treatment [30]. The ozone treatment was reportedly so effective that an activated-carbon polishing unit designed to follow the ozonator was not required.

Activated carbon use is also well established for removal of trace organics as a tertiary treatment process, and is equally effective for moderate strength and dilute phenolic wastes (see Table 20.7). Activated carbon has been reported to have a phenol capacity of 0.5–25 lbs/100 lbs carbon [30, 67]. The process is most effective at lower pH, for phenol and other weak organic acids.

Use of macroreticular polystyrene resin columns for removal of phenol from waste brine has been reported [80]. A three-column series unit in continuous treatment accomplished 99% removal overall for influent levels of 5 mg/l phenol. Regeneration was achieved with 1% NaOH. Relatively few full-scale resin adsorption plants are in use. One plant in Indiana is reported to have installed a facility in 1976 for phenol recovery, and another has been installed at a coal liquification plant in West Virginia for removal of phenol and polycyclic hydrocarbons [41].

Recycle Practice

Recycle of phenolic wastewaters, with or without phenol removal, has been reported. Several reports [15–17] have detailed the reuse of rinsewater to wash continuous formation belts in fiberglass manufacturing. The water cycles in the rinse system and is used as formation water for the phenolic resin binder sprayed onto the glass. Therefore, soluble phenolics are recovered and reused. From 20 to 40% blowdown is necessary to prevent excessive solids buildup. Recycle water characteristically contains 240 mg/l phenol, 16,000 mg/l dissolved solids, and 200 mg/l suspended solids [16]. It has been reported that total recycle is practical for this particular industry, and 32% of existing plants have already achieved this status [53]. Generalized costs based upon a 0.15 MGD blowdown were reported (in 1974) to be $5,233/1,000 gpd and $9.31/1,000 gal [53]. Wastewater recycle has also been recommended for the plywood and the wood and wood-preserving industries [48].

One refinery [39] has incorporated biological treatment directly into its recycle system. Phenol wastewaters are employed as makeup to the cooling water system at 800–1,000 gpm. The forced-draft and induced-draft cooling towers function as biological reactors. The blowdown of 200–500 gpm typically contains 0.08–9.8 mg/l phenol, representing 99.4–99.9+% removal. Nine years' operating experience has substantiated that limited clogging and corrosion problems exist.

SUMMARY

Phenol removal to very low residual levels is technologically feasible and in practice. Removals below the mg/l level are accomplished by multiple treatment sequences, since increasing efficiency for any one process usually is associated with a substantial increase in cost. Processes are available for 99% or greater removals for the 500–10,000 mg/l phenol level. Such processes are typically product recovery oriented. Biological and physicochemical processes have been shown capable of up to 99% phenol removal in the 5–500 mg/l range. Specialized biological systems are capable of treating even higher phenol concentrations. Although not favorable for direct treatment of concentrated or intermediate level phenolic wastes, chemical oxidation processes such as chlorination and ozonation demonstrate favorable economics when employed as polishing processes after biological systems.

Carbon adsorption is in use for both intermediate and dilute phenol levels, but displays a wide range of capital and operating costs depending upon the specificity of contaminant removal.

REFERENCES

1. McKee, J.E. and H.W. Wolf. *Water Quality Criteria*, 2nd ed. California State Water Quality Control Board, Publication 3-A (1963).
2. "Development Document for Effluent Limitations Guidelines and New Source Performance Standards for the Steel Making Segment of the Iron and Steel Manufacturing Point Source Category," USEPA 440/1 74-024-a (June 1974).
3. Adams, C.E., Jr., R.M. Stein, and W.W. Eckenfelder, Jr. "Treatment of Two Coke Plant Wastewaters to Meet EPA Effluent Criteria," *Proceedings of the 29th Industrial Waste Conference, Purdue University* (May 1974), p 29.
4. Pearce, A.S. and S.E. Punt. "Biological Treatment of Liquid Toxic Wastes," *Effluent Water Treatment J.* 15 (1975): 32–39.
5. Reid, G.W., R. Daigh, and L. Wortman. "Phenolic Wastes from Aircraft Maintenance," *J. Water Poll. Control. Fed.* 32 (1960): 353–391.
6. Reid, W. and Roscoe W. Libby. "Phenolic Waste Treatment Studies," *Proceedings of the 12th Annual Purdue Industrial Waste Conference, Purdue University*, West Lafayette, IN (May 1957), pp 250–258.
7. Barzler, R.P., D.J. Giffels, and E. Willoughby. "Pollution Control in Foundry Operations," *Industrial Pollution Control Handbook*, Herbert F. Lund, ed. (New York: McGraw-Hill, 1971).
8. Schlesinger, H.A., E.F. Dul, and T.A. Fridy, Jr. "Pollution Control in Textile Mills," *Industrial Pollution Control Handbook*, Herbert F. Lund, ed. (New York: McGraw-Hill, 1971).

9. Ross, R.D. *Industrial Waste Disposal* (New York: Reinhold, 1968).
10. Parsons, W.A. *Chemical Treatment of Sewage and Industrial Wastes* (Washington, DC: National Lime Association, 1965).
11. Sechrist, W.D. and N.S. Chamberlin. "Chlorination of Phenol Bearing Rubber Wastes," *Proceedings of the 6th Industrial Waste Conference, Purdue University* (May 1951), pp 396–412.
12. Wurm, H.J. "The Treatment of Phenolic Wastes," *Proceedings of the 23rd Industrial Waste Conference, Purdue University* (May 1968), pp 1054–1073.
13. Ide, T. "Formal Discussion to Biczysko and Suschka, 'Investigations on Phenolic Wastes Treatment in an Oxidation Ditch,'" *Advances in Water Pollution Research Conference, Munich* (New York: Pergamon Press, 1967), vol. 2, pp 285–295.
14. "Development Document for Effluent Limitations Guidelines and New Source Performance Standards for the Textile Mills Point Source Category," USEPA 440/1-74-022-a (June 1974).
15. Baloga, J.M., F.B. Hutto, Jr., and E.I. Merrill. "A Solution to the Phenolic Problem in Fiberglass Plants," *Water Sew. Works* 118 (1971): 7–13.
16. Fletcher, G.W., S.H. Thomas, and D.E. Cross. "Development and Operation of a Closed Wastewater System for the Fiberglass Industry," paper presented at the 45th Annual Water Pollution Control Federation Meeting, Atlanta, GA (1972).
17. "Phenolic Waste Reuse by Diatomite Filtration," Johns-Manville Products Corporation, USEPA 12080EZF 09/70 (1970).
18. Kostenbader, P.D. and J.W. Flecksteiner. "Biological Oxidation of Coke Plant Weak Liquor," *J. Water Poll. Control. Fed.* 41 (1969): 199–207.
19. Fisher, C.W. "Coke and Gas," *Chemical Technology,* Vol 2. *Industrial Wastewater Control*, F. Fred Gurnham, ed. (New York: Academic Press, 1965).
20. Carbone, W.E., R.N. Hall, H.R. Kaiser, and C.G. Bazell. "Commercial Dephenolization of Ammonical Liquors with Centrifugal Extractors," *Proceedings of the 5th Ontario Industrial Waste Conference* (1958), pp 42–58.
21. *State of the Art: Review on Product Recovery*, Water Pollution Control Research Services 17-7 ODJW 11/69 (Washington, DC: USDI, 1969).
22. Biczysko, J. and J. Suschka. "Investigations of Phenolic Wastes Treatment in Oxidation Ditch," *Advances in Water Pollution Research Conference, Munich,* (New York: Pergamon Press, 1967), vol. 2, pp 294–295.
23. Clough, G.F.S. "Biological Oxidation of Phenolic Waste Liquor," *Chem. Proc. Eng.* 42 (1961): 11–14.
24. Eklund, C.W. and J.W. Irvin. "Alternate Methods of Treating Steel Mill Coke Plant Wastewaters," paper presented at the 53rd Annual Water Pollution Control Federation Meeting, Las Vegas, NV (1980).
25. Lesperance, T.W. "Biological Treatment of Phenols," *Proceedings of the 8th Ontario Industrial Waste Conference* (1961), pp 59–66.
26. Wong-Chong, G. and A. Dequittner. "Physical-Chemical Treatment of Coke Plant Wastewater at Crucible Steel, Midland, PA," paper presented at the Water Pollution Control Federation Conference, Las Vegas, NV (September 1980).
27. Graves, B.S. "Biological Oxidation of Phenols in a Trickling Filter," *Proceedings of the 14th Industrial Waste Conference, Purdue University* (May 1959), pp 1–6.
28. Valiknoc, T. and R.D. Neufeld. "Thiocyanate Toxic Inhibition to Phenol Biooxidation," paper presented at the 33rd Annual Purdue Industrial Waste Conference, West Lafayette, IN (May 1978).

29. Sack, W.A. and W.R. Bokey. "Biological Treatment of Coal Gasification Wastewater," paper presented at the 33rd Annual Purdue Industrial Waste Conference, West Lafayette, IN (May 1978).
30. McPhee, W.T. and A.R. Smith. "From Refinery Waste to Pure Water," *Proceedings of the 16th Industrial Waste Conference, Purdue University* (May 1961), pp 311–312.
31. Benger, M. "The Disposal of Liquid and Solid Effluents from Oil Refineries," *Proceedings of the 21st Industrial Waste Conference, Purdue University* (May 1966), pp 759–767.
32. Steck, W. "The Treatment of Refinery Waste Water with Particular Consideration of Phenolic Streams," *Proceedings of the 21st Industrial Waste Conference, Purdue University* (May 1966), pp 783–790.
33. Short, T.E., Jr., B.L. DePrater, and L.H. Meyers. "Controlling Phenols in Refinery Wastewaters," *Oil Gas J.* 72 (1974): 119–124.
34. Peoples, R.F., P. Krishnan, and R.N. Simonsen. "Nonbiological Treatment of Refinery Wastewater," *J. Water Poll. Control. Fed.* 44 (1972): 2120–2128.
35. Mahmud, Z. and W.C. Thanh. "Biological Treatment of Refinery Wastes," paper presented at the 33rd Annual Purdue Industrial Waste Conference, West Lafayette, IN (May 1978).
36. "Development Document for Effluent Limitations Guidelines and New Source Performance Standards for the Petroleum Refining Point Source Category," USEPA 440/1-74-014-a (April 1974).
37. Lindsay, J.T. and B.V. Prather. "Solving Petroleum Refinery Wastewater Problems," *J. Water Poll. Control. Fed.* 49 (1977): 1779–1785.
38. Wigren, A.A. and F.L. Burton. "Refinery Wastewater Control," *J. Water Poll. Control. Fed.* 44 (1971): 117–128.
39. Mohler, E.F., Jr., H.F. Elkin, and L.R. Kumnick. "Experience with Reuse and Biooxidation of Refinery Wastewater in Cooling Tower Systems," *J. Water Poll. Control. Fed.* 36 (1964): 1380–1392.
40. Carnes, B.A., J.M. Eller, and J.C. Martin. "Reuse of Refinery and Petrochemical Wastewaters," *Ind. Water Eng.* 9 (1972): 25–29.
41. Cawley, W., ed. *Treatability Manual*, vol. III, *Technologies for Control/Removal of Pollutants*, USEPA 600-8-80-0420 (July 1980).
42. Dickenson, B.W. and W.T. Laffey. "Pilot Plant Studies of Phenol Waste from Petrochemical Operations," *Proceedings of the 14th Industrial Waste Conference, Purdue University* (May 1959), pp 780–799.
43. Joyce, T. "Design Criteria for Phenol Treatment by Plastic Media Trickling Filters," *AIChE Symposium Series* 167 (1977): 160–165.
44. Capestany, G.J., J. McDaniels, and J.L. Opgrande. "The Influence of Sulfate on the Biological Treatment of Phenol-Benzaldehyde Wastes," paper presented at the 48th Annual Water Pollution Control Federation Meeting, Miami Beach, FL (October 1975).
45. Henshaw, T.B. "Adsorption/Filtration Plant Cuts Phenols from Effluent," *Chem. Eng.* 78 (1971): 47–49.
46. Sidwell, A.E. "Biological Treatment of Chlorophenolic Wastes," USEPA 12130EGK 06/71 (1971).
47. "Development Document for Effluent Limitations Guidelines and New Source Performance Standards for the Smelting and Slag Processing Segments of the Fer-

roalloy Manufacturing Point Source Category," USEPA 440/1-74-008-a (February 1974).

48. "Development Document for Effluent Limitations Guidelines and New Source Performance Standards for the Plywood, Hardboard and Wood Preserving Segment of the Timber Products Processing Point Source Category," USEPA 440/1-74-023-a (April 1974).
49. Vela, G. and J. Ralston. "The Effect of Temperature on Phenol Degradation in Wastewater," *J. Microbiol.* (Canada) 24 (1978): 1366–1370.
50. *The Cost of Clean Water*, Vol. III. *Industrial Waste Profile No. 10, Plastics Materials and Resins* (Washington, DC: USDI, 1967).
51. Schumaker, T.P. and R.H. Zanitsch. "Physical/Chemical Treatment: A Solution to a Complex Waste Problem," paper presented at the 45th Annual Water Pollution Control Federation Meeting, Atlanta, GA (1972).
52. Jorgensen, S.E. "Recovery of Phenols from Industrial Waste Water," *Prog. Water Technol.* 8 (1976): 65–79.
53. "Development Document for Effluent Limitations Guidelines and New Source Performance Standards for the Insulation Fiberglass Manufacturing Segment of the Glass Manufacturing Point Source Category," USEPA 440/1-74-001-b (January 1974).
54. "Potassium Permanganate Frees Effluent of Phenols," *Chem. Proc.* 36 (1975): 22.
55. Reed, G.D. and L. Hemphill. "Electrochemical Treatment of Tank Truck Wash Water," paper presented at the 33rd Annual Purdue Industrial Waste Conference, West Lafayette, IN (May 1978).
56. Heller, A.N., E.W. Clark, and W. Reiter. "Some Factors in the Selection of a Phenol Recovery Process," *Proceedings of the 12th Industrial Waste Conference, Purdue University* (May 1957), pp 103–122.
57. Van Stone, G.R. "Treatment of Coke Plant Effluent," *Ind. Wastes* 18 (1972): 23–35.
58. Ross, R.D. "Pollution Waste Control," *Industrial Pollution Control Handbook*, Herbert F. Lund, ed. (New York: McGraw-Hill, 1971).
59. Sample, G.E. and R.D. Rea. "Floats Away Refinery Wastes, Oil, Phenols Reduced 98%," *Chem. Proc.* 32 (1969): 41–42.
60. Pinkstaff, E., C.A. Leahu, and G.F. Gurnham. "Phenol Removal from Coke Plant and Petroleum Refinery Wastewaters by Municipal Treatment," *Water Poll. Abstr.* 40: (1967).
61. Robertson, J. and D. Tierney. "Fate of Selected Trace Organics in Canadian Petroleum Refinery Wastewater Treatment Processes," paper presented at the 56th Water Pollution Control Federation Conference, Atlanta, GA (October 1983).
62. Nakashio, M. "Phenolic Wastes Treatment by Activated Sludge Process. The Operation Condition and Actual Operation," *J. Ferment. Technol. Osaka.* 47 (1969): 389–393; *Water Poll. Abstr.* 43: (1970).
63. Heath, H., Jr. "Full-Scale Demonstration of Industrial Wastewater Treatment Utilizing du Pont's PACT Process," USEPA 600/52-81-159 (December 1981).
64. Hentschel, M.L. and T.L. Cox. "Effluent Water Treatment at Charter International Oil Company's Houston Refinery," *Water Reuse*, AIChE 70: (1973): 638–644.
65. Johnson, L. and J. Young. "Inhibition of Anaerobic Digestion by Organic Priority Pollutants," paper presented at the 54th Water Pollution Control Federation Conference, Detroit, MI (October 1981).

66. "Development Document for Effluent Limitations Guidelines and New Source Performance Standards for the Synthetic Resins Segment of the Plastics and Synthetic Manufacturing Point Source Category," USEPA 440/1-74-010-a (March 1974).
67. Hager, D.G. "Industrial Wastewater Treatment by Granular Activated Carbon," *Ind. Water Eng.* (January/February 1974): 14–28.
68. "Development Document for Effluent Limitations Guidelines and New Source Performance Standards for the Major Inorganic Products Segment of the Inorganic Chemicals Manufacturing Point Source Category," USEPA Publication 440/1-74-007-a (March 1974).
69. Edwards, A.W., W.W. Eckenfelder, Jr., and J.A. Roth. *Industrial Water Cost Studies*, vol. 1, chapters I-IV, Office of Research and Development, USEPA Contract 68-01-2485 (1975).
70. Cleary, E.J. and J.E. Kinney. "Findings from a Cooperative Study of Phenol Waste Treatment," *Proceedings of the 6th Industrial Waste Conference, Purdue University* (May 1951), pp 158–170.
71. Chamberlin, N.S. and A.E. Griffin. "Chemical Oxidation of Phenolic Wastes with Chlorine," *Sew. Ind. Wastes* 24 (1952): 750–760.
72. Strunk, W.G. "Hydrogen Peroxide for Industrial Wastewater Pollution Control," *Proceedings of the National Conference on Treatment and Disposal of Industrial Wastewaters and Residues*, Houston, TX (April 1977).
73. Deinzer, M., F. Schaumburg and E. Klein. "Environmental Health Sciences Center Task Force Review on Halogenated Organics in Drinking Water," *Environ. Health Perspectives* 24 (1978): 209–239.
74. Anderson, G. "Ozonation of High Levels of Phenol in Water," *AIChE Symposium Series* 166 (1977): 265–271.
75. Joshi, M. and R. Shambaugh. "The Kinetics of Ozone-Phenol Reaction in Aqueous Solutions," *Water Res.* (Br.) 16 (1982): 933–938.
76. Besselievre, E.B. *The Treatment of Industrial Wastes* (New York: McGraw-Hill, 1969).
77. "Phenols in Refinery Wastewater Can Be Oxidized with Hydrogen Peroxide," *Oil Gas J.* (January 20, 1975): 84–86.
78. Klemetson, S. and M. Scharbow. "Removal of Phenolic Compounds in Coal Gasification Wastewaters Using a Dynamic Membrane Filtration Process," *Proceedings of the 32nd Industrial Waste Conference, Purdue University* (May 1977), pp 786–796.
79. Schutt, H.C. and J. Loftus. "Waste Water Conditioned by Carrier Gas," *Oil Gas J.* 64 (1966): 70–72.
80. Malinowski, M. and R. Heropolitanski. "Studies on the Removal of Phenol from Waste Brine," *Przem. Chem.* 55 (1976): 479–482; *Chem. Abstr.* 86 (1977): 47034e.

21 Selenium

Industries using selenium include paint, pigment and dye formulating, electronics and glass manufacturing, and the insecticide industry. Relatively little information is available in the water pollution literature on levels of selenium in industrial wastewaters, treatment methods for selenium wastes, or costs associated with removal of selenium from industrial wastewaters. Table 21.1 presents data on industrial sources and wastewater concentrations of selenium. Raible [9] has reported the presence of selenium in a pharmaceutical wastewater, but did not indicate concentrations.

It has been reported that selenium is present in almost all types of paper, and it might be concluded from this information that pulp and paper mill wastes could contain selenium [5]. One primary concern has been the release of selenium as an air pollutant, upon incineration of paper products. Selenium wastewater levels have been reported as follows. For incinerator fly-ash quench water, soluble selenium was measured at 5–23 μg/l, while incinerator residue quench water contained 3 μg/l soluble selenium. The incinerator handled a solid waste containing approximately 55–69% paper [5].

Selenium also occurs as a trace impurity in coal. In one study, a selenium content of 2.2 μg/g coal was reported [10]. Upon combustion in a coal-fired steam electric power plant, the selenium partitioned as follows: 0.3% in the slag, 68.1% in the fly ash, and 18.6% in the vapor phase. The remaining 13% was unaccounted for. All effluent selenium was in the elemental form. Reduction of selenium dioxide (SeO_2) by sulfur dioxide to yield elemental selenium and sulfur trioxide was postulated as an on-site reaction which yielded the effluent elemental selenium observed [10]. As a result of its accumulation in coal fly ash, selenium is normally present in the effluent from fly ash ponds. Selenium concentrations below 10 μg/l appear typical, however. Much higher levels may be observed in power-plant air-pollution scrubber water, up to 2.5 mg/l [2, 3].

Selenium occurs as an impurity in the form of selenide, Se^{-2}, in metallic sulfide ores [11]. When the sulfide ores are roasted in air (to convert the sulfide and drive off SO_2), the selenide is oxidized to selenium dioxide. This may react with the sulfur dioxide by the mechanism postulated above, to yield elemental selenium. If the flue gas is quenched, residual selenium dioxide reacts with the quenchwater to form selenious acid, H_2SeO_3.

Table 21.1 Industrial Sources and Wastewater Concentrations of Selenium

Industry	*Selenium Concentration (μg/l)*	*Reference*
Coal mining	2–50	1
Coal pile drainage	1–30	2
Power plant scrubber waste	1–2,200	3
	<20–2,500	4
Power plant ash pond	1–170	2
	3–42	1
Incinerator ash quench water	5–23	5
Petroleum refining	25–62	1
Iron and steel		1
Continuous casting	220	
Basic oxygen furnace	37	
Iron ore milling	20	1
Copper production		
Ore milling	200	1
Milling and smelting	320	1
Milling, smelting, and refining	220	1
Smelting and refining	700	6
Lead/zinc ore milling	20–140	1
Molybdenum ore milling	40	1
Titanium ore milling	15	1
Zinc smelting	7,000	7
Hydrofluoric acid production	63	1
Copper sulfate production	200	8

A selenium level of 0.7 mg/l has been reported in wastewaters from copper smelting and electrolytic refining [6]. Electrolytic refining of copper appears to be one of the best characterized industrial sources of selenium wastewater [7, 12]. A major portion of the selenium impurity in crude copper settles to the bottom of the electrolytic cells as slime during the refining process. These slimes are typically processed to solubilize and recover copper and selenium. Selenium is recovered by precipitation of elemental selenium, upon addition of sulfuric acid and sulfur dioxide [7]. Selenium wastewater levels reported for the electrolytic copper refining industry are presented in Table 21.2. Highest selenium levels appear to result from the electrolytic cell slime by-product recovery process.

The selenite ion (SeO_3^{-2}) appears to be the most common form of selenium in wastewaters, except for solid forms of the selenide resulting from mining, ore milling and pigment and dye manufacture (e.g., yellow cadmium selenide). Other forms decompose in water to yield the selenite ion.

Table 21.2 Selenium Levels in Electrolytic Copper Refining Wastewaters [7]

Source	*Selenium (mg/l)*	*pH*
By-product recovery	1.7	3.8–4.4
Refining plus by-product recovery	2.5	7.0
Electrolytic refining	0.40	7.5
Electrolytic refining	0.14	4.6
Electrolytic refining	0.20	6.7
Arsenic recovery	0.04	7.0

TREATMENT TECHNOLOGY

The efficiencies of various treatment technologies for selenium depend upon its physical state (particulate or soluble) and its oxidation state. Particulate selenium is effectively removed by physical processes such as sedimentation or filtration. Precipitation, coagulation/coprecipitation, activated carbon, ion exchange, and reverse osmosis have been applied for removal of soluble selenium.

Physical Treatment Processes

The predominant form of selenium in certain types of industry wastewater is particulate. These include coal mining and preparation, and metallic-ore mining and milling. Applicable technologies for these wastewaters include sedimentation and filtration. Other industry wastes contain predominantly soluble selenium, usually as the selenite ion. Physical treatment processes are ineffective here. As one example, filtration applied at pilot scale to petroleum-refining wastewaters resulted in only 0–10% selenium removal [1].

Sedimentation has been applied widely, at full scale, to wastewaters of the coal-mining and metallic ore-milling industries. Treatment results are summarized in Table 21.3. Removal efficiencies generally exceed 90%, except at influent selenium levels of 20 μg/l or less.

Chemical Treatment Processes

For selenium in the soluble form, various chemical and physicochemical treatment processes have been described. Lime precipitation appears to be ineffective for selenium removal, as would be expected for a pollutant which exists in solution primarily as an anion. Lime precipitation treatment of steam electric power-plant ash pond effluent, initially at 8–41 μg/l selenium, resulted in no

Table 21.3 Efficiency of Sedimentation Treatment for Selenium [after 1]

	Selenium (μg/l)		
Industry	*Influent*	*Effluent*	*Removal*
Coal mining	4	3	25
	50	<5	>90
	34	3	91
Iron and steel[a]	220	5	98
Iron and steel[b]	37	31	16
Copper ore milling	200	<5	>97
Copper ore milling, smelting	320	7	98
Copper ore milling, smelting, and refining	220	12	94
Iron ore milling	20	<5	>75
Lead/Zinc ore milling	140	<10	>93
	20	<5	>75
Molybdenum ore milling	40	<5	>87
Titanium ore milling	15	<5	>67

[a] Continuous casting.
[b] Basic oxygen furnace.

removal [1]. Lime treatability studies on copper sulfate-manufacturing wastewaters containing soluble selenium at 110–180 μg/l, at treatment pH 9.8–10.4, was also ineffective [8]. Identical results have been reported for lime treatment of selenium in a hydrofluoric acid production wastewater [1].

Selenium has been reported at 7.0 mg/l in the effluent from one zinc smelter. This was reduced to 1.8 mg/l upon lime treatment to pH 8.2. A probable mechanism for the selenium removal is coprecipitation, or adsorption into the zinc hydroxide precipitate. Zinc was reduced in the lime treatment step from 513 to 50 mg/l [7]. Pilot studies on ferrous sulfate coagulation/coprecipitation of selenium in ash pond effluent resulted in up to 24% removal [1]. It is likely that the pilot process was not optimized for selenium removal, however.

Results of pilot studies on coagulation/coprecipitation of selenite, employing ferric sulfate and alum, are presented in Table 21.4. Ferric sulfate efficiency of selenium removal is higher at lower pH, although this influence is somewhat offset at higher coagulant dosages. Ferric sulfate removal efficiency for selenium drops off rapidly at pH above 7.0 [14]. Alum treatment efficiency is low, but increases with increased coagulant dosage. This benefit is offset by increasing chemical cost and sludge yield. Alum coagulation treatment efficiency appears to be relatively unaffected by treatment pH, over the range of 6–10 [14].

Activated carbon treatment of a secondary treated municipal wastewater at pH 7.6 reduced selenium from 9.32 to 5.85 μg/l. This represents a 37% removal efficiency [15]. Logsden et al, however, have reported less than 4%

Table 21.4 Results of Pilot Treatment Studies on Selenite [13]

Treatment and Dosage	*pH*	*% Removal*
Ferric sulfate coagulation		
25 mg/l	5.5	78
	7.0	60
60 mg/l	5.5	83
	7.0	75
100 mg/l	6.0	84
	7.0	81
Alum coagulation		
25 mg/l	6.0	10
100 mg/l	6.0	27
200 mg/l	6.0	45

removal of selenite ion by activated carbon, from well water [13]. Pilot-scale granular activated-carbon treatment of petroleum refining wastewater continuing 25–62 μg/l selenium resulted in only 6–11% removal [1].

Secondary municipal sewage treatment plant effluents containing 2–9 μg/l selenium have been reported [16]. A tertiary sequence of treatment, which included lime treatment to pH 11, sedimentation, mixed media filtration, activated carbon adsorption, and chlorination yielded selenium removal of 0–89%. Best removal was obtained at higher initial selenium concentrations.

Pilot-scale studies were performed to compare lime precipitation with iron and alum coagulation; these processes being followed by sedimentation, filtration and activated carbon adsorption. The results are presented in Table 21.5. As demonstrated by other studies discussed here, lime precipitation plus solids removal was ineffective. However, at the high treatment pH (11.5) in-

Table 21.5 Pilot Treatability Results for Selenium [17]

Treatment	*Initial Se (μg/l)*	*Cumulative % Removal*		
		Sedimentation	*+ Filtration*	*+ Activated Carbon*
Lime at 415 mg/l to pH 11.5	500	36	35	96
Ferric chloride at 40 mg/l as Fe and pH 6.2	50	68	80	77
Alum at 220 mg/l and pH 6.4	500	53	48	82

Table 21.6 Removal of Selenium by Bench-Scale Advanced Wastewater Treatment Processes [18]

Process	*% Removal*
Lime precipitation-settling (pH 7.6)	16.2
Cation exchange	0.9
Cation plus anion exchange	99.7
Process sequence[a]	
1. sand filtration	9.5
2. activated carbon	43.2
3. cation exchange	44.7
4. anion exchange	99.9

[a] Cumulative removal after lime precipitation plus indicated process sequence.

vestigated, activated carbon efficiently removed the lime-treated selenium. It is possible that the high treatment pH resulted in a change in selenium speciation, to a form more susceptible to carbon treatment. The treatment results for coagulation were consistent with other coagulation results described above, with the exception that carbon treatment improved the removal of the alum-pretreated selenium.

Linstedt et al [18], measured 2.3 μg/l selenium in the effluent from a secondary sewage treatment plant. These workers investigated the efficiencies of removal of selenite by several advanced wastewater treatment processes, and reported the results presented in Table 21.6. The authors concluded that efficient removal of selenium could be achieved with a strong acid-weak base ion-exchange system. Figure 21.1 presents capital and operating costs for such a weak-base system, as a function of dissolved solids content of the wastewater. The nonselective nature of ion exchange resins is a drawback when attempting to remove a low level contaminant in the presence of significant quantities of other ions. The simultaneous removal of other ions rapidly increases ion exchange costs.

The pilot study reported in Table 21.6 utilized a weak-base anion exchange resin. Selenious acid is a weak acid, fully dissociated at pH above 7 [20]. Therefore, most effective anion exchange treatment of the selenite ion would occur at pH 7 or greater. Use of a strong-base resin could enhance selectivity and capacity [20].

The selenite ion can be oxidized to selenate (SeO_4^{-2}), and selenate is more effectively removed by anion exchange than is selenite [21]. Selenic acid (H_2SeO_4) is a strong acid, and therefore fully dissociated even at pH below 7.0. Chemical oxidants tested to convert selenite to selenate include oxygen, permanganate, hydrogen peroxide, and chlorine. The latter was the most ef-

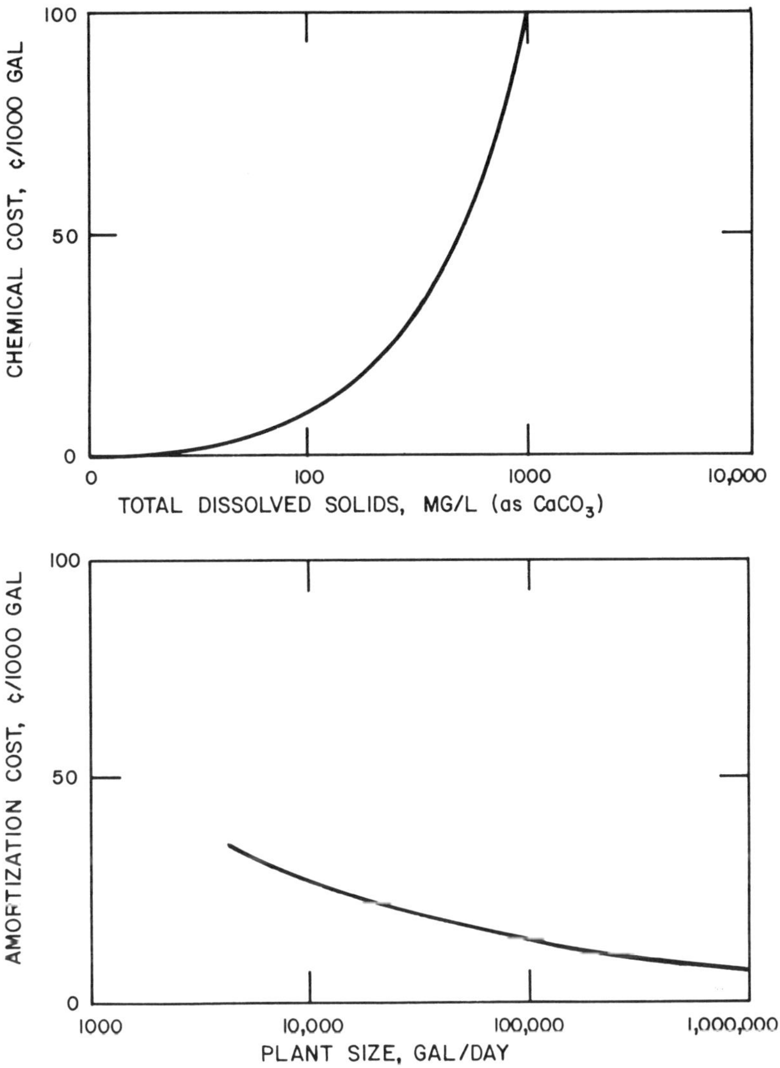

Figure 21.1 Chemical and amortization costs (in 1969) for a weak-base ion exchange system. (From Ahlgren [19], courtesy of *Industrial Water Engineering*.)

fective. At an optimum oxidation pH of 7–8, and 5 mg/l chlorine dosage, oxidation was essentially completed within 15 minutes [21]. Anion-exchange column comparisons of selenite versus selenate removal demonstrated much higher exchange capacity for the selenate. However, Calmon et al [20] have cautioned that anion exchange selectivity for sulfate and selenate are nearly identical. Thus, ion exchange treatment of a selenate wastewater also containing appreciable sulfate could result in low, and uneconomical, anion exchange capacity for the selenium.

One other process has been reported applied to selenium wastewater, at pilot scale. Reverse osmosis has been tested on steam-electric power-plant ash pond effluents [1]. For influent selenium levels of 3–42 μg/l, reverse osmosis achieved effluent concentrations of 1–6.2 μg/l, and removal efficiencies up to 85%.

SUMMARY

Selenium is associated primarily with wastewaters from coal mining, processing, and combustion, plus wastewaters from sulfide ore milling and refining of metals. Petroleum-refining effluents are also reported to contain selenium. Depending upon the industrial source, selenium occurs in either the particulate or the soluble form. For the former, physical treatment processes such as sedimentation or filtration are quite effective.

Lime precipitation of soluble selenium has been shown to be ineffective. Alum coagulation/coprecipitation is somewhat effective, with removal efficiencies approaching 50% achievable. Iron coagulation/coprecipitation is capable of above 80% selenium removal, with efficiency increasing with decreasing pH. Activated carbon adsorption is not effective, except in one reported instance involving high pH pretreatment. The most efficient process for soluble selenium appears to be anion exchange. Preoxidation of the selenite to selenate is reported to enhance exchange capacity.

REFERENCES

1. Cawley, W., ed. *Treatability Manual*, vol. III. *Technologies for Control/Removal of Pollutants*, USEPA 600-8-80-042-c.
2. Chu, T., R. Ruane, and P. Krenkel. "Wastewater Recycling in Steam-Electric Power Plants," *Proceedings of the 3rd Annual Conference on Treatment and Disposal of Industrial Wastewaters and Residues*, Houston, TX (April 18–20, 1978).
3. Bornstein, L. "Reuse of Power Plant Desulfurization Wastewater," USEPA 600/2-76-024 (1976).
4. Rossoff, J. "Disposal of By-Products from Non-Regenerable Flue Gas Desulfurization Systems—a Status Report," *Proceedings of the Symposium on Flue Gas Desulfurization* (1974).
5. Johnson, H. "Determination of Selenium in Solid Waste," *Environ. Sci. Technol.* 4 (1970): 850–853.
6. Hallowell, J.B., J.F. Shea, G.R. Smithson, Jr., A.B. Tripler, and B.W. Gearser. *Water-Pollution Control in the Primary Nonferrous Metals Industry*, vol. I. *Copper, Zinc and Lead Industries*, USEPA EPA-R2-73-247-a (September 1973).
7. "Development Document for Interim Final Effluent Limitations Guidelines and Proposed New Source Performance Standards for the Primary Copper Smelting

Subcategory and the Primary Copper Refining Subcategory of the Copper Segment of the Nonferrous Metals Manufacturing Point Source Category," USEPA 440/1-75/032-b, Group I, Phase II (February 1975).

8. *Treatability Studies for the Inorganic Chemicals Manufacturing Point Source Category*, USEPA 440/1-80/103 (July 1980).
9. Raible, J.A. "Fluorometric Determination of Selenium in Effluent Streams with 2,3-Diaminonaphthalene," *Environ. Sci. Technol.* 6 (1972): 621–622.
10. Andren, A.W., D.H. Klein, and Y. Talmal. "Selenium in Coal-Fired Steam Plant Emissions," *Environ. Sci. Technol.* 9 (1975): 856–858.
11. Hutchinson, E. *Chemistry: The Elements and Their Reactions* (Philadelphia: Saunders, 1959).
12. "Development Document for Interim Final Effluent Limitations Guidelines and Proposed New Source Performance Standards of the Significant Inorganic Products Segment of the Inorganic Chemicals Manufacturing Point Source Category," USEPA 440/1-75/037, Group I, Phase II (May 1975).
13. Logsden, G.S., T.J. Sorg, and J.M. Symons. "Removal of Heavy Metals by Conventional Treatment," *Proceedings of the 16th Water Quality Conference, University of Illinois*, Urbana, IL (February 12–13, 1974).
14. "Manual of Treatment Techniques for Meeting the Interim Primary Drinking Water Regulations," USEPA 600/8-77-005 (May 1977).
15. Culp, G.L. and R.L. Culp. *New Concepts in Water Purification* (New York: Van Nostrand Reinhold, 1974).
16. Argo, D.G. and G.L. Culp. "Heavy Metals Removal in Wastewater Treatment Processes: Part 2—Pilot Plant Operation," *Water Sew. Works* 119 (1972): 28–132.
17. Hannah, S.A., M. Jelus, and J.M. Cohen. "Removal of Uncommon Trace Metals by Physical and Chemical Processes," *J. Water Poll. Control. Fed.* 49 (1977): 2297–2309.
18. Linstedt, K.D., C.P. Houck, and J.T. O'Connor. "Trace Element Removals in Advanced Wastewater Treatment Processes," *J. Water Poll. Control. Fed.* 43 (1971): 1507–1513.
19. Ahlgren, R.M. "Membrane vs. Resinous Ion Exchange Demineralization," *Ind. Water Eng.* 8 (1969): 12–14.
20. Calmon, C., J. Casana, and H. Gold. "Ion Exchange and Resin Adsorption Data for the Removal of Toxic Contaminants from Wastewater," Report to the USEPA Industrial Environmental Research Laboratory, Cincinnati, OH (September 1980).
21. "Monthly Report," USEPA Municipal Environmental Research Laboratory, Cincinnati, OH (May 1984).

22
Silver

Silver, as the solid metal, is used in the jewelry, silverware, metal alloy, and food and beverage-processing industries. Little soluble silver waste would be expected to result from use of the solid metal. The only appreciably soluble common silver salt, silver nitrate, is used in the porcelain, mirroring, photographic, electroplating, and ink formulation industries; it has also found application as an antiseptic [1]. Two major sources of soluble silver wastes are the electroplating and photographic industries. Silver cyanide-plating baths may contain from 13,000 to 45,000 mg/l silver [2]. The cyanide-plating bath is by far the most widely used in the silver-plating industry. However, in recent years other baths such as nitrate, iodide, thiourea, thiocyanate, sulfamate, and thiosulfate have come into use [3]. An additional use of silver in plating is in the electroplating of gold-silver alloys. The compositions of these plating baths are generally proprietary. Silver compounds are widely used in the photographic industry. About 75% of the silver used in this industry finds its way into the fixing solution [4]. Table 22.1 summarizes typical industrial sources and levels of silver.

TREATMENT TECHNOLOGY

Recovery of silver from waste streams is more lucrative than is that for most other metals. A recent study on product recovery places a net recovery value of $1.60–9.00/1,000 gal dilute plating rinsewaters containing 50–250 mg/l silver [14]. Because of the high price of the metal, most silver in wastewater is recovered after treatment for refining, and extra precautions are taken to avoid spills and leaks. Dumping of concentrated plating solutions is not practiced.

Basic methods for silver removal from wastewaters are discussed under four categories: precipitation, ion exchange, reductive exchange, and electrolytic recovery. Other processes reported in use include reverse osmosis and electrodialysis. More than one of these categories of process may be combined, in order to achieve high levels of silver recovery. Silver in rinsewaters of the electroplating industry generally is recovered by ion exchange, evaporation, or electrodeposition (electrolytic recovery). Treated rinsewater silver levels of less than 0.01 up to 0.2 mg/l have been reported. The average for six plating

Table 22.1 Industrial Sources and Concentrations of Silver Wastewaters

Industry	*Silver Concentration (μg/l)*	*Reference*
Paint manufacture	20	5
Textile dyeing	50	6
Textile mills	5–130	5
Photographic chemicals	46,000	7
Photographic processing	230,000	8
	1,310,000	9
	200,000	10
	65,000	11
Silver plating	700–40,000	12
	1,500	5
Iron and steel	175	5
Commercial laundries	7–120	5
Steam electric power plants		
Ash pond effluent	1–14	5
Scrubber liquor	5–600	13

plants was 0.5 mg/l [3]. Concentrates as well as unusable baths are returned to salvage companies, for silver recovery by chemical treatment.

Precipitation

Hydroxide precipitation of silver is effective, although not commonly utilized. Minimum silver hydroxide solubility is achieved at pH 11.1, with a residual soluble silver level of 0.2 mg/l achievable [15]. In one study, lime precipitation at pH 11.5 reduced silver from 500 to 21 μg/l. Filtration provided further removal, to 14.5 μg/l [16].

Silver is frequently precipitated from plating wastewaters as silver chloride. In contrast to most metal chlorides, which are relatively soluble, silver chloride is an extremely insoluble salt. The salt will dissolve in water to a maximum concentration of approximately 1.4 mg/l silver ion [17]. Slight excess of chloride ion will reduce this value, but greater excess increases the solubility of silver through the formation of soluble silver chloride complexes [12, 17]. Thus, silver can be selectively recovered as silver chloride from a mixed-metal waste stream without initial waste stream segregation or concurrent precipitation of other metals. In the event that treatment conditions are alkaline, thus resulting in precipitation of hydroxides of other metals along with the silver chloride, acid washing of the precipitated sludge removes contaminant metal ions, leaving the insoluble silver chloride.

Table 22.2 Results of Bench-Scale Experiments on Silver Removal by Chlorine Addition [12]

	Cyanide (mg/l)		*Silver (mg/l)*	
Experiment	*Initial*	*Final*	*Initial*	*Final*
1	102	0	105	3.5
4	208	0	250	2.8
10	156	0	150	1.8

Plating wastes generally contain silver in the form of silver cyanide, which interferes with precipitation of silver chloride. Therefore, cyanide removal is necessary (see Chapter 8) prior to precipitation of silver as the chloride salt. Oxidation of the cyanide with chlorine releases chloride ions into solution, which in turn react to form silver chloride directly [12]. In the event that cyanide is present in great excess over silver, high chloride concentrations resulting from cyanide oxidation may greatly reduce the effectiveness of this silver removal process.

Bench-scale studies, performed in designing a silver waste treatment process, yielded the results presented in Table 22.2. Treatment consisted of addition of chlorine in the ratio of 3.5 mg/mg cyanide, and a 30 min reaction time. Silver reductions of 90–99% were achieved. Modification of the batch treatment process was made, to the sequence described below.

1. Chlorination and 10 min reaction
2. Adjust pH to 6.5 to complete oxidation of cyanide
3. Addition of ferric chloride coagulant
4. Skimming and sedimentation
5. Lime addition to pH 8
6. Settling of floc and decanting supernatant

The treatment plant resulting from the above studies handles silver recovery in batch processes of 10,000 gal. Operating data from the full-scale recovery system are presented in Table 22.3. Best silver recovery was achieved at lowest residual cyanide concentration, indicating that incomplete cyanide destruction interfered with silver recovery.

The role of ferric chloride in the above process is as a coagulant/coprecipitant, rather than to introduce additional chloride into the treatment process. Hannah et al [16] have reported that either ferric chloride or alum coagulation, at pH 6.2–6.4, can reduce silver by 95% from an initial 500 μg/l level. Iron salt coagulation efficiency for silver increases slightly with increasing pH, although alum coagulation efficiency declines rapidly at pH above 8 [19].

Table 22.3 Results of Operation of Batch Cyanide Destruction-Silver Recovery System [18]

	Cyanide (mg/l)		*Silver (mg/l)*	
Batch	*Initial*	*Final*	*Initial*	*Final*
1	642	70	564	8.2
2	829	86	585	4.4
3	509	35	401	2.7
4	348	30	331	6.5
5	389	54	364	2.7
6	672	56	533	5.1
7	431	40	440	0.7
8	159	15	130	0.0

Other silver precipitation processes are mentioned in the literature. One, a patented process of Eastman Kodak Co. [20], precipitates silver from waste photographic solutions containing high levels of organic acids by adding magnesium sulfate and lime. The silver likely precipitates as a mixed sulfate-oxide, and is recovered from the sludge.

Precipitation of silver by joint lime-ozone treatment has been examined in bench-scale studies [21]. An initial silver concentration of 91 mg/l was reduced to the levels given in Table 22.4, after treatment. The principal oxidation state of silver in aqueous solution is +1. Ozone apparently converts the silver to a higher oxidation state (+2 or +3), which forms a less soluble precipitate than does Ag^+.

Use of sulfide for precipitating silver from photographic solutions as the extremely insoluble silver sulfide is also reported [22], but no operating data or costs are provided. From a solubility standpoint, sulfide precipitation would appear to be an attractive method, although solids separation and addition of sulfide to the wastewater are considered disadvantages. In a review of methods of silver recovery in the photographic film-processing industry, Schreiber [23]

Table 22.4 Precipitation of Silver by Lime Plus Ozone

	Residual Soluble Silver (mg/l)	
pH	*Lime Only*	*Lime + Ozone*
9	37	1.4
10	12	0.1
11	0.4	0.1

has pointed out that the use of sulfide is one of the oldest methods for silver precipitation. Use of hydrosulfite for precipitation yields both free silver and the sulfide, as well as a more compact precipitate with good settling characteristics. Heat is required, however, and chemical costs are high.

Ion Exchange

Ion exchange has been widely utilized for silver recovery. One application, for a plating rinsewater, achieved complete removal of silver initially at 1.5 mg/l [5].

Silver can exist as a positively charged ion, or as a negatively charged anion in the presence of cyanide, thiosulfate, or other plating bath stabilizers. Depending upon the nature and charge of the ion, silver removal is possible by either cationic or anionic exchange resins [24]. For the positively charged silver ion, weak cationic exchangers are reported to achieve excellent results. However, in their application, treatment pH must be maintained high since weak resins have a high affinity for the hydrogen ion [24]. Such systems should be operated on the hydrogen cycle, since the sodium cycle with attendant sodium chloride regeneration will cause silver chloride precipitation within the column.

A patented process involves the recovery of silver from dilute photographic wash water by passage through basic ion exchange resins [25]. Recovery is by elution of silver salts from the resin, or by direct resin incineration and recovery of pure silver. Above 85% removal of trace levels of silver were achieved from an extremely dilute secondary sewage effluent by cation exchange, and 91.7% removal by combined cation-anion exchange [26]. Treatment efficiencies in excess of 90% have been reported for ion exchange treatment for silver in photographic-processing wastes [10, 11].

A series cation-anion exchanger combination has been used for recovery of silver cyanide from plating rinsewaters [27]. The recovered sodium silver cyanide may be returned to the plating bath, or converted to silver chloride by the precipitation process. Chemical costs of recovery are given (in 1958) as 18¢/lb silver recovered.

A 5-column ion exchange recovery system has been reported for silver cyanide rinsewaters [28]. The first column contains a cation exchange resin (hydrogen cycle) which converts the silver cyanide to a silver hydrogen cyanide complex. This in turn passes through the second column (an anion exchange resin) where silver cyanide is removed. The additional three columns function to effect further cyanide removal. The eluant from the second column is treated with potassium or sodium cyanide to recover the silver as potassium or sodium silver cyanide.

The cost of silver cyanide waste treatment by ion exchange was reported (in 1968) as $1.75/1,000 gal [29]. However, the value of silver recovered resulted in a net profit of $27.20/1,000 gal of wastewater treated. Approximately 280

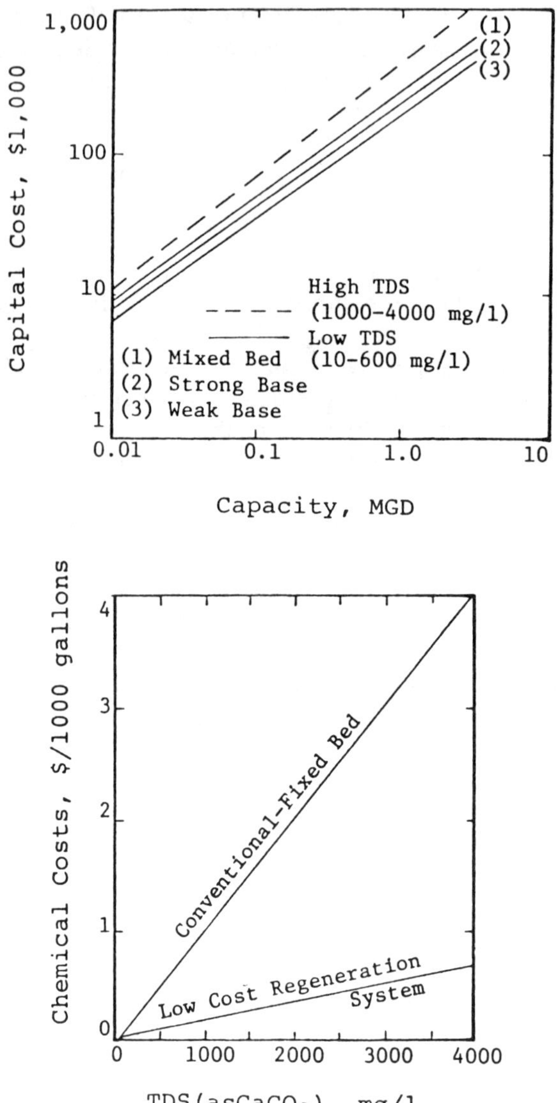

Figure 22.1 Capital and chemical regenerant costs (in 1974) for ion exchange treatment [30].

mg silver were recovered from each liter of wastewater [29]. Recovery costs would be somewhat higher if the silver chloride precipitation method of recovery was practiced, as cyanide destruction would be a prerequisite to formation of the silver chloride. Approximate capital and chemical regenerant costs of ion exchange units are given in Figure 22.1. These costs are for complete units, including valving and resins. The economics described above are typical in that recovery values of silver normally more than pay for treatment costs [31].

Problems in recovering silver from ion exchange resin may make ion exchange uneconomical for some wastewaters. Ion exchange in conjunction with electrolytic recovery for thiosulfate developer baths has been suggested [32]. In that case, fixing-solution silver is recovered by plating out the silver. Washout silver waste is accumulated by ion exchange, and the resin regenerant containing the concentrated silver is passed through an electrolytic process, thereby removing silver and providing developer-bath makeup water.

The development of a silver-specific ion exchange resin, similar to that available for gold recovery, is reported under way [33]. The application of the projected resin is for silver recovery from spent fixer solutions.

Reductive Exchange

This method involves cementation of silver onto a sacrificial metal. Usually, zinc or iron is used because these metals are relatively inexpensive. The net result is a substitution of iron or zinc ion for silver in solution. This process had been in operation at one plant prior to installation of a cyanide chlorination-silver chloride recovery system [34]. Although the process was effective for silver recovery, replacement of the reductive exchange process was necessary for effective cyanide treatment.

Using zinc or steel wool under proper conditions to maintain sufficient metal surface exposed, 95% removal of silver can be accomplished with column operation cementation. Silver ions are replaced with equivalent quantities of zinc or iron ions.

The potential use of activated carbon as a treatment method for silver has been reported [35]. The mechanism is described as one of reductive recovery by formation of elemental silver at the carbon surface. Laboratory results indicate that the carbon is capable of retaining silver to 9% of carbon weight at pH 2.1, and 12.5% at pH 5.4. However, other full- and pilot-scale studies on the removal of silver on activated carbon indicate that the process is very inefficient [5, 16].

Electrolytic Recovery

This process has been reported in conjunction with treatment of photographic developer solutions [36] and for electroplating rinsewaters [3]. As with any electrolytic process, high wastewater ionic concentrations are necessary for the method to function effectively. If sufficient metallic ions are in solution such an operation is feasible; however, most silver wastewaters do not appear to contain sufficient concentrations of silver ion alone. Many electrolytic units will operate at silver concentrations of 500 mg/l, and some at as low as 100 mg/l. Usually operation begins at 5,000 mg/l, however, and reduces bath silver concentrations to 500 mg/l.

Closed-loop operation of the first rinse after silver-plating baths is economically feasible for electrolytic recovery, using laminar or packed-bed recovery cells [37]. High electrode efficiency and value of metal recovered makes the process worthwhile for dilute silver wastes.

Other Processes

Both electrodialysis and reverse osmosis have been reported to remove silver from wastewater. Electrodialysis is employed by Oneida, Ltd (Sherrill, NY) for a silver cyanide rinsewater. The concentrate is recovered at twice the plating-bath strength. The process replaced a cyanide destruction-silver chloride precipitation system [38]. Silver plating bath solution is reported to cost \$47.93/gal [39].

Reverse osmosis has been tested at pilot scale on textile mill wastewaters and power-plant ash pond effluent [5]. For the latter waste, an initial silver level of 14 μg/l was reduced to 1.1 μg/l. Textile mill wastewaters ranging from 42 to 120 μg/l were reduced on average only 29%. One large photo-processing facility has recently installed at a cost of approximately \$1 million, a reverse osmosis system which allows recycle of 18,000 gpd water and, in 1978, recovered \$600,000 worth of silver [8].

SUMMARY

The value of silver makes recovery from process streams attractive. Electrolytic recovery, as long practiced in photographic processing, is operative only down to solution concentrations of 100–500 mg/l. However, packed-bed electrodes may find use to much lower concentrations. Precipitation with chloride ion can remove silver to the milligram-per-liter level. Coprecipitation with other metal hydroxides under alkaline conditions improves silver removal to less than 0.1 mg/l. Silver recovery is possible from mixed-metal precipitates by redissolving the metal precipitates under acidic conditions. Ion exchange for low silver concentrations appears feasible for segregated waste streams, where total ionic strength is not appreciably greater than the silver salt concentration. Very low residual silver concentrations are possible with ion exchange. Sulfide precipitation of silver has been used on a small scale for many years, but large-scale operation appears to pose problems, particularly with respect to solids separation and handling.

REFERENCES

1. McKee, J.E. and H. W. Wolf. *Water Quality Criteria*, 2nd ed. California State Water Quality Control Board Publication 3-A (1963).

2. Lowe, W. "The Origin and Characteristics of Toxic Wastes, with Particular Reference to the Metals Industry," *Water Poll. Control* (London, 1970): 270–280.
3. "Development Document for Interim Final Effluent Limitations Guidelines and Proposed New Source Performance Standards for the Common and Precious Metals Segment of the Electroplating Point Source Category," USEPA 440/1-75/040, Group I, Phase II (April 1975).
4. Kuhn, A.T. "Electrochemical Techniques for Effluent Treatment," *Chem. Ind.* (1971): 946–950.
5. Cawley, W., ed. *Treatability Manual*, vol. III. *Technologies for Control/Removal of Pollutants*, USEPA 600-8-80-042-c (July 1980).
6. Feigenbaum, H.N. "Removing Heavy Metals in Textile Wastes," *Ind. Wastes* (March/April 1977): 32–34.
7. Kawakami, T. "Recovery of Silver," *Chem. Abstr.* 87 (1977): 357.
8. Kohn, P.M. "Photo-Processing Facility Achieves Zero Discharge," *Chem. Eng.* (December 4, 1978): 78–79.
9. Hagiwara, Z. and H. Sato. "Treatment of Photographic Processing Wastewater," *Chem. Abstr.* 87 (1977): 324.
10. Marsh, D. "Removal of Residual Silver from Processing Wastewaters by Ion Exchange," *J. Appl. Photographic Eng.* 4 (1978): 1.
11. Nelson, F., K. Kraus, and J. Gilbert. "Beneficiation of Photographic Wastes," paper presented at the 31st Annual Purdue Industrial Waste Conference, West Lafayette, IN (May 1976).
12. Walker, C.A., W. Zabban, R.W. Southworth, and E.P. Heslin. "Disposal of Electroplating Wastes by Oneida, Ltd. II Development of Treatment Processes," *Sew. Ind. Wastes* 26 (1954): 849–853.
13. Bornstein, L. "Reuse of Power Plant Desulfurization Wastewater," USEPA 600/2-76-024 (1976).
14. *State of the Art Review on Product Recovery*, Water Pollution Control Research Services 17-7 00JW 11/69 (Washington, DC: USDI).
15. Netzer, A., A. Bowers, and J. Norman. "Removal of Heavy Metals from Wastewaters by Lime," *Proceedings of the 16th Conference on Great Lakes Research*, Huron, OH (1973), pp 260–265
16. Hannah, S., M. Jelus, and J. Cohen. "Removal of Uncommon Trace Metals by Chemical Treatment Processes," *J. Water Poll. Control. Fed.* 49 (1977): 2297–2309.
17. Butler, J.N. *Ionic Equilibrium* (Reading, MA: Addison-Wesley, 1964).
18. Eichenlaub, P.W. and J. Cox. "Disposal of Electroplating Wastes by Oneida, Ltd. V. Plant Operation," *Sew Ind. Wastes* 26 (1954): 1130–1135.
19. "Manual of Treatment Techniques for Meeting the Interim Primary Drinking Water Regulations," USEPA 600/8-77-005 (May 1977).
20. Pool, S.C. "U.S. Patent 2,507,175, to Eastman Kodak Co.," *Water Poll. Abstr.* 24 (123) (1951).
21. Netzer, A., A. Bowers, and J.D. Norman. "Removal of Trace Metals from Wastewater by Lime and Ozonation," *Pollution: Engineering and Scientific Solutions* Plenum Publishing Co. (1972).
22. Fusco, F. "Recovery of Silver from Photographic Fixing Baths and Other Sources," *Chem. Ind.* 24 (1942): 356; *Water Poll. Abstr.* 21 (1948).
23. Schreiber, M.L. "Present Status of Silver Recovery in Motion-Picture Laboratories," *J. Soc. Motion Picture Technol. Eng.* 74 (1965): 505–513.

24. Calmon, C., J. Casana, and H. Gold. "Ion Exchange and Resin Adsorption Data for the Removal of Toxic Contaminants from Wastewater," report to USEPA Industrial Environmental Research Laboratory, Cincinnati, OH (September 1980).
25. Permutit Co., Ltd. "Improvements Relating to the Recovery of Silver from Dilute Solution," B.P. 626, 081; *Water Poll. Abstr.* 23 (1950).
26. Linstedt, K.D., C.P. Houck, and J.T. O'Connor. "Trace Element Removals in Advanced Wastewater Treatment Processes," *J. Water Poll. Control. Fed.* 43 (1971): 1507–1513.
27. Heidorn, R.F. and H.W. Keller. "Methods for Disposal and Treatment of Plating Room Solutions," *Proceedings of the 13th Industrial Waste Conference, Purdue University* (May 1958), pp 418–427.
28. Schore, G. "Electronic Equipment for Use in Automated Water Treatment Systems," paper presented at the 27th Annual Purdue Industrial Waste Conference, West Lafayette, IN (May 1972).
29. Ross, R.D., ed. *Industrial Waste Disposal* (New York: Reinhold, 1968).
30. "Development Document for Effluent Limitations Guidelines and New Source Performance Standards for the Major Inorganic Products Segment of the Inorganic Chemicals Manufacturing Point Source Category," USEPA 440/1-74-007-a (March 1974).
31. "Development Document for Interim Final Effluent Limitations Guidelines and Proposed New Source Performance Standards for the Significant Inorganic Products Segment of the Inorganic Chemicals Manufacturing Point Source Category," USEPA 440/1-75/037, Group I, Phase II (May 1975).
32. Arden, T.V. *Water Purification by Ion Exchange* (New York: Plenum Press, 1968).
33. "Winning Heavy Metals from Waste Streams," *Chem. Eng.* 78 (1971): 62, 64.
34. Walker, C.A., B.F. Dodge, and J.F. Madden. "Disposal of Electroplating Wastes by Oneida, Ltd.," *Sew. Ind. Wastes* 26 (1954): 1002–1013.
35. Sigworth, E.A. and S.B. Smith. "Adsorption of Inorganic Compounds by Activated Carbon," *J. Am. Water Works Assoc.* 64 (1972): 386–391.
36. "Disposal of Photographic-Processing Waste," Eastman Kodak Co. Publication J-28, Rochester, New York (1969).
37. Silman, H. "Treatment of Rinse Water from Electrochemical Processes," *Metal Fin.* 69 (1971): 62–66.
38. Millmam, W. and R. Heller. "Some Successful Applications of Electrodialysis," *Proceedings of the 4th Conference on Advanced Pollution Control for the Metal Finishing Industry*, USEPA Report 600/9-82-022 (December 1982).
39. Smith, D. and C. Roy. "Disposal: What Does It Really Cost?" *Proceedings of the 4th Conference on Advanced Pollution Control for the Metal Finishing Industry*, USEPA Report 600/9-82-022 (December 1982).

23
Total Dissolved Solids

Although dissolved solids are to some extent present in any water, natural or waste, most industrial processes, and in fact many industrial waste treatment processes, result in increased quantities of total dissolved solids (TDS). One major exception is where lime is added for anion precipitation. In general, TDS of natural waters consist mainly of carbonates, chlorides, sulfates, phosphates, and nitrates of calcium, magnesium, sodium and potassium, with traces of iron, manganese, and other substances.

The mineral content of natural waters is normally augmented by the addition of chemical wastes, dissolved salts, acids, and alkalis. For example, TDS levels of metal-plating effluents, after waste treatment for acid neutralization and reduction of cyanide and heavy metals, are presented in Table 23.1. A recent study indicates that TDS in inorganic-chemical industry effluents may range from 1,000 to 150,000 mg/l [2]. An effluent TDS of 1,163 mg/l has been reported from a petroleum industry refinery [3], and TDS in a carpet mill waste of 5,850 mg/l, with a raw water supply TDS of 410 mg/l [4]. Source water TDS is quite variable, depending upon geographical location and surface versus groundwater source, and thus can influence wastewater TDS. High TDS values have also been reported for inorganic pigment manufacture, coal preparation and coal mine drainage, and other industries (Table 23.2).

TREATMENT TECHNOLOGY

Removal of dissolved solids from wastewaters is one of the more uncommon and, in many cases, most expensive wastewater treatment procedures. Where high dissolved solids result from heavy metal or hardness ions, reduction may be accomplished by chemical precipitation methods, as discussed elsewhere in this volume. Where the dissolved solids are present, as for example sodium or sulfate ions, total dissolved solids reduction requires more specialized treatment methods, as discussed in this chapter.

The nature of nonprecipitable dissolved solids, and the treatment processes available to remove or reduce such solids, are such that there are, practically speaking, no intermediate levels of treatment. That is to say, treatment for removal of dissolved solids, if applied at all, generally produces a water containing very low residual TDS. Partial treatment of a high-TDS waste stream

Table 23.1 Total Dissolved Solids Concentration in Treated Metal-Plating Wastewaters [1]

Location	*Plant No.*	*TDS (mg/l)*
Arkansas	1	393
California	2	625
	3	435
Illinois	4	555
	5	15,100
	6	1,065
	7	2,260
Indiana	8	166
Kentucky	9	810
Michigan	10	490
New York	11	610
	12	1,580
Ohio	13	4,066
Pennsylvania	14	793
Virginia	15	906

Table 23.2 Total Dissolved Solids Concentrations in Selected Industrial Wastewaters

	TDS Concentration (mg/l)		
Source	*Range*	*Average*	*Reference*
Carpet mill	—	5,850	4
Carbon monoxide plant blowdown	—	2,590	5
Chrome pigments manufacture	7,000–12,000	9,800	5
	—	6,000	
Zinc yellow (pigment) manufacture	—	28,000	5
Titanium oxide (pigment) manufacture	1,660–22,371	—	6
Soda ash manufacture	2,560–76,000	—	6
Blast furnace scrubber water	—	2,721	7
Aluminum anodizing	1,500–3,500	—	8
Nitrocellulose manufacture	2,700–3,600	—	9
Coal mining			10
Coal preparation	636–2,240	1,433	
Acid surface mine drainage	120–8,870	4,060	
Acid deep mine drainage	12–15,572	4,749	
Alkaline surface mine drainage	152–8,358	2,867	
Alkaline deep mine drainage	418–22,158	2,702	

Table 23.3 Capital and Maintenance and Operating (M&O) Treatment Costs (in 1970) for Reduction of TDS [2]

Flow (MGD)	*TDS (mg/l)*	*Type of Process*[a]	*Capital Cost of Process ($1,000s)*	*Total*[b] *Capital Costs ($1,000s)*	*M & O Cost of Process (¢/1,000 gal)*	*Total*[b] *M & O Cost (¢/1,000 gal)*
0.5	3,000	RO	640	1,439	40.0	65.5
		ED	455	1,240	65.0	97.5
		DI	806	1,701	98.0	122.5
		IE	1,700	2,551	25.0	49.5
	30,000	RO	784	1,704	75.0	112.5
		DI	806	1,713	98.0	130.5
1.0	3,000	RO	1,120	2,185	31.0	51.5
		ED	896	1,929	54.0	79.3
		DI	1,610	2,810	85.0	103.0
		IE	3,020	4,700	33.3	48.3
	30,000	RO	1,340	2,550	50.0	77.6
		DI	1,610	2,892	85.0	110.3
10.0	3,000	RO	6,700	10,651	18.5	26.0
		ED	6,400	10,496	38.0	49.0
		DI	9,070	13,724	38.0	44.0
		IE	14,800	21,160	30.0	34.0
	30,000	RO	8,400	13,567	25.0	38.0
		DI	9,070	14,100	38.0	49.0

[a] Process: RO, reverse osmosis; ED, electrodialysis; DI, distillation; IE, ion exchange.
[b] Total costs include prefiltration for removal of suspended solids, and brine waste disposal.

thus may involve splitting the stream, treating one portion to remove essentially all dissolved solids, and recombining the treated and untreated streams in the proportion necessary to achieve the desired final solids concentration. Treatment costs discussed later in this chapter thus do not refer necessarily to the total waste volume produced by an industrial process, but only to that portion of the waste which must be separately treated to yield required effluent quality.

The major processes employed for reduction of TDS are reverse osmosis, electrodialysis, distillation, and ion exchange. Ion exchange and electrodialysis are generally applicable for TDS concentrations up to 5,000 mg/l, while reverse osmosis and distillation are employed for TDS concentrations to 50,000 mg/l [2]. Evaporative ponds may be required at extremely high TDS concentrations, above 50,000 mg/l. Table 23.3 presents comparative capital and operating costs (1970 basis) of TDS treatment processes, in terms of waste stream volume, and TDS concentrations. Capital and operating costs are also presented in Figures

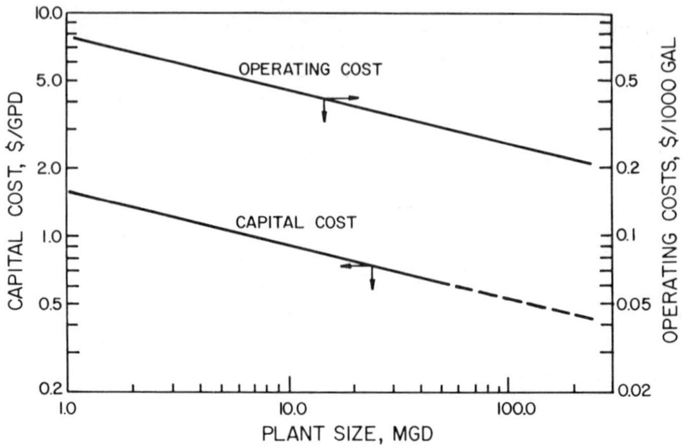

Figure 23.1 Capital and operating costs (in 1970) for multiple effect evaporation/distillation [2].

23.1–23.6. Similar costs to those shown in Figures 23.1–23.6 have also been reported by others [6, 11, 12]. Additional cost curves are presented in Figures 23.7 and 23.8. The lower three curves of Figure 23.8 refer to various ion exchange media. Figure 23.9 presents an energy comparison for distillation, electrodialysis, and reverse osmosis. For wastewaters having TDS of 1,000 to 10,000 mg/l, reverse osmosis is most economical in terms of energy utilization (Figure 23.9) and chemical costs (Figure 23.8). Above a TDS level of 10,000

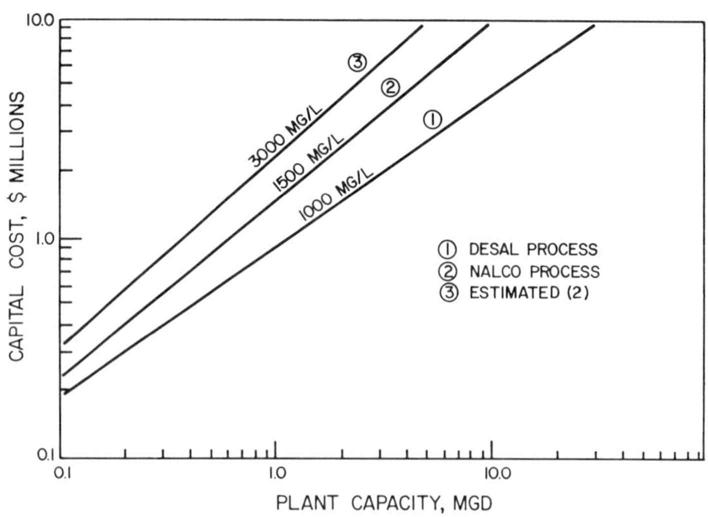

Figure 23.2 Capital cost (in 1970) of dissolved solids reduction by ion exchange treatment [2].

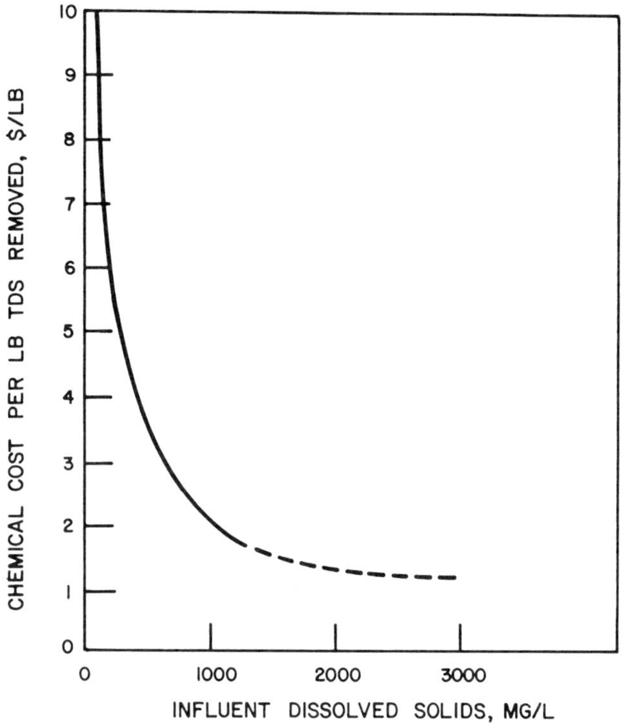

Figure 23.3 Ion exchange resin-regenerating chemicals cost per pound (in 1970) of total dissolved solids removed [2].

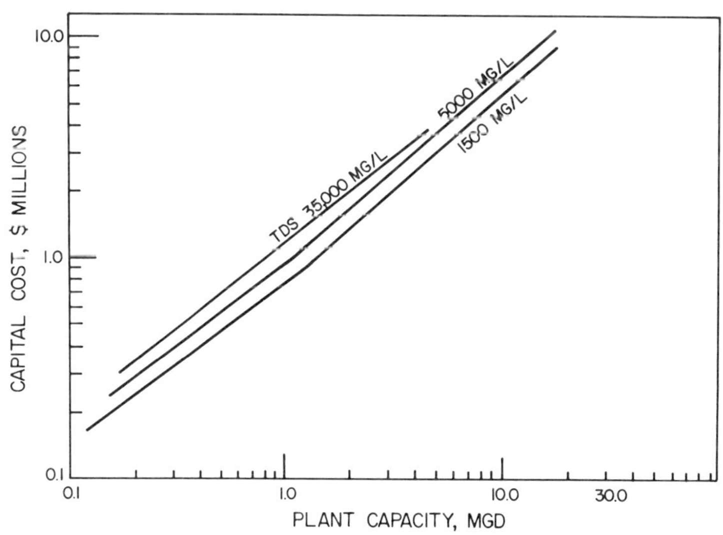

Figure 23.4 Capital costs (in 1970) for reverse osmosis treatment plants [2].

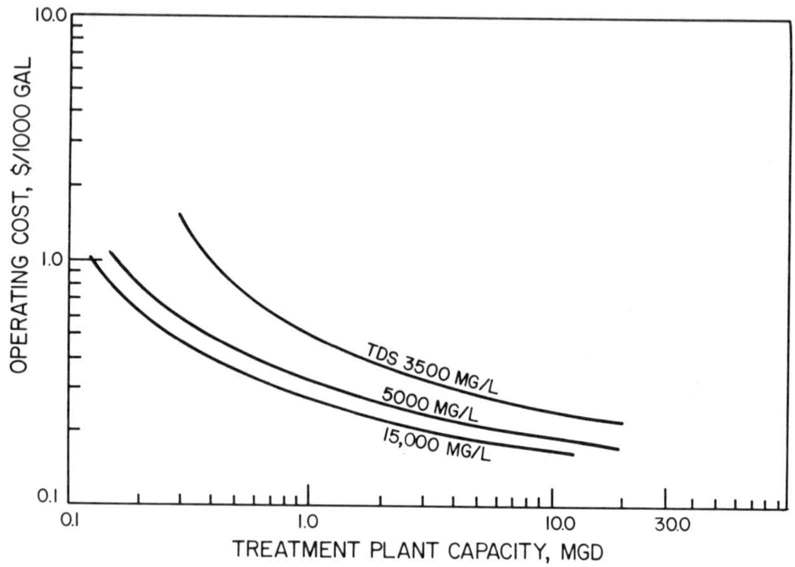

Figure 23.5 Operating costs (in 1970) for dissolved solids reduction by reverse osmosis [2].

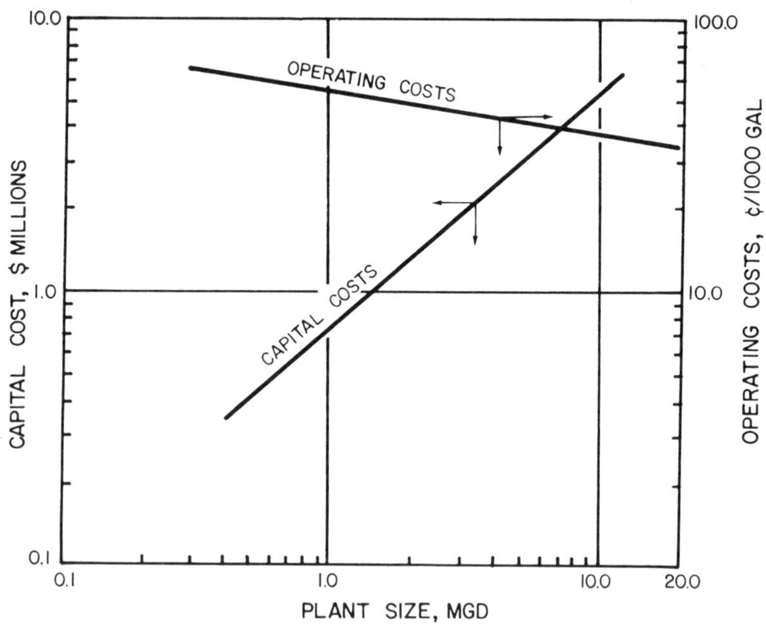

Figure 23.6 Capital and operating costs (in 1970) for an electrodialysis treatment plant, at an influent total dissolved solids concentration of 3,000 mg/l [2].

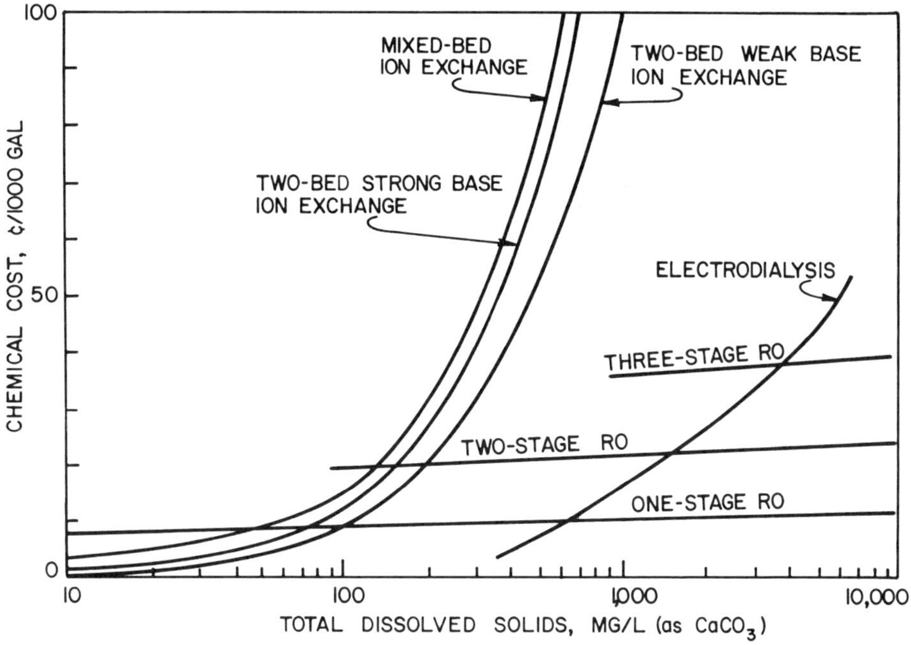

Figure 23.7 Chemical treatment cost (in 1971) versus total dissolved solids for various demineralization systems. RO = reverse osmosis. (From Ahlgren [13], courtesy of *Industrial Water Engineering*.)

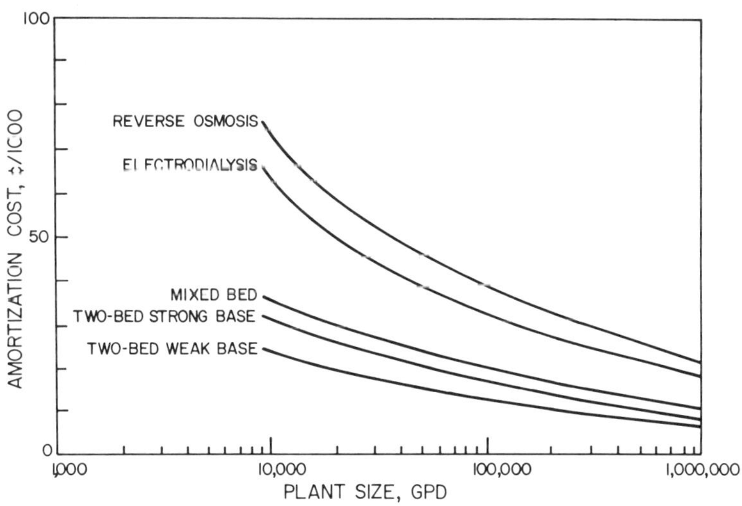

Figure 23.8 Amortization costs (in 1971) versus plant size for various demineralization systems. Costs are based on 7-year write-off, 7% interest rate, and 330 days/year plant operation. (From Ahlgren [13], courtesy of *Industrial Water Engineering*.)

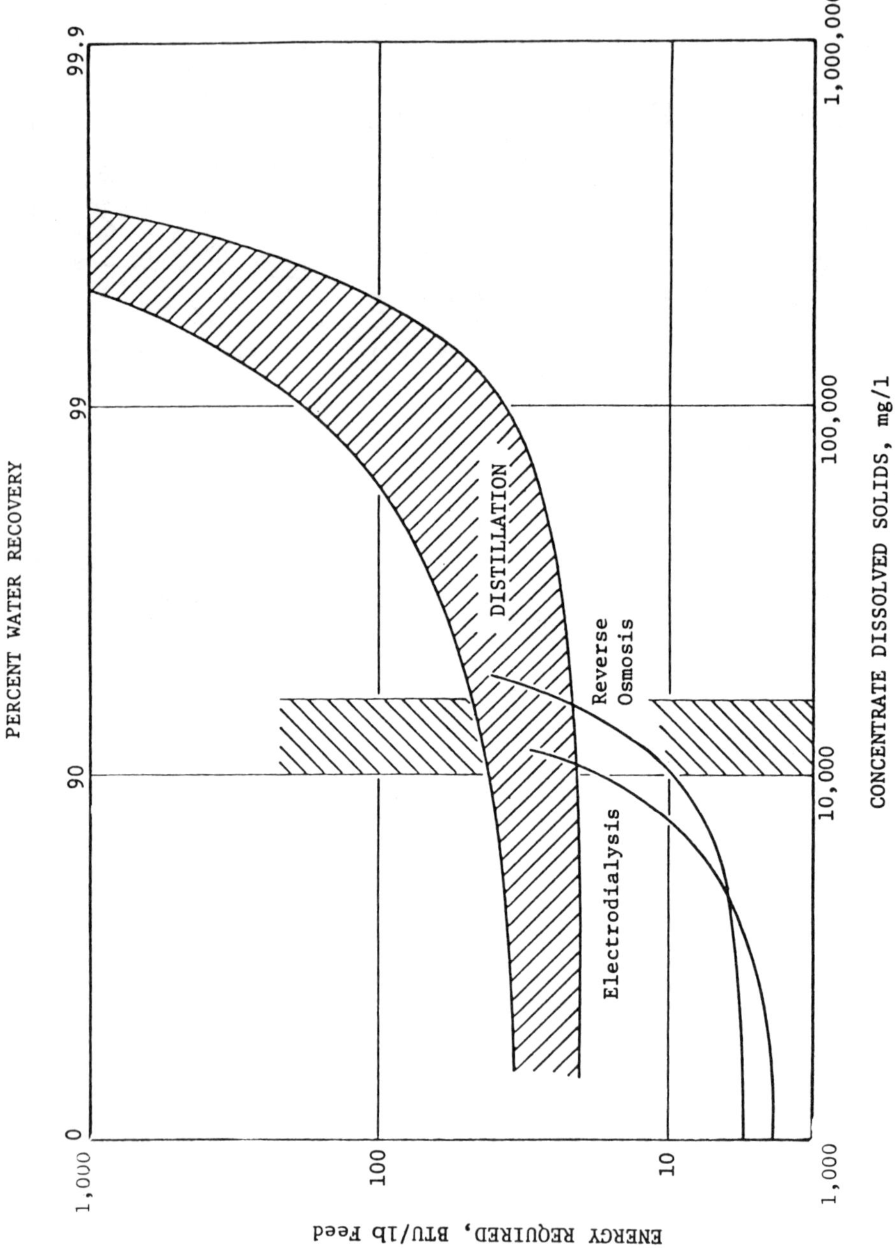

Figure 23.9 Energy comparison for dissolved solids removal [6].

Table 23.4 Operating Costs (in 1974) for Distillation Treatment [6]

Evaporator	*Volume Treated (gpd)*	*Operating Cost ($/1,000 gal)* Steam	*Labor*	*Total*
Single-effect	10,000	—	—	5.64
	50,000	—	—	5.51
	100,000	—	—	5.45
Six-effect	100,000	0.95	0.91	1.86
	250,000	0.95	0.80	1.75
	500,000	0.95	0.71	1.66
	1,000,000	0.95	0.68	1.63

mg/l, distillation is more energy conservative. Single-effect evaporators are often chosen for total flows below 5,000 gpd. Higher flows may require multiple-effect evaporators.

All treatment processes for TDS yield, in addition to the treated product water (effluent), a brine solution containing the dissolved solids removed in the process. Disposal of this brine is typically by deep-well injection or to a solar evaporating lagoon. An important exception exists for those industrial wastes where recovery of a valuable constituent can be combined with waste treatment.

Distillation

Recondensed water vapor from distillation processes contains little inorganic dissolved solids, but may contain volatile organic fractions driven off with the water vapor, and recondensed. A summary of current operating costs for distillation treatment is presented in Table 23.4. For single-effect evaporators, cost of steam is the major factor, while for multiple-effect evaporators, steam costs are comparable to labor and maintenance costs. Single-effect evaporators use much more steam per volume of wastewater evaporated than do multiple-effect evaporators (Figure 23.10). At current high fuel prices, distillation is increasingly falling out of favor as a wastewater treatment process. Most industrial experience with distillation is to purify raw water for boiler operations.

Prescott [14] has described a 50,000 gal/day evaporative process used to treat cooling-tower blowdown near El Paso, Texas. The process requires only 20 BTU/lb water evaporated. The blowdown has a TDS of 2,331 mg/l, and the condensed vapors, which are recycled to the cooling towers, contain 8 mg/l TDS. Scaling in the evaporator is prevented by using a seed slurry of brine crystals. The concentrated waste brine from the process is disposed to an evaporation pond. Capital investment for the unit was $400,000, and operating costs are reported at $2.00/1,000 gallons. This does not include savings in water reuse.

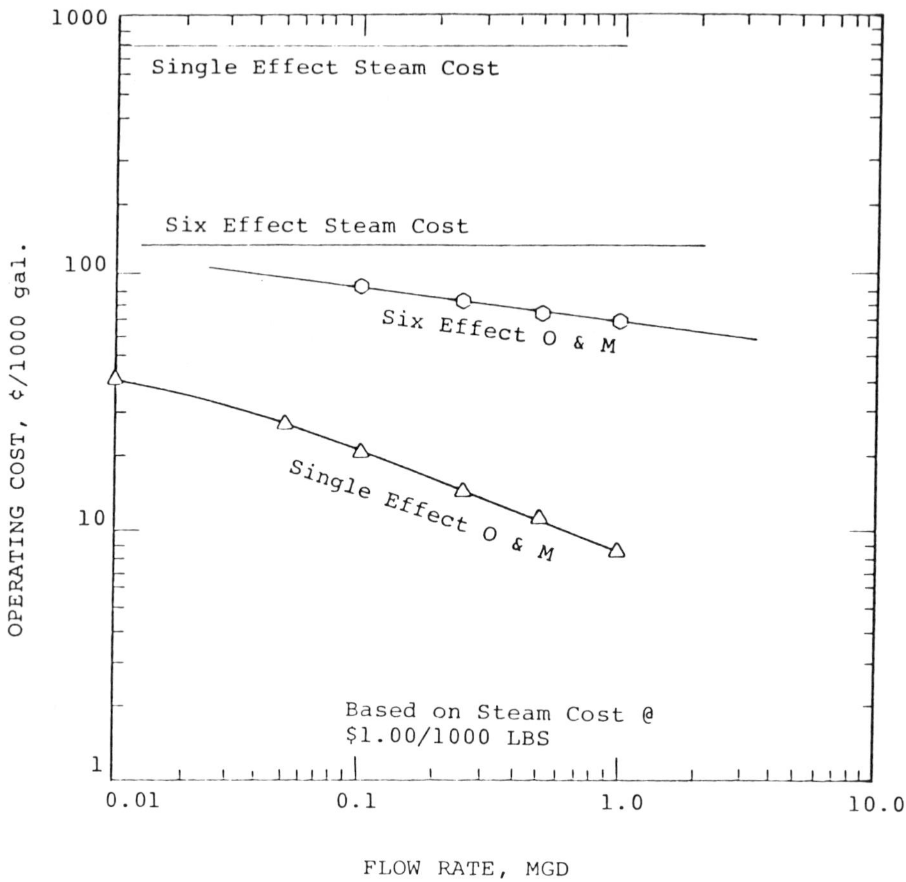

Figure 23.10 Steam cost and operating and maintenance cost (in 1975) for evaporation [11].

Ion Exchange

Ion exchange resins, depending upon their chemical nature, show preferential selectivity for specific ions. For example, weak-base anion resins are more selective for removing chloride than are strong-base resins. Table 23.5 presents data on ion selectivity for various types of exchange resins. The greater the affinity of the resin for the adsorbed ion, the more complete removal is achieved. However, a high affinity also means greater difficulty in regenerating the resin. Because regeneration of an effective resin is difficult, it is never carried to completion. As a result, the operational capacity of an ion exchange resin may be only 50–60% of theoretical capacity. For treatment of concentrated wastes, large resin regenerant volumes result, which may present a considerable disposal problem.

Table 23.5 Ion Exchange Resins Selectivity [15]

Resin	*Resin Selectivity*[a]
Strong acid cation (sulfonic)	$Li^+ > H^+ > Na^+ > NH_4^+ > K^+ > Rb^+ > Cs^+ > Mg^{+2} > Zn^{+2} > Cu^{+2} > Ca^{+2} > Pb^{+2}$
Weak acid cation (carboxylic)	$Na^+ > K^+ > Mg^{+2} > CA^{+2} > Cu^{+2} > H^+$
Strong-base anion (type I)	$F^- > OH^- > H_2PO_4^- > HCO_3^- > Cl^- > NO_2^- > HSO_3^- > CN^- > Br^- > NO_3^- > HSO_4^- > I^-$
Strong-base anion (type II)	$F^- > H_2PO_4^- > OH^- > HCO_3^- > Cl^- > NO_2^- > HSO_3^- > CN^- > Br^- > NO_3^- > HSO_4^- > I^-$
Weak-base anion	$F^- > Cl^- > Br^- > I^- > PO_4^{-3} > NO_3^- > CrO_4^{-2} > SO_4^{-3} > OH^-$

[a] Increasing selectivity left to right.

Ion exchange processes are very sensitive to both clogging and fouling. An ion exchange resin bed is a good filter, and suspended solids in the wastewater will rapidly clog the bed. Fouling results when the resin adsorbs materials which, either because of their affinity or absorption into the resin pores, cannot be removed in the regeneration step [16]. The Mn^{+4} ion, for example, has such high affinity for certain exchangers that it is almost impossible to remove by conventional regeneration. Strong-base anion exchange resins are particularly sensitive to organic fouling. The net result of fouling is loss of resin reactivity and reduced service life.

Ion exchange has been used with wide success for water softening and boiler water deionization. It has had less success, however, in industrial wastewater treatment, with many reports of irreversible fouling, poor removal efficiency, and similar difficulties. Its most successful industrial wastewater use to date has been with metal-plating rinsewaters.

Ion exchange has been used to purify a brackish municipal water supply which contained 1,000 mg/l sulfate and several hundreds of mg/l carbonates [17]. The effluent water contains less than 250 mg/l dissolved solids. Treatment costs (in 1971) were 20¢/1,000 gallons, with a capital investment of $700,000 for the 0.5 MGD plant. This is much less than the capital cost indicated in Table 23.3 for an equivalent waste. The two-stage ion exchange process is regenerated by lime and sulfuric acid. Large amounts of calcium sulfate sludge (gypsum) are produced and disposed of by landfill [17].

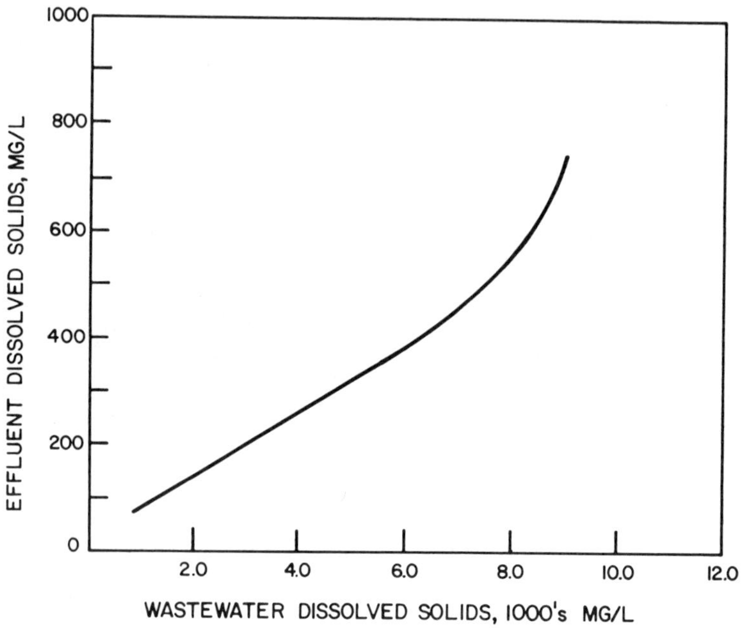

Figure 23.11 Relationship of effluent to influent total dissolved solids concentration for reverse osmosis treatment [20].

Higgins [18] has presented costs of ion exchange purification of a secondary sewage treatment-plant effluent totaling 10 MGD. Capital costs were estimated (in 1965) at \$1,963,000 and amortizated cost at 4.9¢/1,000 gal. Operating costs were 6.87¢/1,000 gal, or \$68.70/day. This plant would effectively treat a waste containing dissolved solids of 683 mg/l. Young has reported ion exchange treatment cost of about 9¢/1,000 gal, while amortizated capital costs were between 15 and 20¢/1,000 gal [19]. Table 23.3 indicates much higher operating costs for ion exchange treatment, at 25–50¢/1,000 gal.

Reverse Osmosis

The reverse osmosis process produces desalted effluent by forcing water through semipermeable membranes at high pressure. These membranes are more permeable to pure water than to dissolved solids. Dissolved solids removal is not complete, as shown in Figure 23.11. Effluent TDS content increases with increasing wastewater dissolved-solids level. In addition, low molecular-weight organic compounds will pass the membrane as effectively as water. Phenol (MW 94) is a good example. Its concentrations may be 10–20% greater in the permeate than feedwater, even for a 97% rejection (of NaCl) membrane [21]. Removals of up to 95% inorganic TDS are obtainable, as long as the membranes are in good condition [22].

Table 23.6 Industrial Process Water Reverse Osmosis Plants [25]

Capacity (gpd)	*TDS (mg/l)*			*% Water Recovery*
	Influent	*Effluent*	*Brine*	
151,000	3,310	212	20,865	85
21,600	453	40	1,692	75
57,500	306	40	1,104	75
40,000	700	—	—	70
50,000	8,000	30	3,110	75
352,000	200	5	785	75
40,000	800	25	3,125	75
320,000	600	45	2,820	80
20,000	2,000	200	7,400	75
1,500	280	25	—	40
44,000	140	10	—	75
24,000	750	75	2,775	75
320,000	750	81	—	77

At acidic wastewater pH, or at pH above 9, the reverse osmosis membranes will hydrolyze and dissolve in a matter of hours. Higher temperature liquids, above 100°F, may also shorten membrane life [20, 23]. Other factors which affect the useful life of reverse osmosis membranes are compaction due to excessive pressure (above 500 psig), membrane fouling, bacterial degradation, and precipitate formation at the membrane surface. Of these six factors, only precipitation is reversible. The adverse effect of pH and temperature, in particular, prevents the use of reverse osmosis on many industrial wastewaters. For example, most effective cyanide removal by reverse osmosis occurs at pH above 10.0. However, at this pH alkaline hydrolysis destroys the membrane. Fouling can occur due to the presence of oil or colloidal organic or inorganic materials in the wastewater. These materials will be trapped in the membrane and can reduce both product water quantity and quality. Common colloidal foulants are iron, manganese, silica, organics, bacteria, and algae [24].

Protection of the reverse osmosis membrane may require extensive pretreatment, including filtration for suspended solids removal, pH adjustment, and activated carbon treatment for removal of organics. Pretreatment requirements increment both the capital and operating costs of reverse osmosis systems.

One of the principal uses of reverse osmosis has been to produce high purity process water for industrial use. Table 23.6 summarizes results for U.S. plants. Typical raw water TDS is less than 1,000 mg/l, with low to moderate percentages of water recovery.

Good results have been obtained in purifying brackish water containing up to 20,000 mg/l dissolved solids, with operating costs estimated at 19–36¢/1,000 gal for 1 to 5 MGD plants, respectively [26]. Operating results are presented in Table 23.7 for three full-scale reverse osmosis plants. Plants A and B, at 24,000 and 150,000 gal/day, treated brackish water for domestic consumption. Plant C treated municipal sewage treatment plant effluent at 10,000 gal/day. The ranges of percent removals for various ions monitored are also contained in Table 23.7. Although efficiency of removal was not consistent for all minerals, TDS reductions exceeding 90% were found. Table 23.8 contains cost information on municipal reverse osmosis treatment. These values, which are more recent than those shown in Figures 23.4 and 23.5, indicate comparable capital and operating costs.

One modern plant produces boiler feed water at TDS below 100 mg/l, from seawater averaging 37,000 mg/l [30]. The 283,700 gpd plant, constructed in 1980, cost $4 million for equipment plus $1.5 million for a building. Equipment cost therefore represented an extremely high $14,100/1,000 gpd capacity. An operating cost analysis is presented in Table 23.9.

Since total dissolved solids have not historically been regulated as a pollutant, there is limited data on the capability of reverse osmosis for wastewater TDS control. Most experience is at the pilot scale and directed toward removal of one or more specific wastewater pollutants, other than TDS. Table 23.10 summarizes available TDS data from both full-scale and pilot experience with wastewater treatment by reverse osmosis. Costs would be expected to be higher than those found in other reverse osmosis applications, due to wastewater pretreatment requirements. This is demonstrated in the cost data presented in Table 23.11, for a 100,000 gpd full-scale ultrafiltration plus reverse osmosis system installed by Alcoa in 1976 [35]. The system is used to treat an oily waste for an aluminum-rolling mill, with water recovery. The total unit operation cost (in 1978) of $10.20/1,000 gal is partially offset by a credit for water recovery of $1.37/1,000 gal.

Electrodialysis

The amount of dissolved solids removed by electrodialysis depends, among other engineering factors, upon the retention time of the waste in the electrodialysis unit. Thus, the degree of purification achieved by this process, unlike those described previously, can be selected to a degree. Electrodialysis uses ion exchangers in membrane form, with an electrical driving force to promote ion transport across the membrane. Ions are thus removed from solution, leaving water behind. In contrast to ion exchange, which must undergo periodic regeneration, electrodialysis is essentially a continuous process. Since electric current must pass through the liquid, high resistance is encountered as the water becomes increasingly purer.

Table 23.7 Operating Results for Full-Scale Reverse Osmosis Plants

	Plant A [27]		*Plant B [27]*		*Plant C [28]*		*Range of %*
Mineral (mg/l)	*Influent*	*Effluent*	*Influent*	*Effluent*	*Influent*	*Effluent*	*Removal*
Calcium	312	10	160	2	68	0.8	96.8–99.8
Magnesium	141	12	49	4	12	1.5	87.5–91.8
Sodium + potassium	60	15	410	41	139	15.3	75.0–90.0
Bicarbonate	183	37	297	59	41	22	46.4–79.8
Chloride	95	59	351	32	260	15	29.5–94.2
Sulfate	1008	21	864	29	85	Nil	96.6–100.0
Iron	0.05	Nil	2.9	0.05	0.02	Nil	98.3–100.0
TDS	1008	155	2100	170	756	24	91.4–96.8

Table 23.8 Costs (in 1975) of Municipal Reverse Osmosis Plants [29]

Location	*Capital Cost ($/1,000 gpd)*	*Operating Costs (¢/1,000 gal)*
Greenfield, Iowa	628.97	86.2
Ocean Reef, Florida	476.19	42.5
Rotunda West, Florida	514.00	65.1
Average	539.72	64.4

Table 23.9 Operating and Maintenance Cost Analysis (in 1981) for a Reverse Osmosis Plant [30]

Item	*Cost ($/1,000 gal)*
Power	1.08
Maintenance and labor	0.69
Chemicals	1.53
Membrane replacement (60% at 5 years)	0.60
Prefilter cartridges	0.08
Equipment replacement (10-year life)	0.77
Total M&O cost	$4.75

Table 23.10 Wastewater Treatment by Reverse Osmosis Plants [31–34]

Type Wastewater	*Capacity (gpd)*	*TDS (mg/l)*		*% Water Recovered*	*Reference*
		Influent	*Effluent*		
Acid mine drainage	10,000[a]	2,491	10	75	31
Textile waste	[a]	764	76	91	32
Municipal sewage	25,000[a]	4,000	500	75	33
Plating rinse	2,000	525	50	60	25
Plating rinse	144,000	750	125	90	34
Rolling mill waste	100,000	2,437	—	—	35

[a] Pilot plant results.

Table 23.11 Economics of Alcoa Ultrafiltration–Reverse Osmosis Waste Treatment System [35]

Cost Factor	*Ultrafiltration*	*Ultrafiltration Plus Reverse Osmosis*
Capital investment ($\$ \times 10^6$)	1.45	1.55
Operating costs ($/1,000 gal)		
Depreciation	2.44	2.69
Power	0.47	0.62
Steam	0.19	0.19
Labor	2.00	2.00
Chemicals	1.48	1.48
Maintenance	0.92	0.92
Membrane replacement	1.61	2.30
Total operating, $/1,000 gal	$9.11	$10.20

The operation of an electrodialysis plant in Buckeye, Arizona has been described [36]. This plant, which has been in operation since 1962, reduces TDS of approximately 2,100 mg/l to about 500 mg/l. The plant is used to produce municipal water for the town of Buckeye, from brackish groundwater which would otherwise be too high in TDS for consumption. Operating costs for this electrodialysis plant of 4.8¢/1,000 gal of treated water produced have been reported [37]. The 1964 cost of a 10–20 MGD electrodialysis plant has been estimated at 9–10¢/1,000 gal [38]. This included amortization plus operating and maintenance. These costs contrasted with 1962 cost data of up to 20¢/1,000 gal.

Deep-Well Injection

Deep-well injection has become increasingly popular over the past 35 years, with an estimated 400 deep-well injection sites permitted today [39]. Most deep wells are greater than 1,000 feet deep, and many extend below 5,000 feet. The principal users of deep wells have been the chemical and petroleum-refining industries. The technology of deep-well injection is well established. However, uncertainties regarding the long-term impact of injected pollutants is bringing this technology under increasing regulatory scrutiny.

Deep-well disposal has been employed for ultimate disposal of a chemical industry wastewater containing above 100,000 mg/l dissolved solids. The waste stream contained sodium chloride, plus dissolved trace metals and trace organics [40]. Other reports of deep-well disposal of high TDS wastages are also in the literature [6, 41]. Figures 23.12 and 23.13 present capital investment and

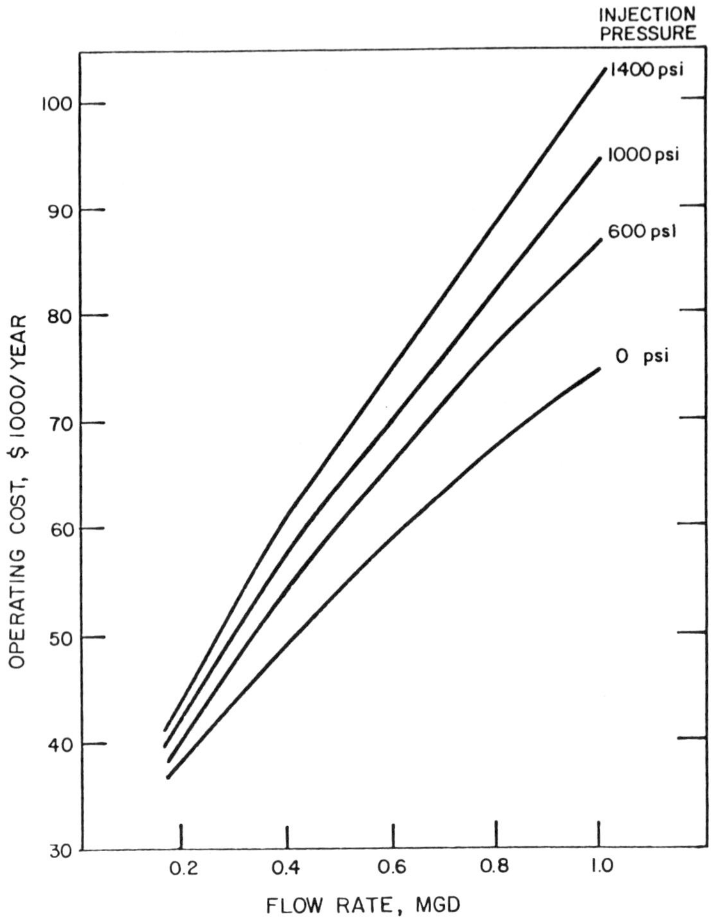

Figure 23.12 Annual operating costs (in 1966) for waste disposal by deep-well injection systems [3].

operating costs for deep-well disposal. Injection pressures of 200–5,000 psi have been reported for high-strength chloride wastes [41]. A comparison of these costs with costs for other processes (see Figures 23.1–23.6), reveal the substantial cost advantage of deep-well injection. In light of the potential hazards and uncertainties associated with subsurface waste disposal, however, deep-well injection may create pollutional problems far more severe than the ones being corrected [42].

SUMMARY

Most TDS treatment processes, with the exception of industrial waste precipitation, have been developed either to purify saline or brackish waters or to pro-

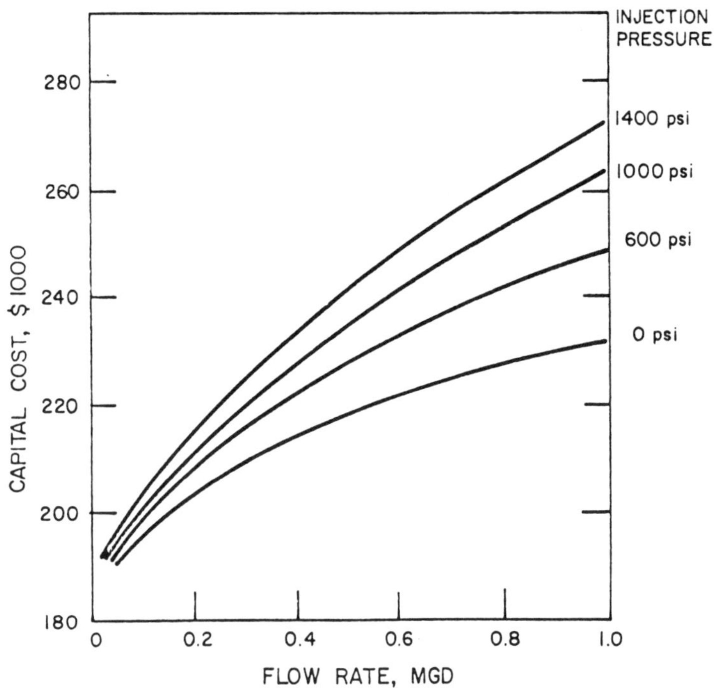

Figure 23.13 Capital costs (in 1966) for deep-well injection waste disposal systems [3].

vide deionized boiler feed and manufacturing process waters. There is little information in the technical literature on industrial wastewater TDS control by processes discussed in this chapter. The inorganic chemicals industries appear to be the major users of nonprecipitation methods such as ion exchange and reverse osmosis to reduce TDS in wastewaters. Costs of TDS removal are high compared to many other types of industrial waste treatment processes. Often, however, only a fraction of the waste stream need be treated in order to achieve intermediate TDS levels in the waste. This may be offset by higher costs required for pretreatment, for some TDS control processes.

REFERENCES

1. Zievers, J.F., R.W. Crain, and F.G. Barclay. "Waste Treatment Metal Finishing: U.S. and European Practices," *Plating* 55 (1968): 1171–1179.
2. *The Economics of Clean Water*, vol. III. *Inorganic Chemicals Industry Profile*, (Washington, DC: USDI, 1970).
3. Benger, M. "The Disposal of Liquid and Solid Effluents from Oil Refineries," *Proceedings of the 21st Industrial Waste Conference, Purdue University* (May 1966), pp 759–767.

4. Stone, R. "Walter Carpet Mill Industrial Waste System," *J. Water Poll. Control. Fed.* 44 (1972): 470–478.
5. "Development Document for Interim Final Effluent Limitations Guidelines and Proposed New Source Performance Standards for the Significant Inorganic Products Segment of the Inorganic Chemicals Manufacturing Point Source Category," USEPA 440/1-75/037, Group I, Phase II (May 1975).
6. "Development Document for Effluent Limitations Guidelines and New Source Performance Standards for the Major Inorganic Products Segment of the Inorganic Chemicals Manufacturing Point Source Category," USEPA 440/1-74-007-a (March 1974).
7. Brower, G., S. Ryckman, and P. Luther. "Control of Water Problems in a Steel Mill's Blast Furnace Gas Wash Recirculation System," paper presented at the 32nd Annual Purdue Industrial Waste Conference, West Lafayette, IN (May 1977).
8. Olver, J., W. Kreye, R. Roberts, and H. Sutton. "Anodizing Wastewater Treatment and Sludge Disposal: A Total Systems Design," *Proceedings of the 3rd Annual Conference on Treatment and Disposal of Industrial Wastewaters and Residues*, Houston, TX (April 18–20, 1978).
9. Wang, L., M. Pressman, W. Schuster, and T. Lynch. "Treatment of Nitrocellulose Manufacturing Wastewater," *Proceedings of the 3rd Annual Conference on Treatment and Disposal of Industrial Wastewaters and Residues*, Houston, TX (April 18–20, 1978).
10. "Development Document for Interim Final Effluent Limitations Guidelines and New Source Performance Standards for the Coal Mining Point Source Category," USEPA 440/1-75/057, Group II (October 1975).
11. Edwards, A.W., W.W. Eckenfelder, Jr., and J.A. Roth. *Industrial Water Cost Studies*, vol. I, Office of Research and Development, USEPA Contract 68-01-2485 (1975).
12. Monti, R.P. and B.T. Silbermann. "Wastewater System Alternatives: What Are They and What Cost?," *Water Eng.* 11 (1974): 52–56.
13. Ahlgren, R.M. "Membrane vs. Resinous Ion Exchange Demineralization," *Ind. Water Eng.* 8 (1971): 12–14.
14. Prescott, J.H. "New Evaporation-Step Entry," *Chem. Eng.* 78 (1971): 30–31.
15. Calmon, C. "Modern Ion Exchange Technology," *Ind. Water Eng.* 9 (1972): 12–15.
16. Polosi, P. and J. McCarthy, "Preventing Fouling of Ion-Exchange Resins," *Chem. Eng.* (August 9, 1982): 75–78.
17. "Ion Exchangers Sweeten Acid Water," *Environ. Sci. Technol.* 5 (1971): 24–25.
18. Higgins, I.R. "A Unique Process for Demineralizing Waste Water," *Ind. Water Eng.* 2 (1965): 26–29.
19. Young, L. "Ion Exchange Softening Process Operation and Maintenance," *Proceedings of the 25th Annual Short Course for Superintendents and Operators of Water, Sewage and Industrial Waste Disposal Systems* (1962), pp 103–106.
20. Morris, D.C., W.R. Nelson, and G. O. Walraven. "Recycle of Papermill Waste Waters and Applications of Reverse Osmosis," USEPA 12040 FUB 01/72 (1972).
21. Spatz, D.D. "Electroplating Waste Water Processing with Reverse Osmosis," *Prod. Fin.* 36 (1972): 79–89.
22. Culp, R.L. and G.L. Culp. *Advanced Wastewater Treatment* (New York: Van Nostrand Reinhold, 1971).

23. "Reverse Osmosis Module Operates Below 450 PSI," *Chem. Eng.* 78 (1971): 72–73.
24. Beach, W.A. and A.C. Epstein. "Summary of Pretreatment Technology for Membrane Processes," *Ind. Water Eng.* 12 (1975): 13–17.
25. "RO,: A Profile," *Ind. Water Eng.* 10 (1973): 10.
26. Channabasappa, K.C. and F.L. Harris. "Economics of Large-Scale Reverse Osmosis Plants," *Ind. Water Eng.* 7 (1970): 20–24.
27. Shields, C.P. "Reverse Osmosis for Municipal Water Supply," *Water Sew. Works* 119 (1972): 64–70.
28. Nusbaum, I., J.H. Sleigh, Jr., and S.S. Kremen. "Study and Experiments in Waste Water Reclamation by Reverse Osmosis," USEPA 17040-05/70 (1970).
29. Mattson, M.E. "Membrane Desalting Gets Big Push," *Water Waste Eng.* 12 (1975): 35–42.
30. Vera, I. "Operating Experiences on a Reverse Osmosis Plant Which Converts Sea Water into Boiler Feed Water," *Ind. Water Eng.* 18 (November/December 1981): 22–29.
31. Nusbaum, I. and R.G. Sudak. "Industrial Applications of Spiral-Wound Reverse Osmosis Modules," undated reprint from Gulf Environmental Systems Co., San Diego, CA.
32. Goodall, R., personal correspondence, ROGA Division, Universal Oil Products Co. (1974).
33. Dodson, R.E. "Pure Water from Sewage," *American City* (February 1971).
34. Hartman, J., personal communication, US Army Rock Island (IL) Arsenal (1975).
35. Sonksen, M.K., F.M. Sittig, and E.F. Maziarz. "Treatment of Oily Wastes by Ultrafiltration/Reverse Osmosis—A Case History," *Proceedings of the 33rd Industrial Waste Conference, Purdue University* (May 1978).
36. Oliver, J.C. "The Port Mansfield Desalting Plant," *Public Works* (June 1965).
37. Gilliland, E.R. "The Current Economics of Electrodialysis," paper presented at the 1st International Symposium of Water Desalination, Washington, DC (October 1965).
38. Smith, J.D. and J.L. Eisenmann. "Electrodialysis in Waste Water Recycle," *Proceedings of the 19th Industrial Waste Conference, Purdue University* (May 1971), pp 738 760.
39. Cawley, W., ed. *Treatability Manual*, vol. III. *Technologies for Control/Removal of Pollutants*, USEPA 600-8-80-042-c (July 1980).
40. Henkel, H.O. "Deep Well Disposal of Chemical Waste Water," *Proceedings of the 5th Annual Sanitary Water Resource Engineers Conference, Vanderbilt University* (1966).
41. Gloyna, E.F. and D.L. Ford. "The Characteristics and Pollutional Problems Associated with Petrochemical Wastes—Summary," (Washington, DC: USDI, 1970).
42. "Subsurface Disposal of Industrial Wastes," Interstate Oil Compact Commission, Oklahoma City, OK (1968).

24 Zinc

Industries discharging waste streams which contain significant levels of zinc include steel works with galvanizing lines, zinc and brass metal works, zinc and brass plating, viscose rayon yarn and fiber production, groundwood pulp production and newsprint paper production. Zinc salts are also used in the inorganic pigments industry (e.g., zinc chromate), and high zinc levels have been reported in acid mine drainage water. In a survey of active mines (deep and surface), zinc values ranged from 0.01 to 5.6 mg/l [1].

The primary source of zinc in wastewaters from plating and metal-processing industries is the dragout solution adhering to the metal product after removal from pickling or plating baths. The metal is rinsed free of this solution, and the contaminants are thus transferred to the rinsewater. The pickling process consists of immersing the metal (zinc or brass) in a strong acid bath to remove oxides from the metal surface. Finished metals are brightened by submergence in a bright dip bath containing strong acid chromate solution.

Plating solutions typically contain 5,000–34,000 mg/l zinc. These concentrated solutions may be discharged periodically, due to contamination. The concentration of zinc in a plating rinsewater will be a function of the bath zinc concentration, plated part drainage time over the bath, and the volume of rinsewater used. Most zinc and all brass plating and rinse solutions contain cyanide.

Waste concentrations of zinc range from less than 1 to more than 48,000 mg/l in various waste streams described in the literature. Average values, however, seem to be between 10 and 200 mg/l. Table 24.1 summarizes values reported for various zinc-bearing wastewaters.

TREATMENT TECHNOLOGY

Treatment processes employed for wastewater zinc removal may involve either chemical precipitation, with disposal of the resultant sludge, or recovery. Recovery may involve ion exchange, evaporative recovery, and other processes including precipitation, where a relatively pure zinc sludge is reclaimed. Recovery of plating wastes may prove to be more economical on an overall basis than is conventional precipitation and sludge disposal.

Table 24.1 Concentrations of Zinc in Process Wastewaters

Industrial Process	*Zinc Concentration (mg/l)*	*Reference*
Metal processing		
Bright dip wastes	0.2–37.0	2
Brass mill wastes	40–1,463	3
	8–10	3
Pickle bath	4.3–41.4	2
	0.5–37	3
	20–35	4
Wire mill pickle	36–374	5
Electrogalvanizing rinse	500	6
Bonderizing baths	1,000–3,000	7
	10	7
Conversion coating rinse	30.7	8
Plating		
General	2.4–13.8	9
	55–120	10
	15–20	11
	5–10	4
	7.0–215	2
	440–930	12
Chrome	245–1,050	13
Nickel	30	13
Zinc	480	13
	20–30	4
	70–150	2
	42	14
	70–350	15
	23.2	16
Brass	11–55	2
	10–60	15
Plating on zinc castings	3–8	4
Galvanizing of cold rolled steel	2–88	17
Anodizing plus plating	0.3–33	12
Silver plating		
Silver-bearing wastes	0–25	2
Acid waste	5–220	2
Alkaline waste	0.5–5.1	2
Rayon wastes		
General	250–1,000	18
	20[a]	19
	20–120	20
Typical	1,000	21

Table 24.1 (*continued*)

Industrial Process	*Zinc Concentration (mg/l)*	*Reference*
Other		
Latex rubber products	33–975	22
Vulcanized fiber	100–300	23
Cooling-tower blowdown	6	24
	2.5	25
	0.5	26
Power-plant boiler-cleaning waste	0.5–15.8	27
Municipal refuse incinerator scrubber water	650	28
Paint-manufacturing wastes	0.3–77.4	29
Ink formulating (tub washwater)	0.1–4.9	29
Dye house waste	7.5	30
Textile dyeing waste	2–6	31
Chrome pigment manufacturing	0–1,702	25
	0–1,000	32
	0.02	33
Hydrofluoric acid production	0.84	33
Sodium bisulfite production	1.3	33
Petroleum refining	0.04–1.84	34
Ferroalloy-smelting scrubber waters	3–89	26
Nonferrous-smelting scrubber water	114	12
Lead smelting	0.4–9.9	35
Lead battery manufacture	54.5	36
Gold ore-milling	1.5	37
Ferrous foundry	36–162	38
Steel-making—open hearth	2.1–1,210	39
Steel-making—degassing	1–23	39
Primary copper smelting and refining		40
Acid plant blowdown	0.8–13.7	
Arsenic plant washdown	37	
Secondary copper manufacturing	3–2,000	41
Zinc smelting		
Combined	11–243	42
Acid plant effluent	25–1,500	
Auxiliary metal reclamation	1,300–48,000	
Scrap steel cupola scrubber water	135–316	43
Coal mine drainage	0.01–5.6	1
Acidic coal-cleaning leachate	16	44
Base metal mine drainage	138	45

[a] After zinc recovery by ion exchange.

Chemical Precipitation

The precipitation process most frequently employed involves adjustment of pH with either lime or caustic to achieve alkaline conditions, and precipitation of zinc hydroxide. Lime addition has been the most widely accepted method for pH adjustment, in spite of the concurrent precipitation of calcium sulfate in the presence of high sulfate levels in some wastes. The precipitation of calcium sulfate along with the zinc hydroxide increases the total amount of sludge to be disposed of.

Table 24.2 is a representative summary of published zinc treatment results for hydroxide precipitation. These values reflect a wide range of industrial systems; generally, treatment is not for zinc alone. In cases where cyanide or chromate is also present in the waste, as frequently occurs in zinc and brass plating, cyanide destruction and chromate reduction must precede metal hydroxide precipitation. Settling efficiency affects effluent concentration, as is evident in Table 24.2 by improvements in effluent zinc levels resulting from filtration of settled effluents. Incomplete cyanide treatment will increase effluent zinc levels, due to complexation, as will improper control of treatment pH.

There is a great deal of confusion in the literature regarding the optimum pH for hydroxide precipitation of zinc. Theoretical considerations, and laboratory studies on pure zinc solutions, indicate optimum treatment at pH 9.0–9.8 [54–57]. However, results from full-scale systems often indicate improved performance at considerably high operating pH. Yost and Scarfi [56] have reported, for example, optimum zinc precipitation treatment at pH 11 and above, for zinc-plating rinses containing alkaline cleaning compounds. These workers have also presented treatability data which suggest that both optimum treatment pH and residual soluble effluent zinc are influenced by initial zinc concentration [56, 57].

The influence of pH on treatment of actual zinc plating wastes is apparently quite complex, particularly when the zinc waste is combined with rinsewaters from other plating operations. Table 24.3 presents zinc treatability data as a function of pH, for four plating wastes from one facility [13]. Three different pH optimas are observed among the four wastewaters, ranging from pH 8.0 to 11.5. Further, treatment efficiency for the zinc-plating line wastewater was quite poor at all pH values tested. The interference of chelating materials was reported for this wastewater [13].

Yost and Scarfi [57], in testing actual plating wastewaters, reported a pH optimum of 10 for zinc cyanide wastes alone, with residual filterable zinc increasing with increasing initial zinc concentration. At pH 10 and initial zinc levels of 1,000, and 50 mg/l, residual filterable zinc levels of about 1.0 and 0.1 mg/l were achieved, respectively. The solubility patterns corresponded to the classical U-shaped metal hydroxide solubility curves described in many textbooks. However, treatment of combined zinc plus other plating wastes (e.g., copper cyanide, cadmium cyanide) yielded quite different results. Depending

Table 24.2 Summary of Hydroxide Precipitation Treatment Results for Zinc Wastewaters

	Zinc Concentration (mg/l)			
Industrial Source	*Initial*	*Final*	*Comments*[a]	*Reference*
Zinc plating	—	0.2–0.5	pH 8.7–9.3	46
General plating	18.4	2	pH 9.0	47
	—	0–6	sand filtration	48
	55–120	1.0	pH 7.5	10
	4.1	0.39		49
	46	2.9	pH 8.5	14
		1.9	pH 9.2	
		2.8	pH 9.5	
		2.9	pH 10.5	
Vulcanized fiber	100–300	1.0	pH 8.5–9.5	23
Brass wire mill	36–374	0.08–1.60	integrated treatment for copper recovery	5
Tableware plant	16.1	0.02–0.23	sand filtration	2
Viscose rayon	20–120	0.88–1.5		20
	70	3–5	pH 5	50
	20	1.0		19
Metal fabrication	—	0.5–1.2	sedimentation	51
		0.1–0.5	sand filtration	
Radiator manufacture	—	0.33–2.37	sedimentation	52
		0.03–0.38	sand filtration	
Blast furnace gas scrubber water	50	0.2	pH 8.8	53
Zinc smelter	744	50		42
	1,500	2.6		42
Ferroalloy wastes	11.2–34	0.29–2.5		26
	3–89	4.2–7.9		26
Ferrous foundry	72	1.26	sedimentation	38
		0.41	sand filtration	
Deep coal mine—acid waters	33–7.2	0.01–10		1

[a] All treatment involved precipitation plus sedimentation. Special or additional aspects of treatment are indicated under *Comments*.

upon initial metal concentrations, pH 10 yielded either minimum solubility (for lower initial concentrations) or maximum solubility observed between pH 8 and 11 (for higher initial concentrations). Thus, the blend of the wastewaters and their initial concentrations influenced both optimum treatment pH and residual filterable zinc concentration.

Table 24.3 Effect of Treatment pH on Caustic Precipitation of Zinc for Plating Wastes [13]

Treatment pH	*Filtered Zinc Concentration (mg/l)*			
	Chrome-1	*Chrome-2*	*Nickel*	*Zinc*
Initial[a]	1,050	245	30	480
7.5	0.03	0.08	0.24	19.0
8.0	0.09	<0.02	<0.02	23.0
8.5	0.04	<0.02	<0.02	32.0
9.0	<0.02	0.08	<0.02	32.0
9.5	<0.02	0.04	<0.02	12.0
10.0	<0.02	<0.02	<0.02	23.0
10.5	<0.02	<0.02	<0.02	16.0
11.0	<0.02	<0.02	<0.02	14.0
11.5	<0.02	<0.02	<0.02	4.0
12.0	<0.02	<0.02	0.04	8.0

[a] Initial pH: Chrome-1 = 2.0; Nickel = 7.4; Chrome-2 = 2.5; Zinc = 9.4.

Such results, particularly in initially alkaline cyanide wastewater of high carbonate content, may reflect changes in the zinc solid phase precipitated. In the presence of carbonate, zinc may precipitate as either zinc hydroxide or zinc hydroxycarbonate, depending upon pH and carbonate level [54]. The results reported by Yost and Scarfi [57] at high initial metal (and carbonate) levels are suggestive of zinc hydroxycarbonate precipitation.

The issue of proper treatment pH is further confused when a wastewater contains other metals, with differing pH optima, in addition to zinc. This situation is demonstrated by the data of Table 24.4, from full-scale treatment of a combined metal-cleaning and -plating waste. Optimum treatment for all metals except zinc, plus optimum clarifier performance, was observed at the highest

Table 24.4 Effect of Operating pH on Precipitation/Sedimentation Treatment Plant Performance [14]

Treatment pH	*Average[a] Effluent Concentration (mg/l)*				
	Suspended Solids	*Chromium(T)*	*Iron*	*Lead*	*Zinc*
8.5	87	0.9	3.1	1.7	2.9
9.2	43	0.6	2.2	1.1	1.9
9.8	27	0.4	3.8	1.0	2.8
10.5	28	0.4	1.6	0.6	2.9

[a] Data are the average of 4 to 9 analyses.

Table 24.5 Comparison of Lime Precipitation versus Two-Stage Lime Precipitation-Sulfide Precipitation Treatment for Zinc [after 12]

Wastewater Source	*Treatment pH*	*Initial Zinc (mg/l)*	*Lime Precipitation*		*Two-Stage Treatment Filtered*
			Clarified	*Filtered*	
Printed circuit board	8.5	0.770	0.430	0.053	0.014
Plating rinsewater	8.75	90.0	1.00	0.210	0.010
	9.0	11.0	2.15	0.167	0.331
	8.5	13.0	0.625	0.010	0.005
	8.75	253.0	0.400	0.295	0.008
	10.0	290	1.20	0.510	0.012
	8.5	2.8	0.044	0.010	0.011
	8.5	440	75.0	71.0	4.7
	9.0	440	37.0	29.0	2.0
	8.5	930	9.6	1.4	0.340
	10.0	930	3.3	1.0	—
Nonferrous smelter	8.5	114	0.511	0.030	0.036

operating pH. However, best zinc removal occurred at pH 9.2. Situations such as this are frequently observed in full-scale systems, and in these cases, operating pH is either a compromise value, or based upon the pH optimum for the wastewater constituent representing the most stringent effluent limitation.

Capital and operating costs for conventional zinc precipitation systems are similar to those for such systems used to treat other metal pollutants. Unit processes can include equalization, chemical feed and rapid mix, reaction vessel, and clarification and filtration. Typical costs for such systems were presented in Table 1.5 (see page 6).

The use of hydrogen sulfide at pH 2 has been reported to achieve "complete removal" of zinc from an electrolytic nickel recovery system [58]. The operating pH would appear to represent cause for concern for evolution of toxic hydrogen sulfide gas, however. Sodium sulfide treatment of a gold ore-refining waste, initially containing zinc at 42 mg/l, achieved precipitation to 1.16 mg/l at a treatment pH of 7.0 [59]. However, treatment of the same waste by caustic at pH 10 reduced zinc to 0.02 mg/l. Comparative costs (in 1975) for lime versus sulfide treatment for a 0.3 MGD zinc smelter plant waste were as follows [42]: capital cost, $1,230,000 (lime), $2,620,000 (sulfide); total annual cost, $503,000 (lime), $777,000 (sulfide). Treatment of acid mine drainage containing zinc at 122–294 mg/l by first-stage lime addition to pH 5.0–5.5, followed by second-stage sulfide addition to pH 5.5–6.4 produced variable effluent zinc levels ranging from 0.2 to 30.0 mg/l [60].

Pilot-scale treatment results are reported in Table 24.5 for lime precipitation alone, followed by clarification and filtration; and lime-precipitation/clarification, followed by second-stage sulfide precipitation/filtration [12]. The results confirm the difficulty of effective precipitation treatment for some zinc wastewaters (see the last two plating rinsewaters listed). Filtration of the lime-treated wastewaters yielded significant improvement in effluent quality in almost all instances, and two-stage treatment normally yielded further slight improvement. The greatest improvement by second-stage sulfide polishing resulted from the two difficult-to-treat plating rinsewaters. Except in such special instances, however, it is doubtful that the increased capital and operating costs (see Table 4.11, page 47) warrant the two-stage process.

The proprietary Sulfex process (Permutit Co.) has been applied to zinc wastes. The process involves addition of ferrous sulfide, which gradually releases sulfide to precipitate the zinc and other metals. Iron is coprecipitated in the process. The process is claimed to overcome the hazards of hydrogen sulfide generation and high residual sulfide in the treated effluent, both of which are associated with the use of more soluble sulfide treatment salts. In one comparison of the Sulfex process with conventional hydroxide precipitation of a waste initially at pH 7.1 and containing 3.2 mg/l zinc, hydroxide treatment to pH 9 reduced zinc to 0.11 mg/l while the Sulfex process at pH 8.5 reduced zinc to 0.09 mg/l [61].

However, the Sulfex process produces about three times greater quantities of sludge than does conventional precipitation [61, 62], which can lead to sludge-handling problems [55]. In addition, the Sulfex process requires fresh ferrous sulfide precipitate, produced on-site by blending ferrous sulfate, a soluble sulfide salt, and lime [61]. These latter two chemicals are required by normal sulfide treatment, and the Sulfex process therefore involves, in essence, an intermediate step of ferrous sulfide production requiring a third treatment chemical. When only a portion of the ferrous sulfide reacts with the wastewater, larger amounts of the sulfide salt and alkali are required in this process. Scott [62] has suggested the application of the Sulfex process as a second, polishing, stage for zinc after conventional precipitation/clarification in a primary stage. He reports that the additional sludge produced is only 1–2% of the sludge produced in the primary stage, but that operating costs will approximately double.

The patented Integrated treatment system has been claimed to achieve effluent zinc levels below 1.0 mg/l in plating and pickling systems [63]. The process involves immersion of a freshly plated part in a concentrated soda ash solution, prior to rinsing of the part. In one reported application of the process in a small plating shop, effluent rinsewater zinc was reduced from 2.0–5.0 mg/l to 0.29–0.44 mg/l, after installation of the process [64]. The treatment bath contained hydrazine, soda ash, and caustic, and was maintained at pH 9.5–10.5. Hydrazine was present to reduce trace amounts of cupric ion and allow precipitation of cuprous oxide.

Inherent in the Integrated process is individualized treatment for a specific metal, with the potential for recovery. Even without recovery, treatment of zinc waste separately may be justified on technical grounds where the zinc process waste represents a small portion of the total plant flow [6]. Separate treatment normally reduces both the size of treatment facility, and the quantity of treatment chemicals used.

A process has been developed by DuPont primarily for small zinc and cadmium cyanide-plating operations [65]. A proprietary formulation called Kastone (composed of a catalyst, hydrogen peroxide, and stabilizers) oxidizes the cyanide and promotes the formation of metal oxide instead of hydroxide precipitate. The former are readily removed by filtration apparatus, which most small plating plants already have in operation. Effluent treatment levels of 0.29–2.9 mg/l zinc have been reported for zinc-plating rinsewaters [66].

Ion Exchange

Although considered a recovery process, ion exchange may be used as a polishing process for high final effluent quality, or as a means for water recovery and reuse, while providing a more concentrated waste via regeneration for final treatment. Ion exchange effluent zinc levels below 0.05 mg/l are achievable [13]. However, many cation exchange resins are more selective for the common divalent ions calcium and magnesium than for zinc [67]. Such lack of selectivity results in early resin exhaustion and unattractive operating economics.

Unless associated with other materials which are costly to treat and replace, or a requirement for water reuse, zinc treatment by ion exchange for recovery of the metal value alone is usually not economical, except at high zinc concentrations. Under conditions of high levels of relatively pure zinc, precipitation and acid regeneration of the zinc hydroxide is satisfactory. This application has been reported in viscous rayon manufacturing [68]. Zinc recovery by ion exchange in viscous rayon manufacturing has also been used [18, 19], with the acid regenerant reused directly without purification. High residual zinc concentrations have been experienced. In one case, residual zinc of 20 mg/l in the ion exchange effluent necessitated further treatment [19]. Ion exchange treatment was considered for one plant, due to a water volume discharge limitation which required recycle of 90% of a 53,000 gpd stream [8]. However, the ion exchange process was rejected in favor of reverse osmosis treatment, due to the high total dissolved solids concentration of the wastewater discharge which would have resulted from the resin regeneration step.

Laboratory studies have indicated that at flow rates of 2.5 gal/hr/ft^2 ion exchange was not as effective as lime precipitation for zinc removal [69]. A published cost comparison of the treatment of zinc and nickel wastes by ion exchange versus lime treatment indicated ion exchange to be more than twice as costly as lime treatment for a waste flow of 35,000 gal/day. Precipitation costs were $0.34/1,000 gal. Recovery value from the ion exchange process was credited for water reuse only [70]. Pilot plant studies have been reported to reduce zinc in cooling-tower blowdown from 6 to 0.4 mg/l [71]. Ion exchange has been investigated for rinsewater recycle in the zinc-bonderizing process

[7], as well as for zinc cyanide-electroplating rinses [16], electrolytic refinery effluents [72], and polyvinyl fiber-manufacturing wastes [73].

Other Processes

A major problem in treating dilute wastewater is the large volumes involved, and the attendant capital and operation costs. Direct recycle or evaporative recycle of pickling- and plating-bath rinsewater is made feasible by waste volume reduction techniques such as installation of countercurrent rinses. Rinsewater volumes can be reduced to one-third or one-fourth of normal volumes by the use of cascade or countercurrent rinse practices [63]. This reduction in volume yields higher zinc concentrations in the rinsewater. One plant was able to reduce zinc cyanide-plating rinse requirements from 310 to 12 gal/hr for two units, by employing countercurrent rinse [9]. The final rinsewaters were of sufficient zinc concentration to allow evaporative concentration and recovery of the plating dragout. Final rinse dragout required no further treatment, due to the low zinc levels present. The cost (in 1982) of zinc cyanide-plating bath makeup solution is reported to be $1.02/gal [74]. Complete elimination of dragout by evaporative recovery results in impurity buildup in the plating bath, however, and consequent need to bleed or periodically dump the bath solution.

Treatment of a zinc- and brass-plating wastewater by evaporative recovery has been reported [75]. Plant modifications reduced process water from 3,000 gal/hr to 50 gal/hr and resulted in process chemical savings of $18,000/yr, plus equal savings in chlorine and other treatment chemicals. Capital and operating costs associated with such a closed-loop system were reported to be approximately the same as for precipitation. Application of evaporative recovery has been reported for another plating plant. A 300 gallon/hr double-effect evaporator was installed on the zinc cyanide-plating line. Cost analysis, based on 4,000 hr/year operation, yielded operating cost of $10,800/year. For a 10 gal/hr plating-bath dragout, savings of $18,000 per year in chemical costs were claimed. The annual net savings over operation was thus $7,200 [76]. Evaporative recovery of viscose rayon acid-bath wash, for 0.768 MGD capacity, was reported (in 1974) at $1,237/1,000 gpd capital costs and 77.7¢/1,000 lb of recycled solution [21].

Evaporative treatment using packed-tower atmospheric evaporators has been estimated (in 1974) to cost $1.00/1,000 gal [77], based upon chemical and rinsewater savings. The process is reported not economical for processes such as chromating baths and other small-scale operations [77].

Reverse osmosis has been proposed for treatment for recovery of zinc-plating wastes (zinc cyanide and acid zinc chloride). To protect the reverse osmosis membranes, feed solution pH must be maintained between 5.0 and 11.0 [78]. Further, it has been demonstrated that reverse osmosis alone will not achieve complete closed-loop recovery and reuse of the zinc solutions, and

Table 24.6 Average Performance of Reverse Osmosis System [8]

Parameter	*RO Feed*	*RO Permeate*
pH	5.1	4.9
Zinc (mg/l)	7.3	2.2
Iron (mg/l)	0.5	<0.2
Phosphate (mg/l as P)	414	109
TDS (mg/l)	750	238

RO = reverse osmosis.

evaporative concentration of the reverse osmosis permeate is required as an intermediate step [78].

A full-scale reverse osmosis system has recently been installed to treat a 53,000 gpd zinc phosphate conversion coating waste which, due to effluent volume limitations, required 90% water volume recovery and recycle [8]. The remaining concentrated brine (5,000 gpd) is batch treated by conventional precipitation, and discharged. Pretreatment required prior to the reverse osmosis unit includes equalization, dual media filtration, pH adjustment, cooling (from 140° to 95°F to prevent membrane compaction), and dispersant and antiprecipitant additions (to prevent membrane fouling). Table 24.6 presents average data on the reverse osmosis process performance. Capital and operating costs have not been reported. Reverse osmosis has been tested at pilot scale on various industry wastewaters. Operating results are presented in Table 24.7. Removal efficiencies are quite high, although permeate quality is comparable to that achieved by conventional precipitation plus effluent filtration.

Electrolytic treatment has also been proposed for treatment of zinc wastes. The process has particular application for wastewaters which also contain cyanide, with simultaneous cyanide destruction. Hillis [80] has reported electrolytic reduction of zinc in a cyanide solution from 27.5 mg/l to a residual 0.3 0.4 mg/l, within 5 hours reaction. The zinc can be directly recovered from the electrode, in pure form. Table 24.8 presents zinc and cyanide data for a carbon fiber electrolytic bed. The electrolytic process has not been widely utilized in full-scale treatment of zinc wastes, however.

SUMMARY

In the case where zinc removal is the only consideration and recovery is not warranted, removal by precipitation can be accomplished by standard pH adjustment through lime addition, precipitation and flocculation, and sedimentation, employing standard waste treatment equipment. Operating data for existing chemical precipitation units indicate that levels of 1 mg/l or less of zinc are readily obtainable with lime precipitation [41, 42], although assurance of consistent removal of precipitated zinc to lower levels from the effluent stream

Table 24.7 Pilot Scale Results for Reverse Osmosis Treatment of Zinc Wastewater [after 79]

	Zinc Concentration (μg/l)		
Industry Source	*Feed*	*Permeate*	*% Removal*
Zinc-cyanide plating rinse	1,700	30	98
Steam electric power plant	300	53	82
	780	3	99
Textile mill	7,200	140	98
	5,400	6,600	(−20)
	460	250	46
	520	360	31
	7,200	360	95
	1,400	30	98
	4,100	180	96
	1,200	22	98
	24,000	430	98
	9,700	37	>99
Cooling-tower blowdown	10,000	300	97

may require filtration. A summary of reported effluent zinc concentrations for various treatment systems in the plating industry is presented in Table 24.9.

Other treatment alternatives exist and may be more economical although not necessarily more efficient. Among these are process modification for evaporative and ion exchange recovery. Existing process layout and current treatment facilities may cause costs to vary widely with industrial site and nature of product. Evaporative recovery, in concert with countercurrent rinsing to reduce waste volumes, has proved successful on both economic and zinc removal efficiency bases.

Table 24.8 Electrolytic Treatment of Zinc Cyanide Wastes [81]

		Concentration (mg/l)	
Waste	*Parameter*	*Initial*	*Final*
A	Zinc	352	0.7
	Cyanide	258	12.0
B	Zinc	117	0.3
	Copper	842	0.5
	Cyanide	1,230	<0.1

Table 24.9 Reported Effluent Zinc Concentrations in the Plating Industry [82]

Type of Treatment	*Effluent Zinc Concentration (mg/l)*
Evaporation	0.15
	18.4
	0.12
Continuous chemical precipitation	0.4
	0.6
	0.25
	0.08
	0.87
	0.35
	0.8
	0.5
	0.5
	0.82
Batch chemical precipitation	0.03
	5.0
	0.14
Integrated process	0.20
	0.45
Electrolytic recovery	0.32
	7.9
	0.05
	0.05

REFERENCES

1. "Development Document for Interim Final Effluent Limitations Guidelines and New Source Performance Standards for the Coal Mining Point Source Category," USEPA 440/1-75/057, Group II (October 1975).
2. Nemerow, N.L. *Theories and Practices of Industrial Waste Treatment* (Reading, MA: Addison-Wesley, 1963).
3. McGarvey, F.X., R.E. Tenhoor, and R.P. Nevers. "Brass and Copper Industry: Cation Exchangers for Metals Concentration from Pickle Rinse Waters," *Ind. Eng. Chem.* 44 (1952): 534–541.
4. Lowe, W. "The Origin and Characteristics of Toxic Wastes with Particular Reference to the Metal Industries," *Water Poll. Control* (London, 1970): 270–280.
5. *Brass Wire Mill Process Changes and Waste Abatement, Recovery and Reuse*, Water Pollution Control Research Series 12010 DFP, (Washington, DC: USEPA, 1971).

6. Pilot, J. "Treatment of Pickling Plant Effluent at B.S.C. Whitehead Works," *Effluent Water Treatment J.* 14 (1974): 260–263.
7. McGarvey, F.X. and S.A. Fisher. "Effluent Control in Zinc Bonderizing Processes. I. The Recycle of Rinse Waters via an Ion Exchange Process," *Plating* 62 (1975): 231–234.
8. Marino, M., M.E. Terril, B. Burke, and A. Simon. "Application of the Reverse Osmosis Process for Achieving Industrial Wastewater Reuse," paper presented at the 51st Annual Conference of Water Pollution Control Federation, Anaheim, CA (October 2, 1978).
9. Barnes, G.E. "Disposal and Recovery of Electroplating Wastes," *J. Water Poll. Control. Fed.* 40 (1968): 1459–1470.
10. Nyquist, O.W. and H.R. Carroll. "Design and Treatment of Metal Processing Wastewaters," *Sew. Ind. Wastes* 31 (1959): 941–948.
11. Pinkerton, H.L. "Waste Disposal," *Electroplating Engineering Handbook*, 2nd ed., A. Kenneth Graham, ed. (New York: Reinhold, 1962).
12. Robinson, A. and J. Sum. "Sulfide Precipitation of Heavy Metals," USEPA 600/2-80-139 (June 1980).
13. Yeats, A. "Ion Exchange Selectively Removes Heavy Metals from Mixed Plating Wastes," paper presented at the 32nd Annual Purdue Industrial Waste Conference, West Lafayette, IN (May, 1977).
14. McVaugh, J. and W.T. Wall, Jr. "Optimization of Heavy Metals Wastewater Treatment Effluent Quality versus Sludge Treatment," *Proceedings of the 31st Industrial Waste Conference, Purdue University* (May 1976), pp 17–25.
15. *State of the Art: Review on Product Recovery*, Water Pollution Control Research Services 17-7 ODJW 11/69 (Washington, DC: USDI, 1970).
16. Moore, F.L. "An Improved Ion Exchange Resin Method for Removal and Recovery of Zinc Cyanide and Cyanide from Electroplating Wastes," *J. Environ. Sci. Health* 7 (1976): 459–467.
17. Donovan, E.J., Jr. "Treatment of Wastewater for Steel Cold Finishing Mills," *Water Wastes Eng.* 7 (1970): F22–F25.
18. McGarvey, F.X. "The Application of Ion Exchange Resins to Metallurgical Waste Problems," *Proceedings of the 17th Industrial Waste Conference, Purdue University* (May 1952), pp 289–304.
19. Sharda, C.P. and K. Namwannan. "Viscose Rayon Factory Wastes and Their Treatment," *Technology, Sindii* 3 (1956): 58–60; *Water Poll. Abstr.* 41 (1968).
20. *Zinc Precipitation and Recovery from Viscose Rayon Wastewater*, Water Pollution Control Research Series, 12090 ESG, (Washington, DC: USEPA, 1971).
21. "Development Document for Effluent Limitations Guidelines and New Source Performance Standards for the Synthetic Resins Segment of the Plastics and Synthetic Materials Manufacturing Point Source Category," USEPA No. 440/1-74-001-a (March 1974).
22. "Development Document for Effluent Limitations Guidelines and New Source Performance Standards for the Fabricated and Reclaimed Rubber Segment of the Rubber Processing Point Source Category," USEPA 440/1-74/030-a, Group I, Phase II (December 1974).
23. "Reclaiming Zinc from an Industrial Waste Stream," *Environ. Sci. Technol.* 6 (1972): 880–881.
24. Culotta, J.M. and W.F. Swanton. "Recovery of Plating Wastes: Selection of Lowest Cost Evaporator," *Plating* 57 (1970): 1121–1123.

25. "Development Document for Interim Final Effluent Limitations Guidelines and Proposed New Source Performance Standards for the Significant Inorganic Products Segment of the Inorganic Chemicals Manufacturing Point Source Category," USEPA 440/1-75/037, Group I, Phase II (May 1975).
26. "Development Document for Effluent Limitations Guidelines and New Source Performance Standards for the Smelting and Slag Processing Segment of the Ferroalloy Manufacturing Point Source Category," USEPA 440/1-74-008-a (February 1974).
27. Kuppusamy, N. "Copper Removal from Power Plant Boiler Cleaning Waste," *Ind. Wastes* (March/April 1977): 43–45.
28. Iammartino, N.R. "Mercury Cleanup Routes—II," *Chem. Eng.* 82 (1975): 36–37.
29. "Development Document for Effluent Limitations Guidelines and New Source Performance Standards for the Oil Base Solvent Wash Subcategories of the Paint Formulating and the Ink Formulating Point Source Category," USEPA 440/1-75/050a, Group II (July 1975).
30. Rouse, J.V. "Removal of Heavy Metals from Industrial Effluents," *ASCE J. Environ. Eng. Div.* (October 1976): EE5 929–936.
31. Brandon, C., J. Ponter, and T. Sargent. "Reuse of Total Composite Wastewater Renovated by Hyperfiltration in Textile Dyeing Operations," *Proceedings of the National Conference on Management and Disposal of Residues from the Treatment of Industrial Wastewaters*, Washington, DC (February 1975).
32. Campbell, H.J., Jr., N.C. Scrivner, K. Batzar, and R.F. White. "Evaluation of Chromium Removal from a Highly Variable Wastewater Stream," *Proceedings of the 32nd Industrial Waste Conference, Purdue University* (May 1977), pp 102–115.
33. *Treatability Studies for the Inorganic Chemicals Manufacturing Point Source Category*, USEPA 440/1-80/103 (July 1980).
34. "Development Document for Effluent Limitations Guidelines and New Source Performance Standards for the Petroleum Refining Point Source Category," USEPA 440/1-74-014-a (April 1974).
35. "Development Document for Interim Final Effluent Limitations Guidelines and Proposed New Source Performance Standards for the Lead Segment of the Nonferrous Metals Manufacturing Point Source Category," USEPA 440/1-75/032-a, Group I, Phase II (February 1975).
36. Liu, H. and J. Blacklidge. "Cross Flow Filtration Technology for Metal Finishers," *Proceedings of the 4th Conference on Advanced Pollution Control for the Metal Finishing Industry*, USEPA Report 600/9-82-022 (December 1982).
37. Schmidt, J., L. Simovic, and E. Shannon. "Development Studies for Suitable Technologies for the Removal of Cyanide and Heavy Metals from Gold Milling Effluents," paper presented at the 36th Annual Purdue Industrial Waste Conference, West Lafayette, IN (May 1981).
38. Osantowski, R. and J. Ruppersberger. "Upgrading Foundry Wastewater Treatment," paper presented at the 39th Annual Purdue Industrial Waste Conference, West Lafayette, IN (May 1984).
39. "Development Document for Effluent Limitations Guidelines and New Source Performance Standards for the Steel Making Segment of the Iron and Steel Manufacturing Point Source Category," USEPA 440/1-74-024-a (June 1974).
40. "Development Document for Interim Final Effluent Limitations Guidelines and Proposed New Source Performance Standards for the Primary Copper Smelting

Subcategory and the Primary Copper Refining Subcategory of the Copper Segment of the Nonferrous Metals Manufacturing Point Source Category," USEPA 440/1-75/032-b, Group I, Phase II (February 1975).

41. "Development Document for Interim Final Effluent Limitations Guidelines and Proposed New Source Performance Standards for the Secondary Copper Subcategory of the Copper Segment of the Nonferrous Metals Manufacturing Point Source Category," USEPA 440/1-75/032-c, Group I, Phase II (February 1975).
42. "Development Document for Interim Final Effluent Limitations Guidelines and Proposed New Source Performance Standards for the Zinc Segment of the Nonferrous Metals Manufacturing Point Source Category," USEPA 440/1-75-032, Group I, Phase II (February 1975).
43. McGuire, M.E., D.R. Williams, and D.A. Fausch. "Pollution Control Through Water Conservation at a Gray Iron Foundry," paper presented at the 33rd Annual Purdue Industrial Waste Conference, West Lafayette, IN (May 1978).
44. Wangen, L. and J. Williams. "Control by Alkaline Neutralization of Trace Elements in Acidic Coal Cleaning Waste Leachate," *J. Water Poll. Control. Fed.* 54 (1982): 1302–1310.
45. Huck, P.M., K.C. Murphy, C. Reed, and B.P. LeClair. "Optimization of Polymer Flocculation of Heavy Metal Hydroxides," *J. Water Poll. Control. Fed.* 49 (1977): 2411–2418.
46. "Effluent Treatment at B.S.R.," *Metal Fin. J.* 17 (1971): 248.
47. Chalmers, R.K. "Trade Effluent Treatment at the Cannock Factory of Joseph Lucas Limited," *J. Proc. Inst. Sew. Purif.* (1965): 357–359.
48. Chalmers, R.K. "Pretreatment of Toxic Wastes," *Water Poll. Control* (London, 1970): 281–291.
49. "Development Document for Interim Final Effluent Limitations Guidelines and Proposed New Source Performance Standards for the Common and Precious Metals Segment of the Electroplating Point Source Category," USEPA 440/1-75/040, Group I, Phase II (April 1975).
50. Rock, D.M. "Hydroxide Precipitation and Recovery of Certain Metallic Ions from Waste Waters," paper presented at the Annual Meeting, American Institute for Chemical Engineers, Chicago, IL (November/December 1970).
51. Stone, E.H.F. "Treatment of Nonferrous Metal Process Waste at Kynoch Works, Birmingham, England," *Proceedings of the 25th Industrial Waste Conference, Purdue University* (May 1967), pp 848–865.
52. Knapp, R. and E. Paulson. "Gravity Filtration Reduces Suspended Metals in a Lime Precipitation Treatment System," paper presented at the 37th Annual Purdue Industrial Waste Conference, West Lafayette, IN (May 1982).
53. Harrison, J.L. "Iron and Steel Works Pollution Control: Water and Effluents," *Steel Times* 202 (1974): 557–567.
54. Patterson, J.W., H.E. Allen, and J.J. Scala. "Carbonate Precipitation for Heavy Metals Pollutants," *J. Water Poll. Control. Fed.* 49 (1977): 2397–2410.
55. Metzner, A.V. "Removing Soluble Metals from Wastewater," *Water Sew. Works* (April 1977): 98–101.
56. Yost, K.J. and A. Scarfi. "Effects of Alkaline Cleaner on Metal Solubility in Treated Cyanide Plating Solutions," *Fin. Manag.* (July 1978): 19–24.
57. Yost, K. and A. Scarfi. "Factors Affecting Zinc Solubility in Electroplating Waste," *J. Water Poll. Control. Fed.* 51 (1979): 1878–1886.

58. Banerjee, N.G. and T. Banerjee. "Recovery of Nickel and Zinc from Refinery Waste Liquors; Part I—Recovery of Nickel by Electrodeposition," *J. Sci. Ind. Res.* 11B (1952): 77–78.
59. Rosehart, R. and J. Lee. "Effective Methods of Arsenic Removal from Gold Mine Wastes," *Can. Min. J.* (June 1972): 53–57.
60. Larsen, H.P., J.K.P. Shou, and L.W. Ross. "Chemical Treatment of Metal-Bearing Mine Drainage," *J. Water Poll. Control. Fed.* 45 (1973): 1682–1695.
61. Feigenbaum, H.N. "Removing Heavy Metals in Textile Waste," *Ind. Wastes* (March/April 1977): 32–34.
62. Scott, M. "Sulfide Process Removes Metals, Produces Disposable Sludge," *Ind. Wastes* (July/August 1979): 34–38.
63. Lancy, L.E. "An Economic Study of Metal Finishing Waste Treatment," *Plating* 54 (1967): 157–161.
64. Martin, J.J.,Jr. "Chemical Treatment of Plating Waste for Removal of Heavy Metals," USEPA EPA-R2-73-044 (May 1973).
65. "New Process Detoxifies Cyanide Wastes," *Environ. Sci. Technol.* 5 (1971): 496–497.
66. Lawes, B.C., L.B. Fournier, and D.B. Mathre. "A Peroxygen System for Destroying Cyanides in Zinc and Cadmium Electroplating Rinse Waters," *Plating* 60 (1973): 902, 909.
67. Calmon, C., J. Casansa, and H. Gold. "Ion Exchange and Resin Adsorption Data for the Removal of Toxic Contaminants from Wastewaters," Report to the USEPA Industrial Environmental Research Lab, Cincinnati, OH (September 1980).
68. Saxena, K.L. and R.N. Chakraboryt. "Viscose Rayon Wastes and Recovery of Zinc Therefrom," *Technology, Sindii* 3 (1964): 29–33; *Water Poll. Abstr.* 41 (1968).
69. Kantawala, D. and H.D. Tomlinson. "Comparative Study of Recovery of Zinc and Nickel by Ion Exchange Media and Chemical Precipitation," *Water Sew. Works* III (1964): R281–R286.
70. Ross, R.D. *Industrial Waste Disposal* (New York: Reinhold, 1968).
71. Chamberlain, D.G. and R.E. Anderson. "Selective Removal of Zinc from Tower Blowdown by Ion Exchange," *Ind. Water Eng.* 8 (1971): 33.
72. Dixon, D.R. and D.B. Hawthorne. "Continuous Ion Exchange Using Magnetic Shell Resins, III. Treatment of Effluents Containing Heavy Metal Ions," *J. Applied Chem. Biotechnol.* 28 (1978): 10–16.
73. Uejima, H., M. Hirai, and T. Ishibashi. "Recovery of Zinc from Industrial Waste Water Using Chelating Resin," *Prog. Water Technol.* 9 (1977): 871–880.
74. Smith, D. and C. Roy. "Disposal: What Does It Really Cost?" *Proceedings of the 4th Conference on Advanced Pollution Control for the Metal Finishing Industry*, USEPA 600/9-82-022 (December 1982).
75. Culotta, J.W. and W.F. Swanton. "The Role of Evaporation in the Economics of Waste Treatment for Plating Operations," *Plating* 55 (1968): 957–961.
76. Gallo, B.G. and J.M. Culotta. "Save on Plating System," *Water Wastes Eng.* 9 (1972): A18–A19.
77. Kolesar, T.J. "The Use of Atmospheric-Evaporative Towers for Recycle and Waste Elimination," *Plating* 61 (1974): 571–574.
78. Crampton, P. and R. Wilmoth. "Reverse Osmosis in the Metal Finishing Industry," *Metal Fin.* (March 1982): 21–27.
79. Cawley, W., ed. *Treatability Manual*, vol. III. *Technologies for Control/Removal of Pollutants*, USEPA Report 600-8-80-042-c (July 1980).

80. Hillis, M. "Treatment of Cyanide Wastes by Electrolysis," *Trans. Inst. Metal Fin.* 53 (1975): 65–73.
81. Kennedy, I.F.T. and S.D. Gupta. "Electrochemical Removal of Trace Metals for Metal Wastes with Simultaneous Cyanide Destruction," *Fin. Manag.* (July 1978): 7–18.
82. "Development Document for Effluent Limitations Guidelines and New Source Performance Standards for the Copper, Nickel, Chromium and Zinc Segment of the Electroplating Point Source Category," USEPA 440/1-74-003-a (March 1974).

Index